Modelling Methodology for Physiology and Medicine

Second Edition

T0329115

Modelling Methodology for Physiology and Medicine

Second Edition

Ewart Carson

Centre for Health Informatics
City University London
London, UK

Claudio Cobelli

Department of Information Engineering
University of Padova
Padova, Italy

ELSEVIER

AMSTERDAM • BOSTON • HEIDELBERG • LONDON • NEW YORK • OXFORD
PARIS • SAN DIEGO • SAN FRANCISCO • SINGAPORE • SYDNEY • TOKYO

Elsevier
32 Jamestown Road, London NW1 7BY
225 Wyman Street, Waltham, MA 02451, USA

First edition 2001
Second edition 2014

Notices
Knowledge and best practice in this field are constantly changing. As new research and experience
broaden our understanding, changes in research methods, professional practices, or medical treatment
may become necessary.

Practitioners and researchers must always rely on their own experience and knowledge in evaluating and
using any information, methods, compounds, or experiments described herein. In using such information
or methods they should be mindful of their own safety and the safety of others, including parties for
whom they have a professional responsibility.

To the fullest extent of the law, neither the Publisher nor the authors, contributors, or editors, assume
any liability for any injury and/or damage to persons or property as a matter of products liability,
negligence or otherwise, or from any use or operation of any methods, products, instructions, or ideas
contained in the material herein.

British Library Cataloguing-in-Publication Data
A catalogue record for this book is available from the British Library

Library of Congress Cataloging-in-Publication Data
A catalog record for this book is available from the Library of Congress

ISBN: 978-0-12-411557-6

For information on all Elsevier publications
visit our werbsite at store.elsevier.com

This book has been manufactured using Print On Demand technology. Each copy is produced to order
and is limited to black ink. The online version of this book will show color figures where appropriate.

Working together
to grow libraries in
developing countries

www.elsevier.com • www.bookaid.org

Contents

3   Deconvolution   45

Giovanni Sparacino, Giuseppe De Nicolao, Gianluigi Pillonetto and Claudio Cobelli

4   Structural Identifiability of Biological and Physiological Systems   69

Maria Pia Saccomani

Preface

Mathematical modelling is now widely adopted in physiology and medicine to support the life scientist and clinical worker. However, good modelling practice must be based upon sound methodology. This is the focus of this book. It builds upon the basic idea of an integrated methodology for the development and testing of mathematical models. It covers many specific areas of methodology in which important advances have taken place over recent years and illustrates the application of good methodological practice in key areas of physiology and medicine.

Over the past few decades, several books have been written on mathematical modelling in physiology and medicine. Some have focused on methodology, while others have centred around a specific area of physiology and medicine. Over the past 20 years, we ourselves have contributed extensively to this field, including our volume from the early 1980s entitled *Mathematical Modelling of Metabolic and Endocrine Systems: Model Formulation, Identification and Validation*, which combined methodological detail with a demonstration of its applicability in relation to metabolics and endocrinology.

This present volume follows suit by combining advances in methodology with demonstration of its applicability. It is one of two volumes on the theme of modelling included in this Biomedical Engineering series. The other one, which is currently in production, provides an introduction to modelling in physiology. The essence of our other volume is summarized in the first chapter of this book. This book serves as both a stand-alone volume and a complementary work. For the reader who has some experience in modelling, this volume will provide an accessible account of recent advances in the field. For the reader who has absorbed the messages of the introductory volume, the chapters herein build logically upon those insights.

This book has been designed to appeal to all those who wish to advance their knowledge of good modelling practice. It will be useful to postgraduate students and those in the final year of study who have chosen modelling specialities as part of biomedical engineering or medical or health informatics courses. It is equally designed to meet the needs of advanced practitioners and researchers in the field of modelling as it applies to physiology and medicine.

Although formally an edited text, this volume is the collaborative work of two teams in London and Padova who together have extensive experience in communicating these ideas and concepts to a wide range of audiences, including undergraduate and postgraduate students, and researchers and professionals across a spectrum of disciplines from engineering and informatics to medicine and related clinical professions. Hence, this book has been produced as an integrated work, meant as tutorial in style and containing references at the end of the volume.

In writing this volume, we thank those of our colleagues in our teams who have chosen to work with us on this project. Their support and encouragement has been greatly appreciated, and without their efforts this volume would not exist. We also wish to thank our friends and colleagues, who over many years have encouraged us to develop our modelling ideas, whether from their perspectives as fellow engineers and computer scientists or from their invaluable viewpoints as physiologists and clinicians. There are many that we would wish to recognize, including Riccardo Bonadonna, Derek Cramp, Ludwik Finkelstein, Antonio Lepschy, and Peter Sönksen.

Finally, we thank Joseph Bronzino, Editor-In-Chief of this Biomedical Engineering Series, and Joel Claypool, Jane Phelan, and colleagues at Academic Press for their encouragement, support, and tolerance in working with us to see our ideas come to fruition.

July 2000

Ewart Carson and Claudio Cobelli
London, England and Padova, Italy

Preface to the Second Edition

In the 12 years that have passed since the first edition was published, considerable progress has occurred in this field of modelling in physiology and medicine. Advances have taken place in modelling methodology, availability of new data, and increased application, both in terms of numbers of studies and new areas of application. In light of this progress we have been invited to produce this second edition. Accordingly, this volume represents a significant expansion of the first edition (from 13 to 23 chapters) in order that coverage can be given to these important developments. In addition, almost all of the material is new.

We should first like to thank all of the authors who have chosen to work with us on this project. In addition, we express our appreciation to Sarah Lay, Erin Hill-Parks, Cari Owen, and Fiona Geraghty of the Elsevier editorial team for their technical expertise and support.

September 2013

Ewart Carson and Claudio Cobelli
Ludlow, England and Padova, Italy

List of Contributors

Steen Andreassen Center for Model-Based Medical Decision Support (MMDS), Aalborg University, Denmark

James B. Bassingthwaighte Department of Bioengineering, University of Washington, Seattle, WA, USA

Riccardo Bellazzi Department of Electrical, Computer and Biomedical Engineering, University of Pavia, Pavia, Italy

B. Wayne Bequette Department of Chemical & Biological Engineering, Rensselaer Polytechnic Institute, Troy, NY, USA

Alessandra Bertoldo Department of Information Engineering, University of Padova, Padova, Italy

Roberto Burattini Previously at the Department of Information Engineering, Polytechnic University of Marche, Italy; Department of Veterinary and Comparative Anatomy, Pharmacology and Physiology, Washington State University, Pullman, Washington, USA

Ewart Carson Centre for Health Informatics, City University London, London, UK

Andrea Caumo Dipartimento di Scienze Biomediche per la Salute, Universitá di Milano, Milano, Italy

Claudio Cobelli Department of Information Engineering, University of Padova, Padova, Italy

Filippo Cona Department of Electrical, Electronic, and Information Engineering, University of Bologna

Chiara Dalla Man Department of Information Engineering, University of Padova, Padova, Italy

Giuseppe De Nicolao Dipartimento di Informatica e Sistemistica, University of Pavia, Pavia, Italy

Barbara Di Camillo Department of Information Engineering, University of Padova, Padova, Italy

Francis J Doyle, III Department of Chemical Engineering, University of California Santa Barbara, Santa Barbara, CA; Institute for Collaborative Biotechnologies, University of California Santa Barbara, Santa Barbara, CA, USA

Fulvia Ferrazzi Institute of Human Genetics, Friedrich-Alexander-Universität Erlangen-Nürnberg, Erlangen, Germany

Dan S. Karbing Respiratory and Critical Care Group (rcare), MMDS, Aalborg University, Denmark

Søren Kjærgaard Anaesthesia and Intensive Care, Aalborg University Hospital, Denmark

Boris Kovatchev Center for Diabetes Technology, University of Virginia, Charlottesville, VA; Department of Systems and Information Engineering, University of Virginia, Charlottesville, VA, USA

Paolo Magni Dipartimento di Ingegneria Industriale e dell'Informazione, Università degli Studi di Pavia, Pavia, Italy

Elisa Magosso Department of Electrical, Electronic, and Information Engineering, University of Bologna

Lorenzo Pasotti Dipartimento di Ingegneria Industriale e dell'Informazione, Università degli Studi di Pavia, Pavia, Italy

Stephen Patek Center for Diabetes Technology, University of Virginia, Charlottesville, VA; Department of Systems and Information Engineering, University of Virginia, Charlottesville, VA, USA

Morten Gram Pedersen Department of Information Engineering, University of Padova, Padova, Italy

Maria Pia Saccomani Department of Information Engineering, University of Padova, Padova, Italy

Gianluigi Pillonetto Department of Information Engineering, University of Padova, Padova, Italy

Italo Poggesi Model-Based Drug Development, Janssen R&D, Via Michelangelo Buonarroti, Cologno Monzese(MI), Italy

Alberto Redaelli Dipartimento di Elettronica, Informazione e Bioingegneria (DEIB), Politecnico di Milano, Milan, Italy

Stephen E. Rees Respiratory and Critical Care Group (rcare), MMDS, Aalborg University, Denmark

Maurizio Rocchetti Independent Consultant, Rho, Milan, Italy

Maria Rodriguez-Fernandez Department of Chemical Engineering, University of California Santa Barbara, Santa Barbara, CA; Institute for Collaborative Biotechnologies, University of California Santa Barbara, Santa Barbara, CA, USA

Francesco Sambo Department of Information Engineering, University of Padova, Padova, Italy

Zimi Sawacha Department of Information Engineering, University of Padova, Padova, Italy

Bernhard Schrefler Department of Civil, Environmental and Architectural Engineering, University of Padova, Padova, Italy

Monica Simeoni Biopharm and ImmunoInflammation, Clinical Pharmacology, Quantitative Sciences, GlaxoSmithKline, Stockley Park West, Uxbridge, Middlesex, UK

Giovanni Sparacino Dipartimento di Ingegneria dell'Informazione, Università di Padova, Padova, Italy

Peter C. St. John Department of Chemical Engineering, University of California Santa Barbara, Santa Barbara, CA, USA

Gianna Toffolo Department of Information Engineering, University of Padova, Padova, Italy

Mauro Ursino Department of Electrical, Electronic, and Information Engineering, University of Bologna

Paolo Vicini Pharmacokinetics, Dynamics and Metabolism, Pfizer Worldwide Research and Development, San Diego, CA; Pharmacokinetics, Dynamics and Metabolism, Pfizer Worldwide Research and Development, Science Center Drive, San Diego, CA, USA

Emiliano Votta Dipartimento di Elettronica, Informazione e Bioingegneria (DEIB), Politecnico di Milano, Milan, Italy

Ruoting Yang Institute for Collaborative Biotechnologies, University of California Santa Barbara, Santa Barbara, CA, USA

Susanna Zucca Dipartimento di Ingegneria Industriale e dell'Informazione, Università degli Studi di Pavia, Pavia, Italy

1 An Introduction to Modelling Methodology

Claudio Cobelli[a] and Ewart Carson[b]

[a]Department of Information Engineering, University of Padova, Padova, Italy, [b]Centre for Health Informatics, City University London, London, UK

1.1 Introduction

The aim of this second edition is to describe more recent developments in modelling theory and practice in relation to physiology and medicine. The chapters that follow offer detailed accounts of several facets of modelling methodology (Chapters 2−12), as well as demonstrations of how such methodological development can be applied in areas of physiology and medicine. This application material, contained in Chapters 13−23, is not intended to be comprehensive. Rather, a number of specific topics have been chosen to illustrate the methodology: respiration (Chapter 13), metabolism (Chapters 15−17), clinical imaging (Chapter 18), oncology (Chapter 19), circulation (Chapters 20−21), musculoskeletal biomechanics (Chapter 22), and synthetic biology (Chapter 23). The authors of the respective chapters have very considerable expertise in these areas of physiology and medicine.

Before moving to more advanced areas of methodology, it is appropriate to review the fundamentals of the modelling process, which put simply can be viewed as a mapping or transforming of a physiological system into a model as shown in Figure 1.1. The process has now reached substantial maturity, and the basic ingredients are well established. The fundamentals of the overall modelling framework are described in detail elsewhere [1]. In this chapter, we provide a distillation of that framework and revisit the fundamentals upon which the later, more detailed chapters are built.

1.2 The Need for Models

1.2.1 Physiological Complexity

Complexity is what characterizes much of physiology, and we must have a method to address this. Complexity manifests itself through elements that comprise any physiological system through the nature of their connectivity, in terms of hierarchy, and through the existence of nonlinear, stochastic, and time-varying effects.

Modelling Methodology for Physiology and Medicine. DOI: http://dx.doi.org/10.1016/B978-0-12-411557-6.00001-X

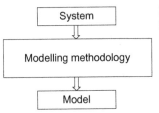

Figure 1.1 Modelling methodology: transforming a system into a model.

Complexity is exhibited at each level of the hierarchy and across levels within the physiological system.

What do we mean by some of these concepts? First, the physiological hierarchy will include the levels of molecule, cell, organ, and organism. Complex processes of regulation and control are evident at each level. Feedback is another key feature that occurs in many forms. It is inherent in chemical reactions within the organism. There are explicit hormonal control mechanisms superimposed upon metabolic processes. The physiological organ systems exhibit explicit control mechanisms. In many instances, there is negative feedback, although examples of positive feedback also exist. Feedback offers examples of control action being taken not only in relation to changes in the value of a physiological variable *per se* but also in response either to its rate of change or to the integral of its value over a period of time. Some of these concepts of feedback and control are examined in more detail in Chapter 2[1].

As a result of this physiological complexity, it is not often possible to measure directly (*in vivo*) the quantities of interest. Only indirect measures may be feasible, implying the need for some model to be able to infer the value of the quantity of real interest. Measurement constraints usually mean that it is only possible to obtain readings of blood values of a metabolite when the real interest lies in its value in body tissue. Equally, it is not generally possible to measure the secretions of the endocrine glands.

Overall, this complexity—coupled with the limitations that are imposed upon the measurement processes in physiology and medicine—means that models must be adopted to aid our understanding.

1.2.2 Models and Their Purposes

What do we mean by the term *model*? In essence, it is a representation of reality involving some degree of approximation. Models can take many forms. They can be conceptual, mental, verbal, physical, statistical, mathematical, logical, or graphical in form. For the most part, this volume focuses on mathematical modelling.

Given that a model provides an approximate representation of reality, what is the purpose of modelling activity? As is shown in Figure 1.2, the purpose is a key driver of good modelling methodology. In classic scientific terms, modelling can be used to describe, interpret, predict, or explain. A mathematical expression, for example, a single exponential decay, can provide a compact description of data that approximate to a first order process. A mathematical model can be used to interpret data

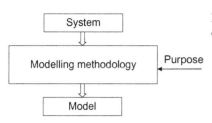

Figure 1.2 The purpose of modelling: a key driver of modelling methodology.

collected as part of a lung function test. A model of renal function—which, for instance, includes representations of the dynamics of urea and creatinine—can be used to predict the time at which a patient with end-stage renal disease should next undergo hemodialysis. A model of glucose and insulin can be used to gain additional insight into, and explanation of, the complex endocrine dynamics in the diabetic patient.

Rather more specific purposes for modelling can be identified in the physiological context. These include aiding understanding, testing hypotheses, measuring inferences, teaching, simulating, and examining experimental design. For example, competing models—constituting alternative hypotheses—can be examined to determine which are compatible with physiological or clinical observation. Equally, a model of the relevant metabolic processes, when taken together with measurements of a metabolite made in the bloodstream, can be used to infer the value of that metabolite in the liver. Models are also increasingly used as a medium in teaching and learning processes, where by means of simulation, the student can be exposed to a richer range of physiological and pathophysiological situations than would be possible in the conventional physiological laboratory setting. Models can also play a powerful role in experimental design. For instance, if the number of blood samples that can be withdrawn from a patient is limited in a given period of time, models can be used to determine the times at which blood samples should be withdrawn to obtain the maximum information from the experiment, for example, in relation to pharmacokinetic or pharmacodynamic effects.

Considering what is meant by a model and its purposes, we now focus on the nature of the process itself. As already indicated, this is the process of mapping from the physiological or pathophysiological system of interest to the model, as shown in Figure 1.1. The essential ingredients are model formulation, including determination of the degree to which the model is an approximation of reality; model identification, including parameter estimation; and model validation. These are discussed in the following sections.

1.3 Approaches to Modelling

In developing a mathematical model, two fundamental approaches are possible. The first is based on experimental data and is essentially a data-driven approach. The other is based on a fundamental understanding of the physical and chemical

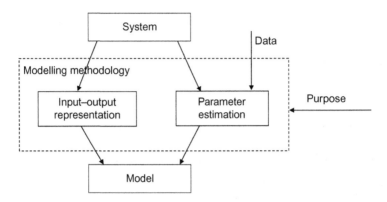

Figure 1.3 Modelling the data: a methodological framework.

processes that give rise to the resultant experimental data. This can be referred to as *modelling the system*.

1.3.1 Modelling the Data

Models that are based on experimental data are generally known as *data-driven* or *black box models*. Fundamentally, this means seeking quantitative descriptions of physiological systems based on input−output (I/O) descriptions derived from experimental data collected on the system. Simply put, these are mathematical descriptions of data, with only implicit correspondence to the underlying physiology.

Why should we use such data models? First, they are particularly appropriate where there is a lack of knowledge of the underlying physiology, whether *a priori* knowledge or knowledge acquired directly through measurement. Equally, they are appropriate when an overall I/O representation of the system's dynamics is needed, without specifying how the physiological mechanisms gave rise to such I/O behavior.

The methodological framework for modelling data is depicted in Figure 1.3. Several specific methods are available for formulating such data models, including time series methods, transfer function analysis, impulse response methods, and con-volution−deconvolution techniques.

1.3.2 Modelling the System

In contrast to data modelling, when modelling the system there is an attempt to explicitly represent the underlying physiology, albeit at an appropriate level of approximation and resolution. The degree of approximation will be largely deter-mined by the availability of *a priori* knowledge and the nature of the assumptions that can be made. The basic framework in this case is shown in Figure 1.4.

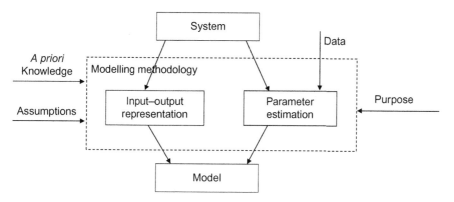

Figure 1.4 Modelling the system: a methodological framework.

Models of the system, that is, models that are physiologically based, may be categorized in a number of ways according to the included attributes. This classification, corresponding to the approaches that can be adopted, includes static versus dynamic models, deterministic versus stochastic, time-invariant versus time-varying, lumped versus distributed, linear versus nonlinear, and continuous versus discrete models. All combinations are possible so one could envisage a dynamic, deterministic, time-invariant, lumped, linear, continuous model in one case or a dynamic, stochastic, time-varying, distributed, nonlinear, discrete model at the other end of the spectrum.

A first classification is into static models and dynamic models. Clearly, static models are restricted to steady-state conditions and do not attempt to capture the richness of a system's dynamics. However, in the circulatory or respiratory context, for example, static models can provide useful relationships between mean pressure, flow, and impedance (or resistance in the linear case). Simple dynamic models have been employed in fields such as cellular dynamics and metabolic compartmental descriptions (e.g., one-compartment elimination), and as descriptions of circulatory and respiratory dynamics.

In more complex formulations, distributed modelling enables spatial effects to be incorporated as well as the basic system dynamics. For example, a distributed model of the arterial system enables the blood pressure patterns along the length of any blood vessel to be analyzed as well as variations of blood pressure over time.

Nonlinear modelling reflects the fact that almost all physiological phenomena are truly nonlinear. In some situations, however, it might be appropriate to assume that linearity applies, for example, if one is simply interested in the dynamics resultant from a small perturbation of the system from some steady-state operating condition. However, if it is the response to a large perturbation that is required (e.g., blood glucose dynamics following a meal), a full nonlinear model of the relevant portions of the carbohydrate metabolism is needed.

Many physiological systems can be treated as if they were time-invariant; this means that system parameters are assumed not to vary with time. However, there are some circumstances in which this assumption would not hold. One example might be the change in the elastic properties of blood vessels that occurs over a long period of time. A model that was intended for use over such an extended period would incorporate elastic parameters of the relevant blood vessels that did vary over time. At the other end of the timescale, a model designed to explore the rapid dynamics of the cardiovascular system that occur during the pumping cycle of the heart would incorporate time-varying representations of those parameters corresponding to the elastic properties of the heart chambers.

A further degree of approximation applies whenever it is assumed that a physiological system can be treated as deterministic. In reality there is usually some stochastic component present. If this component is small, it may be acceptable to treat the system as if it were deterministic. Where the stochastic effects are more dominant, as is the case when examining cellular dynamics, then stochasticity must be incorporated into the model, either as stochastic variation of key variables or as stochastic variation in the parameters. Issues relating to probabilistic and stochastic modelling are discussed in Chapters 11 and 12.

Another facet of modelling relates to the physiological/biological level at which the model is to be formulated. Recent developments in system biology and genomic/genetic data are considered in Chapters 8 and 9.

A number of issues relating to model validation (validation is considered in more detail later in the chapter) must be considered at this stage. These include verification of the compatibility of the proposed model with the relevant physiology, the validity of the assumptions, and the degree to which the complexity of the physiology has been reduced for the particular modelling purpose in question. The model should also be checked to ensure it is logically complete and consistent.

1.4 Simulation

Simulation is the process of solving the model (i.e., the equations that are the realization of the model) to examine its output behavior. Typically, this process involves examining the time course of one or more of the variables, in other words, performing computer experiments on the model.

When is simulation required? It can be used either during the process of model building or once the model is completed. During model building, simulation can be performed to clarify aspects of system behavior to determine whether a proposed model representation is appropriate. This is done by comparison of the model response with experimental data from the same situation. When carried out on a complete, validated model, simulation yields output responses that provide information on system behavior. Depending on the modelling purpose, this information assists in describing the system, predicting behavior, or yielding additional insights (i.e., explanations).

Why use simulation? Simulation offers a way forward in situations in which it might not be appropriate, convenient, or desirable to perform particular experiments on the system. Such situations could include those in which experiments cannot be done at all, are too difficult, are too dangerous, are not ethical, or would take too long to obtain results. Therefore, we need an alternative way to experiment. Simulation offers an alternative that can overcome the preceding limitations. Such experimenting can provide information that, depending on the modelling purpose, aids description, prediction, or explanation.

How do we perform simulation? First, we need a mathematical model that is complete; that is, all its parameters are specified and initial conditions are defined for all of the variables. If the model is not complete (i.e., has unspecified parameter values), parameter estimation techniques must be employed. Once a complete model is available, it is solved numerically, yielding the time course of the system variables.

1.5   Model Identification

1.5.1   A Framework for Identification

To complete the transformation from system to model as depicted in Figure 1.1, we must have both a model structure and fully determined parameters corresponding to that structure. In other words, we need a complete model. However, we may not have such a model. We should by this stage have at least one candidate model, and we may have more than one to choose from. If a single model is incomplete, it will be due to some unknown parameter values. This is true whether the modelling approach adopted has been driven by the data or by the physiology of the system. We may be dealing with the whole model or just part of it. In either case, an integrated identification framework is needed. A schematic representation of this process is shown in Figure 1.5.

The solution of this problem requires data. Data sometimes occur from the intrinsic dynamics of the system (e.g., spontaneous oscillations or noise). Electrophysiological signals would be instances of such output dynamics as they relate to brain, muscle, or cardiac function. Usually, however, we must design experiments. The question is then what experiments must be designed to yield appropriate data. Clearly, the I/O data from the experiment must contain information for that part of the model with unknown parameter values. Tracer methods, as discussed in Chapter 9, offer one approach to data generation convenient for the identification process.

In the identification process, data are mapped into parameter values by the model, where errors can occur in both the data and the model. The first arises as a consequence of measurement errors. The second involves errors relating to model structure. There are other types of errors, including noise on test signals and disturbances. However, one cannot address more than one type of noise at a time. In fact, rigorously we can only address measurement errors. Errors in model structure cannot be dealt with explicitly. They can only be solved by considering each competing model structure in turn. Therefore, it is customary to focus on a single model

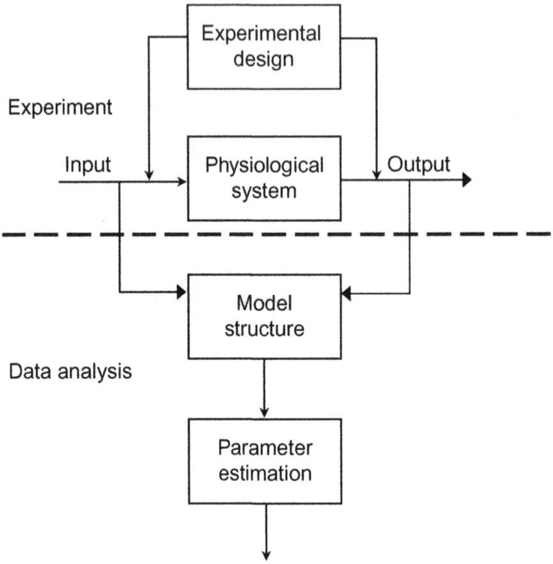

Figure 1.5 Model identification.

and concentrate on the impact of measurement errors that are assumed to be additive. The available approaches can be divided into two groups: situations with parametric models and those with nonparametric models.

1.5.2   Identification of Parametric Models

The first issue to be addressed is that of identifiability. In essence, this is asking whether it would be theoretically possible to make unique estimates of all the unknown parameters assuming that the experimental data were complete and noise-free. In other words, the experimental data must be rich enough to estimate all of the unknown parameters. Problems of identifiability arise when there is a mismatch between the complexity of the model and the richness of the data. That is, the model is too complex (too many unknown parameters) for the available data, or the data are not sufficient for the model provided. In such cases, one must explore whether the model might, in a manner that retains validity, be reduced or whether the experimental design might be enriched, for example, by making measurements of an additional variable. Issues of identifiability are considered in Chapter 4.

If the model is uniquely identifiable, assuming perfect data, it is possible to proceed directly to estimating the parameters. In some situations, multiple solutions (finite, greater than one) may be theoretically possible for some parameters. In some cases, it might at the stage of validating the complete model be possible to select between these alternatives, such as on the basis of physiological plausibility. Where an infinite number of values is theoretically possible for one or more parameters, remedy must be sought to the mismatch between model and data outlined previously.

The next step is parameter estimation for which a number of techniques exist, including nonlinear least squares, maximum likelihood, and Bayesian estimation. These are addressed in Chapter 5, while in Chapter 6 nonparametric linear system identification methods are considered. These approaches relate to data on individual subjects. When the available data are sparse, population parameter estimation techniques can be used, as discussed in Chapter 7.

Some effort has also been directed to the problem of optimal experimental design. This has largely focused on the interaction between the features of an experiment in which data are obtained as a set of discrete values over the experimental period following the application of an input test signal and the information content of the experiment in relation to the quality of the parameter values obtained by the estimating process.

1.5.3   Identification of Nonparametric Models

Nonparametric models arise from some of the situations described earlier in which a data modelling approach has been adopted. In other words, the overall I/O model description has been obtained, such that it is specified as an integral equation. Such a description has essentially three ingredients: the input, the output, and the impulse response that provides the connection between them. Two of these are known, and the third is to be determined. The most usual situation is when the output is specified. To solve this problem, deconvolution techniques can be employed as discussed in Chapter 3.

1.6   Model Validation

Validating a model is essentially examining whether it is good enough in relation to its intended purpose. This assumes that, in a Popperian sense, it can be tested. Clearly, no model can have absolute unbounded validity given that, by definition, a model is an approximation of reality. If one is working with a set of competing candidate models, the validation process involves determining which of them is best in relation to its intended purpose. A valid model is one that has successfully passed through the validation process.

Validation is integral to the overall modelling process. It is an activity that should take place during model building and upon model completion. The issue of validity testing during the process of model formulation was addressed in Section 1.3.2. At this stage, we shall assume that the model is complete; that is, it has no unspecified parameters. The essence of the validation process at this stage is shown in Figure 1.6.

It cannot be stressed too strongly that in examining the validity of the complete model, the process is dependent upon model purpose; that is, it is problem-specific. The implication is that we are testing whether appropriate components are contained in the model. For example, we would be testing the appropriateness of

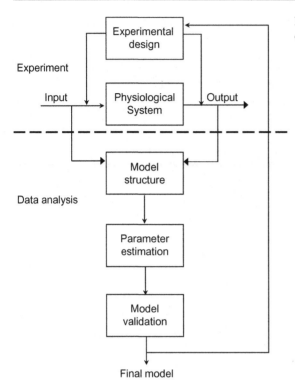

Figure 1.6 Validation of the complete model.

model structure in relation to any intended use for the hypothesis testing parameters that would be meaningful in the context of a specific disease (i.e., a change in parameter values could correspond to the change from a healthy to a diseased state). Dependent upon the purpose, usually some features of the model and system output (i.e., experimental output data) must correspond sufficiently for the same input (an acceptably small difference between them). In other words, within the necessary domain of validity, we are testing whether the model is credible. The model performance also may be tested out with its nominal specific domain of validity to define the effective boundary of the actual domain within which it may be considered valid.

The basic approach in validating a single model is to compare the model and system behavior, based on appropriate output features of response. Any mismatch between the system and the model output should be analyzed for plausibility of behavior.

For cases in which formal parameter estimation procedures have been employed for the model, additional quantitative tools are available in the validation process. These include examining the residuals of the mismatch and the plausibility of the parameter estimates where the parameters have a clear physiological counterpart. In the case of competing models, choice can be aided by examining the parsimony of the models (by using the Akaike criterion in the case of linear, dynamic models) and the features of response yielded by the mode, again testing the plausibility.

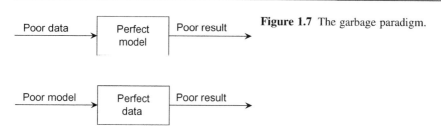

Figure 1.7 The garbage paradigm.

In conclusion, one should always remain critical of a model and not love it too much! All models are approximations; hence, one should always be prepared to include new facts and observations. A good model (in a Popperian sense) is one that is clearly falsifiable and thus is readily capable of bringing about its own downfall. One also should remember that in situations of complexity, it may be appropriate to think of a set of models, in which each would have its own (distinct) domain of validity. It then becomes a case of choosing the most appropriate in relation to, for instance, the level in the physiological hierarchy that is being represented and the timescale of interest in relation to dynamic response (e.g., short-term or long-term). A successful outcome to the modelling process is critically dependent on both the quality of the model and the quality of experimental data (Figure 1.7).

Following this brief tour through basic ingredients of good modelling methodology, subsequent chapters elaborate on methodological issues of current importance and demonstrate their applicability in selected domains of physiology and medicine.

Reference

[1] Cobelli C, Carson ER. An introduction to physiological modeling. London: Academic Press; 2008.

2 Control in Physiology and Medicine

B. Wayne Bequette

Department of Chemical & Biological Engineering, Rensselaer Polytechnic Institute, Troy, NY, USA

2.1 Introduction

The notion of feedback control or regulation is prevalent in physiology and medicine. Any textbook on physiology provides many examples of variables that are regulated by changing or manipulating one or more other variables. These innate physiology systems are commonly called *endogenous* control systems. A prime example is blood glucose concentration regulation by the pancreas, as shown in Figure 2.1 [1]. High blood glucose stimulates the beta cells of the pancreas to produce insulin, which then converts glucose in the liver into glycogen (storage); the insulin also stimulates the uptake of glucose from the blood into tissue cells. The net result is a lowering of the blood glucose concentration. If the blood glucose is low, the alpha cells of the pancreas are stimulated to produce glucagon, which converts the glycogen that has been stored in the liver into glucose, thus raising the blood glucose concentration.

The field of control systems has a standard nomenclature. In this example, the desired blood glucose concentration (usually around 80−90 mg/dL) is called the set point. The output is the actual blood glucose concentration. The manipulated inputs are the insulin and glucagon production rates. The method that the pancreas uses to "decide" on the proper insulin and glucose production rates is termed a control algorithm.

Endogenous control of physiological variables is obviously very important, and much research has been conducted to better understand and mathematically model these regulatory systems. The focus of this chapter, however, is on exogenous control; that is, the engineering use of physical and/or chemical devices to regulate a physiological variable. To better understand exogenous control, we will use an example from type 1 diabetes that is related to Figure 2.1. The pancreas of an individual with type 1 diabetes can no longer produce sufficient insulin to regulate the blood glucose concentration. It is necessary to administer insulin using either injections or a continuous insulin infusion pump. The decision on insulin injection amounts or continuous infusion rates is based on knowledge of the blood glucose concentration and expected meals or exercise. An example of an automated, closed-loop artificial pancreas to regulate blood glucose is shown in the form of a control block diagram in Figure 2.2. More details about control system design for an artificial pancreas are presented in Section 2.7.

Modelling Methodology for Physiology and Medicine. DOI: http://dx.doi.org/10.1016/B978-0-12-411557-6.00002-1

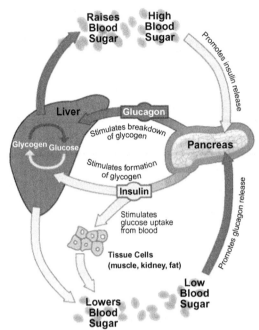

Figure 2.1 Regulation of blood glucose by two main actions of the pancreas. A high glucose concentration causes insulin to be released, resulting in a decrease in glucose concentration. A low glucose concentration causes glucagon to be released, resulting in an increase in glucose concentration.
Source: Image reproduced from Ref. [1], with permission.

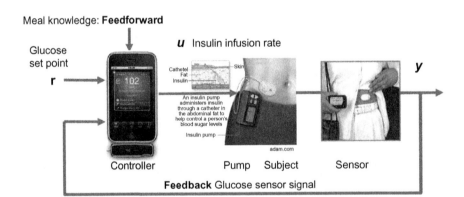

Figure 2.2 Example of an exogenous control system. The sensor measures the interstitial fluid glucose (related to blood glucose) and sends a signal to the controller. The controller, based on the sensed glucose, desired (set point) glucose, and possibly additional information, such as a meal or exercise, calculates an insulin infusion rate. The infusion rate signal is sent to an insulin pump, which then changes the rate of insulin delivered subcutaneously. The physiological "process" then responds to the insulin infusion, resulting in a changing blood glucose. The blood glucose effects the interstitial glucose level, which is sensed and sent to the controller. A sample time of 5−15 min is often used.

Engineers often use the block diagram structure shown in Figure 2.2 to analyze control systems. Each block, while representing a physical component of the system, can be represented by mathematical expressions (i.e., modelling equations) that represent the dynamic behavior of that particular component, as discussed in Section 2.3.

Feedback control has a long history [2] and there are numerous undergraduate textbooks (Ref. [3], for example) that provide an introduction to the topic. The main objectives of this chapter are to provide a concise overview of the mathematical models that are used to describe dynamic systems behavior, and to present some standard control algorithms that are used to regulate these systems.

2.2 Modelling for Control System Design and Analysis

Models can be developed for a number of reasons, and the end use of the model should dictate the form of the model. Since this chapter is focused on control, the models discussed have an end use of control system design and analysis.

2.2.1 Sets of Ordinary Differential Equations

Dynamic models used for control system design are most often based on *lumped parameter systems*, which can be described by a set of initial value ordinary differential equations (ODEs), using the following standard notation:

$$\dot{x} = f(x, u)$$
$$y = g(x) \tag{2.1}$$

where x is a vector of n states ($\dot{x}$ represents a vector of the derivatives of the states with respect to time), u is a vector of m inputs, y is a vector of r outputs (often measured), and p is a vector of q parameters.

$$x = \begin{bmatrix} x_1 \\ x_2 \\ \vdots \\ x_n \end{bmatrix} \quad \dot{x} = \begin{bmatrix} \dfrac{dx_1}{dt} \\ \dfrac{dx_2}{dt} \\ \vdots \\ \dfrac{dx_n}{dt} \end{bmatrix} \quad u = \begin{bmatrix} u_1 \\ u_2 \\ \vdots \\ u_m \end{bmatrix} \quad y = \begin{bmatrix} y_1 \\ y_2 \\ \vdots \\ y_r \end{bmatrix} \quad p = \begin{bmatrix} p_1 \\ p_2 \\ \vdots \\ p_q \end{bmatrix} \tag{2.2}$$

The inputs can further be classified as manipulated or disturbance. A manipulated input can be adjusted, either automatically by a controller or through manual intervention. A disturbance input is sometimes measured and could provide additional feedforward information to a controller.

2.2.2 Linear State Space Models

For control system design, it is most common to use a linear state space model

$$\dot{x}' = Ax' + Bu'$$
$$y' = Cx' + Du' \tag{2.3}$$

where the $(')$ notation represents a perturbation from the steady-state value (the sub-script s denotes a steady-state value). A state vector, in deviation (perturbation) variable form is represented as

$$x' = \begin{bmatrix} x_1 - x_{1s} \\ x_2 - x_{2s} \\ \vdots \\ x_n - x_{ns} \end{bmatrix} \tag{2.4}$$

and the constant coefficient matrices are found from a Taylor series expansion around the steady-state solution

$$A_{ij} = \frac{\partial f_i}{\partial x_j}\bigg|_{xs,us} \qquad B_{ij} = \frac{\partial f_i}{\partial x_j}\bigg|_{xs,us} \qquad C_{ij} = \frac{\partial g_i}{\partial x_j}\bigg|_{xs,us} \qquad D_{ij} = \frac{\partial g_i}{\partial u_j}\bigg|_{xs,us} \tag{2.5}$$

When working with linear state space models, it is more common to assume perturbation variables and to drop the $(')$ notation:

$$\dot{x} = Ax + Bu$$
$$y = Cx + Du \tag{2.6}$$

where the matrices and vectors have the following structure:

$$\begin{bmatrix} \dot{x}_1 \\ \vdots \\ \dot{x}_n \end{bmatrix} = \begin{bmatrix} a_{11} & \cdots & a_{1n} \\ \vdots & \ddots & \vdots \\ a_{n1} & \cdots & a_{nn} \end{bmatrix} \begin{bmatrix} x_1 \\ \vdots \\ x_n \end{bmatrix} + \begin{bmatrix} b_{11} & \cdots & b_{1n} \\ \vdots & \ddots & \vdots \\ b_{n1} & \cdots & b_{nm} \end{bmatrix} \begin{bmatrix} u_1 \\ \vdots \\ u_m \end{bmatrix}$$

$$\begin{bmatrix} y_1 \\ \vdots \\ y_r \end{bmatrix} = \begin{bmatrix} c_{11} & \cdots & c_{1n} \\ \vdots & \ddots & \vdots \\ c_{r1} & \cdots & c_{rn} \end{bmatrix} \begin{bmatrix} x_1 \\ \vdots \\ x_n \end{bmatrix} + \begin{bmatrix} d_{11} & \cdots & d_{1m} \\ \vdots & \ddots & \vdots \\ d_{r1} & \cdots & r_{rm} \end{bmatrix} \begin{bmatrix} u_1 \\ \vdots \\ u_m \end{bmatrix} \tag{2.7}$$

The n eigenvalues, λ, of the A matrix can be found by solving

$$\det(\lambda I - A) = 0 \tag{2.8}$$

where det represents the determinant of an $n \times n$ square matrix. The eigenvalues of A determine the stability of the model. If all n eigenvalues have negative real portions, then the model is stable, and if any single eigenvalue has a positive real portion, the model is unstable. An otherwise stable model with one or more zero eigenvalues has "integrating" behavior.

2.2.3 Transfer Functions

When designing and analyzing control systems, it is common to work with transfer functions, which are developed based on the notion of the Laplace transform, which transforms the time-domain function, $f(t)$, into the Laplace domain function, $F(s)$:

$$L[f(t)] = F(s) = \int_0^\infty f(t)e^{-st}\,dt \tag{2.9}$$

While Laplace transforms can be used to find analytical solutions of dynamic systems for specified inputs, the main usefulness is for controller synthesis and dynamic system analysis.

Transfer function models have the following input−output form:

$$y(s) = G(s)u(s) \tag{2.10}$$

where the individual input−output transfer functions are shown more clearly as

$$\begin{bmatrix} y_1(s) \\ \vdots \\ y_r(s) \end{bmatrix} = \begin{bmatrix} g_{11}(s) & \cdots & g_{1m}(s) \\ \vdots & \ddots & \vdots \\ g_{r1}(s) & \cdots & g_{rm}(s) \end{bmatrix} \begin{bmatrix} u_1(s) \\ \vdots \\ u_m(s) \end{bmatrix} \tag{2.11}$$

That is, the transfer function relating input u_j to output y_i is

$$y_i(s) = g_{ij}(s)u_j(s) \tag{2.12}$$

A continuous state space model can be used to develop the transfer function model using

$$G(s) = [C(sI - A)^{-1}B + D] \tag{2.13}$$

where the D matrix often has 0 in all of the elements.

Individual transfer functions relating a particular input to output have the following form:

$$g_p(s) = \frac{b_m s^m + b_{m-1} s^{m-1} + \cdots + b_1 s + b_0}{a_n s^n + a_{m-1} s^{n-1} + \cdots + a_1 s + a_0} \tag{2.14}$$

where the m roots of the numerator polynomial are called *zeros* and the n roots of the denominator polynomial are called *poles*. When the transfer functions are

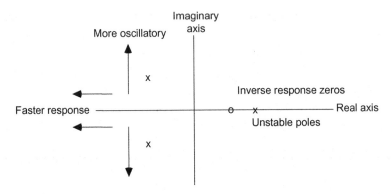

Figure 2.3 Effect of the location of poles and zeros in the complex plane.

obtained directly from a state space model, then the n poles of the transfer function are the same as the n eigenvalues of the A matrix. Occasionally, there will be pole-zero cancellation, causing the number of poles of the transfer function to be less than the number of eigenvalues of the state space model. Indeed, unstable systems can appear to be stable if unstable poles are cancelled by right-half-plane zeros. The plane refers to the complex plane formed by the real (x-axis) and imaginary (y-axis) components of the poles, zeros, or eigenvalues, as shown in Figure 2.3.

2.2.3.1   Pole-Zero Cancellation

As an example of pole-zero cancellation, consider the following state space model of a bioreactor (from Ref. [3]):

$$A = \begin{bmatrix} 0 & 0.9056 \\ -0.75 & -2.5640 \end{bmatrix} \qquad B = \begin{bmatrix} -1.5301 \\ 3.8255 \end{bmatrix}$$
$$C = \begin{bmatrix} 1 & 0 \end{bmatrix} \qquad\qquad D = [0] \tag{2.15}$$

which has the transfer function representation

$$g_p(s) = \frac{-1.5302s - 0.4590}{s^2 + 2.564s + 0.6792} \tag{2.16}$$

or, when written in factored form

$$g_p(s) = \frac{-1.5302(s + 0.3)}{(s + 0.3)(s + 2.2640)} = \frac{-1.5302}{s + 2.2640} \tag{2.17}$$

so a second-order system has become first-order, by virtue of a pole-zero cancellation. Engineers often think in terms of "time constants" which are generally inversely related to the poles. That is, for distinct, real poles, the ith process time constant (here it is also assumed that the pole p_i is negative and, thus, stable) is

$$\tau_{pi} = -1/p_i$$

For the example shown above, the transfer function written in gain-time constant form is

$$g_p(s) = \frac{-1.5302}{s + 2.2640} = \frac{-0.6758}{0.4417s + 1} = \frac{k_p}{\tau_p s + 1} \tag{2.18}$$

where k_p is the process gain and τ_p is the process time constant.

2.2.3.2 Right-Half-Plane Zeros and Time Delays

Dynamic characteristics that are particularly challenging for control systems are right-half-plane zeros and time delays. The primary characteristic of a system with a right-half-plane zero is that a step input change yields an output that has short- and long-term responses in different "directions." A system with a time delay has an output response that starts an appreciable amount of time later, and has a Laplace transfer function representation of $e^{-\theta s}$, where θ is the time delay.

An example of a system with both a time delay and a right-half-plane zero is

$$g_p(s) = \frac{2(-3s + 1)e^{-4s}}{(5s + 1)(6s + 1)} \tag{2.19}$$

which has a gain of 2, a right-half-plane zero at $+1/3$, a time delay of 4, and time constants of 5 and 6. Here, we assume that the timescale is minutes. The corresponding output response to a unit step input change is shown in Figure 2.4.

The discussion thus far has focused on continuous models, which form the core theory for most undergraduate systems and control courses. In practice, when working with experimental data, it is easier to develop discrete-time models.

2.2.4 Discrete-Time State Space Models

A discrete-time, linear state space model has the following form:

$$\begin{aligned} x_{k+1} &= \Phi x_k + \Gamma u_k \\ y_k &= C x_k + D u_k \end{aligned} \tag{2.20}$$

where the subscript on each of the vectors represents the sample time. In this formulation, it is assumed that the input is held constant over the time between samples k and $k + 1$. A discrete-time model can be found directly from the continuous-time model, assuming a *zero-order hold* on the input (again, constant between sample times) as

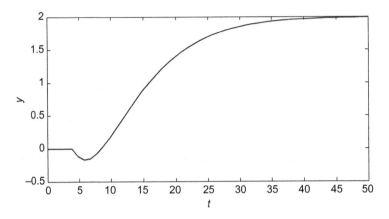

Figure 2.4 Response to a unit step input to the example transfer function, at $t = 0$.

$$\Phi = e^{A\Delta t}$$
$$\Gamma = (e^{A\Delta t} - I)A^{-1}B \tag{2.21}$$

where Δt is the sample interval.

2.2.5 Discrete Auto-Regressive Models

An auto-regressive, moving average with exogenous input model has the following form:

$$y_k = -a_1 y_{k-1} - a_2 y_{k-2} - \cdots - a_n y_{k-n}$$
$$+ b_0 u_k + b_1 u_{k-1} + b_2 u_{k-2} + \cdots + b_m u_{k-m} \tag{2.22}$$

Similar to the use of Laplace transforms to analyze continuous system behavior, z-transforms can be used to analyze discrete-time behavior. The z-transform is defined as

$$y(z) = Z[y_k] \tag{2.23}$$

Or, in terms of the backward shift operator (z^{-1})

$$Z[y_{k-1}] = z^{-1}y(z) \tag{2.24}$$

A discrete-time transfer function can be developed for discrete controller design and analysis

$$y(z) = g_p(z)u(z) \tag{2.25}$$

Also, a discrete state space model is converted to an input–output transfer function model using

$$G(z) = \lfloor C(zI - A)^{-1}B + D \rfloor \tag{2.26}$$

We show the process of going from a continuous state space model to forms of discrete-time models by way of example. Consider the continuous two-state model

$$\begin{bmatrix} \dot{x}_1 \\ \dot{x}_2 \end{bmatrix} = \begin{bmatrix} -0.1 & 0 \\ 0.04 & -0.04 \end{bmatrix} \begin{bmatrix} x_1 \\ x_2 \end{bmatrix} + \begin{bmatrix} 0.1 \\ 0 \end{bmatrix} u$$

$$y = \begin{bmatrix} 0 & 1 \end{bmatrix} \begin{bmatrix} x_1 \\ x_2 \end{bmatrix} \tag{2.27}$$

which has the continuous, input–output transfer function

$$g_p(s) = \frac{1}{(10s + 1)(25s + 1)} \tag{2.28}$$

with a gain of 1 (units of the output divided by units of the input), and time constants of 10 and 25 (here, assume that the time units are minutes). The discrete state space model matrices, assuming a sample time of $\Delta t = 3$ min, are

$$\Phi = e^{A\Delta t} = \exp\begin{bmatrix} -0.3 & 0 \\ 0.12 & -0.12 \end{bmatrix} = \begin{bmatrix} 0.7408 & 0 \\ 0.0974 & 0.8869 \end{bmatrix}$$

$$\Gamma = (\Phi - I)\begin{bmatrix} -0.1 & 0 \\ 0.04 & -0.04 \end{bmatrix}^{-1}\begin{bmatrix} 0.1 \\ 0 \end{bmatrix} = \begin{bmatrix} 0.2592 \\ 0.0157 \end{bmatrix} \tag{2.29}$$

That is,

$$\begin{bmatrix} x_1 \\ x_2 \end{bmatrix}_{k+1} = \begin{bmatrix} 0.7408 & 0 \\ 0.0974 & 0.8869 \end{bmatrix}\begin{bmatrix} x_1 \\ x_2 \end{bmatrix}_k + \begin{bmatrix} 0.2592 \\ 0.0157 \end{bmatrix}u_k$$

$$y_k = \begin{bmatrix} 0 & 1 \end{bmatrix}\begin{bmatrix} x_1 \\ x_2 \end{bmatrix}_k \tag{2.30}$$

and the discrete process transfer function is

$$g_p(z) = \frac{0.0157z + 0.0136}{z^2 - 1.6277z + 0.65702} = \frac{0.0157z^{-1} + 0.0136z^{-2}}{1 - 1.6277z^{-1} + 0.65702z^{-2}} \tag{2.31}$$

which corresponds to the discrete input–output model

$$y_{k+1} = 1.6277y_k + -0.65702y_{k-1} + 0.0157u_k + 0.0136u_{k-1} \tag{2.32}$$

2.2.6   Step and Impulse Response Models

Other forms of discrete-time models include step and impulse response models. A step response model has the following form:

$$y_k = \sum_{i=1}^{\infty} s_i \Delta u_{k-i} \tag{2.33}$$

$$= s_1 \Delta u_{k-1} + \cdots + s_N \Delta u_{k-N} + s_{N+1} \Delta u_{k-N-1} + \cdots + s_{N|\infty} \Delta u_{k-\infty}$$

where $\Delta u_k = u_k - u_{k-1}$. For stable systems, the step response coefficients approach a constant value (thus, $s_{N+j} = s_N$)

$$y_k = s_1 \Delta u_{k-1} + \cdots + s_{N-1} \Delta u_{k-N+1} + s_N \Delta u_{k-N} + \cdots + s_N \Delta u_{k-\infty}$$

$$= s_1 \Delta u_{k-1} + \cdots + s_{N-1} \Delta u_{k-N+1} + s_N \underbrace{(\Delta u_{k-N} + \cdots + \Delta u_{k-\infty})}_{u_{k-N}}, \tag{2.34}$$

or

$$y_k = s_N u_{k-N} + \sum_{i=1}^{N-1} s_i \Delta u_{k-i} \tag{2.35}$$

Similarly, single-input, single-output impulse response models have the following form:

$$y_k = \sum_{i=1}^{\infty} h_i u_{k-i} \tag{2.36}$$

$$= h_1 u_{k-1} + \cdots + h_N u_{k-N} + h_{N+1} u_{k-N-1} + \cdots + h_{+\infty} u_{k-\infty}$$

Step and impulse response models can be found from state space models

$$H_i = C\Phi^{i-1}\Gamma \tag{2.37}$$

$$S_k = \sum_{i=1}^{k} C\Phi^{i-1}\Gamma = \sum_{i=1}^{k} H_i \tag{2.38}$$

where the H and S represent impulse and step response matrices, for the more general multivariable case.

2.2.7   System Identification

While discrete-time models can be developed from given continuous-time models, they are most often developed based on experimental data by using parameter estimation (or model identification). Step and impulse response models can be developed by perturbing an input, as a step or impulse, from steady state and observing the change in output. Care must be taken to smooth out the effects of measurement noise or other artifacts that may occur during the experiment. A disadvantage to the step response approach is that the measured outputs may be perturbed significantly from their desired steady-state values, while a disadvantage to the impulse response approach, particularly for small pulses, is that measurement noise may dominate the output response. More often, optimization-based methods are used to estimate model parameter values that provide a best fit to experimental data, and great care is taken to assure that the input signal provides an appropriate level of "excitation." One common input signal is a pseudorandom binary sequence, in which the input moves between specified minimum and maximum values, with the input changes occurring at random intervals. This discussion has been based on off-line analysis of data from the experiment, but real-time methods, such as recursive least squares, can be used, particularly in the context of adaptive control, as discussed later in this chapter. The reader is encouraged to consult system identification textbooks, such as Ljung [4], for more details.

2.3   Block Diagram Analysis

Transfer functions, whether continuous or discrete, are often used for the design and analysis of closed-loop systems.

2.3.1   Continuous-Time Block Diagram Analysis

Consider a simplified form of the block diagram in Figure 2.2, by lumping the dynamic contributions of the sensor and actuator into the process. The continuous controller and process (insulin−glucose dynamics) transfer functions are represented by $g_c(s)$ and $g_p(s)$, respectively. The signals for the set point, manipulated input (insulin delivery rate), and output (measured blood glucose), are represented by $r(s)$, $u(s)$, and $y(s)$, respectively. The relationship between the measured output and the set point is

$$y(s) = \frac{g_c(s)g_p(s)}{1 + g_c(s)g_p(s)} \cdot r(s) \tag{2.39}$$

The closed-loop stability is determined by solving for the roots of the characteristic polynomial, $1 + g_c(s)g_p(s)$. If the roots are negative, then the closed-loop system is stable.

2.3.2   Discrete-Time Block Diagram Analysis

Similar to the continuous-time case, the stability of a closed-loop discrete-time system can be checked by analyzing the poles of the closed-loop transfer function

$$y(z) = \frac{g_c(z)g_p(z)}{1 + g_c(z)g_p(z)} \cdot r(z) \tag{2.40}$$

The closed-loop system is stable if the roots of the $1 + g_c(z)g_p(z)$ polynomial have magnitudes less than 1.

The continuous and discrete system closed-loop analyses assume that transfer functions are available for both the controller and process. In Section 2.4, we develop example continuous and discrete controller transfer functions.

2.4 Proportional-Integral-Derivative Control

The most common form of control law is known as proportional-integral-derivative (PID) control. In continuous time, the control law is based on the error, or difference between the set point and measured output

$$e(t) = r(t) - y(t)$$

$$u(t) = u_0 + k_c \left[e(t) + \frac{1}{\tau_I} \int_0^t e(t)dt + \tau_D \frac{de(t)}{dt} \right] \tag{2.41}$$

where the three tuning parameters are (i) k_c, proportional gain, (ii) τ_I, integral time, and (iii) τ_D, derivative time.

The relationship between the error and the manipulated input in transfer function form is

$$u(s) = g_c(s)e(s) \tag{2.42}$$

where the ideal continuous PID transfer function is

$$g_c(s) = \frac{k_c(\tau_D\tau_I s^2 + \tau_I s + 1)}{\tau_I s} \tag{2.43}$$

There are at least two disadvantages to the ideal PID algorithm shown above: (i) a perfect derivative cannot actually be implemented and (ii) step set point changes cause discontinuous derivatives in the error and, therefore, manipulated inputs that cannot actually be implemented. Often, the derivative term will be based on the measured output, rather than the error, resulting in

$$u(t) = u_0 + k_c \left[e(t) + \frac{1}{\tau_I} \int_0^t e(t)dt - \tau_D \frac{dy(t)}{dt} \right] \tag{2.44}$$

Also, since the derivative of the output measurement can be sensitive to measurement noise, a filtered derivative is often used. For example, if a first-order filter is used, the transfer function representation is

$$y_f(s) = \frac{1}{\tau_f s + 1} \cdot y(s) \tag{2.45}$$

which, written as an ODE, is

$$\frac{\mathrm{d}y_f}{\mathrm{d}t} = \frac{1}{\tau_f} y_f + \frac{1}{\tau_f} y \tag{2.46}$$

with the PID equation expressed as

$$u(t) = u_0 + k_c \left[e(t) + \frac{1}{\tau_I} \int_0^t e(t)\mathrm{d}t - \tau_D \frac{\mathrm{d}y_f(t)}{\mathrm{d}t} \right] \tag{2.47}$$

2.4.1 PID Tuning Techniques

Many different techniques can be used to tune PID controllers; some are based on the use of a process model, while others are response-based.

2.4.1.1 Ziegler–Nichols Closed-Loop Oscillations

The Ziegler–Nichols [5] closed-loop oscillations tuning approach is developed by starting with a proportional-only controller and increasing the controller gain until a continuous oscillation results. The controller gain causing this is known as the critical gain, k_{cu}, and the other parameter is the period of oscillation, P_u. P, PI, and PID parameters are then found as a function of the critical gain and period of oscillation. The primary disadvantage to this method is that it tends to result in underdamped closed-loop responses that are too sensitive to uncertainty. Tyreus and Luyben modified these relationships to reduce the closed-loop oscillations [6].

2.4.1.2 Frequency Response

Frequency response-based techniques were initially developed in the 1930s and 1940s to understand and improve the behavior of telephone systems [7], which used negative feedback amplifiers to reduce the effect of noise. The basic concepts of gain and phase margins remain one of the best ways to understand the effect of uncertainty in feedback control; PID controller parameters can be adjusted to obtain desired gain and phase margins.

2.4.1.3 Cohen–Coon

The Cohen–Coon method of selecting tuning parameters is based on a first-order + dead time model. The tuning parameters are then functions of the process gain, time constant, and time delay. Tables of these relationships can be found in standard textbooks, including Ref. [3].

2.4.1.4 Internal Model Control-Based PID

Some model-based control techniques can result in feedback controllers that are equivalent to PID control. Internal model control (IMC), when based on low-order models with approximations to time delays and developed in standard feedback form, results in PID control [3,8]. The PID tuning parameters are functions of the process model parameters and an IMC tuning parameter (which is related to the desired closed-loop response time).

2.4.1.5 Ad hoc

The most common PID tuning approach is to make *ad hoc* adjustments to the parameters by increasing the controller gain to speed up the closed-loop response and increasing the integral time to reduce closed-loop oscillations. Because of sensitivity of derivative action to measurement noise, sometimes only the proportional-integral (PI) terms are used.

2.4.2 Discrete-Time PID

Most often, the control algorithm is implemented in discrete time. Assuming a finite-differences approximation to the derivative of the error, a discrete PID equation has the following form:

$$u(k) = u_0 + k_c \left[e(k) + \frac{\Delta t}{\tau_I} \sum_{i=0}^{k} e(i) + \frac{\tau_D}{\Delta t}(e(k) - e(k-1)) \right] \tag{2.48}$$

where Δt represents the constant sample time. It is more common to implement the control law using the following velocity form, based on changes in the manipulated input

$$u(k) = u(k-1) + k_c \left[\left(1 + \frac{\Delta t}{\tau_I} + \frac{\tau_D}{\Delta t} \right) e(k) + \left(-1 - \frac{2\tau_D}{\Delta t} \right) e(k-1) + \frac{\tau_D}{\Delta t} e(k-2) \right]$$

$$\tag{2.49}$$

or

$$u(k) - u(k-1) = b_0 e(k) + b_1 e(k-1) + b_2 e(k-2) \tag{2.50}$$

where

$$b_0 = k_c \left(1 + \frac{\Delta t}{\tau_I} + \frac{\tau_D}{\Delta t} \right), \quad b_1 = -k_c \left(1 + \frac{2\tau_D}{\Delta t} \right), \quad b_2 = \frac{k_c \tau_D}{\Delta t}$$

Note that the corresponding z-domain (discrete-time) controller transfer function is

$$u(z) = g_c(z)e(z) = \frac{(b_0 + b_1 z^{-1} + b_2 z^{-2})}{1 - z^{-1}} e(z) \tag{2.51}$$

and the closed-loop stability can be analyzed by finding the poles of the closed-loop equation (2.40).

2.5 Model Predictive Control

The most widely applied advanced control technique used in the process industries is known as model predictive control (MPC) (see Ref. [9] for early work in the field); this basic approach has recently been used in a number of biomedical systems control problems. The basic concept of MPC is shown in Figure 2.5 for the blood glucose control (artificial pancreas) example presented earlier. At the current sample time, there is a history of previous measured outputs (glucose) and manipulated inputs (insulin) that were applied. A model is used to predict the future output values based on a proposed set of current and future manipulated input changes. An optimizer adjusts these control moves until an objective function is minimized. Most often, the objective function is based on the sum of the squares of future errors (differences between set point and predicted output). An example objective function, shown for a system with one manipulated input and one controlled output, is

$$J = \sum_{i=1}^{P} (r_{k+i} - \hat{y}_{k+i})^2 + w \sum_{i=0}^{M-1} \Delta u_{k+i}^2 \qquad (2.52)$$

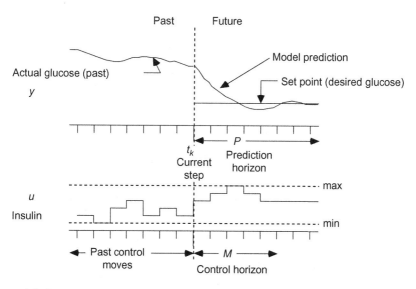

Figure 2.5 Schematic of the optimization-based computation of current and future control moves (control horizon = M) to minimize an objective function over the prediction horizon (P).

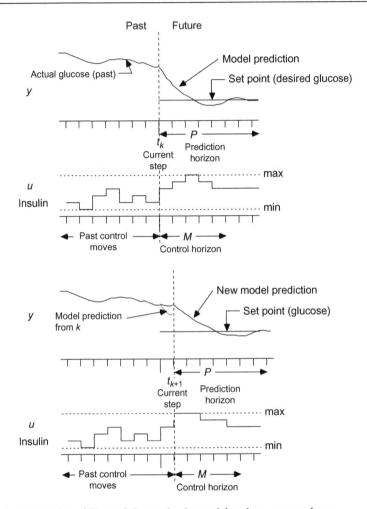

Figure 2.6 Expansion of Figure 2.5 to make the model update process clearer.

where P and M are the prediction and control horizons, respectively, w is a weight on the control moves (Δu), r is the set point, and $\hat{y}$ is the model predicted output.

Although a sequence of future control moves is calculated, only the first move is implemented, with a new measurement obtained at the next sample time, as shown in Figure 2.6.

In Figure 2.6A, the controller has calculated a set of current and future control moves at time step k. The first control move is implemented at time step k, then a new measurement is obtained at step $k + 1$, as shown in Figure 2.6B. Since the model predicted output at step $k + 1$ (which was based on information at step k) is different than the actual measured output, it is important to compensate for this error in the predictions that start with the new time step $(k+1)$, as shown in Figure 2.6B. The simplest compensation method to assume a constant additive disturbance term to correct for the model prediction and measurement difference at the current time step

$$\hat{p}_{k|k} = y_k - \hat{y}_{k|k-1}$$
$$\hat{y}_{k|k} = \hat{y}_{k|k-1} + \hat{p}_{k|k} \tag{2.53}$$

where the subscript $k|k-1$ indicates the output prediction at step k based on measurements at step $k-1$. The prediction at the next time step, when a linear discrete-time model is used, is

$$\hat{x}_{k+1|k} = \Phi\hat{x}_{k|k} + \Gamma u_k$$
$$\hat{p}_{k+1|k} = \hat{p}_{k|k}$$
$$\hat{y}_{k+1|k} = C\hat{x}_{k+1|k} + \hat{p}_{k+1|k} = C\Phi\hat{x}_{k|k} + C\Gamma u_{k-1} + C\Gamma\Delta u_k + \hat{p}_{k+1|k} \tag{2.54}$$

and the predictions P steps into the future are found similarly. The output predictions can be written in matrix-vector form as

$$\begin{bmatrix} \hat{y}_{k+1|k} \\ \hat{y}_{k+2|k} \\ \vdots \\ \hat{y}_{k+P|k} \end{bmatrix} = \underbrace{\begin{bmatrix} C\Phi \\ C\Phi^2 \\ \vdots \\ C\Phi^P \end{bmatrix} \hat{x}_{k|k} + \begin{bmatrix} I \\ I \\ \vdots \\ I \end{bmatrix} \hat{p}_{k|k} + \begin{bmatrix} C\Gamma \\ C\Phi\Gamma + C\Gamma \\ \vdots \\ \sum_{i=1}^{P} C\Phi^{i-1}\Gamma \end{bmatrix} u_{k-1}}_{\text{``free'' or ``unforced response'' (if no more control moves are made)}}$$
$$+ \underbrace{\begin{bmatrix} C\Gamma & 0 & \cdots & 0 \\ C\Phi\Gamma + C\Gamma & C\Gamma & & 0 \\ \vdots & \vdots & & \\ \sum_{i=1}^{P} C\Phi^{i-1}\Gamma & \sum_{i=1}^{P-1} C\Phi^{i-1}\Gamma & \cdots & \end{bmatrix} \begin{bmatrix} \Delta u_k \\ \Delta u_{k+1} \\ \vdots \\ \Delta u_{k+M-1} \end{bmatrix}}_{\text{``forced'' response}} \tag{2.55}$$

where the first terms on the right-hand side represent the *free response* (change in the predicted output if no further control moves are made), and the final term represents the *forced response* (based on control moves to be computed). This can be written in compact form as

$$\hat{Y} = f + S_f \Delta u_f \tag{2.56}$$

It should be noted that other methods of compensating for the difference between model predictions and measured outputs can be used [10], but these methods generally result in the same form as Eq. (2.56).

Writing the objective function in the following form:

$$J = \sum_{i=1}^{P} (r_{k+i} - \hat{y}_{k+i})^T W^y (r_{k+i} - \hat{y}_{k+i}) + \sum_{i=0}^{M-1} \Delta u_{k+i}^T W^u \Delta u_{k+i} \qquad (2.57)$$

and using matrix-vector notation, the optimization problem is stated

$$\min_{\Delta u_f} J = \hat{E}^T W^Y \hat{E} + \Delta u_f^T W^U \Delta u_f \qquad (2.58)$$

where $\hat{E} = r - \hat{Y} = r - f - S_f \Delta u_f$ and the unforced error, E, is written $E = r - f$. The manipulated input vector is

$$\Delta u_f = \begin{bmatrix} \Delta u_k \\ \vdots \\ \Delta u_{k+M-1} \end{bmatrix}$$

and the weighting matrices are

$$W^Y = \begin{bmatrix} W^y & 0 & 0 \\ 0 & \ddots & 0 \\ 0 & 0 & W^y \end{bmatrix} \qquad W^U = \begin{bmatrix} W^u & 0 & 0 \\ 0 & \ddots & 0 \\ 0 & 0 & W^u \end{bmatrix}$$

The quadratic objective function with a linear model results in the analytical unconstrained solution for the vector of current and future control actions as a function of the unforced error, E:

$$\Delta u_f = (S_f^T W^Y S_f + W^U)^{-1} S_f^T W^Y E \qquad (2.59)$$

If there are constraints, a quadratic program (QP) is used to solve the problem [11]; there are many efficient QP codes for this purpose. Whether constrained or unconstrained solutions are used, the first element of Δu_f (i.e., Δu_k) is implemented and the optimization is performed again at the next sample time.

It should be noted that MPC is a broad general approach and that many different types of models (including nonlinear), objective functions, and methods to compensate for model error can be used. This section has served as a basic introduction to MPC by focusing on linear discrete state space models, quadratic objective functions, and an "additive disturbance" assumption for model error compensation.

2.6   Other Control Algorithms

PID and MPC are perhaps the two most frequently used algorithms for biomedical (and other) control problems, but a wide range of other algorithms and approaches have also been used. These other approaches include fuzzy logic, expert systems, and even simple on−off control, as summarized below.

2.6.1   Fuzzy Logic

The field of fuzzy logic and fuzzy systems theory is based on the notion that some input−output relationships are not "crisp." Consider a process where a particular manipulated input change may result in possibly three different magnitudes of changes in an output: low, medium, and high. Fuzzy logic would provide some smoothing to indicate that the output might be a mix of low and medium, for example.

2.6.2   Expert Systems

Expert systems are basically rule based, with rules provided by "experts" with knowledge of the system at hand. These types of models are often used as protocols for insulin delivery in critical care, for example. Here, the clinician would specify rules, such as, if the glucose value is between X and Y, then deliver Z units of insulin. This type of strategy can often be implemented in a fuzzy logic-based framework.

2.6.3   Artificial Neural Networks

Artificial neural networks (ANNs) evolved from a physiological description of the function of neurons and neural networks in animals. An ANN is now more generally used to provide a nonlinear relationship between inputs and outputs. An ANN is first *trained* by providing known input and output data and optimizing parameters in the ANN to provide a best fit to the data. Model verification, or validation, is performed by testing the performance on input−output data that were not used for training.

2.6.4   On−Off

The simplest example of an on−off controller is a simple household thermostat, in which the heating system is switched on when the room temperature is below the set point, and turned off when above the set point. In practice, there is a small deadband to prevent rapid on−off cycling; for example, the heater may be switched on 0.5°C below the set point and switched off 0.5°C above the set point.

2.7 Application Examples

The previous sections have provided an overview of modelling for control, and control algorithms, with a focus on linear systems theory. In this section, we provide a concise review of specific applications of control to physiological systems, with a focus on drug infusion.

2.7.1 Type 1 Diabetes: Blood Glucose Control

An individual with type 1 (juvenile) diabetes must frequently test blood glucose levels and inject insulin several times each day. Most often, they are on multiple daily injection therapy, providing a bolus of long-acting insulin each day, and boluses of rapid-acting insulin at mealtime, or to correct for high blood glucose values. They serve as a *human-in-the-loop* controller by providing feedforward control with an insulin bolus based on their estimate of the amount of carbohydrates in a meal (using an individualized carb/insulin ratio), and feedback control by providing a correction bolus of insulin based on their blood glucose values (and an individualized correction factor, CF).

The development of a closed-loop artificial pancreas, as shown in Figure 2.2, has been an active research area for nearly 50 years. Initial studies were based on intravenous (IV) blood sampling and IV insulin infusion. Kadish [12] developed an on−off strategy, operating much like a household thermostat; insulin was delivered if the blood glucose was above 150 mg/dL, and glucose was delivered if blood glucose was below 50 mg/dL. There are obvious limitations to the on−off approach of Kadish, and the next generation of devices, such as the Biostator produced by Miles Laboratories [13], incorporated more smoothly varying insulin and glucose infusions as a nonlinear function of the measured glucose values. These were bedside devices that could not be used in an ambulatory setting as would be necessary for a truly closed-loop artificial pancreas.

More recent efforts are based on subcutaneous measurements and insulin infusion, using commercially available sensors and pumps, which are the focus of this section. Recent artificial pancreas review articles include Kumareswaran et al. [14], Cobelli et al. [15], and Bequette [16].

2.7.1.1 Models for Simulation

A realistic simulation environment with a wide variety of simulated subjects enables the development of control strategies that are robust and reliable. Indeed, the US Food and Drug Administration (FDA) accepted the UVa−Padova simulator [17] for use in simulated clinical trials, enabling investigators to skip the animal trial stage; this simulator is based on a model presented by Dalla Man et al. [18]. Patek et al. [19] discuss this approach for simulated closed-loop clinical trials. The UVa−Padova simulator contains 300 subjects, and includes sensor errors representative of two continuous glucose monitors (CGMs) and the discrete resolution from two insulin pumps. Wilinska et al. [20,21] discuss the use of simulation studies, based on the model presented by

Hovorka et al. [22], for evaluating model predictive control strategies in simulated clinical trials; their simulation studies involve 18 different subject parameter sets.

2.7.1.2 Models for Control

Ordinarily control-relevant models are developed by using either experimental data, or data from more detailed simulation-based models. A discrete compartmental model, which is individualized by a CF based on the total daily insulin dose, using the 1800 rule [23], is used by Cameron et al. [24]. An integrating first-order + dead time model relating insulin infusion to blood glucose is proposed by Percival et al. [25].

$$g_p(s) = \frac{K_p \exp(-\theta_p s)}{s(\tau_p s + 1)} \tag{2.60}$$

where the model parameters are based on clinically relevant information, such as total daily insulin, $I{:}C$ ratio, and CF. Similarly, van Heusden et al. [26] develop personalized discrete-time control-relevant models based on extensive simulations using the UVa—Padova simulator discussed in Section 2.7.1.1. The second-order dynamics are constant from subject to subject, but the gains are a function of their CF and a safety factor

$$g_p(z) = \frac{F_s K_i c z^{-3}}{(1 - 0.98z^{-1})(1 - 0.965z^{-1})} \tag{2.61}$$

and where a 5-min sample time is assumed.

2.7.1.3 Control

Algorithms for a closed-loop artificial pancreas can largely be placed into one of four categories: (i) on—off (low glucose suspend, LGS), (ii) PID, (iii) MPC, and (iv) fuzzy logic.

2.7.1.3.1 On—Off

The greatest fear of the parent of a child with type 1 diabetes is overnight hypoglycemia (low blood sugar), which, if occurs for an extended period of time, can result in a coma or even (in rare cases) death. Continuous glucose monitors can be set to alarm of hypoglycemia, but individuals (and their caregivers) often sleep through alarms. It is desirable to take people out of the loop and simply shut off the pump rather than sound an alarm; this approach is often called *low glucose suspend*. Two basic LGS approaches can be used: (i) threshold, where the pump is shut off when glucose goes below a threshold value and (ii) prediction, where the pump is shut off when the glucose is predicted to be below a specified value within a future prediction horizon. Choudhary et al. [27] and Danne et al. [28] report outpatient results based on the Medtronic Paradigm Veo, which can suspend the basal insulin delivery for up to 2 h when hypoglycemia is detected by a CGM; the shut-off threshold can be set between 40 and 70 mg/dL. A Kalman filter-based predictive LGS approach was used by Cameron et al. [29] in clinical studies, preventing hypoglycemia in 73% of subjects that had data sets suitable for analysis. Initial outpatient studies are reported by Buckingham et al. [30].

2.7.1.3.2 Proportional-Integral-Derivative (PID)

The Medtronic external physiological insulin delivery (ePID) system includes a PID controller that has been used in animal [31] and human studies [32]. The recent approach used by Medtronic involves model-based feedback of insulin concentration, creating a cascade type of strategy [33,34], called ePID-IFB. Gopakumaran et al. [35] developed a fading memory proportional derivative controller that is roughly equivalent to PID. Castle et al. [36] manipulate both insulin and glucagon in human studies. A strategy by El-Youssef et al. [37] adapts to changing insulin sensitivity that is induced by hydrocortisone administration. Steil [38] reviews a number of PID studies and makes a strong case for the use of PID in a closed-loop artificial pancreas.

2.7.1.3.3 Model Predictive Control (MPC)

Cameron et al. [24] develop a multiple model probabilistic predictive control (MMPPC) approach, with meal probabilities continuously estimated to detect unannounced meals; extensions to the meal modelling approach are presented by Cameron et al. [39]. In simulation studies a risk measure is minimized, also considering the uncertainty. A discrete compartmental model is used, which is individualized by a CF based on total daily dose (TDD), using the 1800 rule. The performance of the MMPPC strategy is compared with several other algorithms in Figure 2.7 (where EMPC refers to the MMPPC strategy). Cameron et al. [40] revise the approach used in simulation studies for their clinical studies involving 10 subjects.

A number of other MPC-based strategies have been conducted in clinical trials. Elleri et al. [41] perform 36-h studies in adolescents. The algorithm is similar to Elleri et al. [42], and is initialized using the subject's weight, total daily insulin dose (mean of the previous 3 days), and the 24-h basal insulin profile programmed on the pump. The algorithm is adapted by updating endogenous glucose flux and carbohydrate bioavailability.

A control-to-range type of approach, known as zone MPC, is used in simulation studies by Grosman et al. [43]; prediction and control horizons of 180 and 25 min, respectively, are used. Dassau et al. [44] present clinical results using a multiparametric MPC algorithm, which has the advantage of a fast computation time; 6-h prediction and 30 min control horizons are used.

Turksoy et al. [45] develop an adaptive generalized predictive control (GPC) strategy to regulate the estimated blood glucose levels based on CGM measurements. In an approach that involves dual hormone delivery (glucagon and insulin), El-Khatib et al. [46], use a PD controller for glucagon and an adaptive GPC strategy for insulin delivery. Bequette [47] summarizes a number of MPC-based strategies in this analysis of the artificial pancreas.

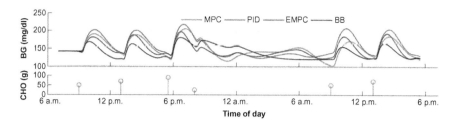

Figure 2.7 Performance of controllers averaged over the nine valid simulated subjects. EMPC (Enhanced MPC) is the MMPPC strategy. The PID controller parameters were adjusted to minimize the blood glucose risk index averaged over the subjects. MPC represents a "standard" MPC strategy with a symmetric objective function. The basal—bolus strategy represents optimal performance and is based on perfect meal knowledge; none of the other strategies used meal anticipation.
Source: Figure reproduced from Cameron et al. [24].

2.7.1.3.4 Fuzzy Logic

A fuzzy logic-based approach that uses a combination of control-to-range and control-to-set point strategies is incorporated into the MD-Logic Artificial Pancreas System [48], with a 5-min sample time; this has been tested in a trial on seven adults, without the use of meal announcement. Mauseth et al. [49] describe a fuzzy logic-based controller with a 15-min sample time that uses blood glucose (BG), its rate of change, and its acceleration as inputs, and is tuned based on a personalization factor. A preliminary version was tested (without a personalization factor) on four subjects before enhancements were made and performance was demonstrated in simulation studies. Overnight studies using this control strategy at a diabetes camp are reported by Phillip et al. [50].

2.7.2   *Intensive Care Unit Blood Glucose Control*

Individuals who are critically ill may suffer from hyperglycemia (high blood glucose) and insulin resistance, even if they do not have diabetes. Current clinical practice in the intensive care unit (ICU) often requires monitoring and control of blood glucose to maintain euglycemia in the face of injury or illness; this practice is called tight glycaemic control.

2.7.2.1   *Models*

A number of different models are available for studies of insulin—glucose dynamics in critical care, including Florian and Parker [51], Chee et al. [52], and Chase et al. [53]. The physiological model presented by Wong et al. [54] is used in the studies performed by Bequette [55].

The equations are

$$\frac{dG}{dt} = -p_G G - S_I(G + G_E)\frac{Q}{1 + \alpha_G Q} + \frac{G_f}{V_G}$$

$$\frac{dQ}{dt} = -kQ + kI$$

$$\frac{dI}{dt} = \frac{-nI}{1 + \alpha_I I} + \frac{u_{ex}}{V_I}$$

where the parameters and variables are defined as (example values are shown in parentheses)

G is the glucose concentration, as a perturbation from G_E
G_E is the equilibrium glucose concentration, with no external glucose feeding or insulin infusion
Q is the insulin concentration that directly affects the glucose
I is the insulin concentration in the insulin compartment
p_G is the glucose clearance rate (0.02 min^{-1})
S_I is the insulin sensitivity (0.002 L/(min*mU))
α_G is a parameter that accounts for saturation of the insulin effect on glucose ($1/65$ L/mU)
G_f is the glucose feed rate directly into the glucose compartment
V_G is glucose distribution volume (15 L)
k is the rate constant for insulin transfer into the effective compartment (0.0099 min^{-1})
n is a parameter (0.16 min^{-1})
α_I is a saturation parameter (0.0017 L/mU)
V_I is the insulin distribution volume (12 L)
u_{ex} is the exogenous insulin infusion rate.

The steady-state relationship between insulin infusion rate and blood glucose concentration, for various values of G_E are shown in Figure 2.8. Observe that the gain (change in glucose for a given change in insulin infusion rate) is high at higher glucose levels (and lower insulin infusion rates). Due to saturation effects, this gain decreases with increasing insulin infusion rates.

2.7.2.2 Control

Initial closed-loop algorithms for the ICU application were based on directly sampling the blood glucose at intervals ranging from 1 to 4 h, and are studied in detail by Bequette [55]. PID [56], Columnar Insulin Dosing [57], and GRIP [58] algorithms are shown to have similar features and performance. The columnar insulin dosing strategy is a time-varying proportional-only controller (no integral action), while the GRIP algorithm is a nonlinear controller with integral action. A minor modification to the GRIP algorithm is suggested to improve the closed-loop performance. A detailed review of model-based ICU glucose control strategies is provided by Chase [53], while Wong et al. [54] present clinical trial results for a model-based strategy to simultaneously deliver insulin and nutrition.

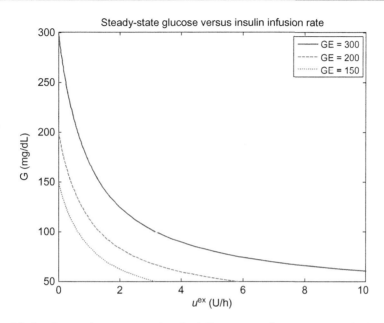

Figure 2.8 Steady-state input–output (insulin delivery rate–glucose concentration) curves for three hypothetical subjects, with identical parameters except for equilibrium glucose concentration (concentration when no exogenous insulin is delivered).

In the algorithms discussed above, the nurse is often in the loop, inputting glucose concentrations into a device containing the algorithm, and programming the infusion pump to provide the rate calculated by the algorithm. More recent research has been based on the use of frequent sensor measurements and infusion pump changes in an automated closed-loop system. In Sun et al. [59], an MPC strategy, based on simultaneously manipulating IV glucose and insulin infusion, is developed to improve blood glucose regulation in ICU patients. In the short term, glucose infusion is used for tighter glucose control, particularly for disturbance rejection, while, in the long term (24-h period), glucose infusion is used to meet nutritional needs. The proposed *habituating control* algorithm is shown to outperform a single-input single-output MPC (which only manipulated insulin) by providing faster set point tracking and tighter glucose control for a patient population, and producing less glucose variability while rejecting disturbances in insulin infusion and insulin sensitivity.

2.7.3   Blood Pressure Control Using Continuous Drug Infusion

It is common for patients that have undergone a cardiac operation to have high blood pressure. These patients are often treated by continuous infusion of sodium nitroprusside (SNP) in the postoperative ICU. It has been estimated that for these hypertensive patients, an ICU nurse spends up to 26% of her/his time monitoring blood pressure and adjusting the SNP infusion rate. Clearly, a reliable automatic closed-loop controller would enable the nurse to spend less time on this tiring task, and to spend more time tending to other responsibilities. Also, during surgery, anesthesiologists monitor the status of patients and regulate blood pressure and cardiac output (CO) (particularly for cardiac surgery) by manipulating the infusion rates of two or more drugs. See Bequette [60] for an overview of drug infusion control.

2.7.3.1 Models

Complex physiological system simulation models have been developed to predict the effect of drugs on hemodynamic variables, such as mean arterial pressure (MAP) and CO. Yu et al. [61] developed a simulation model that includes the (i) circulatory system equations, which describe the effect of specific body parameters on the hemodynamic variables, (ii) drug effect relationships, which describe the influence of the infused drugs on the specific body parameters, and (iii) equations that describe the baroreflex mechanism (the body's blood pressure regulatory mechanism).

While physiological simulation models are good for performing initial controller design and robustness studies, they can be quite limited in predictive capability for any single patient. Often a low-order transfer function model, such as first-order + dead time, is used for control system design and implementation. A major challenge is that drug sensitivity (process gain) can vary tremendously from patient to patient, and as a function of time (and anesthetic level) for any individual patient. This motivated the multiple model-based approach presented, for example, by Yu et al. [62].

2.7.3.2 Control

The poor performance of controllers based on a single fixed model motivated the use of a bank of several models to characterize possible dynamic behavior of the patient, and the formulation of the solution as a multiple model predictive control (MMPC) problem. Based on the recent drug responses of the patient, relative weights are assigned to each of the models, to find the best weighted step response model that describes the behavior. A constrained model predictive controller is then used to determine the infusion rates. Although a constrained optimization (QP) approach is used, the control calculations take less than 2 s, which is minimal for a system with a sample time of 30 s. Detailed multivariable results (simultaneously regulating blood pressure and CO) using the MMPC approach in animal studies are shown by Rao et al. [63,64].

2.7.4   Control of Anesthesia and Sedation

The use of control technology in anesthesia has a long history, with initial studies on feedback control performed by Bickford [65] at the Mayo Clinic. Several different anesthetics were studied as the manipulated input, while the measured variable was the integrated rectified amplitude of the EEG. The control strategy did not gain acceptance because it was not clear what feature of the EEG signal should be used as an indicator of the depth of anesthesia [66]. More recently, the bispectral index (BIS), based on a nonlinear function of the EEG, has been used to infer anesthetic depth [67].

While vapor-phase anesthetics were commonly used for decades, it is becoming more common for anesthesiologists to use liquid phase anesthetics, such as propofol, which are often delivered using infusion pumps. Propofol has a more rapid on−off effect than isoflurane, for example, along with fewer complications such as postoperative nausea.

2.7.4.1 Models

Typically, two- and three-state compartmental-based pharmacokinetic and pharmacodynamics models have been used. A pharmacokinetic model describes the dynamics of drug concentration in the blood, while a pharmacodynamics model describes the dynamics of the drug effect. A pharmacokinetic model of propofol is used to predict the plasma concentration of propofol, while a pharmacodynamics model is used to predict the effect site concentration (which usually cannot be measured, since the site is often the brain). For an overview of modelling and control topics in clinical pharmacology, see Bailey and Haddad [68].

2.7.4.2 Open-Loop Control

The effect of the intravenous anesthetic propofol is directly related to its concentration in the blood. A model-based open-loop strategy to regulate the concentration of a drug in the blood, by providing an initial bolus, followed by time-dependent infusion, is known as target controlled infusion (TCI); see Bailey and Shafer [69] for early results, and Schraag [70] for a more recent review. A commercial device, the Diprifusor, has been available throughout much of the world since 1996, with millions of successful propofol infusions administrated [71]. For a variety of reasons, no TCI device has received FDA approval in the United States [72].

2.7.4.3 Closed-Loop Control

The BIS has been successfully used as a measure of depth of anesthesia, and therefore serves as a reliable measured variable in a closed-loop control strategy, usually with propofol infusion rate as the manipulated input. Haddad and Bailey [73] review a number of algorithms that have been used, including PID [74] and adaptive control [75].

2.8 Summary

The goals of this chapter were to concisely review classic modelling and control techniques used in biomedicine and physiology, and to review specific control applications. Continuous-time models in the original nonlinear ODE form were first presented, followed by linearization into state space form. The Laplace transform is used to create transfer function models. An overview of continuous-time PID control was then discussed. Discrete-time models were presented, and the discrete-time forms of PID controllers were discussed. MPC was developed, and a summary of fuzzy logic and ANNs-based methods was provided. Finally, a range of biomedical control applications were reviewed, including the closed-loop artificial pancreas, ICU blood glucose control, and drug infusion in anesthesiology and clinical pharmacology.

References

[1] Freudenrich C. How diabetes works, < http://science.howstuffworks.com/life/human-biology/diabetes1.htm > ; [accessed 03.05.13].

[2] Bequette BW. A critical assessment of algorithms and challenges in the development of an artificial pancreas. Diabetes Technol Ther 2005;7(1):28−47.

[3] Bequette BW. Process control: modeling, design and simulation. Upper Saddle River, NJ: Prentice Hall; 2003.

[4] Ljung L. System identification: theory for the user. 3rd ed. Upper Saddle River, NJ: Prentice Hall; 1999.

[5] Ziegler JG, Nichols NB. Optimum settings for automatic controllers. Trans ASME 1942;64:750−68.

[6] Luyben ML, Luyben WL. Essentials of process control. New York, NY: McGraw-Hill; 1997.

[7] Kline R. Harold Black and the negative-feedback amplifier. IEEE Control Syst Mag 1993;13(4):82−5.

[8] Rivera DE, Morari M, Skogestad S. Internal model control 4. PID Controller Design Ind Eng Chem Process Des Dev 1986;25:252−65.

[9] Cutler CR, Ramaker BL. Dynamic matrix control—a computer control algorithm. In: Proceedings of the joint automatic control conference. San Francisco, CA; 1980. Paper WP5-B.

[10] Muske KR, Badgwell TA. Disturbance modeling for offset-free linear model predictive control. J Process Control 2002;12:617−32.

[11] Garcia CE, Morshedi AM. Quadratic programming solution of dynamic matrix control (QDMC). Chem Eng Commun 1986;46:73−87.

[12] Kadish AH. Automation control of blood sugar. I. A servomechanism for glucose monitoring and control. Am J Med Electron 1964;3:82−6.

[13] Clemens AH. Feedback control dynamics for glucose controlled insulin infusion systems. Med Prog Technol 1979;6:91−8.

[14] Kumareswaran K, Evans ML, Hovorka R. Artificial pancreas: an emerging approach to treat type 1 diabetes. Expert Rev Med Dev 2009;6(4):401−10.

[15] Cobelli C, Renard E, Kovatchev B. Artificial pancreas: past, present and future. Diabetes 2011;60(11):2672−82.

[16] Bequette BW. Challenges and progress in the development of a closed-loop artificial pancreas. Annu Rev Control 2012;36:255−66.

[17] Kovatchev BP, Breton M, Dalla Man C, Cobelli C. In silico preclinical trials: a proof of concept in closed-loop control of type 1 diabetes. J Diabetes Sci Technol 2009;3 (1):44−55.

[18] Dalla Man C, Raimondo DM, Rizza RA, Cobelli C. GIM, simulation software of meal glucose—insulin model. J Diabetes Sci Technol 2007;1(3):323—30.

[19] Patek SD, Bequette BW, Breton M, Buckingham BA, Dassau E, Doyle III FJ, et al. In silico preclinical trials: methodology and engineering guide to closed-loop control. J Diabetes Sci Technol 2009;3:269—82.

[20] Wilinska ME, Budiman ES, Taub MB, Elleri D, Allen JM, Acerini CL, et al. Overnight closed-loop insulin delivery with model predictive control: assessment of hypoglycemia and hyperglycemia risk using simulation studies. J Diabetes Sci Technol 2009;3:1109—20.

[21] Wilinska ME, Chassin LJ, Acerini CL, Allen JM, Dunger DB, Hovorka R. Simulation environment to evaluate closed-loop insulin delivery systems in type 1 diabetes. J Diabetes Sci Technol 2010;4:132—44.

[22] Hovorka R, Canonico V, Chassin LJ, Haueter U, Massi-Benedetti M, Fedrici MO, et al. Nonlinear model predictive control of glucose concentration in subjects with type 1 diabetes. Physiol Meas 2004;25(4):905—20.

[23] Walsh J, Roberts R, Bailey T. Guidelines for insulin dosing in continuous subcutaneous insulin infusion using new formulas from a retrospective study of individuals with optimal glucose levels. J Diabetes Sci Technol 2010;4(5):1174—81.

[24] Cameron F, Bequette BW, Wilson DM, Buckingham BA, Lee H, Niemeyer GA. Closed-loop artificial pancreas based on risk management. J Diabetes Sci Technol 2011;5(2):368—79.

[25] Percival MW, Bevier WC, Wang Y, Dassau E, Zisser H, Jovanovic L, et al. Modeling the effects of subcutaneous insulin administration and carbohydrate consumption on blood glucose. J Diabetes Sci Technol 2010;4(5):1214—28.

[26] van Heusden K, Dassau E, Zisser HC, Seborg DE, Doyle III FJ. Control-relevant models for glucose control using a priori patient characteristics. IEEE Trans Biomed Eng 2012;59(7):1839—49.

[27] Choudhary P, Shin J, Wang Y, Evans ML, Hammond PJ, Kerr D, et al. Insulin pump therapy with automated insulin suspension in response to hypoglycemia. Reduction in nocturnal hypoglycemia in those at greatest risk. Diabetes Care 2011;34:2023—5.

[28] Danne T, Kordonouri O, Holder M, Haberland H, Golembowski S, Remus K, et al. Prevention of hypoglycemia by using low glucose suspend function in sensor-augmented pump therapy. Diabetes Technol Ther 2011;13(11):1129—34.

[29] Cameron F, Wilson DM, Buckingham BA, Arzumanyan H, Benzsi K, Chase HP, et al. In-patient studies of a Kalman filter based predictive pump shut-off algorithm. J Diabetes Sci Technol 2012;6(5):1142—7.

[30] Buckingham BA, Cameron F, Calhoun P, Maahs DM, Wilson DM, Chase HP, et al. Outpatient safety assessment of an in-home predictive low-glucose suspend system with T1D subjects at elevated risk of nocturnal hypoglycemia. Diabetes Tech Ther 2013;15(8):622—7.

[31] Loutseiko M, Voskanyan G, Keenan DB, Steil GM. Closed-loop insulin delivery utilizing pole placement to compensate for delays in subcutaneous insulin delivery. J Diabetes Sci Technol 2011;5(6):1342—51.

[32] Weinzimer SA, Steil GM, Swan KL, Dziura J, Kurtz N, Tamborlane WV. Fully automated closed-loop insulin delivery versus semiautomated hybrid control in pediatric patients with type 1 diabetes using an artificial pancreas. Diabetes Care 2008;31(5):934—9.

[33] Palerm CC. Physiologic insulin delivery with insulin feedback: a control systems perspective. Comp Meth Prog Biomed 2011;102(2):130—7.

[34] Steil GM, Palerm CC, Kurtz N, Voskanyan G, Roy A, Paz S, et al. The effect of insulin feedback on closed loop glucose control. J Clin Endocrinol Metab 2011;96 (5):1402−8.

[35] Gopakumaran B, Duman HM, Overholser DP, Federiuk IF, Quinn MJ, Wood MD, et al. A novel insulin delivery algorithm in rats with type 1 diabetes: the fading memory proportional-derivative method. Artif Organs 2005;29(8):599−607.

[36] Castle JR, Engle JM, El-Youssef J, Massoud RG, Kagan R, Ward WK. Novel use of glucagon in a closed-loop system for prevention of hypoglycemia in type 1 diabetes. Diabetes Care 2010;33:1281−7.

[37] El-Youssef J, Castle JR, Branigan DL, Massoud RG, Breen ME, Jacobs PG, et al. Controlled study of the effectiveness of an adaptive closed-loop algorithm to minimize corticosteroid-induced stress hyperglycemia in type 1 diabetes. J Diabetes Sci Technol 2011;5(6):1312−26.

[38] Steil GM. Algorithms for a closed-loop artificial pancreas: the case for PID control. J Diabetes Sci Technol 2013;7(6): [in press]

[39] Cameron F, Niemeyer G, Bequette BW. Extended multiple model prediction with application to blood glucose regulation. J Proc Cont 2012;1422−32.

[40] Cameron F, Niemeyer G, Wilson DM, Bequette BW, Buckingham BA. Clinical trials of a closed-loop artificial pancreas rejected large unannounced meals. Presented at the American diabetes association annual meeting, Chicago, IL; June 2013.

[41] Elleri D, Allen JM, Kumareswaran K, Leflarathna L, Nodale M, Caldwell K, et al. Closed-loop basal insulin delivery over 36 hours in adolescents with type 1 diabetes. Diabetes Care 2013;36(4):838−44.

[42] Elleri D, Allen JM, Nodale M, Wilinska ME, Mangat JS, Larsen AMF, et al. Automated overnight closed-loop glucose control in young children with type 1 diabetes. Diabetes Technol Ther 2011;13(4):419−24.

[43] Grosman B, Dassau E, Zisser HC, Jovanovic L, Doyle III FJ. Zone model predictive control: a strategy to minimize hyper- and hypoglycemic events. J Diabetes Sci Technol 2010;4:961−75.

[44] Dassau E, Zisser H, Harvey RA, Percival MW, Grosman B, Bevier W, et al. Clinical evaluation of a personalized artificial pancreas. Diabetes Care 2013;36(4):801−9.

[45] Turksoy K, Bayrak ES, Quinn L, Littlejohn E, Cinar A. Multivariable adaptive closed-loop control of an artificial pancreas without meal and activity announcement. Diabetes Technol Ther 2013;15(5):386−400.

[46] El-Khatib FH, Russell SJ, Nathan DM, Sutherlin RG, Damiano ERA. Bihormonal closed-loop artificial pancreas for type 1 diabetes. Sci Transl Med 2010;2: 27ra27

[47] Bequette BW. Algorithms for a closed-loop artificial pancreas: the case for model predictive control (MPC). J Diabetes Sci Technol 2013;7(6): [in press]

[48] Atlas E, Nimri R, Miller S, Grunberg EA, Phillip M. MD-logic artificial pancreas systems. Diabetes Care 2010;33(5):1072−6.

[49] Mauseth R, Wang Y, Dassau E, Kircher R, Matheson D, Zisser H, et al. Proposed clinical application for tuning fuzzy logic controller of artificial pancreas utilizing a personalization factor. J Diabetes Sci Technol 2010;4:913−22.

[50] Phillip M, Battelino T, Atlas E, Kordonouri O, Bratina N, Miller S, et al. Nocturnal glucose control with an artificial pancreas at a diabetes camp. N Engl J Med 2013;368:824−33.

[51] Florian Jr JA, Parker RS. Empirical modeling for glucose control in diabetes and critical care. Eur J Control 2005;11(6):616.

[52] Chee F, Fernando T, Van Heerden PV. Closed-loop glucose control in critically ill patients using continuous glucose monitoring system (CGMS) in real time. IEEE Trans Inf Technol Biomed 2003;7(1):43−53.

[53] Chase JG, Shaw GM, Wong XW, Lotz T, Lin J, Hann CE. Model-based glycemic control in critical care: a review of the state of the possible. Biomed Signal Process Control 2006;1(1):3−21.

[54] Wong XW, Singh-Levett I, Hollingsworth LJ, Shaw GM, Hann CE, Lotz T, et al. A novel model-based insulin and nutrition delivery controller for glycemic regulation in critically ill patients. Diabetes Technol Ther 2006;8(2):174−90.

[55] Bequette BW. Analysis of algorithms for intensive care unit blood glucose control. J Diabetes Sci Technol 2007;1(6):813−24.

[56] Wintergerst KA, Deiss D, Buckingham B, Cantwell M, Kache S, Agarwal S, et al. Glucose control in pediatric intensive care unit patients using an insulin-glucose algorithm. Diabetes Technol Ther 2007;9(3):211−22.

[57] Osburne RC, Cook CB, Stockton L, Baird M, Harmon V, Keddo A, et al. Improving hyperglycemia management in the intensive care unit. Diabetes Educ 2006;32 (3):394−403.

[58] Vogelzang M, Zijlstra F. Nijsten MWM. Design and implementation of GRIP: a computerized glucose control system at a surgical intensive care unit. BMC Med Inform Decis Mak 2005;5:38.

[59] Sun J, Cameron F, Bequette BW. A habituating blood glucose control strategy for the critically Ill. J Process Control 2012;22(7):1411−21.

[60] Bequette BW. A tutorial on biomedical process control. III. Modeling and control of drug infusion in critical care. J Process Control 2007;17(7):582−6.

[61] Yu CL, Roy RJ, Kaufman H. A circulatory model for combined nitroprusside-dopamine therapy in acute heart failure. Med Prog Technol 1990;16:77−88.

[62] Yu CL, Roy RJ, Kaufman H, Bequette BW. Multiple-model adaptive predictive control of mean arterial pressure and cardiac output. IEEE Trans Biomed Eng 1992;39 (8):765−78.

[63] Rao RR, Palerm CC, Aufderheide B, Bequette BW. Experimental studies on automated regulation of hemodynamic variables. IEEE Eng Med Biol Mag 2001;20(1):24−38 [Jan/Feb].

[64] Rao RR, Aufderheide B, Bequette BW. Experimental studies on multiple-model predictive control for automated regulation of hemodynamic variables. IEEE Trans Biomed Eng 2003;50(3):277−88.

[65] Bickford RG. The use of feedback systems for the control of anesthesia. Elect Eng 1951;70:852−5.

[66] Isaka S, Sebald AV. Control strategies for arterial blood pressure regulation. IEEE Trans Biomed Eng 1993;40(4):353−63.

[67] Sebel PS, Lang E, Rampil IJ, White P, Jopling RCM, Smith NT, et al. A multicenter study of bispectral electroencephalogram analysis for monitoring anesthetic effect. Anesth Analg 1997;84(4):891−9.

[68] Bailey JM, Haddad WM. Drug dosing control in clinical pharmacology. IEEE Control Syst Mag 2005;25(2):35−51 [April].

[69] Bailey JM, Shafer SL. A simple analytical solution to the three-compartment pharmacokinetic model suitable for computer-controlled infusion pumps. IEEE Trans Biomed Eng 1991;38(6):522−5.

[70] Schraag S. Theoretical basis of target controlled anaesthesia: history, concept and clinical perspectives. Best Pract Res Clin Anaesthesiol 2001;15(1):1−17.

[71] Egan TD, Shafer SL. Target-controlled infusions for intravenous anesthetics. Anesthesiology 2003;99(5):1039−41.

[72] Manberg PJ, Vozella CM, Kelley SD. Regulatory challenges facing closed-loop anesthetic drug infusion devices. Clin Pharmacol Ther 2008;84(1):166−9.

[73] Haddad WM, Bailey JM. Closed-loop control for intensive care unit sedation. Best Pract Res Clin Anaesthesiol 2009;23:95−114.

[74] Absalom R, Sutcliffe N, Kenny GN. Closed-loop control of anesthesia using bispectral index: performance assessment in patients undergoing major orthopedic surgery under combined general and regional anesthesia. Anesthesiology 2002;96(1):67−73.

[75] Struys MMRF, Smet TD, Versichelen LFM, et al. Comparison of closed-loop controlled administration of propofol using bispectral index as the controlled variable versus "standard practice" controlled administration. Anesthesiology 2001;95:6−17.

3 Deconvolution

Giovanni Sparacino, Giuseppe De Nicolao,
Gianluigi Pillonetto and Claudio Cobelli

Department of Information Engineering, University of Padova, Padova, Italy

3.1 Problem Statement

Many signals of interest for the quantitative understanding of physiological systems are not directly measurable *in vivo*. Some examples include the secretion rate of a gland, the production rate of a substrate, the appearance rate of a drug in plasma after an oral administration. Very often, it is only possible to measure the causally related effects of these signals in the circulation, for example, in (samples of) the time course of plasma concentrations. Thus, there is the need of reconstructing the unknown causes from the measured effects. In the mathematics/physics/engineering literature, this is an *inverse problem*, that is, instead of going along the cause−effect chain, one has to deal with the reversal of this chain. If the unknown signal is the input of a system, the inverse problem is an *input estimation problem* which, in the linear time invariant (LTI), case can be formalized as that of solving the following integral equation:

$$c(t) = \int_0^t g(t - \tau)u(\tau)d\tau \tag{3.1}$$

where $u(t)$ is the (unknown) input and $c(t)$ is the (measurable) output of the system. The function $g(t)$ describes the input−output behavior of the system and is called the *impulse response* of the system, that is, the time course of the output when the system is forced by a unitary Dirac impulse $\delta(t)$. In Eq. (3.1), $c(t)$ is the *convolution* of $u(t)$ with $g(t)$: hence, obtaining $u(t)$ from Eq. (3.1) given $g(t)$ and $c(t)$ is called *deconvolution*. If the system is linear time varying (LTV), the problem becomes that of solving a Fredholm integral equation of the first kind:

$$c(t) = \int_0^t g(t, \tau)u(\tau)d\tau \tag{3.2}$$

which is also often called (albeit improperly) deconvolution.

Modelling Methodology for Physiology and Medicine. DOI: http://dx.doi.org/10.1016/B978-0-12-411557-6.00003-3

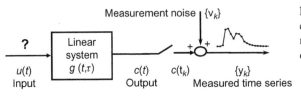

Figure 3.1 The input estimation problem from noisy data for a linear dynamic system.

Deconvolution is a classic problem in many disciplines of engineering, physics, and science, e.g., in spectroscopy, quantum physics, image restoration, geophysics, seismology, telecommunications, astronomy, acoustics, and electromagnetism, where it is often encountered either to remove distortion from an observed signal or to indirectly measure a nonaccessible signal [1,2]. Biomedical applications of deconvolution not only include, as mentioned above, endocrinology and metabolism and pharmacokinetics but also biomechanics, confocal microscopy, blood pressure measurement, ultrasounds, tracer kinetics, nuclear medicine, radiology and tomography, functional imaging, neurophysiology, and evoked potentials. An extended list of bibliographic references can be obtained at http://www.pubmed.com.

The deconvolution problem, schematically illustrated in Figure 3.1, is well known to be *ill-conditioned* (i.e., a small percent error in the measured output can produce a much greater percent error in the estimated input). Moreover, dealing with physiological signals adds to the complexity of the problem, since they are often nonnegative and sampled at a nonuniform and/or infrequent rate. Finally, the impulse response $g(t)$ is a model (often a sum of exponentials) either identified through a specific input—output experiment or obtained from population studies.

To better grasp the specific ingredients of the problem in physiological systems, we report the following example. Suppose we want to reconstruct insulin secretion rate (ISR) from C-peptide concentration data (C-peptide, instead of insulin, is used because they are secreted equimolarly by the pancreas, but C-peptide does not undergo any liver extraction). In normal conditions, the pancreas releases C-peptide (and insulin) in an oscillatory fashion resulting in the so-called ultradian oscillations (UOs) with period between 90 and 150 min. If glucose concentration increases, for example, as an effect of a glucose stimulus, UOs are obscured by the biphasic response of the pancreas: a sudden and large secretory peak (first phase) is followed by a smooth release (second phase). The pancreatic secretion is not directly measurable and the only available information is the plasma concentration of C-peptide. Panel A of Figure 3.2 shows the C-peptide plasma concentration measured every 20 min, for 12 h and under fasting conditions, in a normal hospitalized volunteer whose glycemic levels were maintained relatively stable by infusing glucose in vein by a pump [3]. The UO pattern of the secretion is evident from the measured concentration. Panel B of Figure 3.2 depicts the C-peptide plasma concentrations nonuniformly sampled for 4 h in a normal subject during an intravenous glucose tolerance test (IVGTT) [4]. The time series clearly reflects the biphasic response of the pancreas to the glucose stimulus. For both cases, since C-peptide kinetics is linear, the problem of reconstructing the C-peptide secretion rate (i.e., input in Figure 3.1) from the C-peptide plasma concentrations (i.e., output in

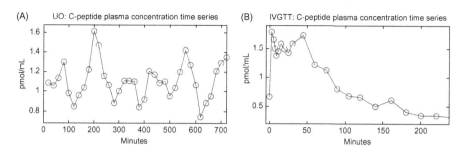

Figure 3.2 Estimation of ISR from C-peptide plasma concentration time series.
Panel A: Spontaneous UOs ($n = 36$ samples). Panel B: IVGTT ($n = 22$).
Source: Data are courtesy of the authors of Refs. [3,4].

Figure 3.1) is a deconvolution problem. In order to solve it, the impulse response g (t) is required. In the C-peptide case, an *ad hoc* experiment can be performed in the same individual of the "deconvolution experiment" on a separate occasion, as documented by one of the case studies reported in Chapter 5.

3.2   Difficulty of the Deconvolution Problem

The deconvolution difficulties are discussed by using a classic conceptual simulated example [1,5,6], hereafter referred to as the Hunt problem. Consider the input given by

$$u(t) = e^{-\left(\frac{t-400}{75}\right)^2} + e^{-\left(\frac{t-600}{75}\right)^2} \quad 0 \le t \le 1025 \tag{3.3}$$

and the impulse response of the system given by

$$g(t) = \begin{cases} 1, & t \le 250 \\ 0, & t > 250 \end{cases} \tag{3.4}$$

From $u(t)$ and $g(t)$, one obtains $c(t)$ from Eq. (3.1). Assume that n samples of $c(t)$, say $\{c_k\}$ where $c_k = c(t_k)$, are measured without error on the uniform sampling grid $\Omega_s = \{kT\}$, $k = 1, \ldots, n$, with $T = 25$ and $n = 41$. The input $u(t)$, the impulse response $g(t)$, and the output $c(t)$ together with the samples $\{c_k\}$ are shown in Panels A, B, and D of Figure 3.3, respectively. Since there are an infinite number of continuous-time functions which, once convoluted with the impulse response, describe perfectly the sampled data $\{c_k\}$, the deconvolution problem is *ill-posed*.

In order to tackle ill-posedness, any deconvolution approach must in some way restrict the set of the functions within which the solution of the problem is sought. For instance, in the so-called *discrete deconvolution* the signal $u(t)$ is assumed

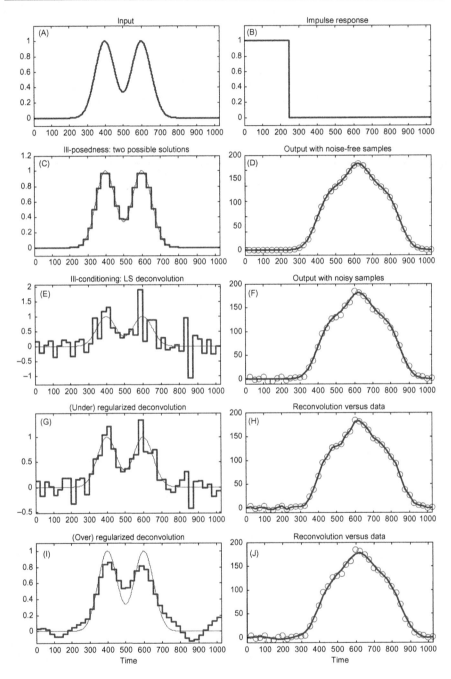

Figure 3.3 The Hunt simulated deconvolution problem. Panel A: True input. Panel B: Impulse response. Panel C: Ill-posedness of the deconvolution problem: the staircase function (thick line) is a solution of the deconvolution problem with error free data which

piecewise constant within each interval of the sampling grid $\Omega_s = \{t_1, t_2, \ldots, t_n\}$, that is, $u(t) = u_i$ for $t_{i-1} < t \le t_{i+1}$, $i = 1, 2, \ldots, n$, where $t_0 = 0$. From Eq. (3.2), which includes Eq. (3.1) as a particular case, it follows that

$$c(t_k) = \int_0^{t_k} g(t_k, \tau) u(\tau) d\tau = \sum_{i=1}^{k} u_i \int_{t_{i-1}}^{t_i} g(t_k, \tau) d\tau \tag{3.5}$$

One may also think of u_i as the mean level of $u(t)$ during the ith sampling interval. By letting:

$$g_{k,i} = \int_{t_{i-1}}^{t_i} g(t_k, \tau) d\tau \tag{3.6}$$

it follows:

$$c(t_k) = \sum_{i=1}^{k} u_i g_{k,i} \tag{3.7}$$

Adopting a matrix notation:

$$c = Gu \tag{3.8}$$

where $c = [c_1, c_2, \ldots, c_n]^T$ is the n-dimensional vector of the sampled output, $u = [u_1, u_2, \ldots, u_n]^T$ lists the levels of the piecewise constant input and G is an $n \times n$ lower triangular matrix, whose entries are

$$G(k, i) = \begin{cases} g_{k,i}, & k \ge i \\ 0, & k < i \end{cases} \tag{3.9}$$

In the LTI case with uniform sampling $G(k,i) = G(k - i)$, so that G is a Toeplitz matrix and Eq. (3.7) represents a discrete-time convolution.

◀ describes perfectly the samples of Panel D just as the true input (thin line). Panel D: True continuous output (thin line) with noise-free samples ($n = 41$). Panel E: Ill-conditioning of the deconvolution problem: solution provided by the Least Squares approach (thick line) and true input (thin line) from the noisy output samples of Panel F. Panel F: True continuous output (thin line) and noisy samples (SD = 3). Panel G: Regularized deconvolution obtained with a too small value of the regularization parameter ($\gamma = 0.5$, $q(\gamma) = 35.12$). Panel H: Reconvolution obtained from the input of Panel G and data. Panel I: Regularized deconvolution obtained with a too large value of the regularization parameter ($\gamma = 400$, $q(\gamma) = 11.31$) and true input (thin line). Panel J: Reconvolution obtained from the input of Panel I and data.

Provided that G is invertible, Problem (3.8) admits a unique solution, i.e., $\hat{u} = G^{-1}c$. For the Hunt problem, this solution is displayed in Panel C of Figure 3.3. Once convoluted with $g(t)$, the staircase function describes perfectly the output samples just as the true input (thin line), of which it could be regarded as a reasonable approximation, whose main drawback is the staircase profile.

The above deals with the noise-free situation. However, output samples are usually affected by some measurement error. Let y_k denote the kth measurement

$$y_k = c_k + v_k, \quad k = 1, 2, \ldots, n \tag{3.10}$$

where v_k is the error. Thus, in vector notation:

$$y = Gu + v \tag{3.11}$$

where $y = [y_1, y_2, \ldots, y_n]^T$ and $v = [v_1, v_2, \ldots, v_n]^T$. Vector v is hereafter assumed a zero-mean random vector with covariance matrix Σ_v given by

$$\Sigma_v = \sigma^2 B \tag{3.12}$$

where B is an $n \times n$ positive definite matrix and σ^2 is a scale factor, possibly unknown. Usually measurement errors are uncorrelated so that B is diagonal. For example, if noise is white with constant variance equal to σ^2, one has $B = I_n$.

The simplest estimate of u obtainable from Eq. (3.11) is

$$\hat{u}_{LS} = G^{-1}y \tag{3.13}$$

The subscript LS stands for "least squares." In fact, Eq. (3.13) is the solution of the LS problem

$$\min_{\hat{u}} (y - G\hat{u})^T B^{-1}(y - G\hat{u}) \tag{3.14}$$

The presence of noise in the measurement vector y of Eq. (3.13) may have a dramatic effect on the quality of the estimate, possibly due to the numerical instability of least squares [7]. In Panel F of Figure 3.3, Gaussian noise (standard deviation, SD = 3) was added to the data of the Hunt problem and LS deconvolution was performed (Panel E): note that wide, spurious, and unrealistic oscillations contaminate the estimated input, which also takes on some negative values. The reason for this deterioration is that deconvolution is not only an ill-posed but also an *ill-conditioned* problem: small errors in the observed data can be amplified, thus yielding much larger errors in the estimate.

One could think that increasing the number of samples is beneficial to the solution of the problem. On the contrary, both theory and practice show that increasing the sampling rate worsens ill-conditioning. In addition, the "smoother" the system kernel is, the worse the ill-conditioning of the deconvolution problem.

For example, the longer the hormone half-life is and the higher the sampling rate, the more difficult it is to reconstruct the secretion rate of a hormone by deconvolution [8]. Indices that measure the degree of ill-conditioning of a deconvolution problem as a function of sampling rate and kernel smoothness are available in Refs. [9,10].

3.2.1 Dealing with Physiological Systems

Dealing with physiological signals adds to the complexity of the deconvolution problem. For instance, to cope with technical and cost limitations as well as for the patient's comfort, data are very often collected with *infrequent* and *nonuniform* sampling schedules (Figure 3.2). Among other things, nonuniform sampling hinders the possible use of frequency domain techniques such as Wiener filtering. Furthermore, physiological inputs are often intrinsically *nonnegative*, e.g., a hormone secretion or a substrate production rate. Thus, negative input estimates due to ill-conditioning (Figure 3.3, Panel E) are physiologically unplausible. Finally, physiological systems are sometimes *time varying*, for example, the glucose−insulin system during a glucose perturbation [11].

3.2.2 A Classification of the Deconvolution Approaches

In the literature, many methods have been developed to circumvent ill-conditioning. Broadly speaking, they can be divided into two categories: *parametric deconvolution* assumes the analytic expression of the input to be known except for a small number of parameters, so that the deconvolution problem becomes a parameter estimation problem (see Chapter 5), while *nonparametric deconvolution* does not postulate an analytic form for the input. The best known nonparametric approach is the regularization method, which is described in detail in the next section. Some other deconvolution approaches, both parametric and nonparametric, will be briefly reviewed in Section 3.4.

3.3 The Regularization Method

3.3.1 Deterministic Viewpoint

The regularization method (sometimes also referred to as damped or penalized least squares) is a nonparametric approach that has been extensively exploited since the 1960s [12−14]. The idea behind the method is to look for a solution that provides a good data fit and enjoys, at the same time, a certain degree of "smoothness." This is done by solving the optimization problem

$$\min_{\hat{u}} (y - G\hat{u})^{\mathrm{T}} B^{-1}(y - G\hat{u}) + \gamma \hat{u}^{\mathrm{T}} F^{\mathrm{T}} F\hat{u} \tag{3.15}$$

where B is an $n \times n$ matrix as in Eq. (3.12), F is an $n \times n$ penalty matrix (see below) and γ is a real nonnegative parameter (see below). Problem (3.15) is quadratic and its solution

$$\hat{u} = (G^T B^{-1} G + \gamma F^T F)^{-1} G^T B^{-1} y \qquad (3.16)$$

linearly depends on the data vector y. Note that if $\gamma = 0$, Eq. (3.16) coincides with Eq. (3.14) and the LS solution is obtained. When $\gamma > 0$, the cost function of Eq. (3.16) is made up of two terms. The first one penalizes the distance, weighted by the inverse of B, between the model predictions $G\hat{u}$ (the *reconvolution* vector) and the data. The second contribution, i.e., $\hat{u}^T F^T F \hat{u}$, is a term which penalizes the "roughness" of the solution. The standard choice is to penalize the energy of the mth order time derivatives, m being an integer parameter. The parameter m is usually adjusted by trials. Its choice is usually not considered a major issue and typical choices are $m = 1$ or $m = 2$. Hence, F is a square lower triangular Toeplitz matrix (size n) whose first column is $F = [1, -1, 0, \ldots, 0]^T$ or $F = [1, -2, 1, 0, \ldots, 0]^T$, respectively. In general, one can penalize the energy of the mth time derivatives by letting

$$F = \Delta^m \qquad (3.17)$$

Δ being a square lower triangular Toeplitz matrix (size n) whose first column is $[1, -1, 0, \ldots, 0]^T$.

The relative weight given to data fit and solution regularity is governed by the so-called *regularization parameter* γ. By raising γ, the cost of roughness increases and the data match becomes relatively less important. Conversely, by decreasing the value of γ, the cost of roughness gets lower and the fidelity to the data becomes relatively more important. Notably, too large values of γ will lead to very smooth estimates of $\hat{u}$ that may not be able to explain the data (oversmoothing), while too small values of γ will lead to ill-conditioned solutions $\hat{u}$ that accurately fit the data, but exhibit spurious oscillations due to their sensitivity to noise (for $\gamma \to 0$ the LS solution is approached), see Panels G and I of Figure 3.3 for the Hunt problem (Panels H and J display how well the estimated input, once convoluted with the impulse response, matches the data).

In the literature, several criteria have been proposed for the choice of the regularization parameter. Two of the most popular are described below.

3.3.1.1 The Choice of the Regularization Parameter

A widely used criterion [14], which goes under the name of the *discrepancy*, suggests to compute the residuals vector

$$r = y - G\hat{u} \qquad (3.18)$$

and then adjust γ until the residual sum of squares equals the sum of the measurement error variances. In mathematical terms, the condition to be satisfied can be expressed as

$$\text{WRSS} = (y - G\hat{u})^T B^{-1}(y - G\hat{u}) = n\sigma^2 \tag{3.19}$$

Since the residuals vector can be interpreted as an estimate of the measurement error vector v, the discrepancy criterion has a very intuitive motivation. For instance, in the case $B = I_n$ it is "logical" to expect that

$$r^T r \approx E[vTv] = \sum_{i=1}^{n} \text{var}(v_k) = n\sigma^2 \tag{3.20}$$

Unfortunately, this intuitive rationale does not have a solid theoretical foundation. In particular, as will be discussed in the following, the discrepancy criterion is at risk of oversmoothing [1,15].

Another popular regularization criterion is *cross-validation*[16]. The so-called *ordinary cross-validation* (OCV) exploits a "leave-one-out" strategy, that is, each of the output samples is, in turn, left out. Deconvolution is obtained from the remaining $n - 1$ samples and the corresponding "leave-one-out" prediction error is computed: the best γ is the minimizer of the sum of the squared prediction errors. In order to reduce the computational burden required by OCV, an approximation called *generalized cross-validation* (GCV) is most commonly used. Accordingly, the regularization parameter γ is selected as the minimizer of the cost function:

$$\text{GCV}(\gamma) = \frac{\text{WRSS}}{\text{trace}[I_n - \psi]^2} \tag{3.21}$$

where Ψ is the so-called hat matrix

$$\Psi = G(G^T B^{-1} G + \gamma F^T F)^{-1} G^T B^{-1} \tag{3.22}$$

and WRSS is the weighted residuals sum of squares defined as in Eq. (3.19).

For other classic regularization criteria and for an asymptotic/analytical comparison, we refer the reader to examples in the relevant references [15,17−20].

3.3.1.2 The Virtual Grid

The regularization method is based on the discrete model (Eq. (3.7)), which was derived assuming that the unknown input is constant during each sampling interval, no matter how long. In the infrequent sampling case, this results in a poor approximation of the signal. For instance, consider the problem of estimating ISR. Panels A and C of Figure 3.4 show the C-peptide secretory profiles obtained by deconvoluting the data of Panels A and B of Figure 3.2, respectively, using an

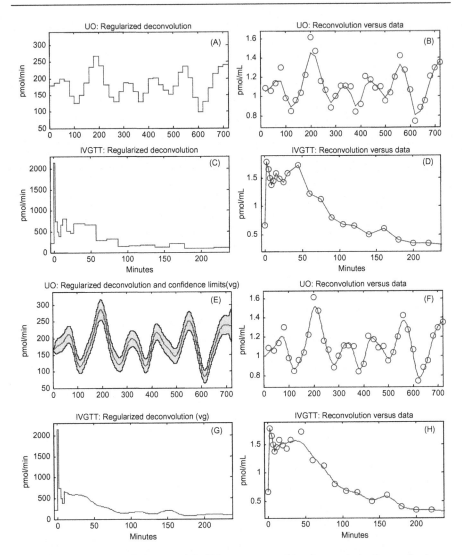

Figure 3.4 Estimation of ISR. Panel A: UO reconstructed by the regularization method ($\gamma = 0.0012$, $q(\gamma) = 19.29$, discrepancy criterion). Panel B: Reconvolution obtained from the input of Panel A and data. Panel C: ISR during IVGTT reconstructed by the regularization method ($\gamma = 0$, $q(\gamma) = 22$). Panel D: Reconvolution obtained from the input of Panel C and data. Panel E: UO reconstructed using the virtual grid with 95% confidence limits ($\gamma = 0.25$, $q(\gamma) = 21.07$, maximum likelihood criterion). Panel F: Reconvolution obtained from the input of Panel E and data. Panel G: ISR during IVGTT reconstructed using the virtual grid ($\gamma = 2.2e - 05$, $q(\gamma) = 9.16$, maximum likelihood criterion). Panel H: Reconvolution obtained from the input of Panel F and data.

individually tuned impulse response. Due to the infrequent sampling rate the staircase approximation is hardly acceptable. Panels B and D show the data and how well the estimated input $\hat{u}(t)$ predicts them once reconvoluted with the impulse response.

The roughness of the staircase approximation can also be appreciated by looking at the deconvoluted profiles obtained for the Hunt simulated problem in both the ideal (Figure 3.3, Panel C) and noisy (Figure 3.3, Panels E, G, and I) cases.

Such an unsatisfactory performance is due to the fact that the number of levels of the unknown vector u is assumed to be equal to the number n of measurements. In order to remove this assumption, a different discretization grid can be used for the input and the output [1]. Let Ω_s be the (experimental) sampling grid and $\Omega_v = \{T_1, T_2, \ldots, T_k, \ldots, T_N\}$ a finer $(N \geq n)$ grid (possibly uniform) over which the unknown input $u(t)$ is described as a piecewise constant function. Ω_v must contain Ω_s but, apart from this, it is arbitrary and does not have an experimental counterpart. For this reason, Ω_v is called the *virtual grid*. Let $c_v(T_k)$ denotes the (noise-free) output at the virtual sampling times T_k. Assuming that $u(t)$ is a piecewise constant within each time interval of the virtual grid, it follows that

$$c_v(T_k) = \int_0^{T_k} g(T_k, \tau)u(\tau)\mathrm{d}\tau = \sum_{i=1}^{k} u_i \int_{T_{i-1}}^{T_i} g(T_k, \tau)\mathrm{d}\tau \qquad (3.23)$$

where $T_0 = 0$. Adopting the usual matrix notation one has $c_v = G_v u$, where c_v and u are N-dimensional vectors obtained by sampling $c(t)$ and $u(t)$ on the virtual grid, and G_v is an $N \times N$ lower triangular matrix. Times belonging to the virtual grid Ω_v, but not present in the sampling grid Ω_s, have no counterpart in the sampled output data. We can regard them as (virtually) missing data. Denote by G the $n \times N$ matrix obtained by removing from G_v those rows that do not correspond to sampled output data.

The measurement vector is thus

$$y = Gu + v \qquad (3.24)$$

where v is the n-dimensional vector of the measurement error, u is the N-dimensional vector of the input discretized over the virtual grid, and G is the $n \times N$ matrix obtained by removing suitable rows of G_v. If the system is LTI and Ω_v is uniform, G has a near-to-Toeplitz structure, meaning that it misses some of the rows of the Toeplitz matrix G_v.

The estimate $\hat{u}$ is obtained by solving Eq. (3.15), where G and u are those of Eq. (3.24) and F has size $N \times N$. This method, provided that Ω_v has a fine time detail, yields a stepwise estimate that is virtually indistinguishable from a continuous profile.

Panel A of Figure 3.5 shows the results obtained with a 1-min virtual grid and employing the discrepancy criterion for the Hunt simulated problem of Panel F in Figure 3.3 (noisy data). The estimate is able to describe the true continuous-time

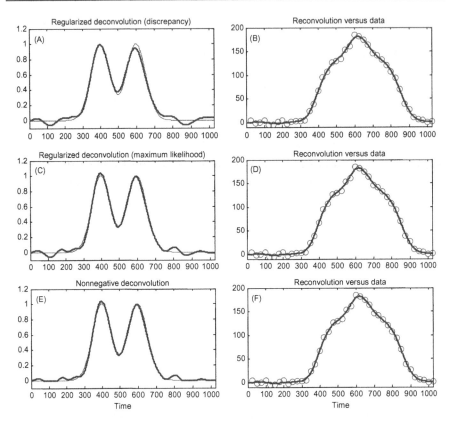

Figure 3.5 The Hunt simulated problem (virtual grid). Panel A: Regularized deconvolution using the discrepancy criterion ($\gamma = 1.24$e06, $q(\gamma) = 13.84$) and true input (thin line). Panel B: Reconvolution obtained from the input of Panel A and data. Panel C: Regularized deconvolution using the maximum likelihood criterion ($\gamma = 6.18$e05, $q(\gamma) = 15.55$) and true input (thin line). Panel D: Reconvolution obtained from the input of Panel C and data. Panel E: Nonnegative deconvolution, obtained with the same γ of Panel C, and true input (thin line). Panel F: Reconvolution obtained from the input of Panel E and data.

input much better than the staircase functions of Figure 3.3. Panel B displays the reconvolution fit to the data.

Remark 1 Consider a virtual grid $\Omega_v = \{kT_v\}$, $k = 1, 2, \ldots, N$, with $NT_v = t_s$. By letting $T_v \to 0$ and $N \to \infty$, the deconvoluted input tends toward a continuous-time function, continuous up to the $(2m - 2 + p)$th time derivative, where p is the difference between the degree of the denominator and that of the numerator of the Laplace transform of $g(t)$ [1]. This result offers a guideline for choosing m. For instance, for an LTI system with $p = 1$, it will be sufficient to let $m = 1$ in order to have an asymptotically continuous estimate together with its first time derivative.

Remark 2 In Eqs. (3.1) and (3.2), it is implicitly assumed that $u(t) = 0$ for $t < 0$. In several cases this is not true, for example, a basal spontaneous hormone secretion also occurs for $t < 0$. A possible solution is to estimate the input on a virtual grid starting at a negative time sufficiently far from 0, e.g., $-100T_v$ (the input estimate in the negative portion of the time axis is then discarded).

3.3.1.3   Assessment of Confidence Limits

Since deconvolution provides an indirect way of measuring a nonaccessible variable, it is important to assess the reliability of such a measurement. For the estimation error $\tilde{u} = u - \hat{u}$, the following expression is easily derived as

$$\tilde{u} = -\Gamma v + [I_N - \Gamma]u \tag{3.25}$$

where $\Gamma = (G^T B^{-1} G + \gamma F^T F)^{-1} G^T B^{-1}$. Given the variance of v, it is possible to obtain the variance of the first term in the right-hand side of Eq. (3.25). However, the second term, which is a bias term (its statistical expectation is nonzero if $\gamma > 0$), cannot be computed since it depends on the true (unknown) vector u. Unless a probabilistic description of u is provided, confidence intervals accounting for the bias error cannot be obtained. This suggests to embed the deconvolution problem within a stochastic setting.

3.3.2   Stochastic Viewpoint

Consider Eq. (3.24), possibly derived by discretizing equation (3.2) on a virtual grid, and assume that u and v are zero-mean random vectors whose covariance matrices Σ_u and Σ_v are known. It is assumed that $\sum_v = \sigma^2 B$, see Eq. (3.12), and Σ_u (size $N \times N$) is factorized as

$$\Sigma_u = \lambda^2 (F^T F)^{-1} \tag{3.26}$$

In this stochastic setting, the deconvolution problem, i.e., estimating u from y through Eq. (3.24), can be stated as a *linear minimum variance estimation problem*: find the estimate $\hat{u}$, linearly depending on the data vector y, such that $E[||u - \hat{u}||^2]$ is minimized, that is, minimize the expectation E of the squared Euclidean norm of the estimation error. Let $\gamma^\circ = \sigma^2/\lambda^2$. If u and v in Eq. (3.24) are uncorrelated, the estimate $\hat{u}$ coincides with the solution of the optimization problem(3.15), provided that $\gamma = \gamma^\circ$[1,21]. When u and v are jointly Gaussian, the estimator (3.16) with $\gamma = \gamma^\circ$ has minimum error variance among all estimators, either linear or nonlinear, of u given y.

To solve deconvolution as a linear minimum variance estimation problem, the *a priori* covariance matrix of the input vector u, i.e., $\Sigma_u = \lambda^2 (F^T F)^{-1}$, is required. However, we only know that the input u is a smooth function of time. A simple *a priori* probabilistic model of a smooth signal on a uniformly spaced grid is to describe it as the realization of a stochastic process obtained by the cascade of m integrators driven by a zero-mean white noise process $\{w_k\}$ with variance λ^2. For instance, for $m = 1$ this corresponds to a random-walk model:

$$u_k = u_{k-1} + w_k, \quad k = 1, 2, \ldots, N, \quad u_0 = 0 \tag{3.27}$$

In a Gaussian setting, Eq. (3.27) tells us that, given u_k, then u_{k+1} will be in the range $u_k \pm 3\lambda$ with probability 99.7% [6]. It is easily demonstrated that the covariance matrix of the random vector u whose components are obtained from m integrations of a white noise process of variance λ^2 is given by Eq. (3.26), with F as in Eq. (3.17).

Both regularization and minimum variance estimation determine the estimate by solving Eq. (3.16). This allows the establishment of an insightful analogy between the two approaches. In particular, penalizing the mth time derivative energy in the regularization method equals to model the unknown input by a $(m - 1)$-fold integrated random-walk process in the stochastic approach. In view of this analogy, $\gamma^o = \sigma^2/\lambda^2$ represents, in some sense, the "optimal" value of the regularization parameter. Such a value is, however, unknown since λ^2 and, possibly, also σ^2 are unknown. Obviously, the lower λ^2, the smoother $\{u_k\}$. In the following section, we will introduce some statistical criteria for estimating γ^o.

3.3.2.1 Confidence Limits

If all vectors involved in Eq. (3.25) are *stochastic*, the calculation of the variance is possible. By means of the matrix inversion lemma, one obtains

$$\text{var}[\tilde{u}] = \sigma^2 \Gamma B \Gamma^T + \gamma \lambda^2 (G^T B^{-1} G + \gamma F^T F)^{-1} F^T F (G^T B^{-1} G + \gamma F^T F)^{-1} \tag{3.28}$$

It is easily verified that the contribution of noise to the error variance, that is, the first term in the right-hand side of Eq. (3.28), is a monotonically decreasing function (in the matrix sense) of γ, whereas the contribution of bias, that is, the second term of Eq. (3.28), is monotonically increasing. Not surprisingly, the *minimum* value of $\text{var}[\tilde{u}]$ is obtained for the *optimal* value of γ that is $\gamma = \gamma^o = \sigma^2/\lambda^2$:

$$\text{var}[\tilde{u}] = \sigma^2 (G^T B^{-1} G + \gamma^o F^T F)^{-1} \tag{3.29}$$

If a reliable estimate of γ^o is available, this covariance matrix can be used to compute the confidence intervals for the entries of $\hat{u}$.

3.3.2.2 Statistically Based Choice of the Regularization Parameter

Let $\text{WRSS} = (y - G\hat{u})^T B^{-1}(y - G\hat{u})$ and $\text{WESS} = \hat{u}^T F^T F \hat{u}$ denote the weighted residuals sum of squares and the weighted estimates sum of squares, respectively. Both of these quantities depend on $\hat{u}$ and thus on the value of γ. In the stochastic setting,

WRSS and WESS are random variables. For the linear minimum variance estimate, the following two properties hold, see Ref. [22] for a proof:

$$E[\text{WESS}(\gamma^o)] - \lambda^2 q(\gamma^o) \tag{3.30}$$

$$E[\text{WRSS}(\gamma^o)] = \sigma^2 \{n - q(\gamma^o)\} \tag{3.31}$$

where

$$q(\gamma^o) = \text{trace}(B^{-1/2}G(G^T B^{-1}G + \gamma^o F^T F)^{-1}G^T B^{-}) \tag{3.32}$$

Observe the analogy of Eq. (3.31) with a well-known property of linear regression models, where the averaged sum of squared residuals is a biased estimator of the error variance with the bias depending on the (integer) number of degrees of freedom of the model. For this reason, $q(\gamma)$ defined by Eq. (3.33) is named *equivalent degrees of freedom* associated with γ[1]. The quantity $q(\gamma)$ is a real number varying from 0 to n: if $\gamma \to 0$ then $q(\gamma) \to n$ whereas if $\gamma \to \infty$ then $q(\gamma) \to 0$. The fact that $q(\gamma)$ is a real number is in agreement with the nature of the regularization method, where the flexibility of the model (its degree of freedom) can be changed with continuity through the tuning of γ.

By dropping the expectations in Eqs. (3.30) and (3.31) and recalling that $\gamma^o = \sigma^2/\lambda^2$, two "consistency" criteria can be intuitively derived that allow the choice of γ when either λ^2 or both λ^2 and σ^2 are unknown. It is worth noting that the same criteria can be obtained on a firmer statistical ground under Gaussianity assumptions by determining necessary conditions for λ^2 and σ^2 to maximize the likelihood of the data vector y[1]. The two criteria are formulated as follows:

Criterion ML1. When λ^2 is unknown (σ^2 is assumed to be known), tune γ until

$$\text{WESS} = \lambda^2 q(\gamma) \tag{3.33}$$

with $\lambda^2 = \sigma^2/\gamma$.

Criterion ML2. When both σ^2 and λ^2 are unknown, tune γ until

$$\frac{\text{WRSS}}{n - q(\gamma)} = \gamma \frac{\text{WESS}}{q(\gamma)} \tag{3.34}$$

and then estimate σ^2 as

$$\hat{\sigma}^2 = \frac{\text{WRSS}}{n - q(\gamma)} \tag{3.35}$$

according to Eq. (3.31).

Panel C of Figure 3.5 shows the input profile of the Hunt problem estimated with the virtual grid and using the ML regularization criterion. By comparing this profile with that of Panel A, obtained using the discrepancy criterion, one notes that the latter is oversmoothed. In fact, by comparing Eq. (3.18) with Eq. (3.31), it is easily seen that the discrepancy criterion leads, on the average, to oversmoothing.

To complete the examples, Panel E of Figure 3.4 displays the reconstruction of UOs from the time series of the Panel A of Figure 3.2, obtained by the stochastic approach employing Criterion ML1. The 95% confidence intervals obtained from Eq. (3.29) are also reported. Panel G shows the reconstruction of ISR during IVGTT obtained by the stochastic approach using Criterion ML2 (in this case, Σ_u was partitioned in two blocks in order to express the prior information concerning the biphasic response of the pancreatic beta cells to the glucose stimulus, see Ref. [22] for details). In both cases, the use of the virtual grid renders these profiles more plausible than those of Panels A and C.

Remark 3 Some regularization criteria, e.g., discrepancy and minimum risk, are restricted to the case of known σ^2. In contrast, some others, e.g., L-curve and GCV, do not require the knowledge of σ^2, but do not use it when it is available. For instance, for a given data set, GCV always selects the same γ, no matter whether the variance of the measurement error is known to be, say, 1 or 100. The two ML criteria presented above deal with both cases of known and unknown σ^2.

Remark 4 The absolute value achieved for γ also depends on a number of ingredients of the problem, such as sampling rate, virtual grid, noise variance, impulse response, and even units adopted for the signals under study. For instance, in the Hunt problem, $\gamma = 400$ leads to oversmoothing in Panel I of Figure 3.3, but $\gamma = 6.18\text{e}05$ (fixed in agreement with Criterion ML1 in presence of the virtual grid) leads to suitable regularization in Panel C of Figure 3.5. A better indicator of the amount of regularization is the degrees of freedom $q(\gamma)$, since it is a real number varying from 0 to n. For instance, the degrees of freedom in the above two cases were $q(\gamma) = 11.31$ and $q(\gamma) = 15.55$, respectively, suggesting that less regularization (in spite of a higher value of γ) was used in the determination of the latter input estimate.

3.3.3 Numerical Aspects

In the regularization approach, the computation of the solution via Eq. (3.15) or (3.16) would require $O(N^3)$ memory occupation and $O(N^3)$ operations to accomplish matrix inversion (the notation $O(f(N))$ means "of the same order of magnitude as $f(N)$"). This computational burden can be reduced by applying the matrix inversion lemma to Eq. (3.16), thus obtaining:

$$\hat{u} = F^{-1}F^{-T}G^{T}(GF^{-1}F^{-T}G^{T} + \gamma B)^{-1}y \tag{3.36}$$

In this way, an $n \times n$ matrix must be inverted at the price of $O(n^3)$ operations. Note that if F is as in Eq. (3.17), its inverse admits an easy to derive analytic expression.

In the LTI system case with uniform sampling, matrices G and F in Eq. (3.15) exhibit a Toeplitz structure so that only their first column needs to be stored. Then, an efficient numerical technique to compute the regularized estimate is available, first presented in Ref. [5] and subsequently refined in Refs. [1,6]. In particular, Eq. (3.15) can be solved by the iterative conjugate gradient (CG) method, whose basic iteration can be performed in $O(N \log N)$ operations through the Fast Fourier Transform [6] or in $O(N)$ operations by the use of recursive difference equations [1]. The Toeplitz structure of the matrices can also be exploited in order to devise suitable "preconditioners" that improve the rate of convergence of the algorithm [6]. Since theory guarantees the convergence of the CG algorithm in N iterations at most, the overall complexity of the algorithm is $O(N^2 \log N)$ or $O(N^2)$. In the LTV case, these methods do not apply equally well because G does not have a Toeplitz structure.

However, the bottleneck of numerical algorithms for deconvolution is given by the need of computing several trial solutions of Eq. (3.15). In fact, the tuning of the regularization parameter γ (according to any criterion) requires a trial and error procedure. The strategy of Ref. [1] first puts Eq. (3.24) in diagonal form in $O(n^3)$ operations through a singular value decomposition (SVD). This dramatically speeds up the trial and error procedure for the determination of the optimal regularization parameter since only $O(n)$ scalar operations are required to compute $\hat{u}$ for each trial value of γ. The strategy also allows the computation of the confidence intervals with $O(N^2)$ complexity, with a significant improvement over the use of Eq. (3.16), which would require $O(N^3)$ operations. The overall complexity of the algorithm is $O(n^3 N)$ which, at least when $n \ll N$, is better than the above-mentioned method based on the CG. Of note is that, since the SVD procedure is unaffected by the Toeplitz structure of G, this numerical strategy applies also to the time-varying case. An alternative numerical strategy that takes into account the need of calculating Eq. (3.16) for several values of the regularization parameters is the QR factorization approach proposed in Ref. [23].

For "large" values of n, further refinements can be obtained in the LTI case when the sampling is uniform. In particular, a spectral factorization can be exploited to calculate, with a computational burden not dependent on the complexity of the problem, the degrees of freedom $q(\gamma)$. Some explicit formulas to compute $q(\gamma)$, under appropriate assumptions, can also be obtained [24]. In the unconstrained case, using state−space methods, it has been shown [25] that the regularized estimate $\hat{u}(t)$ can be seen as a regularization network whose output is the weighted sum of N suitable basis functions. Interestingly, the neural network weights are computable in $O(n)$ operations via a Kalman filtering approach [26]. In the absence of nonnegativity constraints, this approach is particularly convenient when deconvolution must be performed on several data sets that share the same statistical parameters and impulse response (for instance, a population model): in fact, the basis functions remain unchanged across the data sets and one has only to compute the individual weights.

3.3.4   Constrained Deconvolution

In a number of physiological cases, the input $u(t)$ is known to be intrinsically non-negative (e.g., hormone secretion rates, drug absorption rates). Nevertheless, due to measurement errors and impulse response model mismatch, the solution provided by Eq. (3.16) may take on negative values (Figure 3.5, Panels A and C). To obtain nonnegative estimates, the regularization method can be reformulated as a constrained optimization problem [1,6]:

$$\min_{\hat{u} \geq 0} (y - G\hat{u})^{\mathrm{T}} B^{-1} (y - G\hat{u}) + \gamma \hat{u}^{\mathrm{T}} F^{\mathrm{T}} F \hat{u} \tag{3.37}$$

where $\hat{u} \geq 0$ stands for $\hat{u}_k \geq 0$, $\forall k$. This problem does not admit a closed-form solution and must be solved by an iterative method such as the constrained CG algorithm.

Remarkably, the incorporation of nonnegativity constraints as in Eq. (3.37) makes the estimator nonlinear. This impairs the use of some regularization criteria, e.g., GCV, ML. In addition, nonnegativity contradicts Gaussianity so that the computation of confidence intervals by exploiting analytic approaches is not possible. Empirical strategies are most often used to cope with these problems, see Ref. [1] for details. Panel E of Figure 3.5 shows, for the Hunt problem, the input profile estimated with the nonnegativity constraint (the value of the regularization parameter is the same adopted for the unconstrained estimate of Panel C).

A method to deal with constrained regularized deconvolution on a more theoretically sound basis has been proposed in Ref. [27]. Briefly, entries of u are the exponential of the entries of vector s, the latter being an m-fold integrated white Gaussian noise $\{w_k\}$ with variance λ^2. In this way, the covariance matrix of s is $\Sigma_s = \lambda^2 (F^{\mathrm{T}} F)^{-1}$. If v in Eq. (3.11) is Gaussian, the maximum *a posteriori* (MAP) estimate of s can be obtained by solving

$$\min_{\hat{s}} (y - G \exp(\hat{s}))^{\mathrm{T}} B^{-1} (y - G \exp(\hat{s})) + \gamma \hat{s}^{\mathrm{T}} F^{\mathrm{T}} F \hat{s} \tag{3.38}$$

where $\gamma = \sigma^2/\lambda^2$. From $\hat{s}$, a nonnegative estimate $\hat{u}$ is straightfordly obtained. An approximate expression for the covariance matrix of the error affecting $\hat{s}$ can also be obtained. In practice, instead of facing Eq. (3.38), it can be convenient to resort to a Markov chain Monte Carlo (MCMC) approach [28], which jointly handles computation of confidence intervals (analytically intractable due to nonlinearities) and tuning of γ. As example of application, Figure 3.6 (top) shows the secretion rate of luteinizing hormone (LH) in a healthy subject and its confidence interval (left), obtained from samples of its concentration in plasma collected every 5 min in spontaneous conditions (right). It is apparent that the confidence interval is not credible since it includes negative values. Results using the approach of Ref. [27] are depicted in Figure 3.6 (bottom), which shows that the 95% confidence intervals are now realistic. The posterior mean of u exhibits a peak around 90 min, whose amplitude is larger than that obtained from the linear estimator, thus

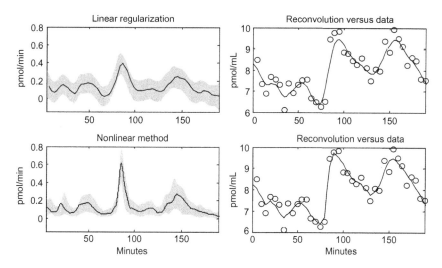

Figure 3.6 Reconstruction of secretion rate of LH in spontaneous conditions. Top left: Regularization using maximum likelihood criterion (thick line) with confidence interval (shadowed area). Top right: Reconvolution versus data. Bottom left: Deconvolution computed by the nonlinear stochastic approach (thick line) with confidence interval (shadowed area). Bottom right: Reconvolution versus data.

introducing less bias in the residuals (see right panels). A further development of this nonlinear stochastic approach for estimating hemodynamic parameters from MRI perfusion images was documented in Ref. [29].

3.4 Other Deconvolution Methods

A number of deconvolution techniques, often referred to as *parametric deconvolution* methods, circumvent ill-posedness and ill-conditioning by making functional assumptions on the input. The analytic expression of the input is assumed to be known except for a small number of parameters, so that the deconvolution problem becomes a parameter estimation problem with more data than unknowns. This guarantees the uniqueness (at least locally) and regularity of the solution. In Ref. [30], for example, a linear combination of M exponentials is used, so that deconvolution turns into the problem of determining the amplitudes and the rates ($2M$ parameters). Parametric approaches were also developed for more specific problems. For instance, in Ref. [31] episodic hormone release is described by assuming secretory spikes to be Gaussian-shaped (with each pulse characterized by location, amplitude, and width), while in Ref. [32] lagged-normal models have been used to approximate the transport function of an indicator into an organ.

The required heavy assumptions on the shape of the unknown input constitute in general a major drawback. Moreover, one has also to deal with model-order selection problems, for example, selecting M in Ref. [30] or the number of pulses that form the secretory profile in Ref. [31]. Remarkably, model-order selection resembles the problem of tuning the smoothing parameter in the regularization method and some of the criteria defined for the regularization method, for example, cross-validation, discrepancy, and L-curve can be extended to this purpose. In some parametric methods, it is difficult to impose nonnegativity constraints and to handle time-varying systems. Finally, the intrinsic nonlinearity of parametric deconvolution methods exposes them to the problem of correctly selecting the initial guess of the parameters in order to avoid local minima in parameter estimation [33].

A parametric approach which allows a larger flexibility is based on *regression splines*[34]. Having a fixed number M of spline knots and vector τ of their location, the reconvolution vector linearly depends on a vector μ which can be estimated by least squares. However, addressing the problem of selecting M involves the usual fit versus smoothness trade-off, while locating the spline knots is even more complex (intuitively, the density of knots should be higher where fast changes in the input are expected, see the quoted paper for empirical strategies). An appealing feature of the regression splines method is the possibility of incorporating monotonicity and nonnegativity constraints on the input by adding suitable inequality constraints on vector μ. However, these constraints complicate the model-order selection problem.

Turning back to nonparametric methods, it is worth pointing out that, in addition to regularization, there are also other approaches to deconvolution. For instance, *truncated singular value decomposition* (TSVD) methods [35] first perform a singular value expansion of the algebraic system in Eq. (3.11) and then determine an (unconstrained) estimate by truncating such an expansion before the small singular values start to dominate. The (integer) number k of the eigenvectors that are left in the expansion determines the regularity of the estimate. *Conjugate gradient regularization* (CGR) is based on the fact that, when the CG algorithm is used to solve the system $y = Gu$, the low-frequency components of the solution tend to converge faster than the high-frequency components. Hence, the CG has some inherent regularization effect where the number of CG iterations tunes the regularity of the solution [36]. Another iterative method based on the same principle is based on an algorithm developed by Landweber for the solution of an algebraic system, a modified version being available in order to deal with nonnegativity constraints [2]. Finally, *maximum entropy* (ME) methods can be viewed as a variant of the regularization method where the term $u^T \log u$ replaces the quadratic term $u^T F^T F u$ in Eq. (3.15)[37]. In this way, for a fixed γ, the ME estimator provides (strictly) positive signals, but it is no more linear in the data, so that a closed-form solution does not exist and an iterative method is required to compute the estimate. The structure of the cost function makes ME methods particularly suitable for solving problems where the unknown input is essentially zero, for example, in a nearly black image, in the vast majority of its domain.

In all the deconvolution approaches mentioned in this section, the computation of the confidence intervals is, however, difficult due to the nonlinearity of the estimator and/or the deterministic nature of the underlying setting.

3.5 Conclusions

In this chapter, we have introduced the deconvolution problem critically, reviewing the available methods against the real-world challenges arising in the analysis of physiological systems, including ill-conditioning, treatment of infrequently sampled data sets, computation of confidence intervals, nonnegativity constraints, and efficiency of the numerical algorithms. In this respect, the regularization method makes only mild assumptions about the unknown input and has some important advantages over the other deconvolution approaches, especially in its stochastic formulation. An extension of the method solves the deconvolution problem subject to nonnegativity constraints, preserving the statistical interpretation of the estimate and its confidence intervals.

Among recent research issues, there is the development of algorithms that explicitly account for the uncertainty in the impulse response, which has been neglected in this chapter, but is an important feature, given that $g(t,\tau)$ is often uncertain, being either identified from experimental data or an "average" population model that ignores interindividual variability. In such a case, an analytic solution being not available, a first approach relies on Monte Carlo techniques [1,22], but it is also possible to pursue a fully Bayesian approach via MCMC methods [38] at the cost of some increase in algorithmic complexity.

Another significant problem, especially in pharmacokinetics, is embedding the deconvolution problem in a population modelling context in which M individuals undergo the same experiment whose constraints (e.g., several subjects but few measurements per subject) prevent one from obtaining acceptable estimates. In such situations, the M subjects can be simultaneously analyzed so as to improve each individual estimate by taking advantage of the relevant information in the other $M - 1$ subjects. For instance, in the semiparametric spline-based approach proposed in Ref. [39], the input is seen as the distortion of a template common to all of the population that is estimated exploiting the data of all subjects. In a nonparametric context, it is natural to adopt a Bayesian paradigm and model the inputs to each individual as realizations of stochastic processes sharing the same prior distribution, typical of the population. The two key points are the definition of the most suitable prior distribution and the development of efficient algorithms for the derivation of the posterior distribution of each individual input conditional on all the M data sets. These problems have already been successfully addressed in the function reconstruction problem [40,41] that can be seen as a particular deconvolution problem having a Dirac delta as impulse response. In particular, the computational burden involved by the need of jointly processing the M data sets has been substantially reduced by exploiting sampling grids common to all subjects [42] or just the presence of coincident sampling times among subjects [43].

On the computational side, recent research has focused on the development of numerically efficient algorithms that solve the regularization problem without any discretization grid and express the estimated input as the linear combination of continuous-time basis functions [25,44]. There are also numerically efficient MCMC algorithms that yield discretization-free continuous-time solutions [28].

Finally, although this chapter focused mostly on the estimation of the unknown input of a linear system, deconvolution can also be used to estimate the impulse response of a linear system from input and output measurements. Many considerations and techniques of this chapter apply equally well to this case, with the possible important exception of the choice of the regularization penalty, or equivalently the prior distribution of the unknown impulse response. In fact, imposing only regularity on the derivatives of the signal does not account for the possible exponential decay of the impulse response. A recent investigation has demonstrated the benefit of incorporating a stability constraint within the statistical prior on the impulse response [45,46], a result important in the context of physiological systems.

References

[1] De Nicolao G, Sparacino G, Cobelli C. Nonparametric input estimation in physiological systems: problems, methods, case studies. Automatica 1997;33:851−70.

[2] Bertero M, Boccacci P. Introduction to inverse problems in imaging. Bristol: IOP Publishing; 1998.

[3] Sturis J, Van Cauter E, Blackman JD, Polonsky KS. Entrainment of pulsatile insulin secretion by oscillatory glucose infusion. J Clin Invest 1991;87:439−45.

[4] Shapiro ET, Tillil H, Rubenstein AH, Polonsky KS. Peripheral insulin parallels changes in insulin secretion more closely than C-peptide after bolus intravenous glucose administration. J Clin Endocrinol Metab 1988;67:1094−9.

[5] Hunt BR. Biased estimation for nonparametric identification of linear systems. Math Biosci 1971;10:215−37.

[6] Commenges D. The deconvolution problem: fast algorithms including the preconditioned conjugate-gradient to compute a MAP estimator. IEEE Trans Automatic Control 1984;29:229−43.

[7] Cohen A, Davenport MA, Leviatan D. On the stability and accuracy of the least squares. Found Comput Math 2013;13:819−34.

[8] De Nicolao G, Liberati D. Linear and nonlinear techniques for the deconvolution of hormone time-series. IEEE Trans Biomed Eng 1993;40:440−55.

[9] Hunt BR. A theorem on the difficulty of numerical deconvolution. IEEE Trans Audio Electroac 1972;20:94−5.

[10] Ekstroem MP. A spectral characterization of the ill-conditioning in numerical deconvolution. IEEE Trans Audio Electroac 1973;21:344−8.

[11] Caumo A, Cobelli C. Hepatic glucose production during the labelled IVGTT: estimation by deconvolution with a new minimal model. Am J Physiol 1993;264:E829−41.

[12] Phillips DL. A technique for the numerical solution of certain integral equations of the first kind. J Ass Comput Mach 1962;9:97−101.

[13] Tikhonov AN. Solution of incorrectly formulated problems and the regularization method. Soviet Math Dokl 1963;4:1624.

[14] Twomey S. The application of numerical filtering to the solution of integral equations of the first kind encountered in indirect sensing measurements. J Franklin Inst 1965;279:95–109.

[15] Hall P, Titterington DM. Common structure of techniques for choosing smoothing parameters in regression problems. J Roy Stat Soc Ser B 1987;49:184–98.

[16] Golub GH, Heath M, Wahba G. Generalized cross-validation as a method for choosing a good ridge parameter. Technometrics 1979;21:215–24.

[17] O'Sullivan F. A statistical perspective on ill-posed inverse problems. Stat Sci 1986;1:502–27.

[18] Hansen PC, O'Leary DP. The use of the L-curve in the regularization of discrete ill-posed problems. SIAM J Sci Comput 1993;14:1487–503.

[19] Rice JA. Choice of the smoothing parameter in deconvolution problems. Contemp Math 1986;59:137–51.

[20] Kay J. Asymptotic comparison factors for smoothing parameter choices in regression problems. Stat Probab Lett 1992;15:329–35.

[21] Beck JV, Arnold KJ. Parameter estimation in engineering and science. New York: Wiley; 1977.

[22] Sparacino G, Cobelli C. A stochastic deconvolution method to reconstruct insulin secretion rate after a glucose stimulus. IEEE Trans Biomed Eng 1996;43:512–29.

[23] Hanke M, Hansen PC. Regularization methods for large scale problems. Surv Math Ind 1993;3:253–315.

[24] De Nicolao G, Ferrari-Trecate G, Sparacino G. Fast spline smoothing via spectral factorization concepts. Automatica 2000;36:1733–9.

[25] De Nicolao G, Ferrari-Trecate G. Regularization networks for inverse problems: a state-space approach. Automatica 2003;39:669–76.

[26] De Nicolao G, Ferrari-Trecate G. Regularization networks: fast weight calculation via Kalman filtering. IEEE Trans Neural Netw 2001;12:228–35.

[27] Pillonetto G, Sparacino G, Cobelli C. Handling non-negativity in deconvolution of physiological signals: a nonlinear stochastic approach. Ann Biomed Eng 2002;30:1077–87.

[28] Pillonetto G, Bell B. Bayes and empirical Bayes semi-blind deconvolution using eigenfunctions of a prior covariance. Automatica 2007;43:1698–712.

[29] Zanderigo F, Bertoldo A, Pillonetto G, Cobelli C. Nonlinear stochastic regularization to characterize tissue residue function in bolus-tracking MRI: assessment and comparison with SVD, block-circulant SVD, and Tikhonov. IEEE Trans Biomed Eng 2009;56:1287–97.

[30] Veng-Pedersen P. An algorithm and computer program for deconvolution in linear pharmacokinetics. J Pharmacokin Biopharm 1980;8:463–81.

[31] Veldhuis JD, Johnson M. Deconvolution analysis of hormone data. Methods Enzymol 1992;210:539–75.

[32] Knopp TJ, Dobbs WA, Greenleaf JF, Bassingwaighte JB. Transcoronary intravascular transport functions obtained via a stable deconvolution technique. Ann Biomed Eng 1986;4:44–59.

[33] Verotta D. Comments on two recent deconvolution methods. J Pharmacokinet Biopharm 1990;18:483–9.

[34] Verotta D. Estimation and model selection in constrained deconvolution. Ann Biomed Eng 1993;21:605–20.

[35] Hansen PC. The truncated SVD as a method for regularization. BIT 1987;27:354–553.

[36] Van der Sluis A, Van der Vorst HA. SIRT and CG type methods for iterative solutions of sparse linear least-squares problems. Lin Alg Appl 1990;130:257–302.

[37] Donoho D, Johnstone IM, Hoch J, Stern A. Maximum entropy and the nearly black object. J R Stat Soc Ser B 1992;54:41–81.

[38] Fattinger KE, Verotta D. A nonparametric subject-specific population method for deconvolution. I. Description, internal validation, and real data examples. J Pharmacokinet Biopharm 1995;23:581–610.

[39] Magni P, Bellazzi R, De Nicolao G. Bayesian function learning using MCMC methods. IEEE Trans Pattern Anal Mach Intell 1998;20:1319–31.

[40] Neve M, De Nicolao G, Marchesi L. Nonparametric identification of population models via Gaussian processes. Automatica 2007;43:1134–44.

[41] Neve M, De Nicolao G, Marchesi L. Nonparametric identification of population models: an MCMC approach. IEEE Trans Biomed Eng 2008;1:41–50.

[42] Pillonetto G, De Nicolao G, Chierici M, Cobelli C. Fast algorithms for nonparametric population modeling of large data sets. Automatica 2009;45:173–9.

[43] Pillonetto G, Dinuzzo F, De Nicolao G. Bayesian online multi-task learning using regularization networks. In: Proceedings of the American control conference 2008; p. 4517–22.

[44] Bell B, Pillonetto G. Estimating parameters and stochastic functions of one variable using nonlinear measurement models. Inver Probl 2004;20:627–46.

[45] Pillonetto G, De Nicolao G. A new Kernel-based approach for linear system identification. Automatica 2010;46:81–93.

[46] Chen T, Ohlsson H, Ljung L. On the estimation of transfer functions, regularizations and Gaussian processes − Revisited. Automatica 2012;48:1525–35.

4 Structural Identifiability of Biological and Physiological Systems

Maria Pia Saccomani

Department of Information Engineering, University of Padova, Padova, Italy

4.1 Introduction

Mathematical modelling of biological systems is becoming a standard tool to investigate complex dynamic, nonlinear interaction mechanisms in cellular processes, like signal transduction pathways and metabolic networks [1−3]. These mechanisms are often modelled by ordinary differential equations involving parameters such as reaction rates. For example, the Michaelis−Menten equation is often used to describe the internal structure of the biochemistry of the system, assuming that diffusion is fast, compared to reaction rates. The system parameters contain key information, but in general they can only be measured indirectly, as it is usually not possible to measure directly the dynamics of every portion of the system. The recovery of parameter values can then only be approached indirectly as a parameter estimation problem starting from external, input−output measurements [4]. In this context, the first question is whether the parameters of the model can be determined, at least for suitable input functions, assuming that all observable variables are error-free. This is the property called *structural identifiability* of the model [5−9].

The answer to this question is clearly a necessary prerequisite for the parameter estimation problem to be well-posed. Although necessary, *a priori* identifiability is obviously not sufficient to guarantee an accurate model identification from real data or, even more importantly, model validity. In fact, an *a priori* identifiable model may have to be rejected for several reasons, for example, it cannot explain the data or the precision with which its parameters can be estimated is very poor either because of a too complex structure for the data or simply because of the paucity of the data. However, these aspects should in no way detract from the necessity of satisfying the *a priori* identifiability requirement for any model of interest that is to have any biological meaning at all. It is also worth emphasizing that *a priori* identifiability cannot be tested when estimating the parameters from the data, for example, by nonlinear least squares with software like SAAMII or MATLAB.

Modelling Methodology for Physiology and Medicine. DOI: http://dx.doi.org/10.1016/B978-0-12-411557-6.00004-5

In fact, these packages can only assess identifiability numerically, thus they cannot distinguish whether nonidentifiability arises from *a priori* or *a posteriori* reasons (structure too complex or paucity of data), and may also interpret as globally identifiable, models that are only locally identifiable [10]. This last issue is particularly critical when dealing with physiological systems where a different numerical estimate can characterize a pathological from a normal state. *A priori* global identifiability is also crucial in qualitative experiment design [11] that studies the input–output configuration necessary to ensure unique estimation of the unknown parameters. In fact, it distinguishes among those experiments that cannot succeed and those that might succeed and, among those last ones, it determines the minimal input–output configuration to ensure estimation of the unknown parameters.

Thus, the need to investigate the identifiability properties of nonlinear systems is unquestionable. Referring to the more recent literature, different approaches have been proposed which can be grouped in three different classes: the revisited Taylor Series approach [12], the methods based on the local state isomorphism theorem [5,6,13,14], and the differential algebra based methods [9,15–17]. Whatever the method used, it requires solving a system of nonlinear algebraic equations where the number of unknowns increases very rapidly with the number of terms, the nonlinearity degree, and the model order. Thus, the solution of the *a priori* identifiability problem is severely limited by computational bounds.

It should be stressed that identifiability depends also on the experimental conditions. More precisely, for a fixed model structure and measurement schedule, identifiability does in general depend on the class of admissible input functions acting on the system. Input functions which do not "excite" the system properly may render some parameters invisible from the external output. Structural identifiability analysis is performed under the assumption that the input is *persistently exciting*, see Refs. [15,18] for a precise definition of this condition. Of course the admissible inputs class must contain persistently exciting functions.

For completeness, one should mention also the so-called *practical identifiability*, which is a data-dependent property, a property that can be tested based on simulated model data coupled with parameter estimation routines. Techniques for testing practical identifiability have been proposed in the literature [1]. Although in principle these methods may give an answer in case of local identifiability, practical identifiability tests based on a specific data set cannot give exact answers about structural identifiability and one should therefore resort to heuristics and extensive simulations. Moreover, it may be impossible to distinguish between nonidentifiability and lack of convergence of the iterative optimization algorithm used for parameter estimation.

Certainly, the introduction of concepts of differential algebra in control and system theory, mainly due to Fliess [19], has led to a better understanding of the nonlinear identifiability problem. In particular, Ollivier [16] and Ljung and Glad [15] have shown that the concept of a *characteristic set* of a differential ideal, introduced by Ritt [20], is a very useful tool in identifiability analysis. The main goal of the research in this area has become the construction of an efficient algorithm, based on differential algebra, to test global identifiability of nonlinear systems

[9,15]. Recently a differential algebra algorithm, based on several conceptual improvements on the methods existing in the literature, has been implemented in a computer program making use of the symbolic language *Reduce* [21]. The algorithm has been tested on several complex linear and nonlinear models used in the biological and biomedical literature.

Ideally, one would like to establish rigorously the domain of validity of the identifiability algorithm in terms of model structure complexity, such as the number of states, number of parameters, number of essential nonlinearities, number of measured variables, and so forth. This is, however, very difficult, if not impossible. It would, in fact, require defining the limits of applicability of the Buchberger algorithm [22] for solving the *exhaustive summary* of the model, which does not only depend on the number of parameters, but also, and in a complicated way, on the other variables. For example, a rather complicated model with 20 state variables, but with 20 observed outputs, may turn out to be easier to analyze than a very interconnected model with 4 states and only 1 measured output [23].

Like other algorithms based on differential algebra, the DAISY (Differential Algebra Identifiability of SYstems) algorithm of Ref. [21] analyzes identifiability of systems assuming generic initial conditions. It has been pointed out that it may give wrong answers in special cases when the initial condition is fixed to some special value, a situation frequently encountered in identification of biological and medical systems [18]. Thus, work has been devoted to extend the applicability of the algorithm to systems started at specific initial conditions [24].

In the next chapter, after briefly reviewing some fundamentals of differential algebra, we will present the theoretical and algorithmic aspects of a method for testing structural global identifiability of nonlinear models starting at specific initial conditions. An example will also be presented.

4.2   Background and Definitions

4.2.1   The System

Consider a parameterized nonlinear system described in state-space form:

$$\begin{cases} \dot{\mathbf{x}}(t) = \mathbf{f}[\mathbf{x}(t), \mathbf{p}] + \mathbf{G}[\mathbf{x}(t), \mathbf{p}]\mathbf{u}(t) \\ \mathbf{y}(t) = \mathbf{h}[\mathbf{x}(t), \mathbf{u}(t), \mathbf{p}] \end{cases} \tag{4.1}$$

where x is the n-dimensional state variable; $\mathbf{u}$ is the m-dimensional input vector of smooth functions; $\mathbf{y}$ is the r-dimensional output; $\mathbf{p} \in \mathscr{P}$ is the ν-dimensional parameter vector. If initial conditions are specified, the relevant equation $\mathbf{x}(t_0) = \mathbf{x}_0$ is added to the system. Although this is not strictly necessary, we have assumed the system affine in the control variable. This simplifies some technical steps of the algorithm. The essential assumption here is that the entries of $\mathbf{f}$, $\mathbf{G} = [\mathbf{g}_1, \ldots, \mathbf{g}_m]$, and $\mathbf{h}$ are *polynomial functions* of their arguments.

4.2.2 Structural Identifiability

Different definitions have been given in the literature [5,6,9,15]. Here we adopt the one used in Ref. [9]. Let $\mathbf{y} = \Phi(\mathbf{p}, \mathbf{u})$ be the input–output map of the system (4.1). Although $\Phi(\mathbf{p}, \mathbf{u})$ obviously depends also on the initial state of the system (4.1), we shall not write this dependence explicitly.

Definition: We say that system (4.1) is *a priori globally (or uniquely) identifiable* if and only if, for at least a generic set of points $\mathbf{p}^* \in \mathscr{P}$, the equation

$$\Phi(\mathbf{p}^*, \mathbf{u}) = \Phi(\mathbf{p}, \mathbf{u}) \tag{4.2}$$

has only one solution $\mathbf{p}^* = \mathbf{p}$ for at least one input function $\mathbf{u}$;
the system (4.1) is *locally (or nonuniquely) identifiable* if and only if, for at least a generic set of points $\mathbf{p}^* \in \mathscr{P}$, the equation (4.2) has more than one, but at most a finite number of solutions, for all input functions $\mathbf{u}$;
the system (4.1) is *nonidentifiable* if, for at least a generic set of points $\mathbf{p}^* \in \mathscr{P}$, the equation (4.2) has an infinite number of solutions for all input functions $\mathbf{u}$.

As we shall review in detail later in this chapter, the use of differential algebra permits the writing of the input–output relation of the system in *implicit form*, that is, as a set of m polynomial differential equations in the variables $(\mathbf{y}, \mathbf{u})$. The coefficients of these differential equations will depend, in general polynomially, on the parameter $\mathbf{p}$ [16]. The availability of these r input–output differential equations greatly simplifies the analysis. In fact, in order to analyze the *a priori* identifiability of the model (4.1), one just has to define a proper "canonical" set of coefficients of the polynomial differential equations, say $\mathbf{c}(\mathbf{p})$. One refers to this family of functions of $\mathbf{p}$ as the *exhaustive summary* of the model [5,16]. After the exhaustive summary is found, to study *a priori* global identifiability of the model, one has to see if the map $\mathbf{c}(\mathbf{p})$ is *injective*, that is, to see if the system of nonlinear algebraic equations:

$$\mathbf{c}(\mathbf{p}) = \hat{\mathbf{c}}$$

where $\hat{\mathbf{c}}$ is a family of given values, has (generically) a unique solution in $\mathbf{p}$.

4.3 Identifiability and Differential Algebra

4.3.1 The Problem

With the basic definitions at hand, we can now return to the dynamic system (4.1). This can be looked upon as a set of $n + r$ differential polynomials:

$$\dot{\mathbf{x}}(t) - \mathbf{f}[\mathbf{x}(t), \mathbf{p}] + \mathbf{G}[\mathbf{x}(t), \mathbf{p}]\mathbf{u}(t) \tag{4.3}$$

$$\mathbf{y}(t) - \mathbf{h}[\mathbf{x}(t), \mathbf{u}(t), \mathbf{p}] \tag{4.4}$$

Polynomials (4.3) and (4.4) are the generators of a differential ideal I in a differential ring. It is known that the state-space description ensures the primality of the ideal generated by polynomials (4.1) [15]. The *characteristic set* of the ideal I is a finite set of $n + r$ nonlinear differential equations that describes the same solution set of the original system. Its special structure allows the construction of the so-called *exhaustive summary* of the model used to test identifiability.

The problems now are (i) to construct, in an algorithmic way, the characteristic set starting from the model equations and (ii) to solve the algebraic nonlinear equations of the exhaustive summary. This can be solved by a computer algebra method, for example, the Buchberger algorithm [22] to calculate the Gröbner basis. To solve the first problem, we have observed [9,18] that the computational complexity is strongly influenced by the choice of the ring of multipliers of the differential ring $K[\mathbf{z}]$ and by the ranking of the variables of the differential polynomials.

4.3.2 The Characteristic Set

The characteristic set of the differential polynomials (4.3) and (4.4) will be considered in the ring $R(\mathbf{p})[\mathbf{u}, \mathbf{y}, \mathbf{x},]$, where $R(\mathbf{p})$ is the field of rational functions of the parameter $\mathbf{p}$. The variables are the states, inputs, and outputs. In this way, once the characteristic set is obtained, the coefficients of the differential polynomials are rational functions (or polynomials) in $\mathbf{p}$ and the differential polynomials, themselves, depend only on $\mathbf{u}$, $\mathbf{y}$, and $\mathbf{x}$ variables and, possibly, on their derivatives. To calculate the characteristic set, one has to use Ritt's algorithm, which is analogous to the Gauss elimination algorithm used to solve linear algebraic equations. This requires the introduction of a ranking among the variables; in particular, the higher ranked variables are eliminated first. In the context of parameter identifiability, the unknown state variables and their derivatives are ranked highest so they are preferentially eliminated. Normally the following ranking is chosen:

$$u_1 < \dot{u}_1 < \ldots < u_2 < \dot{u}_2 < \ldots < y_1 < \dot{y}_1 < \ldots \\ < y_2 < \dot{y}_2 < \ldots < x_1 < \dot{x}_1 < \ldots < x_2 < \dot{x}_2 < \ldots \tag{4.5}$$

Then the polynomials defining the systems in Eqs. (4.3) and (4.4) are reduced using Ritt's pseudodivision algorithm [20]. Once the system can no longer be reduced, the characteristic set has been obtained. For explanations of technical terms, one may consult Ref. [21]. It has been shown that if the original dynamic system is in state-space form, as in Eq. (4.1), the characteristic set has a diagonal structure, and the first r (number of outputs) polynomials of the characteristic set contain only input and output variables. With respect to the ranking in Eq. (4.5), the characteristic set of the polynomials (4.3) and (4.4) has the following form:

$$A_1(\mathbf{u}, \mathbf{y}) \ldots A_r(\mathbf{u}, \mathbf{y}) \\ A_{r+1}(\mathbf{u}, \mathbf{y}, x_1) \\ A_{r+2}(\mathbf{u}, \mathbf{y}, x_1, x_2) \\ \vdots \\ A_{r+n}(\mathbf{u}, \mathbf{y}, x_1, \ldots, x_n) \tag{4.6}$$

We will refer to the corresponding first r differential polynomial equations

$$A_1(\mathbf{u}, \mathbf{y}) = 0 \quad A_2(\mathbf{u}, \mathbf{y}) = 0. \ldots A_r(\mathbf{u}, \mathbf{y}) = 0 \tag{4.7}$$

of Eq. (4.6) as the *input—output relations*. These polynomial differential equations are the implicit description of the input—output map of the system (4.1) that we were referring to in Section 4.2. In fact, these polynomials are obtained after elimination of the state variables x from the set (4.3) and (4.4) and, hence, represent exactly the pairs $(\mathbf{u}, \mathbf{y})$ that are described by the original system. After a suitable normalization, the input—output polynomials can be rendered monic and their coefficients provide a set of rational functions of the unknown parameter $\mathbf{p}$ that forms the so-called *exhaustive summary* of the model.

Remark 1 The triangular form of the characteristic set allows for the easy extraction of information from the equations containing the state variables. For example, if the derivatives of the state components do not appear in the last n equations, the dynamic system (4.1) is *algebraically observable* [15]. Thus, algebraic observability of the system is easy to check from the characteristic set.

Remark 2 If the system is of high dimension, the calculation of the characteristic set can become very complex. In this case, to decrease the computational complexity, a different suitable ranking may be chosen.

4.3.3 A Benchmark Model

In this section, we shall analyze the identifiability of a nonlinear model that describes the kinetics of a drug in the human body. This model has been widely employed as a benchmark model for studying global identifiability [6]. The drug is injected into the blood where it exchanges linearly with the tissues; the drug is irreversibly removed with a nonlinear saturative characteristic from the blood and with a linear one from the tissue. Thus, the model incorporates a Michaelis—Menten type nonlinearity and is mathematically described by the following rational nonlinear differential equations:

$$\begin{cases} \dot{x}_1 &= -(k_{21} + V_M/(K_m + x_1))x_1 + k_{12}x_2 + b_1 u \\ \dot{x}_2 &= k_{21}x_1 - (k_{02} + k_{12})x_2 \\ y &= c_1 x_1 \end{cases} \tag{4.8}$$

where x_1, x_2 are drug masses in blood and tissues, respectively, u is the drug input, y the measured drug output in the blood, k_{12}, k_{21}, and k_{02} are the constant rate parameters, V_M and K_m are the classical Michaelis—Menten parameters, and b_1 and c_1 are the input and output parameters, respectively. The question is whether the unknown vector $\mathbf{p} = [k_{21}, k_{12}, V_M, K_m, k_{02}, c_1, b_1]$ is globally identifiable from the input—output experiment. Let the ranking of the variables be $u < y < x_1 < x_2$.

The reduction procedure is started and the characteristic set is calculated. Here only the input–output relation is reported.

$$
\begin{aligned}
&\ddot{y}y^2 + k_{21}k_{02}y^3 - (k_{21}c_1b_1 + k_{02}c_1b_1)y^2u + \\
&+ (k_{21} + k_{12} + k_{02})\dot{y}y^2 - c_1b_1y^2\dot{u} - K_mc_1{}^3b_1\dot{u} + \\
&+ (2k_{21}K_mk_{02} + k_{12}V_Mk_{02}V_M)c_1y^2 + \\
&- 2(k_{12} + k_{02})c_1{}^2b_1K_myu + \\
&+ 2(k_{21}K_mc_1 + k_{12}K_mc_1 + k_{02}K_mc_1)y\dot{y} + 2K_mc_1y\ddot{y} + \\
&- 2K_mc_1{}^2b_1y\dot{u} + K_mc_1{}^2(k_{21}k_{02} + k_{12}V_M + k_{02}V_M)y + \\
&- (k_{12}K_m{}^2b_1c_1{}^3 + k_{02}K_m{}^2b_1c_1{}^3)u + K_m{}^2c_1{}^2\ddot{y} + \\
&+ (k_{21}K_m{}^2c_1{}^2 + k_{12}K_m{}^2c_1{}^2 + V_MK_mc_1{}^2 + k_{02}K_m{}^2c_1{}^2)\dot{y}
\end{aligned}
\tag{4.9}
$$

Coefficients are extracted (these are the exhaustive summary of the model) and evaluated at a numerical point $\hat{p}$ randomly chosen in the parameter space $\mathscr{P}$. Each coefficient, in its polynomial form, is then set equal to its corresponding numerical value. The Gröbner basis is calculated by Buchberger's algorithm [22] and applied to solve the obtained equations. The obtained system has an infinite number of solutions, thus the model is *a priori* nonidentifiable. Note that if the input parameter is assumed to be known, for example, if $b_1 = 1$, the model becomes *a priori* globally identifiable.

4.4   The Question of Initial Conditions

As it has been shown, the construction of the characteristic set ignores the initial conditions. In particular, the input–output relations (4.7) represent the input–output pairs of the system for "generic" initial conditions. Often, however, physical systems have to be started at special initial conditions, for example, all radiotracer kinetics experiments in humans [9] are necessarily started at the initial state $\mathbf{x}(0) = 0$. Thus, the problem arises if some specific initial conditions can change the input–output relations. To better understand the problem, a very simple example follows. The example will be linear, since identifiability of linear system can be checked by standard transfer function methods. The system will be a three-compartment model describing the dynamics of a drug in a tissue. For this model, the initial conditions have to be set to zero, since no drug quantities are present before the exogenous injection u.

4.4.1   An Example

Consider the system

$$
\begin{cases}
\dot{x}_1 = p_{13}x_3 + p_{12}x_2 - p_{21}x_1 + u & x_1(0) = 0 \\
\dot{x}_2 = -p_{12}x_2 + p_{21}x_1 & x_2(0) = 0 \\
\dot{x}_3 = -p_{13}x_3 & x_3(0) = 0 \\
y = x_2
\end{cases}
\tag{4.10}
$$

where $\mathbf{x} = [x_1, x_2, x_3]$ is the state vector, for example, x_1, x_2, x_3 are drug masses in compartment 1, 2, and 3 respectively; u is the drug input; y is the measured drug output; $\mathbf{p} = [p_{12}, p_{21}, p_{13}]$ is the rate parameter vector (assumed constant). The question is: are all the unknown parameters p_{12}, p_{21}, p_{13} globally identifiable from the input–output experiment?

With the standard ranking from Eq. (4.5), the following characteristic set is calculated:

$$
\begin{aligned}
A_1 &\equiv \dot{u}p_{21} + \dddot{y} + \ddot{y}(p_{12} + p_{21} + p_{13}) + \\
    &\quad + \dot{y}(p_{12}p_{13} + p_{21}p_{13}) + up_{21}p_{13} \\
A_2 &\equiv \dot{y} + yp_{12} - x_1 p_{21} \\
A_3 &\equiv y - x_2 \\
A_4 &\equiv \ddot{y} + \dot{y}(p_{12} + p_{21}) + up_{21} - x_3 p_{13} p_{21}
\end{aligned}
\tag{4.11}
$$

Note that only the differential polynomial A_1 contains information on model identifiability, in fact, it does not include as variables either $\mathbf{x}$ or its derivatives. This polynomial represents the input–output relation of the model. By extracting the coefficients from A_1 (which is already monic) and setting them equal to known symbolic values, the following exhaustive summary equations are obtained

$$
\begin{aligned}
p_{21} &= c_1 \quad p_{12} + p_{21} + p_{13} = c_2 \\
p_{12}p_{13} &+ p_{21}p_{13} = c_3 \quad p_{21}p_{13} = c_4
\end{aligned}
\tag{4.12}
$$

Equations (4.12) have a unique solution, hence all of the parameters p_{12}, p_{21}, and p_{13} seem to be globally identifiable. However, with zero initial conditions, it is immediately apparent that the transfer function of system (4.10) depends only on the parameters p_{12} and p_{21}, so that p_{13} cannot be identifiable. What goes wrong? Three observations are in order.

1. The characteristic set does not take into account the specific initial conditions.
2. The reachability subspace of system (4.10) is the two dimensional subspace $\{x_3 = 0\}$. Said in other language, the system is *nonaccessible* from initial states of the form $\mathbf{x}(0) = [x_1 \ x_2 \ 0]^{\mathrm{T}}$.
3. The exhaustive summary is the one provided by the transfer function method without performing the cancellations on the transfer function due to nonminimality.

Since $x_3(0) = 0 \Rightarrow x_3(t) = 0$ for all times (as the third compartment is evolving autonomously in time), the state variable x_3 can be set equal to zero in the polynomials of the characteristic set (4.11) where A_4 becomes

$$
\bar{A}_4 \equiv \ddot{y} + \dot{y}(p_{12} + p_{21}) + up_{21} = 0 \quad \forall t
\tag{4.13}
$$

Note that the input–output relation A_1 of (4.11) can be written as:

$$
A_1 \equiv \dot{\bar{A}}_4 + p_{13}\bar{A}_4 = 0
\tag{4.14}
$$

Thus, the set formed by A_1, A_2, A_3 of (4.11) and $\overline{A}_4$ is no longer a characteristic set. To take account of initial conditions, the ideal generated by the original differential polynomial system *plus* $\overline{A}_4$ has to be computed. The new characteristic set is

$$
\begin{aligned}
\overline{A}_4 &\equiv \ddot{y} + \dot{y}(p_{12} + p_{21}) + up_{21} \\
A_2 &\equiv \dot{y} + yp_{12} - x_1 p_{21} \\
A_3 &\equiv y - x_2 \\
A_4 &\equiv -x_3 p_{13} p_{21}
\end{aligned}
\tag{4.15}
$$

Only p_{12} and p_{21} are uniquely identifiable, while p_{13} has disappeared. This is in agreement with the transfer function method. Thus, we have observed that the ideal generated by polynomials describing the dynamic system (4.1) may change when initial conditions are taken into account. The above example gives a hint as to why the method may fail, in particular, the lack of reachability from the initial state (i. e., accessibility) may be the cause of problems.

4.4.2 The Role of Accessibility

In the following, we shall refer to a concept of geometric nonlinear control theory [25−27] called *accessibility*. In particular, the *accessibility* can be viewed as a weak counterpart of the concept of reachability (from an arbitrary initial state).

Definition 1 *The system (4.1) is* accessible from $\mathbf{x}_0$ *if the set of states reachable from* $\mathbf{x}_0$ *(at any finite time) has a nonempty interior, that is, it contains an open ball in* $\mathbb{R}^n$.

To study accessibility, one looks at the *Control Lie Algebra*, that is, the smallest Lie algebra $\mathscr{C}$ containing the vector fields $\mathbf{f}, \mathbf{g}_1, \ldots, \mathbf{g}_m$ of Eq. (4.3) and invariant under Lie bracketing with $\mathbf{f}, \mathbf{g}_1, \ldots, \mathbf{g}_m$. To the Lie algebra $\mathscr{C}$ we associate the distribution Δ_C, mapping each $\mathbf{x} \in R^n$ into the vector space

$$
\Delta_C(\mathbf{x}) = \text{span}\{\tau(\mathbf{x}) : \tau \in \mathscr{C}\}
$$

We recall from the literature [25−27] the *accessibility rank condition*.

Theorem 1 *For analytic, in particular polynomial, systems, a necessary and sufficient condition for accessibility from* $\mathbf{x}_0$ *is that* $\dim \Delta_C(\mathbf{x}_0) = n$.

A full understanding of the identifiability problem with specific initial conditions requires the study of the role of accessibility in the structure of the characteristic set.

It has been shown that, when the system is accessible from $\mathbf{x}_0$, adding the specific initial condition $\mathbf{x}(0) = \mathbf{x}_0$ as a constraint cannot change the characteristic set. This is so since the variety where the evolution of the system takes place has the same dimension as the initial variety [25], and the order of the system cannot drop. In the benchmark example presented in Section 4.3.3, the accessibility rank condition is satisfied since $\dim\Delta_C(\mathbf{x}_0) = 2$, that is the system order, and the system is accessible from every point.

Conversely, suppose that the system is algebraically observable, generically accessible (i.e., accessible from all points except from a "thin" subvariety) and assume that $\mathbf{x}_0$ belongs to the "thin" set from which the system is nonaccessible. There is an invariant subvariety where the evolution of the system takes place when started at the initial condition $\mathbf{x}_0$ [26]. This subvariety can be calculated by a construction based on the accessibility Lie Algebra, see Ref. [27, p. 154]. Let $\phi(\mathbf{x}) = 0$ be the equation of the invariant subvariety of nonaccessible states. This algebraic equation must be added to the characteristic set in order to get a reduced representation of the system dynamics. In order to compute the new characteristic set, we note that the polynomials representing the state dynamics alone in Eq. (4.3), with a suitable ranking, are a characteristic set and that $\phi(\mathbf{x})$ is reduced with respect to it. For a known property [20, p. 5] the characteristic set of the ideal generated by polynomials from Eq. (4.3) plus $\phi(\mathbf{x})$ is of a *lower order than the original one*. Once the output polynomial from Eq. (4.4) is added, a new characteristic set is calculated that still remains of the (lower) order obtained in the previous elimination step. Note that in the elimination steps of the procedure described above, the primality of the ideal needs to be checked. If the ideal is not prime, a factorization needs to be performed [20].

4.5   Identifiability of Some Nonpolynomial Models

The differential algebraic method, in principle, deals with polynomial or rational systems. However, it is possible to adopt this method also for testing structural global identifiability of systems involving nonpolynomials, for example, exponential or logarithmic, functions.

As an elementary example, consider a (nonpolynomial) system like

$$\dot{x}_1 = a\,exp(-x_2) + u$$
$$\dot{x}_2 = -bx_1$$

where a and b are the unknown parameters, u is the input, and x_1 and x_2 are the state variables. This system can be rendered polynomial by introducing a new state $x_3 = a\,exp(-x_2)$ and by differentiating it the following additional equation $\dot{x}_3 = -\dot{x}_2 x_3$ is provided. This differential equation will turn it into a third-order system of the following form:

$$\dot{x}_1 = x_3 + u$$
$$\dot{x}_2 = -bx_1$$
$$\dot{x}_3 = bx_1 x_3$$

which is indeed polynomial (and time-invariant).

Furthermore, in many biological and physiological applications, very often in the differential equations describing the phenomena, time-varying coefficients appear with a known functional form, but they depend on some unknown parameters. Consider for example a system like

$$\dot{x}_1 = a \exp(-bt)x_1 + u$$

This system also can be rendered polynomial by introducing a new state $x_2 = \exp(-bt)$ and an additional equation $\dot{x}_2 = -bx_2$, which will turn it into the following second-order polynomial (and time-invariant) system

$$\dot{x}_1 = ax_1x_2 + u$$
$$\dot{x}_2 = -bx_2$$

With this technique one can handle many classical situations where time-varying coefficients of known functional form appear. Note that nonalgebraic nonlinearities lead to an augmented model that is trivially *globally* nonaccessible since the evolution of the system obtained by adding the new state variable is constrained to take place in some invariant submanifold. For details on this issue see Ref. [24].

4.6 A Case Study

In this section, we shall analyze the identifiability of a fifth-order model based on *in vitro* experiments to study the kinetics of homoacetogenesis by human-colon bacteria. Due to its complexity, it is considered in Ref. [28] to be a challenging example to analyze. The model is mathematically described by the following rational nonlinear differential equations:

$$\begin{cases}
\dot{x}_1(t) = \dfrac{\mu_{\max}x_1x_5}{k + x_5} - k_dx_1 \\[2mm]
\dot{x}_2(t) = k_dx_1 - k_1x_2 + k_5x_2x_3 + k_5x_2x_4 \\[2mm]
\dot{x}_3(t) = k_{la}(x_5 - k_{hrt}x_3)V_l/V_g \\[2mm]
\dot{x}_4(t) = \dfrac{(1 - y_h)}{y_h}\dfrac{\mu_{\max}x_1x_5}{k + x_5} - k_5x_2 \\[2mm]
\dot{x}_5(t) = -\dfrac{\mu_{\max}\dot{x}_1x_5(k + x_5) + k_{la}y_h\dot{x}_3(k+x_5)^2}{\mu_{\max}x_1(k + x_5) - \mu_{\max}x_1x_5 - k_{la}k_{hrt}y_h(k+x_5)^2} \\[2mm]
y_1(t) = \alpha(x_1 + x_2) \\[1mm]
y_2(t) = x_3 \\[1mm]
y_3(t) = x_4
\end{cases} \qquad (4.16)$$

where x_i, $i = 1, ..., 5$ are concentrations, k_{la}, k_{hrt}, and α are physical constants known from the literature, see the referenced paper for the details. Following Ref. [28], the ratio V_l/V_g is assumed to be known and is not included in the list of unknown parameters so that the unknown parameter vector turns out to be $\mathbf{p} = [\mu_{max}, k, k_d, k_i, y_h]$. The variables y_1, y_2, and y_3 are the measured outputs of the system. The initial conditions are unknown.

The *a priori* identifiability of this model was first analyzed in Ref. [28] by using a sufficient condition for global identifiability due to Ref. [14]. This condition holds for uncontrolled models started at known initial conditions. To verify this condition (which is only sufficient) several calculations are required.

We test the identifiability of this model by using the software DAISY [21]. With the standard ranking of the input, output, and state variables (i.e., $y < x_1 < x_2 < x_3 < x_4 < x_5$) the program starts all of the required calculations. Due to space limitations, we do not report here the exhaustive summary of this model, which is very long, but only its Gröbner basis solution:

$$\{\mu_{max} - 4, \quad k - 25, \quad k_d - 27, \quad i - 3, \quad y_h - 18 \tag{4.17}$$

which shows that all the parameters μ_{max}, k, k_d, i, and y_h are uniquely identifiable from input−output experiments. Note that in this case, the knowledge of the initial conditions would be redundant to identifiability.

In practice, to check the global identifiability of this model with DAISY, the user has to write the input file in a given format. In the following, the input file for the model (4.16) with the simplification above presented to eliminate redundancy is reported:

Input File of DAISY

```
WRITE "A case study"$
% B_ IS THE VARIABLE VECTOR
B_: = {y1,y2,y3,x3,x4,x1,x2,x5}$
FOR EACH EL_ IN B_ DO DEPEND EL_,T$
%B1_ IS THE UNKNOWN PARAMETER VECTOR
B1_: = {numax,k,kd,ki,yh}$
%NUMBER OF STATE(S)
NX_: = 5$
%NUMBER OF OUTPUT(S)
NY_: = 3$
C_: = {df(x1,t) = numax*x5*x1/(k + x5)-kd*x1,
    df(x2,t) = kd*x1-ki*x2,
    df(x3,t) = kla*(x5-khrt*x3),
    df(x4,t) = ((1-yh)/yh)*numax*x1*x5/(k + x5),
    df(x5,t) = -(numax*df(x1,t)*x5*(k + x5) +
    kla*yh*df(x3,t)*(k + x5)^2)/(numax*x1*(k + x5)-numax*x1*x5-
    kla*khrt*yh*(k + x5)^2),
    y1 = alpha*(x1 + x2),
    y2 = x3,
    y3 = x4}$
LET kla = 6,khrt = 2,alpha = 4$
SEED_: = 125$
DAISY()$
END$
```

Due to space limitations the output file is not reported here, but the reader can directly run the above input file and see that DAISY provides the required structural identifiability answer in less than 1 min. This computer algebra tool does not require expertise in mathematical modelling by the experimenter.

4.7 Conclusion

The goal of this chapter is to make researchers in system biology aware of the relevance of checking structural identifiability of the dynamic model under study. A differential algebra method based on the characteristic set of the ideal generated by the polynomials defining the system can be successfully used in testing the identifiability of many nonlinear systems. This method has been implemented in the software tool DAISY [21] and is able to check structural global identifiability in a fully automatic way.

References

[1] Hengl S, Kreutz C, Timmer J, Maiwald T. Data-based identifiability analysis of nonlinear dynamical models. Bioinformatics 2007;23(19):2612−8.

[2] Roper RT, Saccomani MP, Vicini P. Cellular signaling identifiability analysis: a case study. J Theor Biol 2010;264(2):528−37.

[3] Evans ND, Moyse HAJ, Lowe D, Briggs D, Higgins R, Mitchell D, et al. Structural identifiability of surface binding reactions involving heterogeneous analyte: application to surface plasmon resonance experiments. Automatica 2013;49(1):48−57.

[4] Ljung L. System identification—theory for the user. 2nd ed. Upper Saddle River, NJ: PTR Prentice Hall; 1999.

[5] Walter E, Lecourtier Y. Global approaches to identifiability testing for linear and nonlinear state space models. Math Comput Simul 1992;24:472−82.

[6] Chappell MJ, Godfrey KR. Structural identifiability of the parameters of a nonlinear batch reactor model. Math Biosci 1992;108:245−51.

[7] Godfrey KR, DiStefano III JJ. Identifiability of model parameters. In: Walter E, editor. Identifiability of Parametric Models, 1. Oxford: Pergamon Press; 1987. pp. 1−20

[8] Cobelli C, DiStefano III JJ. Parameter and structural identifiability concepts and ambiguities: a critical review and analysis. Am J Physiol 1980;239:R7−24.

[9] Audoly S, Bellu G, D'Angiò L, Saccomani MP, Cobelli C. Global identifiability of nonlinear models of biological systems. IEEE Trans Biomed Eng 2001;48(1):55−65.

[10] Cobelli C, Saccomani MP. Unappreciation of *a priori* identifiability in software packages causes ambiguities in numerical estimates. Letter to the Editor. Am J Physiol 1990;21:E1058−9.

[11] Saccomani MP, Cobelli C. Qualitative experiment design in physiological system identification. IEEE Control Syst 1992;12(6):18−23.

[12] Joly-Blanchard G, Denis-Vidal L. Some remarks about identifiability of controlled and uncontrolled nonlinear systems. Automatica 1998;34:1151−2.

[13] Chapman MJ, Godfrey KR, Chappell MJ, Evans ND. Structural identifiability of nonlinear systems using linear/non-linear splitting. Int J Control 2003;76(3):209−16.

[14] Denis-Vidal L, Joly-Blanchard G. Equivalence and identifiability analysis of uncontrolled nonlinear dynamical systems. Automatica 2004;40:287–92.

[15] Ljung L, Glad ST. On global identifiability for arbitrary model parameterizations. Automatica 1994;30(2):265–76.

[16] Ollivier F. Le Problème de l'Identifiabilité Structurelle Globale: Étude Théorique, Méthodes Effectives et Bornes de Complexité. Thèse de Doctorat en Science, École Polytéchnique, Paris, France; 1990.

[17] Margaria G, Riccomagno E, Chappell MJ, Wynn HP. Differential algebra methods for the study of the structural identifiability of rational function state-space models in the biosciences. Math Biosci 2001;174:1–26.

[18] Saccomani MP, Audoly S, D'Angiò L. Parameter identifiability of nonlinear systems: the role of initial conditions. Automatica 2004;39:619–32.

[19] Fliess M, Glad ST. An algebraic approach to linear and nonlinear control. Essays on Control: Perspectives in the Theory and its Applications, Progress in Systems and Control Theory, vol. 14. Boston, MA: Birkhuser; 1993. pp. 223–267

[20] Ritt JF. Differential algebra. Providence, RI: American Mathematical Society; 1950.

[21] Bellu G, Saccomani MP, Audoly S, D'Angiò L. DAISY: a new software tool to test global identifiability of biological and physiological systems. Comp Meth Prog Biomed 2007;88:52–61.

[22] Buchberger B. An algorithmical criterion for the solvability of algebraic system of equation. Aequationes Math 1988;4(3):45–50.

[23] Saccomani MP, Audoly S, Bellu G, D'Angiò L. Examples of testing global identifiability of biological and biomedical models with the DAISY software. Comp Biol Med 2010;40:402–7.

[24] D'Angiò L, Saccomani MP, Audoly S, Bellu G. Identifiability of nonaccessible nonlinear systems. Positive Systems, Lecture Notes in Control and Information Sciences, 389. Berlin/Heidelberg, Germany: Springer-Verlag; 2009. pp. 269–277

[25] Hermann R, Krener AJ. Nonlinear controllability and observability. IEEE Trans Autom Control 1977;AC-22(5):728–40.

[26] Isidori A. Nonlinear control systems. 3rd ed. London: Springer; 1995.

[27] Sontag ED. Mathematical control theory. 2nd ed. Berlin, Germany: Springer; 1998.

[28] Munoz-Tamayo R, Laroche B, Leclerc M, Walter E. Modelling and identification of *in vitro* homoacetogenesis by human-colon bacteria. In: Proceedings of the 16th IEEE Mediterranean Conference on Control and Automation, France: Ajaccio; 2008. pp. 1717–22.

5 Parameter Estimation

Paolo Magni[a] and Giovanni Sparacino[b]

[a]Dipartimento di Ingegneria Industriale e dell'Informazione, Università degli Studi di Pavia, Pavia, Italy, [b]Dipartimento di Ingegneria dell'Informazione, Università di Padova, Padova, Italy

5.1   Problem Statement

Let us suppose that, thanks to available physiological knowledge and/or assumptions, we have been able to build the equations that define the model of a physiological system, for example, an endocrine-metabolic system at either the organ or whole body level. In order to complete the model, it is necessary to assign numerical values to its parameters. In practical cases, this can be done by exploiting the availability of one, or more, measurement signals in response to a known input. For the sake of simplicity, in this chapter we will consider one measurable signal only, for example, the plasma concentration of the substance under study in response to a known exogenous perturbation, for instance, a bolus injection. In general, by assuming an error-free model structure, the measurable signal $y(t)$ can be expressed as

$$y(t) = g(t, \boldsymbol{p}) \tag{5.1}$$

where the function $g(\cdot)$ depends on the specific model equations and $\boldsymbol{p} = [p_1, \ldots, p_M]^{\mathrm{T}}$ is the vector containing the M unknown model parameters. In practice, only N samples of $y(t)$, collected at (usually sparse and nonuniformly spaced) times $\{t_1, \ldots, t_N\}$, are normally available. Since samples are corrupted by several sources of noise, including measurement error in particular, the following model can be considered to describe the kth measurement z_k:

$$z_k = y(t_k) + v_k = g(t_k, \boldsymbol{p}) + v_k, \quad k = 1, \ldots, N \tag{5.2}$$

where v_k denotes the (additive) error which affects the kth measurement z_k. The error v_k is usually modelable as a random variable with zero mean, and often some statistical information is available on the random process given by the collection $\{v_k\}$. For instance, $\{v_k\}$ can be an uncorrelated stationary process with variance σ^2. In some cases, each sample of $\{v_k\}$ has a different variance level σ_k^2. The variance σ_k^2 is often known, or known apart from a scale factor, from previous studies made

Modelling Methodology for Physiology and Medicine. DOI: http://dx.doi.org/10.1016/B978-0-12-411557-6.00005-7

on the measurement process. Sometimes the measurement error variance σ_k^2 is dependent on the measurement itself, for example, for the constant coefficient of variation (CV) situation one has

$$\sigma_k^2 = [\text{CV } y(t_k)]^2 = [\text{CV } g(t_k, p)]^2 \tag{5.3}$$

where CV is a positive real.

By letting $z = [z_1, \ldots, z_N]^T$, $G(p) = [g(t_1, p), \ldots, g(t_N, p)]^T$ and $v = [v_1, \ldots, v_N]^T$, Eq. (5.2) can be written in vector notation as

$$z = y + v = G(p) + v \tag{5.4}$$

In Eq. (5.4), the covariance matrix of vector v, i.e., $\Sigma_v = E[vv^T]$, can always be described as

$$\Sigma_v = \sigma^2 B \tag{5.5}$$

where σ^2 is a suitable scalar parameter, known or unknown, and B is an N-dimension matrix which, in the remainder of this chapter, will be assumed to be known. For instance, $B = I_N$ if noise variance is constant and equal to σ^2, while B contains the squared measurements when the CV is constant and equal to σ. Sometimes, following Eq. (5.3), B is assumed to contain the squared model predictions $\{y_k\}$ instead of the measurements $\{z_k\}$, with the consequence that the estimation procedure becomes more complex because B depends on the unknown parameter vector p.

The function $g(t, p)$ of Eq. (5.1) reflects the model of the system. The model can be either a structural model, for example, a compartmental model of the kinetics of a substance under study, or an input−output model, for example, a sum of exponentials describing the clearance function of the substance. Let us clarify the point by the example of Figure 5.1A, which shows the concentration in plasma of C-peptide (a substance secreted by the pancreas equimolarly with insulin, but not extracted by the liver) in a human being after an impulse administration (bolus) at time $t = 0$ of a dose of size D (endogenous C-peptide secretion was previously suspended by a somatostatin infusion started 2 h before the bolus administration) [1].

According to the general formulation of Eq. (5.2), the data $\{z_k\}$ in Figure 5.1A can be described by an input−output model given by the sum of q exponentials:

$$y(t) = D \sum_{i=1}^{q} A_i e^{-\alpha_i t} \tag{5.6}$$

In particular, for the data of Figure 5.1A, it is often sufficient to use two exponentials, and the parameter estimation problem becomes that of assigning numerical values to the elements of $p = [A_1 A_2 \alpha_1 \alpha_2]^T$ from the information contained in the time series $\{z_k\}$. The same data of Figure 5.1A can be also interpreted using a

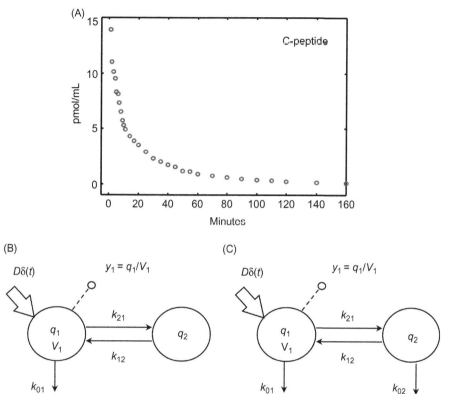

Figure 5.1 (A) C-peptide concentration in plasma after the administration of a Dirac pulse of amplitude D, (B) an *a priori* identifiable kinetics model, and (C) an *a priori* unidentifiable kinetics model.

system of differential equations aiming at describing the kinetics of C-peptide within the organism. For instance, the two compartments model of Figure 5.1B [2] could be used for such a scope, with equations given by

$$\dot{q}_1(t) = -(k_{01} + k_{21})q_1(t) + k_{12}q_2(t) + D\delta(t), \quad q_1(0) = 0$$
$$\dot{q}_2(t) = k_{21}q_1(t) - k_{12}q_2(t), \quad q_2(0) = 0 \tag{5.7}$$
$$y(t) = q_1(t)/V_1$$

In these equations, the masses of C-peptide in the two compartments represent the two-state variables, the system output coincides with the concentration of the substance in the accessible compartment "1" and the system input is the exogenous input (in this case the Dirac pulse of amplitude D). The parameter identification problem is thus that of determining the vector $\boldsymbol{p} = [k_{01} k_{12} k_{21} V_1]^{\mathrm{T}}$ from the data $\{z_k\}$.

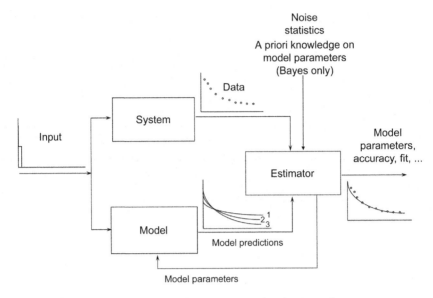

Figure 5.2 Schematic representation of a parameter estimation procedure.

Formally, the *parameter estimation problem* consists in determining an estimate $\hat{p}$ of the true, unknown, vector p from the measurement vector z. The literature on the theoretical and practical aspects of the problem is huge, and several classic textbooks are available [3−6], to which we refer the reader for more insights on the topics covered by the present chapter, which has the aim of providing only the fundamentals to properly handle the most important concepts. As explained in detail in the next subsections, the parameter estimation problem can be attacked by either a Fisher or a Bayes approach. Fisher estimators assume that a certain "deterministic" value of p exists, while Bayes estimators assume p to be a random vector, characterized by its own probability density function (PDF), of which a realization is observed. The parameter estimation procedure is schematically illustrated in Figure 5.2 (the meaning of some terms will become clear during the reading of this chapter).

Before proceeding, it is worthwhile remarking that the kind of model (input−output vs. structural) chosen to describe the system depends, first of all, on the scopes. For instance, to cope with deconvolution (see Chapter 3), the impulse response is needed and an input−output description of the system will be sufficient. For structural models, the choice of number of compartments and interconnections reflects available physiological knowledge, but a compromise must often be established with identifiability properties. For instance, the model of Figure 5.1C is only slightly different from that of Figure 5.1B (only one additional unknown parameter k_{02}), but, as follows from the use of the methodologies described in Chapter 4, the information contained in the set of inputs and outputs accessible to measurement is theoretically insufficient to reconstruct the unknown parameter vector $p = [k_{01}\,k_{12}\,k_{21}\,k_{02}\,V_1]^{\mathrm{T}}$.

5.2   Fisherian Parameter Estimation Approaches

5.2.1   Least Squares

Let us consider a trial numerical value for the unknown parameter vector p. By using the model, it is possible to calculate, at the given sampling times $\{t_k\}$ $k = 1, \ldots, N$, the output of the system in response to any input and finally obtain the N-size vector $G(p)$ of the so-called *model predictions*. The prediction error vector $e(p)$ for the considered trial value p is then defined as

$$e(p) = z - G(p) \tag{5.8}$$

A natural approach to quantify the goodness of the model, comprising equations and the numerical value of p, is to measure the squared norm of $e(p)$

$$J_{LS}(p) = [z - G(p)]^T B^{-1} [z - G(p)] \tag{5.9}$$

is small. In fact, the smaller $J_{LS}(p)$, the "better" the model is expected to be. In Eq. (5.9), the weighting matrix B allows one to take into account that the data $\{z_k\}$ can be affected by error with possibly different variance, that is, the lower the expected noise on a datum, the more the adherence of the model prediction to it should be pursued.

However, p in Eq. (5.9) is unknown. The least squares (LS) parameter estimation technique determines the estimate $\hat{p}_{LS}$ by minimizing $J_{LS}(p)$. Formally

$$\hat{p}_{LS} = \underset{p}{\text{argmin}} \, J_{LS}(p) \tag{5.10}$$

If $G(p)$ linearly depends on p

$$G(p) = Gp \tag{5.11}$$

with G being a suitable $M \times N$ matrix (e.g., $2 \times N$ in the regression line case), the cost function in Eq. (5.10) becomes

$$\hat{p}_{LS} = \underset{p}{\text{argmin}} \, [z - Gp]^T B^{-1} [z - Gp] \tag{5.12}$$

and, thanks to its quadratic shape, an analytic solution exists, given by

$$\hat{p}_{LS} = (G^T B^{-1} G)^{-1} G^T B^{-1} z \tag{5.13}$$

Since, in general, $G(p)$ depends nonlinearly on p, for example, for models of Eqs. (5.6) and (5.7), the solution of Eq. (5.10) cannot be analytically obtained. This solution is termed *nonlinear LS estimator* and iterative descent methods must be used to numerically obtain it. The most widely known goes under the name Gauss−Newton. Some numerically robust implementations, for example, Levenberg−Marquardt, are available. Details can be found in several textbooks [7]. Briefly, the idea of the Gauss−Newton iteration is to linearize the model of Eq. (5.2) by using partial derivatives and first-order Taylor approximation and then compute, by exploiting linear LS in Eq. (5.13), a new approximate solution. For instance, the first iteration of the Gauss−Newton method provides the estimate

$$\hat{p}_1 = \hat{p}_0 + (S^T B^{-1} S)^{-1} S^T (z - G(\hat{p}_0)) \tag{5.14}$$

where $\hat{p}_0$ is the initial guess and S, the sensitivity matrix, is the $N \times M$ matrix given by

$$S = \begin{bmatrix} \dfrac{\partial g(t_1, p)}{\partial p_1}\Big|_{p=\hat{p}_0} & \dfrac{\partial g(t_1, p)}{\partial p_2}\Big|_{p=\hat{p}_0} & \cdots & \dfrac{\partial g(t_1, p)}{\partial p_M}\Big|_{p=\hat{p}_0} \\[2ex] \dfrac{\partial g(t_2, p)}{\partial p_1}\Big|_{p=\hat{p}_0} & \dfrac{\partial g(t_2, p)}{\partial p_2}\Big|_{p=\hat{p}_0} & \cdots & \dfrac{\partial g(t_2, p)}{\partial p_M}\Big|_{p=\hat{p}_0} \\[2ex] \vdots & \vdots & \vdots & \vdots \\[2ex] \dfrac{\partial g(t_N, p)}{\partial p_1}\Big|_{p=\hat{p}_0} & \dfrac{\partial g(t_N, p)}{\partial p_2}\Big|_{p=\hat{p}_0} & \cdots & \dfrac{\partial g(t_N, p)}{\partial p_M}\Big|_{p=\hat{p}_0} \end{bmatrix} \tag{5.15}$$

The generic kth iteration is obtained by replacing 1 and 0 in Eqs. (5.14) and (5.15) with k and $k - 1$, respectively. A delicate point for the user of iterative methods is the definition of the initial guess $\hat{p}_0$, because a bad choice can make the method to converge to a local minimum of the cost function rather than to its global minimum. Figure 5.3 shows, for the simple model $g(t,a) = ae^{-at}$ (with $M = 1$ unknown parameter, i.e., $p = [a]$) fitted against data (open bullets in panels B and C, noise has constant variance), the behavior of $J_{LS}(a)$ (panel A) and how well model predictions $g(t,\hat{a})$ (continuous line) fit the data in correspondence of the global (panel B) and local (panel C) minimum of $J_{LS}(a)$, achieved for 0.14 and 1.2, respectively.

In the literature, the name *weighted LS* is sometimes used to underline the presence of the weighting matrix B in the cost function. In addition, the terminology "absolute weights" is often used when B coincides with Σ_v of Eq. (5.5) and "relative weights" when Σ_v is known apart from the scale factor σ^2. In the latter case, an estimate of σ^2 can be provided *a posteriori* by dividing the value of the cost function $J(\hat{p}_{LS})$ by the degrees of freedom of the model $N - M$:

$$\hat{\sigma}^2 = \frac{J_{LS}(\hat{p}_{LS})}{N - M} \tag{5.16}$$

The LS estimator has a straightforward and intuitive definition. Only limited knowledge and statistical assumptions are needed for its definition. As discussed

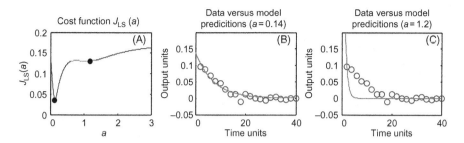

Figure 5.3 The problem of local minima illustrated for parameter estimation of a simple model with only one unknown parameter a. The value $a = 1.2$ is an only local minimum of JLS(a) (panel A) and leads to a fit (panel C) much worse than that obtained for the global minimum $a = 0.14$ (panel B).

later in Section 5.2.4, a measure of the uncertainty affecting the parameter estimates can be easily obtained (this information, together with that brought by the residuals, is key to assessing the reliability of the estimated model parameters). Before entering into these details, however, it is convenient to define the ML estimator. In fact, while ML estimators have statistically insightful properties in general, LS and ML coincide under the Gaussianity hypothesis, a situation of wide practical interest.

5.2.2 Maximum Likelihood

Since v in Eq. (5.4) is a random vector with PDF $f_v(v)$, z is a random vector as well. As, in Fisherian parameter estimation approaches, p is a deterministic vector, the PDF $f_z(z)$ of z depends on the value of p and reflects the randomness of v described by $f_v(v)$. However, p is unknown and its ML estimate is the vector that renders maximum the probability that z takes on, as realization, the values experimentally measured. In formal terms, the ML estimate $\hat{p}_{ML}$ is determined by maximizing with respect to the unknown vector p, the PDF of z, i.e., the *likelihood function* $L(z, p)$, with z known (the measurements):

$$\hat{p}_{ML} = \arg \max_{p} L(z, p) \tag{5.17}$$

The solution of the optimization problem of Eq. (5.17) obviously depends on the expression of the likelihood function $L(z, p)$ which reflects that of $f_v(v)$. A simple but common case is that when the noise vector v is *Gaussian*, with zero mean and a covariance matrix given by Σ_v. In this case, z is *Gaussian* as well with mean and covariance matrix given by $G(p)$ and Σ_v, respectively. The likelihood function $L(z, p)$ is thus

$$L(z, p) = \frac{1}{[(2\pi)^N \det \Sigma_v]^{1/2}} \exp\left(-\frac{1}{2}[z - G(p)]^T \Sigma_v^{-1}[z - G(p)]\right) \tag{5.18}$$

Note that, since p is a deterministic (even unknown) vector, the likelihood function is a deterministic function of p and not a PDF (with z being a known number vector). By computing the logarithm of the likelihood function, it can be easily shown that its maximization with respect to p is equivalent to the minimization of the following cost function, J_{ML}:

$$J_{ML}(p) = [z - G(p)]^T \Sigma_v^{-1} [z - G(p)] + \ln(\det \Sigma_v) \tag{5.19}$$

The ML estimate of p is thus

$$\hat{p}_{ML} = \arg \min_p J_{ML}(p) \tag{5.20}$$

When matrix Σ_v is allowed to be a function of the vector of the unknown parameters, one should think $\Sigma_v = \Sigma_v(p)$. When Σ_v does not depend on p, the second term on the right side of Eq. (5.19) can be neglected and the ML estimate is equivalent to the LS estimate, that is, the one that minimizes the cost function of Eq. (5.9). ML estimator is a very flexible tool and if v is not *Gaussian*, it is still possible to define the ML estimator, even if analytical expressions and maximization can be hard to be derived.

In the case of $N \to \infty$ statistically independent measurements, ML estimators have some interesting asymptotical properties such as unbiasedness, efficiency, normality, and consistency [8]. Hence, in the Gaussian case, the LS estimator enjoys the same properties. Since in physiological model identification problems the number of data N is usually small, these properties cannot be invoked and suitable simulations should be devised to assess the reliability of the estimator in the given application.

5.2.3 Analysis of the Residuals

Let's assume that an estimate $\hat{p}$ of the parameter vector has been obtained. Both equations and numerical values of the model parameters are now known, but, before considering if the model is valid, its quality must be assessed. For such a scope, the first step is the analysis of the residuals vector

$$r = z - G(\hat{p}) \tag{5.21}$$

From this definition, it is interesting to note that both the cost function J_{LS} in Eq. (5.9) and the first term of J_{ML} in Eq. (5.19) are connected to the weighted residuals sum of squares (WRSS)

$$WRSS = r^T B^{-1} r \tag{5.22}$$

which, no matter how $\hat{p}$ is obtained, provides a measure of the goodness of fit, since it represents the distance between data, z, and the model prediction, $G(\hat{p})$, achieved after the termination of the parameter estimation process. In view of Eq. (5.4), vector r can be thought of as a sort of estimate of the vector v. It is then logical to demand that a good model leads to residuals compatible with the statistical description, known or expected, of v. For instance, if v has zero mean and uncorrelated elements, the presence of a significant correlation among the residuals (e.g., a long sequence of residuals with the same sign) and the existence of a non-zero mean is an indicator of either a bad modelling of the system (e.g., too low a number of compartments, errors in the parameterization) or an inefficient optimization of the cost function (e.g., the iterative descent method converged to a local, rather than to a global, minimum). Tests conventionally used to assess the residuals include the Anderson whiteness test and the runs test [9]. Since these tests require the availability of a large number of data, sometimes visual inspection is necessary in practical cases. Residuals assessment can be facilitated by first multiplying the residuals vector for a suitable matrix related to Σ_v. For instance, assuming that errors are uncorrelated and their variance is known, each residual should be divided by the correspondent noise standard deviation and the resulting weighted residuals should mainly lay in the region delimited by -1 and $+1$.

5.2.4 Precision of the Estimates

An important criterion to assess the goodness of model identification is the precision of the estimated parameters. In fact, data $\{z_k\}$ are affected by error and are thus uncertain. As a consequence, the estimated parameter vector, no matter how obtained, is uncertain, too. In order to assess the precision of the estimates, let's define the estimation error as

$$\tilde{p} = p - \hat{p} \tag{5.23}$$

where p is the true, even if unknown, value of the model parameters. Assuming that the estimator is unbiased, that is, $E[\tilde{p}]$ is the null vector, the uncertainty affecting the estimates can be measured by using the covariance matrix of $\tilde{p}$

$$\Sigma_{\tilde{p}} = E[\tilde{p}\tilde{p}^{\mathrm{T}}] \tag{5.24}$$

In particular, the squared root of the diagonal elements of this covariance matrix returns the standard deviations of the estimation error affecting the elements of $\hat{p}$. In the linear case, a closed-form expression can be easily obtained for Eq. (5.13)

$$\Sigma_{\tilde{p}} = (G^{\mathrm{T}}\Sigma_v^{-1}G)^{-1} \tag{5.25}$$

while, in the nonlinear case, the following approximation is often invoked

$$\Sigma_{\tilde{p}} = (S^{\mathrm{T}}\Sigma_v^{-1}S)^{-1} \tag{5.26}$$

where S is the $N \times M$ matrix of the linearized system as in Eq. (5.15) computed for $p = \hat{p}$. Remarkably, the quantity in the right side of Eq. (5.26), in the case of Gaussian measurement error, is the inverse of the Fisher information matrix which, in a ML estimation context, represents a lower bound of the covariance matrix of the estimation error.

It is of utmost importance to take into account that Eq. (5.26) can be used to provide an only approximate estimate of the parameter standard deviation. Furthermore, no confidence intervals can be computed without assuming a normal distribution of the parameter estimation error. In many practical situations, it is advisable to design suitable simulations to correctly propagate the noise from measurements to estimates and to determine the PDF of $\tilde{p}$. For such a scope, methods based on repeated parameter estimations over a large number of synthetic data sets provide a straightforward solution [10]. They are based on two steps: a first one in which hundreds/thousands of synthetic noisy data sets (with the same expected characteristics of the original one) are generated, and a second one in which, from each synthetic data set, the parameter estimator of choice (e.g., LS or ML) calculates the estimates. Finally, simple descriptive statistics (e.g., variance) is used to quantify the precision of the estimates or compute their confidence intervals (e.g., exploiting percentiles). Monte Carlo simulation and the bootstrap represent two popular implementations of these methods. They differ one from the other in the way in which the synthetic data are generated. In both approaches, the parameter estimates $\hat{p}$ are first computed on the experimental data. In Monte Carlo simulations, $\hat{p}$ is used to calculate the noise-free predictions $G(\hat{p})$ in the same time points of the original data set. Then, synthetic measurement errors are generated by randomly extracting numbers from a suitable distribution, in accordance to the adopted error model, and, finally, they are added to the model predictions to obtain the noisy synthetic data sets. Note that Monte Carlo simulations require a complete definition of the PDF $f_v(v)$, including the values of its parameters. Conversely, the bootstrap does not explicitly require the formalization of $f_v(v)$, because it assumes that this information is implicitly included in the experimental data. Therefore, data are not synthetically generated, but perturbed data sets are obtained from the original by removing some data points in a random fashion (e.g., 20−30% of the original points) and randomly replacing each of them with a replicate from the remaining data points. The replacement is carried out to maintain the total number of samples identical to that of the original data set. It is apparent that the bootstrap is specially indicated for analyzing data sets with replicated data points, whereas it is not convenient in time course experiments in which the sampling time schedule should not be modified across data sets.

No matter how it is measured, sometimes the uncertainty of (some of) the estimated parameters are unacceptably large, even if the model is *a priori* identifiable. If residuals are acceptable, a possible reason is the variance of the noise (note how Σ_v plays in Eq. (5.26)) or the too high complexity of the model, that is, the model is over-parameterized. In other cases, the number of the samples is too low to identify the parameter values on solid grounds or the sampling is inefficiently scheduled. Sometimes, a more satisfactory determination of the estimates can be obtained through the use of *a priori* knowledge by Bayesian estimators as discussed in Section 5.3.

5.2.5 Model Selection Among Candidates

Whatever is the approach adopted for parameter estimation, there is often the need of selecting the most suitable model among a number of competing ones. In order to solve this problem, several factors of merit can be considered. For instance, the ability of the model to fit the data is usually the most important criterion. When data are fitted satisfactorily well, e.g., small WRSS and unbiased residuals, by a number of models, other criteria must be considered. For structural models, the relative merit of each model in relation to its adherence to physiological knowledge is also a key ingredient. For input–output models, e.g., sum of exponentials, the selection problem is relatively easier than the structural ones, since only the order is varying among candidate models, that is, one must simply assess whether the improvement in the goodness of fit obtained by increasing model order is the result of a more accurate representation of the data and not only of an increased number of parameters. In fact, it is known that by increasing the model order the data fit always improves, but, at the same time, the precision of the parameter estimates deteriorates.

In the literature, a number of methods are available for the selection of the best model order. Some of these methods are referred to as the *information criteria*. For instance, when parameter estimation is performed by ML, a popular strategy to select the most parsimonious model is based on the Akaike Information Criterion (AIC), according to which the model of choice is that which renders minimum the value of the index:

$$\text{AIC} = 2M + J_{\text{ML}}(\hat{p}) \tag{5.27}$$

where M is the number of unknown parameters and $J_{\text{ML}}(\hat{p})$ is the value of the cost function (5.19) at its minimum value [11]. In Eq. (5.27), the first term penalizes a high number of parameters (linked to parameter estimates uncertainty), while the second one weights the achieved minimum of the cost function J_{ML} (linked to the goodness of fit). Other popular criteria, exploiting the same idea even if in a slightly different formulation, are minimum description length (MDL), final predictor error (FPE), and Bayesian Information Criterion (BIC), also known as Schwartz Criterion (SC). Another interesting criterion, though scarcely known in the biomedical literature, is the Information Measure of Complexity (ICOMP) [12] that directly looks for the best compromise between goodness of fit and precision of the estimates. In particular, ICOMP suggests choosing the model that minimizes

$$\text{ICOMP} = J_{\text{ML}}(\hat{p}) + 2c(F(p)) \tag{5.28}$$

where $c(\cdot)$ is a function of the Fisher informatics matrix F, defined as $c = M/2\log(\lambda_a/\lambda_g)$, with λ_a, the arithmetic mean of the eigenvalues (easy computable as trace$(F)/M$), and λ_g, the geometric mean of the eigenvalues (easy computable as det$(F)^{1/M}$).

5.2.6   Case Study

Let's consider again C-peptide concentration data as those of Figure 5.1. The noise vector v of Eq. (5.4) is assumed Gaussian and independent, that is, the covariance matrix B is diagonal, with zero mean, with constant CV equal to 4.5%, that is, in Eq. (5.5) σ equals 0.045 and B is diagonal with entries given by the squared measurements. In this hypothesis, LS and ML estimators coincide. As shown in Figure 5.4 (top left), data are well modeled by a sum of two-exponential (2E) model, but, in terms of fit, better results are, as expected, obtained by a sum of three-exponential (3E) model (Figure 5.4, top right). This can be better appreciated by the time course of the residuals (lower panels), especially in the minutes immediately after the pulse input.

The numerical values of the parameters of the two models are reported in Table 5.1 together with their uncertainties (in parentheses) expressed as percent standard deviations (Eq. (5.25) was exploited). Of course, price of a lower value of WRSS for 3E is a higher uncertainty. Remarkably, the most uncertain parameters are those related to the fastest mode of the system, for the identification of which relatively less data are available.

The AIC index is lower for the 3E model (ΔAIC $= -4.75$), indicating that, following this criterion, the 3E model represents a better compromise between the two opposite needs: data fit and parameters' precision.

5.3   Bayesian Parameter Estimation Approaches

5.3.1   Fundamentals

In estimating the unknown model parameters, Bayes approaches employ not only the data vector z, indicated as *a posteriori information* because being related to the experiment it is available only after having performed it, but also other information, termed *a priori information* because it is available independently of the experiment (in fact, *a priori* can be intended as standing for "before having seen the data"). In practice, the *a priori* information on p usable by a Bayesian estimator can be obtained by physiological knowledge or, more often, from independent population studies. For instance, if a previous study providing individual model parameters in a population of K subjects homogenous to those under study is available, under Gaussianity assumptions the *a priori* second-order statistical information needed for Bayes estimation can be obtained as

$$\mu = \frac{1}{K} \sum_{i=1}^{K} p^{(i)} \tag{5.29}$$

$$\Omega = \frac{1}{K-1} \sum_{i=1}^{K} (p^{(i)} - \mu)(p^{(i)} - \mu)^{\mathrm{T}} \tag{5.30}$$

where $p^{(i)}$ is the parameter vector of the ith subject of the population.

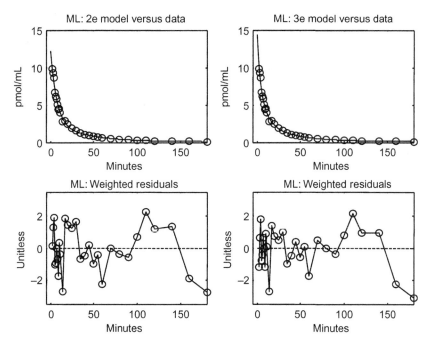

Figure 5.4 The two-exponential (2E) and three-exponential (3E) models identified by ML estimators from C-peptide plasma concentration samples after the administration of a bolus at time 0. (Top) Model predictions (continuous line) versus data (open bullets). (Bottom) Weighted residuals.

Table 5.1 Parameters of the Sum of the 2E and 3E Models Together with Percent Uncertainty (in Parentheses) Estimated by ML or MAP (Section 5.3.2) from the Data of Figures 5.4 and 5.5

	A_1 $(\text{mL})^{-1}$	A_2 $(\text{mL})^{-1}$	A_3 $(\text{mL})^{-1}$	α_1 $(\text{min})^{-1}$	α_2 $(\text{min})^{-1}$	α_3 $(\text{min})^{-1}$	WRSS
2E (ML)	1.894	0.5564		0.1443	0.0262		55.56
	(4.7)	(4.6)		(5.6)	(2.2)		
3E (ML)	1.304	1.093	0.4929	0.2998	0.0998	0.0251	46.83
	(32.0)	(47.1)	(10.2)	(48.2)	(26.4)	(4.0)	
2E (MAP)	1.896	0.5518		0.1436	0.0261		55.58
	(4.3)	(4.3)		(5.0)	(2.1)		
3E (MAP)	1.525	0.9294	0.4593	0.2799	0.0859	0.0245	47.82
	(12.0)	(13.1)	(8.2)	(13.8)	(9.7)	(3.5)	

In the Bayesian approach, the vector p of the unknown model parameters is considered a random vector with an *a priori* PDF $f_p(p) = f_{p_1, p_2, ..., p_M}(p_1, p_2, ..., p_M)$. In the same way, v is a random vector with PDFs $f_v(v) = f_{v_1, v_2, ..., v_N}(v_1, v_2, ..., v_N)$. The measurement vector z of Eq. (5.4) is thus a random vector depending on the randomness of both v and p. In such a context, having the *a posteriori* PDF (also called posterior or posterior distribution) of p given z, that is, the probability distribution of p after seeing the data z

$$f_{p|z}(p|z) = f_{p_1, p_2, ..., p_M|z_1, z_2, ..., z_N}(p_1, p_2, ..., p_M|z_1, z_2, ..., z_N) \qquad (5.31)$$

would allow the derivation of several point estimators by exploiting one of the central tendency measurements. The most straightforward (at least from a conceptual point of view) would be the posterior mean

$$\hat{p}_{MV} = E[p|z] = \int_{\mathcal{R}^M} p f_{p|z}(p|z) \mathbf{dp} \qquad (5.32)$$

which, as far as the estimation error is concerned, enjoys minimum variance (MV) properties. Another popular Bayesian estimator is known as *maximum a posteriori* (MAP) that, being based on the mode, instead of requiring the potentially complex computation of an integral, requires the maximization of the *a posteriori* PDF:

$$\hat{p}_{MAP} = \arg\max_p f_{p|z}(p|z) \qquad (5.33)$$

Notably, by recalling the Bayes theorem, the expression of the posterior required in Eqs. (5.32) and (5.33) can be derived from

$$f_{p|z}(p|z) = \frac{f_{z|p}(z|p) f_p(p)}{f_z(z)} \qquad (5.34)$$

where $f_{z|p}(z|p)$ is the PDF of the data vector z given the parameters vector p, that is, the likelihood of the data, and $f_z(z) = f_{z_1, z_2, ..., z_M}(z_1, z_2, ..., z_N)$ is the PDF of the data vector z. Then, by replacing $f_{p|z}(p|z)$ in Eq. (5.33) with the expression of Eq. (5.34) and ignoring in the maximization problem the denominator of objective function $f_z(z)$, which is independent of p, the MAP estimator con be computed as

$$\hat{p}_{MAP} = \arg\max_p f_{z|p}(z|p) f_p(p) \qquad (5.35)$$

A general expression of the MV/MAP estimator individuated from Eqs. (5.32) and (5.35) does not exist. Indeed, the posterior PDF of p depends on the specific form of both $f_p(p)$ and $f_v(v)$ and, of course, from the model structure (5.4). A tractable analytical expression for this PDF can be derived in only a few cases, one of them is described in the following section.

5.3.2 MAP Estimator in the Gaussian Case

5.3.2.1 Derivation of the MAP Estimator

If p is taken from an *a priori Gaussian* distribution with mean μ and covariance matrix Ω, and v is *Gaussian* with zero mean and covariance matrix Σ_v, it can be easily shown that Eq. (5.35) turns into

$$\hat{p} = \arg \min_{p} J_{\text{MAP}}(p) \tag{5.36}$$

where

$$J_{\text{MAP}}(p) = [z - G(p)]^{\text{T}} \Sigma_v^{-1} [z - G(p)] + \ln(\det \Sigma_v) + (p - \mu)^{\text{T}} \Omega^{-1} (p - \mu) \tag{5.37}$$

In Eq. (5.37), the term $\ln(\det \Sigma_v)$ can be omitted if Σ_v does not depend on p; if this is not the case, one should think $\Sigma_v = \Sigma_v(p)$.

In the case of Eq. (5.37), the formulation of the MAP estimator allows a focus— not on the derivation of the posterior distribution (from which any information about the parameter estimates can be obtained)—but on the simpler task of mini-mizing a cost function as in Eq. (5.19) with a modification reflecting the available prior. In fact, by comparing the cost functions of Eqs. (5.19) and (5.37), it follows that J_{MAP} differs from J_{ML} because of the presence of an additional term expressing the adherence of the estimate to the available *a priori* knowledge on the parameter vector. It is worth noting that both the cost functions of Eqs. (5.19) and (5.37) share the first term, which measures the goodness of fit, that is, the adherence to the *a posteriori* information. Thus, Bayes estimators are said to establish a trade-off between *a priori* and *a posteriori* information, linked to expectations and data, respectively.

It is worth noting that if the *a priori* covariance matrix Ω tends to infinity (i.e., the *a priori* knowledge becomes milder and milder), the last term of Eq. (5.37) can be neglected and the MAP estimator tends to the ML/LS estimator (only *a posteriori* information can be exploited). On the other hand, if the covariance matrix Σ_v of the noise vector v tends to infinity (i.e., the measurements are more and more noisy) the first term of Eq. (5.37) tends to be zero and the optimization problem in Eq. (5.36) is solved by exploiting the *a priori* knowledge only, that is, the MAP estimate of p tends to the *a priori* expected value μ (with uncertainty given by Ω). Finally, of note is that *a priori* information plays a role also dependent on the accuracy of *a posteriori* information. In fact, when Σ_v tends to zero (close-to-perfect measurements, very small parameter of uncertainty) the minimization of only the first term is pursued in Eq. (5.37) and the MAP estimate tends to the ML one.

As in the Fisher approaches, the model selection problem can be stated also if a Bayesian parameter estimation technique is used. For instance, when MAP estima-tion is considered, the Generalized Information Criterion (GIC) can be employed.

This criterion, which can be interpreted as the extension of the AIC to MAP estimation, requires the following index to be minimized:

$$\text{GIC} = \frac{2M}{N} + J_{\text{MAP}}(\hat{p}) \tag{5.38}$$

where N is the number of data and $J_{\text{MAP}}(\hat{p})$ is the value of the cost function in Eq. (5.38) at its minimum. In Eq. (5.38), similarly to Eq. (5.27), the first term weights the number of parameters, while the second term weights the achieved trade-off between data fit and adherence to *a priori* knowledge.

Finally, the problem of determining the precision of the estimates can be approached by computing the matrix $\Sigma_{\tilde{p}} = E[\tilde{p}\tilde{p}^{\text{T}}]$, where $\tilde{p} = p - \hat{p}$ is the difference between two vectors that now are both random. However, analytical (approximate) computations are feasible only under restrictive assumptions. For instance, under Gaussianity assumptions, for the MAP estimator of Eq. (5.37), the following approximation is often used:

$$\Sigma_{\tilde{p}} = (S^{\text{T}}\Sigma_v^{-1}S + \Omega)^{-1} \tag{5.39}$$

where S is as the same of Eq. (5.26) (with $p = \hat{p}_{\text{MAP}}$).

5.3.2.2 Case Study

The bottom rows of Table 5.1 show the estimates obtained by MAP together with their precision for the same data of Figure 5.4 (μ and Ω of the used *Gaussian* prior were obtained from a population study). Model predictions versus data, and weighted residuals, are shown in Figure 5.5 (2E in the left panels, 3E in the right).

In the 2E case, the values of ML and MAP estimates are very similar. WRSS is lower for ML and this is not a surprise since, while the ML estimator only weights the distance of the model predictions from the data (Eq. (5.19)), the MAP estimator also weights the distance of the parameters from their *a priori* expected values (Eq. (5.37)). However, the WRSS difference is very small. In fact, no significant differences can be detected by inspecting the two fits. Turning to precision of parameter estimates (CV, reported in parentheses), one can note that it improves by using Bayes estimation. This is again in line with theoretical expectations, given the incorporation of *a priori* knowledge into the algorithm of Eq. (5.39). However, here the improvement is almost undetectable, since the parameters estimated by ML by exploiting only the *a posteriori* information are already very precise. When the 3E model is considered, only small differences between the values of the parameters estimated by ML or MAP can be detected. Also, WRSS of ML estimation is only slightly better than that of MAP estimation. As with the 2E model, no significant differences can be detected by eye inspection. In contrast with the 2E model, a significant improvement in the precision of parameter estimates now occurs with MAP in comparison with ML estimation. For instance, the CV of the parameter A_2 is 47.1% with ML estimation and 13.1% with MAP estimation. The

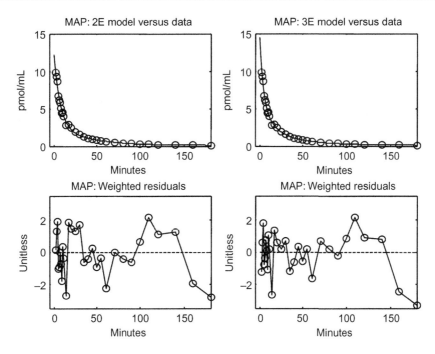

Figure 5.5 As in Figure 5.4 for the MAP estimator.

improvement in the precision obtained by MAP estimation is more significant for the parameters related to the fastest eigenvalues (A_1 and α_1 and A_2 and α_2) than for those related to the slowest one (A_3 and α_3), since their estimation is based on less data (i.e., less *a posteriori* information). This can be explained by recalling that in MAP estimation a trade-off is established between *a priori* and *a posteriori* information. As a consequence, the more the exploitation of *a posteriori* information only is already producing very precise parameter estimates, the less beneficial will be the incorporation into the estimator of *a priori* knowledge. As far as the question of selecting the best model between the two models, GIC is lower for the 3E (ΔGIC $= -6.90$), suggesting that the 3E model is the model of choice also if MAP estimation is considered.

Other results and considerations concerning the C-peptide impulse response identification case study are reported in Ref. [13].

5.3.3 Prior Distribution in Bayesian Analysis

The choice of the prior distribution is one of the intrinsic aspects of the Bayesian analysis. It encodes all of the available information before seeing the data and, as already discussed, can impact the parameter estimates. In defining priors there are two main considerations: first, the choice of the prior distribution, and second, the choice of its parameter values (if we only consider parametric prior, but in some cases priors can be specified by nonparametric distribution through, for example, its sampling distribution [14]). The choice of the prior distribution is not a trivial step, even if often standard conjugate forms (meaning that the conditional posterior distribution will have the same structural form as the prior, but with updated parameter values) are used to have closed-form solutions or more easy and efficient sampling algorithms.

The most popular choices are: normal or multinormal distributions, log-normal distributions for positive parameters, the gamma or Wishart distribution for parameters linked to variances, and the uniform distribution. Independently from that choice, usually the prior distributions are classified as informative or noninformative. However, often it should be more formally correct to use the term "very low informative" priors instead of noninformative. In fact, when, for example, a prior is defined through a normal or log-normal distribution with a very large variance, a very small information is actually encoded. Many Bayesian analyses use the noninformative prior. The main reason is not to affect results by information collected by subjective elicitation methods. In particular, the parameters can be chosen to have, for example, the 95% confidence intervals in a biologically "reasonable" range of values.

On the other hand, in many cases the data collected in the current study may be sparse, such as what may occur when data are collected as a matter of routine clinical care, rather than for model building, which may result in a design that does not allow a full model identification. One option is to fix the nonestimable parameters from previous studies. However, the fixed parameters can bias the estimates of the other parameters. Another option would be to simplify the model, but a more natural option in the Bayesian context is to use appropriately informative priors to aid the modelling process. The setup of this kind of priors is similar to the noninformative one, but the confidence intervals are forced to be within a more narrow range.

5.3.4 Use of Markov Chain Monte Carlo in Bayesian Estimation

Even if MV/MAP estimators use the Bayesian framework to exploit the available *a priori* information, the true goal of the Bayesian approach to the model identification is the computation of the whole posterior distribution, rather than to locate only its mode, as in the special case discussed in Section 5.3.2.1. Then, from the knowledge of the posterior distribution, it is certainly possible to compute point estimates (using one of the measures of central tendency), their precision (computing, e.g., the standard deviation of the posterior distribution) and (even asymmetric) confidence intervals by exploiting the correspondent percentiles of the distribution. As reported in Eq. (5.34), the posterior distribution is fully defined from the knowledge of the prior distribution and of the likelihood, and it is proportional to their product. However, despite this apparent simplicity, in the general case, even just obtaining point estimates (such as MV/MAP) is not trivial and normally requires the use of numerical techniques.

In fact, Bayesian estimation involves the integration over possibly high-dimensional probability distributions. Since this operation is often analytically intractable, it is common to resort to Monte Carlo techniques, which require sampling from the posterior distribution to be integrated. Unfortunately, often it is impossible to extract samples directly from that distribution. Markov chain Monte Carlo (MCMC) methods [15] provide a unified framework to solve this problem.

MCMC denotes a category of sampling methods that can also be used in non-Bayesian settings, essentially based on two steps: a Markov chain generation and a Monte Carlo numerical integration. By sampling from suitable probability distributions, a Markov chain that converges (in distribution) to the target distribution, that is, the distribution to be integrated (in our case the posterior distribution), is generated. Then, the expectation value is calculated through Monte Carlo integration over any desired function of the extracted samples.

The MCMC methods differ from each other in the way the Markov chain is generated. However, all of the different strategies proposed in the literature are special cases of the Metropolis−Hastings [16,17] framework. Also, the well known Gibbs sampler [18] fits into the Metropolis−Hastings scheme.

In the following sections, the application of the Metropolis−Hastings algorithm to Bayesian parameter estimation will be described. Some of its possible variants will also be discussed.

5.3.4.1   The Algorithm

In order to describe the Metropolis−Hastings algorithm, we will use the following notation:

- P^i is the ith sample of the Markov chain;
- $f_{p,z}(p,z) = f_{z|p}(z|p)f_p(p)$ is the target distribution (proportional to the posterior distribution).

The Markov chain, derived by the Metropolis−Hastings method, is obtained through the following steps:

1. at each step $i+1$, a candidate sample P^c is drawn from a proposal distribution $q(\cdot\,|P^i)$, also called transition kernel, that is an arbitrary PDF depending on the current sample of the chain;
2. the candidate point P^c is accepted with probability:

$$\alpha(P^i,P^c) = \min\left(1,\frac{f_{p,z}(P^c,z)q(P^i|P^c)}{f_{p,z}(P^i,z)q(P^c|P^i)}\right) \tag{5.40}$$

   that requires evaluations of both the proposal and the posterior PDF in correspondence of the parameter values P^c and P^i;
3. if the candidate point P^c is accepted, the next sample of the Markov chain is $P^{i+1} = P^c$, else the chain does not move and $P^{i+1} = P^i$.

It is important to remark that the stationary distribution of the chain (i.e., the distribution to which the chain converges) is independent of the proposal distribution and coincides with the target distribution $f_{p,z}(p,z)$ for any starting value P^0.

Although any proposal distribution, in the long run, will deliver samples from the target distribution, the rate of convergence to the stationary distribution of the generated Markov chain crucially depends on the relationships between the proposal and the target distributions. Moreover, the number of samples necessary to perform the Monte Carlo steps depends on the speed with which the algorithm "mixes" (i.e., spans the support of the target distribution).

When the vector of the model parameters is large, it is often convenient to divide p into L components and update, in each iteration, the samples P of these components one-by-one [19]. The components can also be blocked. This situation is suitable when parameters demonstrate high correlation, because generating samples on a component-by-component basis has a high rejection rate for high correlate components. This scheme is called single-component Metropolis−Hastings and it works as follows.

Let $P^{i,j}$ the jth component (or block) of P at the step i and $P^{i,-j} = \{P^{i+1,1}, \ldots, P^{i+1,j-1}, P^{i,j+1}, \ldots, P^{i,L}\}$, the Metropolis−Hastings scheme turns to:

1. at each step $i + 1$, for each component j, the next sample $P^{i+1,j}$ is derived by sampling a candidate point $P^{c,j}$ from a proposal distribution $q^j(\cdot \mid P^{i,j}, P^{i,-j})$;
2. the candidate point $P^{c,j}$ is accepted with probability:

$$\alpha(P^{i,j}, P^{c,j}) = \min\left(1, \frac{f_{p^j,z}(P^{c,j}, z|p^{i,-j})q^j(P^{i,j}|P^{c,j}, P^{i,-j})}{f_{p^j,z}(P^i, z|p^{i,-j})q^j(P^{c,j}|P^{i,j}, P^{i,-j})}\right) \tag{5.41}$$

3. if the candidate point $P^{c,j}$ is accepted, the next sample of the Markov chain is $P^{i+1,j} = P^{c,j}$, else the chain does not move and $P^{i+1,j} = P^{i,j}$.

The Gibbs sampler is just a special case of the single-component Metropolis−Hastings. The Gibbs sampler scheme exploits the full conditional (i.e., the conditional distribution of one component or block, given all of the other ones) as the proposal distribution. In this case, it is easy to verify that the candidate point is always accepted, so that the Markov chain moves at every step. Therefore, the Markov chain generated by the Gibbs sampler does not include repeated elements, but the price is paid by having to obtain in a close form the (potentially expensive) full conditionals. This is analytically possible for linear models and for combinations of conjugate priors and likelihoods. When the full conditionals are standard distributions (easy to sample form), the Gibbs sampler represents a suitable choice. On the contrary, when it is not possible to draw samples directly from the full-conditional distribution, because models are not linear in the unknown parameters or not conjugate priors were chosen, it is convenient to resort to hybrid schemes (Gibbs sampler and Metropolis−Hastings). In this setting, a portion of the model parameters is estimated using the Gibbs sampler, while the others use "*ad hoc*" proposal distributions.

5.3.4.2 *The Choice of the Proposal Distribution*

Different choices of the proposal distribution $q(\cdot, \cdot)$ clearly influence the acceptance rate and result in different algorithms. There is not a best proposal distribution generally applicable, because its selection is likely influenced by the specific parameter estimation problem. However, there are some typical choices proposed in the literature. One of these is the random walk process in which q $(p^c|P^i) = q(p^c - P^i)$ with q a multivariate normal distribution. Another solution is the Metropolis, in which the proposal distribution is symmetrical, i.e., $q(P^c|P^i) = q(P^i|P^c)$. In this case, the acceptance rate α from Eq. (5.40) is equal to the ratio of the posterior distribution computed in the proposed sample P^c and in the current one (P^i). A further method is the independence sampler in which the proposal distribution does not depend on the current sample P^i and, therefore, $q(P^c|P^i) = q(P^c)$. In this case, a further possible solution is to choose the proposal $q(P^c)$ equal to the prior distribution. Consequently, the acceptance rate α from Eq. (5.40) is equal to the ratio of the likelihood computed in the proposed sample P^c and in the current one (P^i).

The success rate (and then a suitable selection of the kernel in relation to the posterior distribution) greatly influences the fast convergence to the posterior distribution. The key aspect is not to have the success rate appear too low or too high. Naively, one would be tempted to make small steps or penalize great deviations in the parameter vector during the generation step to make sure that the success rate is high. This leads, however, to the chain remaining in a small region and moving very slowly to other regions. On the other hand, changes too large in parameters are likely to result in a very low acceptance rate with many repetitions in the chain. Some authors suggest an acceptance rate between 20% and 50%.

5.3.4.3 Assessing the Convergence

One of the problems with MCMC methods is establishing how many samples are necessary for the extraction. There are usually two issues. First, one has to decide whether the Markov chain has reached its stationary distribution. Second, one has to determine the number of iterations to keep after the Markov chain has reached its stationarity. In fact, inferences from the posterior distribution should be made after convergence has been achieved to assure that samples represent the target (posterior) distribution. Then the initial portion of the chain representing the nonconvergent section of the chain (burn-in) has to be discharged and not used in the Monte Carlo integration. Furthermore, it is also only possible to analyze every rth element of the chain to reduce correlation between successive elements. The size of the disregarded chain, the number of skipped elements (r), and indeed the size of the chain depend on how quickly the chain converges to the posterior distribution. In principle, it is not possible to make the decision on successful convergence on a purely theoretical basis, but empirical diagnostic criteria have been developed. They help to address this issue but do not guarantee that convergence has actually occurred. They are designed to verify a necessary, but not sufficient, condition for convergence. So there are no conclusive tests that can tell you when the Markov chain has converged to its stationary distribution, and then it is recommended that several methods are used. Furthermore, it is important to verify the convergence of all parameters, and not just those of interest, before proceeding to make any inference. With some models, certain parameters can appear to have very good convergence behavior, but that could be misleading due to the slow convergence of other parameters. In these cases, you are not allowed to make inference on the posterior never for those parameters that apparently reached the convergence, before all the Markov chain converges.

One of the proposed visual methods, called trace plots, is to visualize the history of the chain against the iteration numbers. These plots indicate both if the chain reached the convergence to its stationary distribution and if the chain mixes well. A chain might have reached stationarity if the distribution of points is not changing as the chain progresses. The aspects of stationarity that are most recognizable from a trace plot are a relatively constant mean and variance. A chain that mixes well traverses its posterior space rapidly and it can jump from one remote region of the posterior to another in relatively few steps. Such plots should look like fuzzy caterpillars. If the appearance of the history is a wiggly snake, it generally indicates that the sampler needs to be run longer and/or that the model needs to be reparameterized. It is associated with serial correlation in the sampling chain.

Other methods recommend running at least two chains simultaneously with overdispersed initial estimates (e.g., the initial estimates of chain 2 are 50% higher than those of chain 1). If the histories of the chains are overlapping and appear to mix with each other, then it is an indication of convergence. Several statistical diagnostic tests can be computed to quantitatively compare the chains. For example, Gelman and Rubin [20] proposed assessing the variance within and between the chains; Geweke [21] proposed comparing values in the early part of the chain to those in the latter to detect failure in convergence.

Finally, the Monte Carlo error can be used to assess how many iterations need to be run after convergence for accurate inference from the posterior distribution. The Monte Carlo error is an estimate of the deviance between the mean of the sample values and the posterior mean: this error can be linked to the standard error. Ideally, in the Monte Carlo integration, all samples from a chain should be independent, that is, free of serial correlation. However, in reality, this is rarely the case. The presence of autocorrelation does not indicate either a lack of convergence or a necessary overparameterization, but is due to the Markov chain sampling mechanism and to the choice of the proposal distribution. It will be necessary, however, to run the chain longer, so that, ultimately, enough independent samples from the chain are kept to ensure that the posterior distribution has been suitably explored by the sampler.

An interesting—more objective—evaluation is provided by the Raftery Criterion [22] that allows the estimation of the number of samples required to approximate with a desired precision several percentiles of the target distribution. The diagnostic was designed to provide the number of iterations and burn-in by first running and testing a shorter pilot chain. However, in practice, we can also just test our normal chain to see if it satisfies the results that the diagnostic suggests. Separate calculations are performed for each variable. This test tends to be conservative in that it will suggest more iterations than necessary.

5.3.5   Case Study

To illustrate the use of the Bayesian parameter estimation approach and, in particular, of MCMC algorithms, in this subsection we summarize a case study related to the identification of a popular model, the minimal model (MM) of glucose kinetics during an intravenous glucose tolerance test (IVGTT), from "poor" data sets comprising much fewer samples than those commonly used. In the identification of physiological systems, the possibility of reducing the number of samples is an appealing issue, for obvious economic and ethical reasons, and thus can have an important impact in clinical practice, especially in large-scale studies. LS/ML estimators are usually unsuccessful in dealing with reduced sampling schedules (RSS) due to unacceptable deterioration of the quality of estimates, while population parameter estimation methods (see Chapter 7) require a sufficiently large number of homogenous subjects. Below, following the procedure described in detail in Ref. [23], we compare the ability of an MCMC Bayesian parameter estimation method to identify the MM and, in particular, its glucose effectiveness (S_G) and insulin sensitivity (S_I) parameters, on a database of 16 normal subjects undertaken in an IVGTT experiment. For each subject, data were originally collected for 4 h on a frequent sampling schedule (FSS) with 30 time measurements that were then artificially lowered to $13-14$ to simulate an RSS situation.

As discussed in Section 5.3.3, one of the intrinsic aspects of the Bayesian estimation is the choice of the prior distribution. In this case, the minimal information we can encode is the nonnegativity of all MM parameters. A natural choice is thus the adoption of a log-normal distribution. However, additional *a priori* knowledge, which is likely useful for overcoming possible numerical identifiability problems in presence of RSS, is obtainable by exploiting the sample distributions of MM parameters estimated by LS in separate groups of subjects undertaken for the IVGTT with FSS. This analysis was performed in an independent study on 50 normal subjects (not including the 16 subjects here considered) and showed that estimates are approximately log-normally distributed. The values of mean and variance of the log-normal prior distribution were determined by fitting the sample distribution of the 50 normal subjects. These values were also found to be stable with respect to changes of the training set, such as the random extraction of subsets of the 50 subjects.

The second aspect concerns the computation of the posterior distribution. It is analytically intractable because of the complex relationships between parameters and data, so an MCMC simulation strategy implemented by the single-component Metropolis−Hastings random walk was used. In particular, the Markov chain was generated by extracting samples (for one MM parameter at a time) from a normal distribution with a mean equal to the current value of the chain and fixed variance (suitably chosen for each parameter to speed up the MCMC algorithm). Convergence was assessed by using the Raftery criterion; 30,000 samples were required in order to have a satisfactory description of the posterior and thus a "robust" assessment of point estimates and confidence intervals.

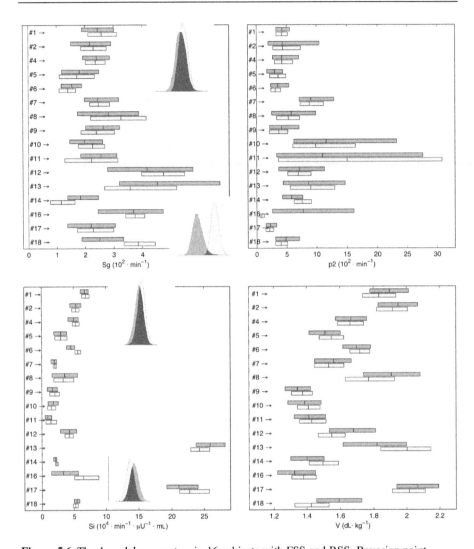

Figure 5.6 The 4 model parameters in 16 subjects with FSS and RSS. Bayesian point estimates (vertical line) and 95% confidence intervals (bars) are shown; for each subject, bottom bar refers to FSS and top RSS. For two subjects, posterior distributions are also shown: light gray FSS, medium gray RSS, and dark gray overlapped area.
Source: Adapted from Ref. [23].

The MCMC Bayesian estimator, as shown in Figure 5.6, provides accurate and precise MM parameter estimates (i.e., no bias with respect to FSS and acceptable CV), operating at the single individual level.

The results reported so far were obtained using a log-normal prior distribution with parameters fitted into a separate population study. However, the sensitivity of

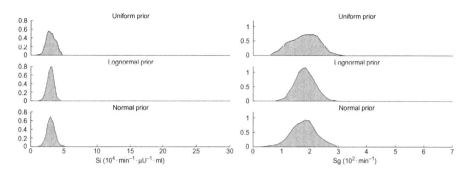

Figure 5.7 Posterior distributions in a subject in presence of RSS with different prior distribution: uniform, normal, and log-normal.
Source: Adapted from Ref. [23].

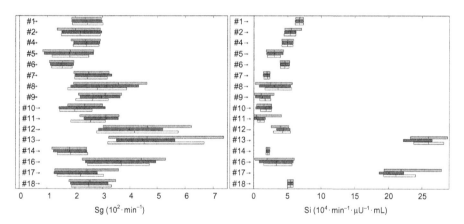

Figure 5.8 S_G and S_I confidence intervals in 16 subjects with RSS obtained by assuming log-normal (bottom bar), normal (middle bar), and uniform (top bar) *a priori* probability distribution.
Source: Adapted from Ref. [23].

Bayesian estimates to the shape of the prior was evaluated. Therefore, the RSS Bayesian estimates were recomputed assuming a normal distribution (whose parameters were fixed to have about the same *a priori* 95% confidence intervals) and a uniform (noninformative) distribution.

Notably, as reported in Figure 5.7 for a representative subject and in Figure 5.8 for all subjects, results are not so sensitive to changes of the shape density function chosen to express *a priori* knowledge.

In fact, in general, the point estimates and confidence intervals obtained are poorly sensitive to changes from a log-normal to a normal prior. Even in the case of the noninformative prior, in several cases, point estimates are very close to the one obtained in the presence of an informative prior, even if some differences are present. Confidence intervals are equal to or larger than those obtained with the log-normal prior. This demonstrates that *a priori* information plays a role in improving the performance of the Bayesian estimator. Moreover, because even using the noninformative prior brings, in most cases, estimates much more precise than those obtained by classical nonlinear LS, we can conclude that the MCMC simulation strategy can avoid the numerical identifiability problems often present with nonlinear least square estimates.

An additional advantage of MCMC Bayesian estimator is that it provides the *a posteriori* probability distribution of the parameter estimates, from which credible confidence intervals can be determined (e.g., asymmetrical and in the range of nonnegative values). This can be crucial in investigations where MM parameters are used in conjunction with other indexes (physiological, clinical, metabolic, genetic) in statistical analyses, such as characterization of a (sub)population, clustering, classification, regression, risk analysis, and so on (see Ref. [24] for more details).

5.4   Conclusions

In this chapter, we introduced the basic elements of model identification following both the Fisherian and Bayesian approaches. The latter are, in general, more complicated to implement, but can often lead to better results. In particular, it is worthwhile to stress that the improvement obtained by Bayes estimation with respect to Fisher estimation is that the more significant, the more the *a priori* information supports *a posteriori* information. Three typical situations in which taking into account *a priori* information can be crucial are when: (i) the model is too complex to be identifiable with satisfactory precision by a Fisherian approach; (ii) data are very noisy (i.e., *a posteriori* information is not reliable enough); and (iii) data are too scarce (i.e., there is only a small amount of *a posteriori* information). The case studies reported in the present chapter explicitly deal with situations (i) and (iii). Many other situations in which *a priori* information can be of determinant help in parameter estimation can be solved by adopting a rationale similar to the one above. For instance, for a situation in which the identification experiment design is not, for some reason, the optimal one and some parameters of the model cannot be identified from available data in an acceptably solid manner [25,26].

For the sake of space, we have not considered many other important aspects of model identification. Among these, we mention what is often indicated under the name of optimal experiment design, that is, to act on design variables such as number of test inputs and outputs, form of test inputs, number of samples and sampling schedule, and measurement errors so as to maximize, according to some criterion, the precision with which model parameters can be estimated. We refer the reader to Refs. [24,27,28] for theoretical results and case studies.

References

[1] Polonsky KS, Licinio-Paixao J, Given BD, Pugh W, Rue P, Galloway J, et al. Use of biosynthetic human C-peptide in the measurement of insulin secretion rates in normal volunteers and type I diabetic patients. J Clin Invest 1986;77:98−105.

[2] Eaton RP, Allen RC, Schade DS, Erickson KM, Standefer J. Prehepatic insulin production in man: kinetic analysis using peripheral connecting peptide behaviour. J Clin Endocrinol Metab 1980;51:520−8.

[3] Beck JV, Arnold KJ. Parameter estimation in engineering and science. New York, NY: Wiley; 1977.

[4] Soderstrom T, Stoica P. System identification. Upper Saddle River, NJ: Prentice Hall; 1989.

[5] Walter E, Pronzato L. Identification of parametric models from experimental data. Berlin: Springer; 1997.

[6] Ljung L. System identification—theory for the user. 2nd ed. Upper Saddle River, NJ: Prentice Hall; 1999.

[7] Nocedal J, Wright SJ. Numerical optimization. 2nd ed. Berlin: Springer; 2006.

[8] Åström KJ. Maximum likelihood and prediction error methods. Automatica 1980;16:551−74.

[9] Landaw E, Di Stefano J. Multiexponential, multicompartmental, and noncompartmental modeling. II. Data analysis and statistical considerations. Am J Physiol 1984;246: R665−77.

[10] Efron B. The jackknife, the bootstrap, and other resampling plans. Philadelphia, PA: SIAM; 1982.

[11] Akaike H. A new look at the statistical model identification. IEEE Trans Automat Contr 1974;19:716−23.

[12] Bozdogan H. Akaike's information criterion and recent developments in information complexity. J Math Psychol 2000;44:62−91.

[13] Sparacino G, Tombolato C, Cobelli C. Maximum likelihood vs maximum *a posteriori* parameter estimation of physiological system models: the C-peptide impulse response case study. IEEE Trans Biomed Eng 2000;47:801−11.

[14] Magni P, Sparacino G, Bellazzi R, Toffolo GM, Cobelli C. Insulin minimal model indexes and secretion: proper handling of uncertainty by a Bayesian approach. Ann Biomed Eng 2004;32:1027−37.

[15] Gilks W, Richardson S, Spiegelhalter DJ. Markov chain Monte Carlo in practice. London: Chapman & Hall; 1996.

[16] Metropolis N, Rosenbluth AW, Rosenbluth MN, Teller NM, Teller AH. Equations of state calculations by fast computing machine. J Chem Phys 1953;21:1087−91.

[17] Hastings WK. Monte Carlo sampling methods using Markov chain and their applications. Biometrika 1970;57:97−109.

[18] Geman S, Geman D. Stochastic relaxation, Gibbs distributions, and the Bayesian restoration of images. IEEE Trans PAMI 1984;6:721−41.

[19] Shephard N, Pitt MK. Likelihood analysis of non-Gaussian measurement time series. Biometrika 1997;84:653−67.

[20] Gelman A, Rubin DB. Inference from iterative simulation using multiple sequences. Stat Sci 1992;7:457−72.

[21] Geweke J. Evaluating the accuracy of sampling-based approaches to calculating posterior moments. In: Bernardo JM, Berger JO, Dawiv AP, Smith AFM, editors. Bayesian statistics, vol. 4. Oxford, UK: Clarendon Press; 1992.

[22] Raftery AE, Lewis SM. The number of iterations, convergence diagnostics and generic Metropolis algorithms. In: Gilks WR, Spiegelhalter DJ, Richardson S, editors. Practical Markov Chain Monte Carlo. London, UK: Chapman and Hall; 1996 [Chapter 7].

[23] Magni P, Sparacino G, Bellazzi R, Cobelli C. Reduced sampling schedule for the glucose minimal model: importance of Bayesian estimation. Am J Physiol Endocrinol Metab 2006;290:E177−84.

[24] Walter E, Pronzato L. Qualitative and quantitative experiment design for phenomenological models: a survey. Automatica 1990;26:195−213.

[25] Magni P, Bellazzi R, Nauti A, Patrini C, Rindi G. Compartmental model identification based on an empirical Bayesian approach: the case of thiamine kinetics in rats. Med Biol Eng Comp 2001;39:700−6.

[26] Pillonetto G, Sparacino G, Cobelli C. Numerical non-identifiability regions of the minimal model of glucose kinetics: superiority of Bayesian identification. Math Biosci 2003;184:53−67.

[27] Cobelli C, Ruggeri A. A reduced sampling schedule for estimating the parameters of the glucose minimal model from a labelled IVGTT. IEEE Trans Biomed Eng 1991;38:1023−9.

[28] Thomaseth K, Cobelli C. Generalized sensitivity functions in physiological system identification. Ann Biomed Eng 1999;27:607−16.

[29] Franceschini G, Macchietto S. Model-based design of experiments for parameter precision: state of the art. Chem Eng Sci 2008;63:4846−72.

6 New Trends in Nonparametric Linear System Identification[1]

Gianluigi Pillonetto[a] and Giuseppe De Nicolao[b]

[a]Department of Information Engineering, University of Padova, Padova, Italy, [b]Dipartimento di Informatica e Sistemistica, University of Pavia, Pavia, Italy

6.1 Introduction

The aim of system identification is the building of mathematical models of dynamical systems from observed input−output data. In the biomedical scenario, the mainstream approaches rely on classical parametric estimation paradigms. First, finite-dimensional models of different orders are postulated (e.g., compartmental descriptions that have important physiological motivations [1−3]). Then, each model is fitted to data using maximum likelihood (ML) or prediction error methods (PEMs), for which a large corpus of theoretical results is available [4,5]. The statistical properties of these techniques are well understood, assuming that the model class is fixed: under certain assumptions, they are in some sense optimal, at least for large samples.

Within this parametric paradigm, a crucial point is the selection of the most adequate model complexity, for example, the number of compartments used to describe the physiological system. In the "classical, frequentist" framework, this is a question of trade-off between bias and variance. It can be faced by adopting several model validation techniques, e.g., cross validation (CV), or complexity measures (e.g., the Akaike's criterion (AIC) [6]). However, some inefficiencies related to these classical approaches have been recently pointed out [7−9]. It has been shown that sample properties of ML and PEM approaches, equipped, for example, with AIC or CV, may lead to unsatisfactory results when tested on experimental data.

In recent work [7], it has been shown that these model selection problems can be successfully faced by a different approach to system identification based on an interaction with the machine learning field [10−12]. Rather than postulating finite-

[1]This research has been partially supported by the European Community under agreement n. FP7-ICT-223866-FeedNetBack, n257462 HYCON2 Network of excellence and by the MIUR FIRB project RBFR12M3AC—Learning meets time: a new computational approach to learning in dynamic systems.

Modelling Methodology for Physiology and Medicine. DOI: http://dx.doi.org/10.1016/B978-0-12-411557-6.00006-9

dimensional hypothesis spaces, e.g., FIR, ARX, ARMAX, or Laguerre models, a new nonparametric paradigm can be followed in which one searches for the unknown model directly in an infinite-dimensional space. In the context of linear system identification, the elements of such space are all the possible impulse responses. The intrinsic ill-posedness of the problem is tackled by resorting to regularization methods admitting a Bayesian interpretation [13,14]. In particular, the impulse response is modeled as the realization of a zero-mean Gaussian process. Hence, prior information is introduced in the identification process just assigning a covariance, named also kernel in the machine learning literature [15−18]. In this regard, a major novelty was the derivation of new kernels that incorporate information on the impulse response regarding its regularity and asymptotic decay to zero [7,9]. These kernels depend on few hyperparameters that are estimated from data using marginal likelihood maximization. This procedure is interpretable as the counterpart of model order selection in the classical ML/PEM paradigm, but it turns out to be much more robust, appearing to be the real reason for the success of these new procedures.

Hence, in view of the increasing importance of these kernel methods also in the general system identification scenario, our aim is to make accessible to the biomedical community some concepts underlying these new learning techniques. The chapter is organized as follows. In Section 6.2, we report the problem statement, while in Section 6.3, the classical approach to system identification is briefly reviewed. Limitations of the classical approach to system identification are then described in Section 6.4 via a Monte Carlo study regarding the assessment of cerebral hemodynamics using magnetic resonance imaging (MRI). The new approach to linear system identification via Gaussian regression is described in Section 6.5. It is then tested in Section 6.6 by considering the same numerical study reported in Section 6.4. Conclusions then end the chapter.

6.2 System Identification Problem

6.2.1 Continuous-Time Formulation

We start considering a continuous-time single-input single-output (SISO) dynamic system. Letting f be the unknown impulse response, we assume that the system is linear, time-invariant, stable, and causal ($f(t) = 0$, $t < 0$) so that

$$y_i = \int_0^{+\infty} u(t_i - \tau)f(\tau)d\tau + e_i, \quad i = 1,\ldots,n \tag{6.1}$$

where u is the input, y_i is the output measurement measured at instant t_i, and e_i is the measurement noise. The goal is to estimate the impulse response f from the knowledge of u and the n measurements y_i.

Under the stated assumptions, it is apparent that system identification corresponds to inverting a convolution integral. This task, also known as a deconvolution problem, is rather ubiquitous. It appears in biology, physics, and engineering with applications, e.g., in medicine, geophysics, and imaging [19–23]. The problem is difficult since convolution is a well-behaved operator in terms of continuity, but its inverse may not exist or may be unbounded [24]. In fact, the estimation of f is an intrinsically ill-posed problem because Eq. (6.1) requires the reconstruction of the whole function from a finite set of observations.

6.2.2 Discrete-Time Formulation

The discretized version of the system identification problem can be obtained assuming the input u piecewise constant and the impulse response f different from zero only on a compact set. Let Δ be a sufficiently small sampling period and m an integer large enough to make the following equations hold with good accuracy:

$$u(t) = u_j, \quad (j-1)\Delta \leq t < j\Delta, \quad j \in Z$$
$$f(t) = 0, \quad t < 0 \text{ and } t > m\Delta$$

For the sake of simplicity, assume also that, for each t_i, there exists $j \in Z$ such that $t_i = j\Delta$. Then, Eq. (6.1) can be rewritten in terms of the following linear algebraic system:

$$Y = \Phi g + E \tag{6.2}$$

where

- Y and E are n-dimensional (column) vectors whose ith components are, respectively, y_i and e_i;
- g is an m-dimensional (column) vector with jth component related to the unknown impulse response as follows:

$$g_j = \int_{(j-1)\Delta}^{j\Delta} f(\tau)d\tau, \quad j = 1,\ldots,m$$

- Φ is the $n \times m$ regression matrix whose ith row is

$$(u_j, u_{j-1}, \ldots, u_{j-m+1})$$

with j such that $t_i = j\Delta$.

Note that, if $m = n$ and Φ is of full rank, the impulse response estimate can be obtained just by inverting the regression matrix:

$$g^* = \Phi^{-1} Y$$

This is the simplest example where discretization allows one to recover the uniqueness of the solution. However, this approach usually leads to an ill-conditioned problem. This means that even small errors e_i in the measurements can lead to a large estimation error $(g^* - g = \Phi^{-1}E)$ [25]. Ill-conditioning is particularly severe when u is a low-pass/smooth signal, as usually happens in biomedical systems, and worsens when the output signal is sampled more frequently. In the case of uniform sampling $(t_i = \Delta \times i)$ these phenomena admit a spectral characterization via the Szegö theorem [26].

6.3   The Classical Approach to System Identification

In the classical approach to system identification, the first step consists of postulating a family of model structures. Each of them is a collection of impulse responses parametrized by a deterministic vector θ, denoted by f_θ in continuous time and g_θ in discrete time. Different structures may contain a different number $\dim(\theta)$ of parameters.

Given the input–output data and one model structure, the most natural approach to estimate θ is the least squares method that dates back to Gauss. The estimate is

$$\theta^* = \arg \min_\theta V(\theta) \tag{6.3}$$

where, in continuous time,

$$V(\theta) = \sum_{i=1}^{n} \left(y_i - \int_0^{+\infty} u(t_i - \tau) f_\theta(\tau) d\tau \right)^2 \tag{6.4}$$

while, in discrete time,

$$V(\theta) = \| Y - \Phi g_\theta \|^2 \tag{6.5}$$

where $\| \cdot \|$ represents the Euclidean norm. The least squares estimate coincides with the ML procedure if the noises e_i are Gaussian, uncorrelated with the same variance.

6.3.1   Model Structures Examples

The impulse response can be modeled as a linear combination of basis functions $h_i(t)$ with coefficients θ_i, i.e.,

$$f_\theta(t) = \sum_{i=1}^{\dim(\theta)} \theta_i h_i(t)$$

For instance, Laguerre functions [27] are often employed. In continuous-time identification, they are easily defined in the Laplace domain by

$$H_i(s) = \frac{(s-\alpha)^{i-1}}{(s+\alpha)^i}, \quad \alpha > 0, \; i = 1, 2, \dots$$

Here, α, that can be possibly included as a further parameter in θ, is associated with the dominant pole of the system: it governs the rate of decrease to zero of the impulse response. Another popular model is given by rational transfer functions [4], in which case the Laplace transform $F(s)$ of f is assumed to be the ratio of two polynomials:

$$F(s) = \frac{s^{d_1} + a_{d_1-1}s^{d_1-1} + \cdots + a_0}{b_{d_2}s^{d_2} + b_{d_2-1}s^{d_2-1} + \cdots + b_0}, \quad d_1 \le d_2$$

whose unknown coefficients a_i and b_i are gathered in $\theta \in \mathbb{R}^{d_1+d_2+1}$. If the roots of the denominator of $F(s)$ are all assumed distinct and real and $d_2 = d_1 + 1$, the impulse response f becomes a sum of exponentials, i.e.,

$$f_\theta(t) = \theta_1 e^{-\theta_2 t}, \qquad \dim(\theta) = 2$$
$$f_\theta(t) = \theta_1 e^{-\theta_2 t} + \theta_3 e^{-\theta_4 t}, \quad \dim(\theta) = 4$$
$$\vdots$$

All of the above models can be easily translated in the discrete-time setting. For instance, the zeta transform $G(z)$ of the impulse response can be assumed to be the ratio of two polynomials:

$$G(z) = \frac{z^{d_1} + a_{d_1-1}z^{d_1-1} + \cdots + a_0}{b_{d_2}z^{d_2} + b_{d_2-1}z^{d_2-1} + \cdots + b_0}, \quad d_1 \le d_2 \tag{6.6}$$

As for the Laguerre basis functions, in the z-transfer domain they are defined by

$$H_i(z) = \frac{(1-\alpha z)^{i-1}}{(z-\alpha)^i}, \quad -1 < \alpha < 1, \; i = 1, 2, \dots$$

where α still regulates the decay rate of the impulse response. In particular, the case $\alpha = 0$ leads to the so-called FIR models [4], where θ contains the impulse response levels, so that $\dim(\theta) = \dim(g_\theta)$:

$$g_\theta = \theta_1, \qquad \dim(\theta) = 1$$
$$g_\theta = (\theta_1, \theta_2)^{\mathrm{T}}, \quad \dim(\theta) = 2$$
$$\vdots$$

(note that no input delay has been included in the model). In the FIR case, the least squares estimate of θ is

$$\theta^* = \arg\min_\theta \| Y - \Phi\theta \|^2 \tag{6.7a}$$

$$\theta^* = (\Phi^T\Phi)^{-1}\Phi^T Y \tag{6.7b}$$

where Φ is assumed of full column rank.

6.3.2 Estimation of Model Dimension

In real applications, the model structure, and hence the dimension of θ, is unknown and must be inferred from data. This step is key and will have a major effect on the quality of the final model.

CV is one of the most widely employed approaches to model selection. Suppose we are given different estimated impulse responses, each characterized by its estimated parameter vector θ^*. Then CV selects that model maximizing an estimate of the prediction capability of the models on future data. Holdout validation is the simplest form of CV: the available data are split in two parts, where one of them (*training set*) is used to train the model and the other one (*validation set*) is used to assess the prediction capability. However, even if CV is a general-purpose approach, in data poor situations (frequently encountered in biomedicine) it may provide unreliable model orders [28].

Other popular techniques are the so-called Akaike-like criteria that assess the total model quality adding a penalty J for model flexibility. To simplify exposition, assume that the output measurement noise is white Gaussian of variance σ^2. Then, if the noise variance is known, the "optimal" model minimizes

$$\left[\frac{V(\theta^*)}{\sigma^2} + J(\dim(\theta), n) \right] \quad \text{known } \sigma^2 \tag{6.8}$$

whereas, if σ^2 is unknown and part of θ, the objective becomes

$$[n \log(V(\theta^*)) + J(\dim(\theta), n)] \quad \text{unknown } \sigma^2 \tag{6.9}$$

The most used penalty

$$J(\dim(\theta), n) = 2\dim(\theta) \quad \text{AIC} \tag{6.10}$$

does not depend on the data size n and leads to the well-known AIC [4,6]. It is derived by information-theoretic arguments: for large samples it provides an approximately unbiased estimator of the Kullback–Leibler divergence (the distance

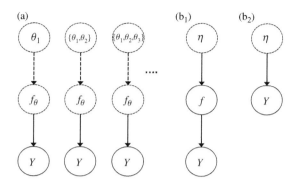

Figure 6.1 Bayesian network describing the classical parametric modelling approach to system identification (case a) and the nonparametric one (case b_1). The latter case with f integrated out (the joint distribution of f and Y is marginalized with respect to f) is also reported (case b_2). In the network, dotted lines denote deterministic variables and relationships, while solid lines denote stochastic variables and relationships.

of a model from the true data generator). There is also a small sample version, known in the literature as corrected AIC (AICc), defined by

$$J(\dim(\theta), n) = \frac{2n\dim(\theta)}{n - \dim(\theta) - 1} \quad \text{AICc} \tag{6.11}$$

Another variant

$$J(\dim(\theta), n) = \log(n)\dim(\theta) \quad \text{BIC, MDL} \tag{6.12}$$

imposes a larger penalty on the model flexibility and is derived following Bayesian arguments. This is also known in the literature as AIC-type B, BIC, or Rissanen's minimum description length (MDL) criterion [4,29].

The classical approach to system identification is graphically depicted by the Bayesian network in Figure 6.1a. Nodes and arrows are either dotted or solid depending on whether they are representative of either deterministic or stochastic quantities/relationships. Thus, one can see that θ is a deterministic parameter vector whose knowledge fully defines the impulse response (e.g., f_θ in continuous time). The model structure is typically unknown. Hence, several model structures need to be estimated from data and then compared using Akaike-like criteria.

6.4 Limitations of the Classical Approach to System Identification: Assessment of Cerebral Hemodynamics Using MRI

Some limitations of the classical approach to system identification are now illustrated via a simulated case study concerning the quantitative assessment of cerebral hemodynamics. This problem is key to understanding brain function in both normal and pathological states. For this purpose, an important technique is bolus-tracking MRI, which relies upon the principles of tracer kinetics for nondiffusible tracers [30,31]. Interestingly, in this scenario, quantification of cerebral hemodynamics corresponds to solving a time-invariant linear system identification problem [32]. In fact, let $C_{VOI}(t)$ and $C_{AIF}(t)$ denote, respectively, the measured tracer concentration within a given tissue volume of interest (the system output) and the measured arterial function (the system input), as a function of the time t. Then, under suitable assumptions, it holds that

$$C_{VOI}(t) = \int_0^t C_{AIF}(t - \tau) f(\tau) d\tau \qquad (6.13)$$

where the impulse response $f(t)$ is proportional to the so-called tissue residue function, which carries fundamental information on the system under study, for example, the cerebral blood flow is given by the maximum of $f(t)$. Hence, the problem is to reconstruct the impulse response from noisy samples of C_{VOI}.

We consider the same simulation described in Ref. [33]. The known system input is a typical arterial function given by

$$C_{AIF}(t) = \begin{cases} 0, & \text{if } t \le 10 \\ (t-10)^3 e^{-2t/3} & \text{otherwise} \end{cases} \qquad (6.14)$$

while the impulse response is the dispersed exponential displayed in the top panel of Figure 6.2 (solid line). It has to be reconstructed from the 80 noisy output samples reported in the bottom panel of the same figure. These measurements are generated as detailed in subsection II.A of Ref. [33], using parameters typical of a normal subject, a signal-to-noise ratio equals to 20 and discretizing the problem using $\Delta = 1$ and $t_i = i$, for $i = 1, \ldots, 80$, as sampling instants.

According to the classical approach to system identification, we proceed to solve the case study assuming that the z-transform of the impulse response g is given by Eq. (6.6) with $d = d_1 = d_2$. For every value of d in the set $1, 2, \ldots, 10$, the model parameters are estimated via least squares using the Matlab function oe.m of the system identification toolbox, equipped with the information that system initial conditions are null ('InitialState', 'Zero'). The values of d chosen by BIC, AIC, and AICc are 2, 10, and 5, respectively. The top panel of Figure 6.2 displays the estimate returned by BIC using a dash–dot line (impulse response coefficients are linearly interpolated and the estimator is denoted by LS + BIC, where LS stays for *least squares*). It is apparent that the reconstructed profile is far from the true one and contains many unphysiological oscillations. The estimates from AIC and AICc (not displayed) are even more distant from the true impulse response.

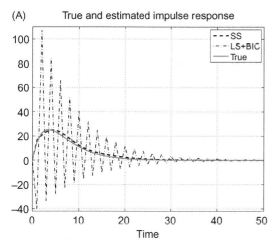

(A) True and estimated impulse response

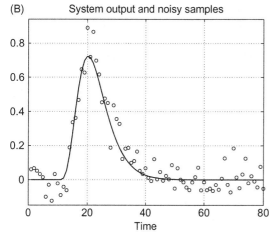

(B) System output and noisy samples

Figure 6.2 Assessment of cerebral hemodynamics using MRI. The top panel shows the true system impulse response (solid line) and the estimates obtained using rational transfer functions, see Eq. (6.6), with order estimated by BIC (LS + BIC, dash−dot) and by exploiting the stable spline estimator (SS, dashed line). The noiseless output (solid line) and the measurements (○) are displayed in the bottom panel.

This kind of result is confirmed by a Monte Carlo study of 1000 runs where independent noise realizations are generated and, given an impulse response estimate $\hat{g}^j$ obtained at the jth run, the error is computed as

$$
\text{err}_j = \sqrt{\frac{\sum_{i=1}^{80} (\hat{g}_i^j - g_i)^2}{\sum_{i=1}^{80} g_i^2}}
$$

Figure 6.3 reports the Matlab boxplots of the 1000 err_j obtained by different estimators. The first boxplot displays the errors obtained by an oracle-based

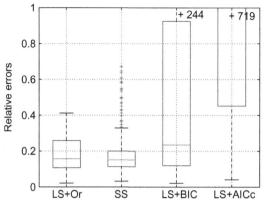

Figure 6.3 Assessment of cerebral hemodynamics using MRI. Boxplot of the 1000 relative errors in the impulse response reconstruction obtained using rational transfer functions, see Eq. (6.6), with order estimated by an oracle (LS + Or), by BIC (LS + BIC) or by AICc (LS + AICc), and by exploiting the stable spline estimator (SS).

procedure (LS + Or). This technique is never implementable in real applications: the oracle knows the true impulse response and, at every run, chooses that model order leading to the least squares estimate minimizing err_j. Hence, LS + Or provides the upper bound on the performance obtainable by the classical system identification approach. The two boxplots on the right part of Figure 6.3 instead show the results obtained using AICc[2] (LS + AICc) and BIC (LS + BIC): it is apparent that their performance is unsatisfactory, much different from that achieved by LS + Or.

To understand the reasons for these findings, it is useful to summarize the limitations of the parametric approaches equipped with AIC-like criteria:

- AIC, AICc, and BIC are all based on an approximation of the likelihood that is only asymptotically exact. This undermines the applicability of the theory when the ratio $n/\dim(\theta)$ is not large enough, a situation routinely encountered in biomedicine;
- all of these criteria select the optimal model without taking into account the uncertainty of the estimated parameters;
- the implementation of these model order selection strategies calls for the solution of several nonlinear least squares problems, one for each model dimension. Hence, computational complexity and local minima can be an issue.

6.5   The Nonparametric Gaussian Regression Approach to System Identification

We start describing the Gaussian regression framework for system identification in continuous time. Under the framework of Gaussian regression [13], instead of postulating a parametric structure, the impulse response is seen as the realization of a

---

[2]The reconstruction errors coming from AIC are larger than those obtained by AICc. For this reason, they are not displayed.

stochastic process f[7]. In particular, f is modeled as a continuous-time zero-mean Gaussian process with autocovariance, also called *kernel*, given by

$$\mathscr{E}(f(t)f(\tau)) = \lambda K(t, \tau), \quad t, \tau \in \mathbb{R}^+$$

where $\mathscr{E}$ denotes mathematical expectation and $\lambda \in \mathbb{R}^+$ is an unknown hyperparameter.[3] The new paradigm is graphically depicted in Figure 6.1b$_1$. In comparison with the parametric scenario on the left side of Figure 6.1, the first notable difference is that the (deterministic) vector θ of unknown size is now replaced by the (deterministic) hyperparameter vector η of fixed dimension. Such a vector contains λ and other possible parameters characterizing K. For example, the Gaussian kernel

$$\mathscr{E}(f(t)f(\tau)) = \lambda e^{-(t-\tau)^2/\rho}$$

is characterized by the hyperparameter vector $\eta = [\lambda, \rho]$.

Note that, while θ fully specifies f in a deterministic way, the hyperparameter vector η specifies only the statistical properties of the impulse response, namely its autocovariance. This explains why, in Figure 6.1, the vector η is connected with f through a stochastic relationship.

6.5.1 Estimate of the Impulse Response for Known Hyperparameters

Given the Bayesian network in Figure 6.1, we first assume that η and the noise variance σ^2 are known. We also assume the noise e_i white and Gaussian, independent of impulse response. According to a Bayesian paradigm, the estimate of the impulse response is taken equal to the posterior expectation, e.g., $\mathscr{E}[f(t)|Y]$ in continuous time.

The key for the derivation is the following well-known result concerning conditioning of jointly Gaussian random variables [34]: letting

$$\begin{bmatrix} x_1 \\ x_2 \end{bmatrix} \sim \mathscr{N}\left(\begin{bmatrix} m_1 \\ m_2 \end{bmatrix}, \begin{bmatrix} P_{11} & P_{12} \\ P_{21} & P_{22} \end{bmatrix} \right) \tag{6.15}$$

it holds that

$$\begin{aligned} x_1|x_2 &\sim \mathscr{N}(m, P) \\ m &= m_1 + P_{12}P_{22}^{-1}(x_2 - m_2) \\ P &= P_{11} - P_{12}P_{22}^{-1}P_{21} \end{aligned} \tag{6.16}$$

---

[3]According to the statistical literature, the term *hyperparameter* is here used to indicate a parameter of a prior distribution.

We start considering the continuous-time setting. For future derivations, given the system input u, it is useful to define the *output kernel matrix* $\mathbf{O}$ as the $n \times n$ positive semidefinite matrix with (i, j) entry given by

$$\mathbf{O}_{ij} = O(t_i, t_j) \tag{6.17}$$

where O is the output kernel induced by K as follows:

$$O(t, \tau) = \int_0^{+\infty} u(t - x) \left(\int_0^{+\infty} u(\tau - a) K(x, a) \mathrm{d}a \right) \mathrm{d}x$$

Then, we have

$$\mathscr{E}[y(t_i) y(t_j)] = \lambda \mathbf{O}_{ij} + \delta_{ij} \sigma^2$$

where δ_{ij} is the Kronecker delta and

$$\mathscr{E}[f(t) y(t_i)] = \lambda \int_0^{+\infty} u(t_i - a) K(t, a) \mathrm{d}a$$

Then, using Eq. (6.16) with the corresponding $x_1 = f(t)$ and $x_2 = Y$, we obtain that the posterior mean is

$$\mathscr{E}[f(t)|Y] = \lambda \sum_{i=1}^{n} c_i \int_0^{+\infty} u(t_i - a) K(t, a) \mathrm{d}a \tag{6.18}$$

where c_i is the ith entry of the vector

$$c = \left(\lambda \mathbf{O} + \sigma^2 I_n \right)^{-1} Y \tag{6.19}$$

with I_n the $n \times n$ identity matrix.

In discrete time, a similar result holds. In particular, let now $\mathbf{O}$ be defined by

$$\mathbf{O} = \Phi \mathbf{K} \Phi^{\mathrm{T}} \tag{6.20}$$

where $\mathbf{K}$ denotes the $m \times m$ *kernel matrix* with (i, j) entry

$$\mathbf{K}_{ij} = K(i, j)$$

Then, under the same assumptions on the noise E stated above and using Eq. (6.16) with $x_1 = g$ and $x_2 = Y$, we obtain

$$\mathscr{E}[g|Y] = \lambda \mathbf{K} \Phi^{\mathrm{T}} (\lambda \mathbf{O} + \sigma^2 I_n)^{-1} Y \tag{6.21}$$

Remark 1 *It can be proved that the estimate $\mathscr{E}[g|Y]$ reported in Eq. (6.21) can also be seen as the solution of a regularized least squares problem. In particular, we have*

$$\mathscr{E}[g|Y] = \arg\min_{g \in \mathbb{R}^m} \| Y - \Phi g \|^2 + \gamma g^{\mathrm{T}} \mathbf{K}^{-1} g \tag{6.22}$$

where $\gamma = \sigma^2/\lambda$. Note that the estimator in Eq. (6.22) controls model complexity not adjusting the dimension m of g, which is fixed and only has to be set to a sufficiently large value to capture the system dynamics, but through the penalty term $g^{\mathrm{T}} \mathbf{K}^{-1} g$ whose importance is regulated by the so-called regularization parameter γ. In that way Eq. (6.22) can be independently interpreted as a standard ℓ_2 regularization of the least squares solution for Eq. (6.2), without any probabilistic connotations [9]. A similar interpretation can be extended to the continuous-time setting: Eq. (6.18) can be seen as the solution of a Tikhonov-type variational problem over a suitable reproducing kernel Hilbert space [7,15].

6.5.2 Hyperparameter Estimation Via Marginal Likelihood Optimization

In the previous subsection, we derived the closed-form expressions for the impulse response estimates assuming known hyperparameters η. However, in real applications, the hyperparameters, possibly including also the noise variance σ^2, are in practice always unknown and need to be estimated from data. Their tuning can be interpreted as the counterpart of model order selection in the classical parametric paradigm. Hence, it has a major impact on the identification performance. In the following, we review one of the most effective methods for their determination.

Consider, just for a moment, the discrete-time case assuming that η contains only the regularization parameter γ. Then, one could be tempted to perform system identification by jointly optimizing equation (6.22) with respect to γ and g. However, one can see that the minimum is obtained for $\gamma = 0$, so that the regularization term vanishes and the problem reduces to least squares. This problem can be circumvented by marginalization, that is, by integrating out the dependence on the unknown impulse response. This leads to the definition of the so-called marginal likelihood $\mathbf{p}(Y|\eta)$. The model after marginalization is also depicted in Figure 6.1b$_2$.

In particular, let η now be a generic hyperparameter vector. Define

$$Z(\eta) = \lambda \mathbf{O} + \sigma^2 I_N$$

where the output kernel matrix $\mathbf{O}$ may also depend on η, being given by Eq. (6.17) in continuous time or by Eq. (6.20) in discrete time. Then, exploiting the Gaussian assumptions on the impulse response and the measurements noise, the ML estimate of η is

$$\eta^* = \arg\max_\eta \mathbf{p}(Y|\eta) = \arg\min_\eta Y^{\mathrm{T}} Z(\eta)^{-1} Y + \log(\det(Z(\eta))) \tag{6.23}$$

By relying upon marginalization, this tuning method is related to the concept of Bayesian evidence. In particular, it embodies the Occam's razor principle, i.e., unnecessarily complex models are automatically penalized, see Refs. [35,36] for nice discussions. Some theoretical results which corroborate its robustness independently of the correctness of the Gaussian assumptions on the impulse response have been also recently obtained in Refs. [37,38].

Once η is determined, according to the Empirical Bayes paradigm [39], its estimate can be inserted into Eq. (6.18) or (6.21), obtaining the impulse response estimate. Full Bayes approaches have also appeared in the literature. In this context, η is also seen as a random variable and stochastic simulation techniques (e.g., Markov chain Monte Carlo [40−42]), are used to reconstruct the posterior of the impulse response and η in sampled form. Typically, such techniques are much more computationally expensive than marginal likelihood optimization, but they return an impulse response estimate, and Bayes intervals around it, able to account also for hyperparameters' uncertainty [43,44].

6.5.3 Covariances for System Identification: The Stable Spline Kernels

The choice of the kernel K greatly influences the quality of the estimate coming from the nonparametric scheme. In continuous time, the typical kernel employed in the Gaussian regression literature reflects the prior knowledge that the unknown function, and possibly some of its derivatives, is continuous with bounded energy. In particular, it is typical to model f as the p-fold integral of white Gaussian noise. Then, the autocovariance of f is proportional to

$$W(s,t) = \int_0^1 G_p(s,u)\, G_p(t,u)\mathrm{d}u \tag{6.24}$$

where

$$G_p(r,u) = \frac{(r-u)_+^{p-1}}{(p-1)!}, \quad (u)_+ = \begin{cases} u & \text{if } u \geq 0 \\ 0 & \text{otherwise} \end{cases} \tag{6.25}$$

This is the autocovariance associated with the Bayesian interpretation of the pth order smoothing splines [45]. In particular, when $p = 1$, one obtains the linear spline kernel:

$$\min\{s,t\} \tag{6.26}$$

while $p = 2$ leads to the cubic spline kernel [45]:

$$\frac{st\min\{s,t\}}{2} - \frac{(\min\{s,t\})^3}{6} \tag{6.27}$$

Note that, if the system input is a Dirac Delta, for p=2, $\mathscr{E}[f(t)|Y]$ defined by Eq. (6.18) is indeed a cubic smoothing spline, that is, a third-order piecewise polynomial. Spline functions enjoy notable numerical properties originally investigated in the interpolation scenario. In particular, piecewise polynomials avoid the Runge's phenomenon [46] (presence of large oscillations in the reconstructed function), which arises, for example, when high-order polynomials are employed. The issue of fit convergence rates is discussed in the classical works [47−49].

In the system identification scenario, the main drawback of the kernel (6.24) is that it does not account for impulse response stability, that is, its decay to zero. In fact, it is immediately seen that if the autocovariance of f is proportional to Eq. (6.26) or (6.27), the variance of $f(t)$ is zero at $t = 0$ and tends to ∞ as t increases. However, if f represents a stable impulse response, we would better have a finite variance at $t = 0$ that goes exponentially to zero as t tends to ∞. How can this problem be overcome? The key idea developed in Ref. [7] to build a kernel having this feature is an exponential change of coordinates to remap $\mathbb{R}^+$ into the unit interval, and then using a spline kernel for functions defined there. This leads to the class of so-called *stable spline kernels* which, by construction, inherit all of the approximation capabilities of the spline curves [49], but differ from them in that they are intrinsically stable. They are defined by

$$S(s, t) = W(e^{-\beta s}, e^{-\beta t}), \quad s, t \in \mathbb{R}^+ \tag{6.28}$$

where β is a positive scalar governing the decay rate of the variance [7]. In practice, β will be unknown so that it is convenient to treat it as a further hyperparameter to be included in the vector η.

Note that, by definition, also the stable spline kernels depend on the integer p entering Eq. (6.24), which is typically set to 1 or 2. In particular, when $p = 1$ in Eq. (6.28), one obtains the impulse response model

$$\mathscr{E}(f(t)f(\tau)) = \lambda e^{-\beta \max(t,\tau)} \tag{6.29}$$

whereas $p = 2$ leads to

$$\mathscr{E}(f(t)f(\tau)) = \lambda \left[\frac{e^{-\beta(t+\tau)}e^{-\beta \max(t,\tau)}}{2} - \frac{e^{-3\beta \max(t,\tau)}}{6} \right] \tag{6.30}$$

The extension of these Bayesian priors for continuous-time system identification to the discrete-time setting is straightforward. In fact, one can describe the vector g containing the m impulse response coefficients as a zero-mean Gaussian random vector given by

$$g \sim \mathcal{N}(0, \lambda \mathbf{S})$$

where the $m \times m$ matrix $\mathbf{S}$ is defined by

$$\mathbf{S}_{ij} = S(i,j)$$

For instance, letting $p = 1$ and $\alpha = e^{-\beta}$, so that $0 \le \alpha < 1$, from Eq. (6.29) one obtains

$$\mathscr{E}(g_i g_j) = \lambda \alpha^{\max(i,j)}, \quad (i,j) \in \{1,\ldots,m\}^2 \tag{6.31}$$

This kernel has been also discussed in a deterministic setting in Ref. [9] and called TC kernel, being derived by an even more sophisticated covariance for system identification called DC kernel.

Finally, if $p = 2$ from Eq. (6.30), one obtains

$$\mathscr{E}(g_i g_j) = \lambda \left[\frac{\alpha^{(i+j)} \alpha^{\max(i,j)}}{2} - \frac{\alpha^{3\max(i,j)}}{6} \right], \quad (i,j) \in \{1,\ldots,m\}^2 \tag{6.32}$$

6.6 Assessment of Cerebral Hemodynamics Using the Stable Spline Estimator

We now reconsider the numerical study discussed in Section 6.4 using the stable spline estimator described in the previous section. In particular, we describe the impulse response g as a zero-mean Gaussian vector of dimension $m = 80$ with autocovariance given by Eq. (6.32). The noise variance is obtained using a low-bias model for the impulse response (as described in Ref. [50]). In particular, data are fitted by an FIR model of order $\dim(\theta) = 50$ and the estimate of σ^2 is the sum of the squared residuals divided by $n - \dim(\theta) = 30$. Then, the hyperparameter vector $\eta = [\lambda, \alpha]$ is determined via marginal likelihood optimization. In particular, Eq. (6.23) is solved setting the regression matrix Φ entering Eq. (6.20) to a lower triangular Toeplitz matrix whose first column is the function $C_{\mathrm{AIF}}(t)$ in Eq. (6.14) sampled on $t_i = 1,\ldots,80$. Finally, the estimate of g is computed via Eq. (6.21).

The top panel of Figure 6.2 displays the estimate returned by the stable spline estimator denoted by SS (dashed line) obtained using the noisy measurements in the bottom panel. In contrast to what happened using BIC, the reconstruction is now close to the true impulse response. This result is confirmed by the Monte Carlo study illustrated in Figure 6.3: the 1000 errors err_j achieved by SS are much lower than those obtained by AICc and BIC. Remarkably, the performance of SS is similar and sometimes also better than that of LS + Or (which is not implementable in practice since it requires the knowledge of the true impulse response). The reason is that the stable spline estimator overcomes the drawbacks of the classical system identification approaches listed at the end of Section 6.4. In fact:

- in the stable spline estimator the marginal likelihood, which is a function of the hyperparameters λ and α, is exact, irrespective of the sample size;
- the stable spline estimator accounts for impulse response uncertainty because the hyperparameter likelihood is obtained after marginalizing with respect to the stochastic impulse response;
- due to marginalization, the domain of the marginal likelihood is a two-dimensional space. Thus, the issue of local minima is far less critical. In fact, instead of solving several nonlinear optimization problems, one is faced with only one optimization problem in a very low-dimensional domain.

6.7 Conclusions

In this chapter, we have first described state-of-the-art linear system identification techniques relying upon parametric models and classical criteria for model order selection. Then, we have illustrated a new nonparametric estimation framework where the impulse response is modeled as realization from a zero-mean Gaussian process. Its covariance embeds information on the smoothness and asymptotic decay to zero (stability) of the unknown function, and depends on few unknown parameters that can be determined from data via marginal likelihood optimization.

The technique here presented establishes a new meeting between modern kernel-based regularized techniques and biomedical systems identification. We believe that this encounter can be especially beneficial and fruitful for the biomedical community. The numerical example, regarding assessment of cerebral hemodynamics using MRI, shows that conventional techniques, based on the difficult choice of model order, may be largely outperformed by carefully tuned regularization. We expect this to happen very frequently in the biomedical scenario, in which data are often not abundant and ill-conditioned problems are routinely encountered.

References

[1] Godfrey K. Compartmental models and their application. New York, NY: Academic Press; 1983.
[2] Jacquez J. Compartmental analysis in biology and medicine. Ann Arbor, MI: The University of Michigan Press; 1985.
[3] Cobelli C, Lepschy A, Jacur GR. Identifiability of compartmental systems and related structural properties. Math Biosci 1979;44(1):1−18.
[4] Ljung L. System identification—theory for the user. 2nd ed. Upper Saddle River, NJ: Prentice-Hall; 1999. p. 505−7
[5] Söderström T, Stoica P. System identification. London: Prentice-Hall; 1989.
[6] Akaike H. A new look at the statistical model identification. IEEE Trans Automat Control 1974;19:716−23.

[7] Pillonetto G, De Nicolao G. A new kernel-based approach for linear system identification. Automatica 2010;46(1):81−93.

[8] Pillonetto G, Chiuso A, De Nicolao G. Prediction error identification of linear systems: a nonparametric Gaussian regression approach. Automatica 2011;47(2):291−305.

[9] Chen T, Ohlsson H, Ljung L. On the estimation of transfer functions, regularizations and Gaussian processes—revisited. Automatica 2012;48(8):1525−35.

[10] Vapnik V. Statistical learning theory. New York, NY: Wiley; 1998.

[11] Evgeniou T, Pontil M, Poggio T. Regularization networks and support vector machines. Adv Comput Math 2000;13:1−150.

[12] Suykens JAK, Gestel TV, Brabanter JD, Moor BD, Vandewalle J. Least squares support vector machines. Singapore: World Scientific; 2002.

[13] Rasmussen CE, Williams CKI. Gaussian processes for machine learning. Cambridge: The MIT Press; 2006.

[14] Hengland M. Approximate maximum a posteriori with Gaussian process priors. Constr Approx 2007;26:205−24.

[15] Aronszajn N. Theory of reproducing kernels. Trans Am Math Soc 1950;68:337−404.

[16] Cucker F, Smale S. On the mathematical foundations of learning. Bull Am Math Soc 2001;39:1−49.

[17] Saitoh S. Theory of reproducing kernels and its applications. Vol. 189 of pitman research notes in mathematics series. Harlow: Longman Scientific and Technical; 1988.

[18] Schölkopf B, Smola AJ. Learning with kernels: support vector machines, regularization, optimization, and beyond (adaptive computation and machine learning). Cambridge, MA: MIT Press; 2001.

[19] Hunt BR. The inverse problem of radiography. Math Biosci 1970;8:161−79.

[20] Bertero M. Linear inverse and ill-posed problems. Adv Electron Electron Phys 1989;75:1−120.

[21] Bertero M, Boccacci P. Introduction to inverse problems in imaging. Bristol: IOP Publishing; 1998.

[22] De Nicolao G, Sparacino G, Cobelli C. Nonparametric input estimation in physiological systems: problems, methods and case studies. Automatica 1997;33:851−70.

[23] Tarantola A. Inverse problem theory and methods for model parameter estimation. Philadelphia, PA: SIAM; 2005.

[24] Phillips DL. A technique for the numerical solution of certain integral equations of the first kind. J Assoc Comput Mach 1962;9:84−97.

[25] Twomey S. Introduction to the mathematics of inversion in remote sensing and indirect measurements. New York, NY: Elsevier; 1977.

[26] Ekstrom MP. A spectral characterization of the ill-conditioning in numerical deconvolution. IEEE Trans Audio Electroac 1973;21(4):344−8.

[27] Wahlberg B. System identification using Laguerre models. IEEE Trans Automat Control 1991;36(5):551−62.

[28] Pillonetto G, De Nicolao G. Pitfalls of the parametric approaches exploiting cross-validation or model order selection. In: Proceedings of the 16th IFAC symposium on system identification (SysId'12); 2012.

[29] Rissanen J. Modelling by shortest data description. Automatica 1978;14:465−71.

[30] Zierler KL. Theoretical basis of indicator-dilution methods for measuring flow and volume. Circ Res 1962;10:393−407.

[31] Zierler KL. Equations for measuring blood flow by external monitoring of radioisotopes. Circ Res 1965;16:309−21.

[32] Calamante F, Thomas DL, Pell GS, Wiersma J, Turner R. Measuring cerebral blood flow using magnetic resonance imaging techniques. J Cereb Blood Flow Metab 1999;19:701–35.

[33] Zanderigo F, Bertoldo A, Pillonetto G, Cobelli C. Nonlinear stochastic regularization to characterize tissue residue function in bolus-tracking MRI: assessment and comparison with SVD, block-circulant SVD, and tikhonov. IEEE Trans Biomed Eng 2009;56 (5):1287–97.

[34] Anderson BDO, Moore JB. Optimal filtering. Englewood Cliffs, NJ: Prentice-Hall; 1979.

[35] Cox RT. Probability, frequency, and reasonable expectation. Am J Phys 1946;14 (1):1–13.

[36] MacKay DJC. Bayesian interpolation. Neural Comput 1992;4:415–47.

[37] Aravkin A, Burke J, Pillonetto G. A statistical and computational theory for robust and sparse Kalman smoothing. In: Proceedings of the 16th IFAC symposium on system identification (SysId'12); 2012.

[38] Carli FP, Chen T, Chiuso A, Ljung L, Pillonetto G. On the estimation of hyperparameters for Bayesian system identification with exponentially decaying kernels. In: Proceedings of the IEEE conference on decision and control (CDC'12); 2012.

[39] Maritz JS, Lwin T. Empirical Bayes method. New York, NY: Chapman and Hall; 1989.

[40] Gilks WR, Richardson S, Spiegelhalter DJ. Markov chain Monte Carlo in practice. London: Chapman and Hall; 1996.

[41] Andrieu C, Doucet A, Holenstein R. Particle Markov chain Monte Carlo methods. J R Stat Soc: Ser B (Stat Method) 2010;72(3):269–342.

[42] Ninness B, Henriksen S. Bayesian system identification via Markov chain Monte Carlo techniques. Automatica 2010;46(1):40–51.

[43] Magni P, Bellazzi R, De Nicolao G. Bayesian function learning using MCMC methods. IEEE Trans Pattern Anal Mach Intell 1998;20(12):1319–31.

[44] Pillonetto G, Bell BM. Bayes and empirical Bayes semi-blind deconvolution using eigenfunctions of a prior covariance. Automatica 2007;43(10):1698–712.

[45] Wahba G. Spline models for observational data. Philadelphia, PA: SIAM; 1990.

[46] Runge C. Uber empirische Funktionen und die Interpolation zwischen aquidistanten Ordinaten. Zeitschrift für Mathematik und Physik 1901;46:224–43.

[47] Ahlberg JH, Nilson EH. Convergence properties of the spline fit. J Soc Ind Appl Math 1963;11:95–104.

[48] Birkhoff G, Boor CD. Error bounds for spline interpolation. J Math Mech 1964;13:827–35.

[49] Atkinson KE. On the order of convergence of natural cubic spline interpolation. SIAM J Numer Anal 1968;5(1):89–101.

[50] Goodwin GC, Gevers M, Ninness B. Quantifying the error in estimated transfer functions with application to model order selection. IEEE Trans Automat Control 1992;37 (7):913–28.

7 Population Modelling

Paolo Magni[a], Alessandra Bertoldo[b], and Paolo Vicini[c]

[a]Dipartimento di Ingegneria Industriale e dell'Informazione, Università degli Studi di Pavia, Pavia, Italy, [b]Department of Information Engineering, University of Padova, Padova, Italy, [c]Pharmacokinetics, Dynamics and Metabolism, Pfizer Worldwide Research and Development, San Diego, CA

7.1 Introduction

In Chapter 5, we covered the parameter estimation problem in a single individual. In that context, the attention was focused on the mathematical formulation of the relationship between a given experimental protocol and the resulting measurements, and in deriving a suitable set of estimates for the model parameters. However, such relationship can differ between individual subjects, even if, in relatively homogenous groups, the differences could be quite limited and there could be many similarities. This is the paradigm of population modelling, a mathematical/statistical modelling methodology that focuses on *a population* of related individuals, in which each individual is represented by a specific instance of model parameter values. The differences in parameter values are due to the so-called *interindividual* (or *between-subject*) variability. As an example, the volume of distribution of a substance changes from one subject to another, and is linked, for example, to the subject's weight, health status, sex, age, and other known or unknown factors. In statistics, we summarize the characteristics of a population by using a small number of statistical moments (in general, the central tendency and the variability); similarly, in population modelling, we are interested in modelling and describing the typical behavior (central or population tendency) of the model and the variability across subjects.

This chapter will consider the situation in which experiments are performed in several elements of a population. The main goal is to describe through suitable mathematical/statistical models both single (typical) subject data and variability between subjects. As in the single subject case, in the population framework, given the model structure and the experimental data, we also have two main approaches to the parameter estimation problem: the maximum likelihood and the Bayesian methods. The maximum likelihood methods are based on the maximization of the likelihood function (or its approximations) derived for the population problem, whereas the Bayesian methods use the Bayesian inference

Modelling Methodology for Physiology and Medicine. DOI: http://dx.doi.org/10.1016/B978-0-12-411557-6.00007-0

approach (and Markov chain Monte Carlo (MCMC) algorithms) to estimate the posterior distribution of the population model parameters. In the following, we will focus on the different formulations of the population problem, mostly on the maximum likelihood estimation approaches.

Although the applicability of the population modelling approach is not limited to a specific field, one of the historic and most popular applications is population pharmacokinetics, i.e., the study of the sources of variability in drug concentrations among individuals who comprise the target patient population receiving clinically relevant doses of a drug of interest [1]. For this reason, some of the examples reported in this chapter come from the pharmacokinetic field.

7.1.1 Problem Statement

Let us consider a sequence of experiments carried out on a set of K individuals. Each experiment can follow its own protocol and can involve measurements in more than one sampling site at predetermined time points (in the following, for simplicity, we will consider the simplest case of only one sampling site). Thus, we will have a collection of pairs of values (t_{ij}, z_{ij}) $i = 1, \ldots, N_j$, and $j = 1, \ldots, K$, where z_{ij} is the ith measurement of the jth subject collected at the t_{ij} time point. As in the single subject case, let us suppose that, thanks to appropriate physiological and/or empirical assumptions, the equations that constitute the mathematical model are known and depend on a set of M (usually unknown) parameters $p = [p^1, p^2, \ldots, p^M]$. Therefore, in our case, we can write that the model prediction of the observed variable in the jth subject is $y_j(t) = f_j(t, p_j)$, where f_j mathematically expresses the relationship between a given experimental protocol and the observed variable in the jth subject, and the vector p_j includes the parameters. Note that the model prediction $y_{ij} = y_j(t_{ij})$ is, in general, different from the measurement z_{ij} for several reasons (often called *residual unknown variability*, RUV), including the presence of measurement error. A common situation is the additive error model in which $z_{ij} = y_{ij} + v_{ij}$, with v_{ij} modelled as an independent zero-mean random variable, normally distributed with standard deviation σ_{ij} (see Chapter 5 for more details). In this chapter, the following alternative, more compact, vector notation will be sometimes adopted: $z_j = y_j + v_j$, $y_j = y_j(t_j) = F_j(t_j, p)$, where z_j, y_j, v_j, t_j, and F_j are the N_j-dimensional vectors of the respective quantities related to the jth subject and Σ_{vj} is the $N_j \times N_j$ variance–covariance matrix of the residual error v_j.

Given this setting, the goals are (i) to derive from experimental data an estimate of the individual parameters p_j, together with their covariance matrixes Σ_{pj}; (ii) to determine their typical (population) value, quantify their variability within the population, and study whether some of this variation is associated with subject characteristics. In the next sections, we will analyze different formulations that, at least partially, try to meet these goals.

7.2 Naïve Data Approaches: Naïve Average and Naïve Pooled Data

The simplest way to approach population studies is to focus the attention only on the typical (population) response, thus avoiding the estimation of the intersubject variability. One way to do this is to "reduce" the problem to the estimation of the parameters p of one reference individual. There are two slightly different approaches: naïve average data (NAD) and naïve pooled data (NPD).

7.2.1 Naïve Average Data

This very simple method consists of the following procedure:

1. Compute the average value of the data of the K subjects at each sample time:

$$\bar{z}_i = \frac{1}{K} \sum_{j=1}^{K} z_{ij} \quad i = 1, \ldots, N \tag{7.1}$$

2. Fit the model $f(t,p)$ against the mean data by using one of the techniques presented in Chapter 5, for example, least squares or maximum likelihood, deriving p_{NAD}, the population parameter estimate. For example, using the weighted least squares approach, minimizing the objective function $O_{NAD}(p) = (\bar{z} - F(t,p))^T \sum_v^{-1} (\bar{z} - F(t,p))$, where the statistical model of a single mean-datum is $\bar{z}_i = f(t_i, p) + v_i$, that provides an estimate p_{NAD}. It is then possible to derive an approximation of the covariance matrix of the estimates through the inverse of the Fisher matrix computed in correspondence of p_{NAD}. As shown in Chapter 5, $\sum_{pNAD} = \left(S_{pNAD}^T \sum_v^{-1} S_{pNAD} \right)^{-1}$, where S_{pNAD} is the sensitivity (or Jacobian) matrix of $F(t,p)$ computed for $p = p_{NAD}$. Note that the covariance matrix provided here is a measure of estimate precision, not of intersubject variability.

This method is attractive because it is simple: a single fitting procedure is sufficient to obtain the typical response. The vector p_{NAD} is often interpreted as the "typical" parameter vector. On the other hand, NAD has a number of caveats. For example, it is suitable only for standardized designs, because one of the intrinsic requisites is that the experimental protocol and the sampling schedule are the same for all of the subjects (in fact, in the previous formulas, $f = f_j$, $t_i = t_{ij}$, and $N = N_j$). Because of the smoothing effect of averaging, mean data generally look more regular than individual data and a better fit is often obtained because the noise is filtered out. However, NAD is suitable only if subjects can be seen as replicates of a typical subject and the interindividual variability, not taken into account, is negligible. This situation might be seen, for example, when experiments are done on standardized laboratory animals of a given strain. On the other hand, if the differences between subjects are significant, the use of NAD can produce a distorted picture. As an example, let us consider a simple monoexponential model and two subjects or group of subjects with very different half-lives. The averaged curve exhibits a nonrepresentative bi-exponential model that would cause model estimation to be wrong. In other cases instead, the averaging operation will tend to obscure peculiarities that can be seen in individual data; the wrong model can also be selected in this case.

Lastly, no estimate of interindividual variability can be obtained with the NAD approach, because it masks variability rather than revealing it. All of the sources of variability disappear with the computation of the mean response. As an example of its application, this approach has been used in Chapter 19 to analyze oncology preclinical data in xenograft mice.

7.2.2 Naïve Pooled Data

This method consists in the simultaneous fitting of the models $f_j(t,p)$ against all of the data from the different subjects to derive the population parameter estimate p_{NPD}. For example, using the least squares approach, the functional to be minimized is $O_{NPD}(p) \sum_{j=1}^{K} (z_j - F_j(t_j, p))^{T} \sum_{v}^{-1}(z_j - F_j(t_j, p))$ being $z_{ij} = f_j(t_{ij}, p) + v_{ij}$.

Also, in this case it is possible to derive an approximation of the covariance matrix of the estimates using the sensitivity matrix.

Similar to NAD, NPD allows one to obtain the parameter estimates after a single fit of all of the data. As NAD, it may perform well when variation between subjects is small. On the other hand, NPD is a more general applicable method than NAD because, in its mathematical formulation, it does not require the same experiment and sampling schedule in all of the subjects (f_j, t_{ij}, and N_j are used in the formulas instead of f, t_i, N). However, when there are many more observations taken from some individuals than others, some bias in the estimates can occur. An example would be a case where eight samples are taken from some individuals, four from others, and one from the rest. If the population is not highly homogenous, the prevalence of samples within one group over other groups can lead to biased estimates.

7.3   Two-Stage Approaches: Standard, Global, and Iterative Two-Stage

Relatively simple ways to approach population studies are to follow one of two opposite strategies: to consider data as coming from a single subject, as in the naïve approaches, or to consider all of the subjects separately as if they were not members of the same population. When the first strategy cannot be followed because the naïve approaches are not applicable, the two-stage approaches can sometimes represent a suitable solution. In particular, two-stage methods can be applied when the number of measurements N_j is such that the individual parameter estimation problem in each individual subject has a reliable solution. In this case, the parameters p_j must be estimated in each individual, providing $\hat{p}_1, \ldots, \hat{p}_K$ parameter estimates. Each estimate is also characterized by its estimated covariance (precision), which is here indicated by $\sum_{\hat{p}_j}$. After having estimated individual parameters in the first stage, in the second stage parameters across individuals are then calculated, thus obtaining a population parameter estimate. As already highlighted, the population parameters of interest are typical values and variability, expressed in these approaches by the (sample) mean and variance/covariance computed on the individual parameters.

7.3.1 Standard Two-Stage

The standard two-stage (STS) approach refers to a well-known, intuitive, and widely used procedure. Population characteristics are estimated as the empirical first (mean) and second-order (covariance) statistical moments of the individual estimates $\hat{p}_j$:

$$p_{STS} = \frac{1}{K}\sum_{j=1}^{K}\hat{p}_j$$
$$\Omega_{STS} = \frac{1}{K-1}\sum_{j=1}^{K}(\hat{p}_j - p_{STS})(\hat{p}_j - p_{STS})^{T}$$

(7.2)

The estimated covariance matrix Ω_{STS} provides a measure of interindividual variability, i.e., how and how much the parameters vary among the different subjects within the population. STS has three intrinsic important limitations: (i) it does not take into account the intraindividual variability summarized in the subject specific covariance matrix $\sum_{\hat{p}_j}$ and this can lead to an overestimation of Ω_{STS}. In fact, the individual parameters are combined in Eq. (7.2) as if they were a K-sample from a multivariate distribution of not uncertain estimates; (ii) no information is gained in the (individual) analysis from the knowledge that the subjects belong to the same population; (iii) no measure of the precision of the population parameters is available. That being said, the main advantage of STS is its simplicity.

7.3.2 Global Two-Stage

The estimate $\hat{p}_j$ can be viewed as an observation of the true individual parameters p_j. It is uncertain because of the v_{ij} random component associated with the measurements z_{ij}. Global two-stage (GTS) uses the covariance matrix $\sum_{\hat{p}_j}$ to take that into account. To understand the procedure, let us consider that

$$\hat{p}_j = p + (p_j + p) + (\hat{p}_j - p_j)$$

(7.3)

where p and p_j are the true population and individual values, respectively, $(p_j - p)$ is the interindividual variability, and $(\hat{p}_j - p_j)$ is the estimation uncertainty of the individual parameters. From this, we can observe that $E[\hat{p}_j] = p$ and $var[\hat{p}_j] = \Omega + \sum_{\hat{p}_j}$, where Ω is the true population variance−covariance matrix.

GTS provides a maximum likelihood estimation of p and Ω assuming that the estimates of the individual parameters are normally distributed. The objective function to minimize is

$$O_{GTS}(p, \Omega) = \sum_{j=1}^{K} (\hat{p}_j - p_j)^T \left(\sum_{\hat{p}_j} + \Omega \right)^{-1} (\hat{p}_j - p) + \ln \left(\det \left(\sum_{\hat{p}_j} + \Omega \right) \right) \quad (7.4)$$

This is the typical objective function for the maximum likelihood estimate of normally distributed data with unknown variance (see Chapter 5). The objective function can be conveniently minimized iteratively by using the following formulas, based on an expectation−maximization (EM) approach that differs from other methods such as Gauss−Newton in that it does not require the calculation of derivatives.

$$p_{GTS}(h+1) = \left(\sum_{j=1}^{K} \left(\sum_{\hat{p}_j} + \Omega_{GTS}(h) \right)^{-1} \right)^{-1} \sum_{j=1}^{K} \left(\sum_{\hat{p}_j} + \Omega_{GTS}(h) \right)^{-1} \hat{p}_j$$

$$\Omega_{GTS}(h+1) = \frac{1}{K} \sum_{j=1}^{K} c_j(h+1) c_j^T(h+1) + \frac{1}{K} \sum_{j=1}^{K} \left(\sum_{\hat{p}_j}^{-1} + \Omega_{GTS}^{-1}(h) \right)^{-1} \quad (7.5)$$

where $c_j(h+1) = \left(\sum_{\hat{p}_j}^{-1} + \Omega^{-1}(h) \right)^{-1} \sum_{\hat{p}_j} (\hat{p}_j - p_{GTS}(h+1))$

The GTS approach, in contrast to STS, provides unbiased estimates of the population mean and variance−covariance matrix under the ideal situation in which Σ_{pj} is exactly known. However, it is well known that in real situations the asymptotic covariance matrix $\Sigma_{\hat{p}_j}$ used in Eq. (7.5) is approximate and that in some cases the approximation can be poor.

7.3.3 Iterative Two-Stage

Another possible solution to the population estimation problem is given by the iterative two-stage (ITS), an iterative algorithm that relies on repeated fitting of individual data, alternatively exploiting maximum *a posteriori* (MAP) estimation and the empirical Bayes approach. In the first stage, all the individual parameters are estimated by using a Bayesian estimator, exploiting the population parameters as prior information; in the second stage, the population parameters are recalculated by using the updated individual parameters, in order to form a new prior distribution. The estimation procedure is repeated until the difference between the new and old prior distributions is close to zero. As for every iterative procedure, ITS has to be initialized with a prior distribution, usually defined using the STS estimated values or the NPD values. More formally, the ITS consists of the following steps:

1. Initialization step (iteration $h = 0$):

$$p_{ITS}(0) = p_{STS} \quad \text{and} \quad \Omega_{ITS}(0) = \Omega_{STS} \quad (7.6)$$

2. For each subject, perform individual MAP estimation, minimizing the following objective function and computing the covariance matrix of the estimates:

$$\hat{p}_j(h+1) = \arg\min O_{ITS}(p_j) - \arg\min(z_j - F_j(t_j, p_j))^T \sum_{v_j}^{-1}(p)(z_j - F_j(t_j, p_j))$$

$$+ \ln\left(\det\left(\sum\nolimits_{v_j}(p)\right)\right) + (p_j - p_{ITS}(h))^T \Omega_{ITS}^{-1}(h)(p_j - p_{ITS}(h)) \qquad (7.7)$$

$$\sum\nolimits_{\hat{p}_j}(h+1) = \left(S_{j,p_j(h+1)}^T \sum\nolimits_{v_j}^{-1} S_{j,p_j(h+1)} + \Omega_{ITS}^{-1}(h)\right)^{-1}$$

3. Compute the new population parameters

$$p_{ITS}(h+1) = \frac{1}{K}\sum_{j=1}^{K}\hat{p}_j(h+1)$$

$$\Omega_{ITS}(h+1) = \frac{1}{K}\sum_{j=1}^{K}\sum\nolimits_{\hat{p}_j}(h+1) + (\hat{p}_j(h+1) - p_{ITS}(h+1))(\hat{p}_j(h+1) - p_{ITS}(h+1))^T$$

$$(7.8)$$

Repeat steps 2 and 3 until a predefined measure of convergence is reached.

The objective function for the individual fitting in this iterative algorithm explicitly takes into account the information on the population mean and variance (as is done in every Bayesian estimator). That can help to fit individual data in the presence of poor informative data (see Chapter 5 for a more detailed discussion). Similarly, computation of the covariance matrix includes available information on the precision of the individual parameter estimates. Even if computationally expensive, ITS can be extensively used in both data-rich and data-poor contexts.

7.4   Nonlinear Mixed-Effects Modelling

A weakness of all two-stage methods is that they can be applied only if enough samples are available for each individual to enable estimation of both p_j and of Σ_{pj}. However, this is often not the case. An estimate of p_j is not available in many situations, such as in clinical studies, where sparse and unbalanced data are routinely collected, and in some nonclinical studies with destructive sampling (e.g., in toxicokinetic studies) where only one data point per subject may be available. On the other hand, the NPD approach, as we know, completely neglects the variability among individuals. That may be unacceptable in several situations, such as in clinical investigation where individual characteristics can be important.

A completely different approach is the explicit modelling of the dependence of the model parameters from individual subjects as a part of a unique mathematical/statistical model to be identified from the whole data set in a one-step procedure. This approach considers the population study sample, rather than the individual, as

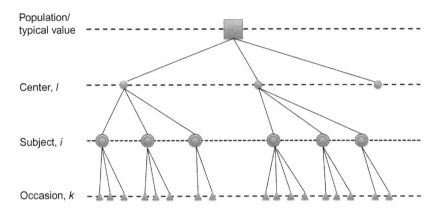

Figure 7.1 Schematic representation of the different levels of variability.

a unit of analysis for the estimation process. It tries to separate the common features from the differences. To do this in a more efficient way, avoiding mixing the different sources of heterogeneity, one can introduce a hierarchy of several levels of variability, which (all together) cause the observed variability in the data. In Figure 7.1, a possible general situation is depicted: a study involves several centers, each center recruited several subjects, each subject is observed on several occasions.

The typical values of the model parameters, common to the investigated population, are at the top of the hierarchy. They can vary in practice because of the difference between centers. Subjects in the same center certainly have common features, but can still have different parameters because of interindividual variability, and observations from the same subject can in turn be different on each different occasion. Lastly, measurements can be affected by residual noise, not included in the previous model features and thus unexplained. The available measurements are the end results of the interaction among all the mentioned effects. These are the basic ideas of the approach to population modelling that will be described in more detail in Section 7.4.1, in which we will consider the simplest two-level hierarchy: overall population and individual subjects.

7.4.1   Basic Definitions: Fixed Effects, Intersubject Variability, Residual Random Effects, Covariates

Population analyses provide an opportunity not only to estimate variability, but also to identify its different sources. Variability is usually characterized in terms of fixed (deterministic) and random effects. The fixed effects are those that do not change across the population of subjects (i.e., they are common to all individuals) such as the parameter population average values, which may in turn be a function of specific subject characteristics such as weight, sex, and health status (called covariates). The random effects quantify the amount of the variability which is not explained by fixed effects. In addition, the variability embedded in the individual parameters can be explicitly modelled by combining parameters that change across the population (random effects) and parameters that do not change (fixed effects). Both of these parameter classes contribute to the description of the population characteristics: mean values are generally derived for fixed effect parameters, whereas variance−covariance values are used to describe random effect parameters. Fixed and random effects can be combined to give rise to very flexible descriptions of population variability. For this reason, the approach is called mixed-effect modelling. We will consider *nonlinear* mixed effects, since the models of interest to biomedicine are most often nonlinear in the parameters.

7.4.2   Formalization: First Stage and Second Stage

Let us consider the widely used two-stage hierarchy. For each level, we will define the general mathematical/statistical model, in order to fully define the nonlinear mixed-effects model.

7.4.2.1   First Stage (Observations)

The observation model can be formalized as in the single subject case, as already done in Section 7.1, even if in this context a more general formulation is usually adopted. Therefore, we assume that the model predictions y of the measurements z can be written as a function of the model parameters p and of a set of design specific variables x, that include all the known design quantities that can change from one observation to another, such as dose and sampling time t. Therefore, $y_{ij} = f_i(x_{ij}, p_j)$. However, measurements usually differ from the model predictions due to different unknown factors. Therefore, at this level, we introduce a random component v to take such factors into account. This random component is often called residual unknown error, because it is not imputed to any kind of structured variability, but is linked to unexplained noise (i.e., unstructured variability). One of the most common examples is the consideration of independent, normally distributed additive error. Therefore, $z_{ij} = f_j(x_{ij}, p_j) + v_{ij}$, or more compactly $z_j = F_j(x_j, p_j) + v_j$, with $\Sigma v_j(p_j, \xi)$ as the covariance matrix, function of the individual parameters, and of other fixed effects ξ.

7.4.2.2   Second Stage (Population)

A population model is a collection of models for individual observations. Various doses, regimens, and/or administration routes are generally used in drug therapy and clinical trials. Correspondingly, the functions F_j will differ across individuals. However, it is realistic to assume that the set of underlying structural parameters is qualitatively the same for all individuals, and that parameters vary quantitatively among individuals. Therefore, we can write $p_j = G(\theta, a_j, \eta_j)$, where G is a known vector function that describes the expected value of p_j as a deterministic function of the fixed effects (population parameters) θ and of the known individual-specific covariates a_j (such as weight, age, disease state). In general, covariates can change over time; however, we will assume them to be constant within an individual for simplicity. The random effect η_j represents the random variation of the individual parameters around the population mean. Usually, the η_js are independent across individuals and they are distributed with covariance matrix Ω (another population parameter). In other words, p_j is assumed to arise from some multivariate probability distribution. The most common choices are the normal and the log-normal distributions. The advantage of the latter is that the individual parameters are guaranteed to be positive, while, if the interindividual variability is large, the Gaussian distribution does not guarantee positivity. The choice of the distribution can be made based on independent physiological knowledge of the system.

7.4.3 An Example: A Population Model of C-Peptide Kinetics

C-peptide (CP) is co-secreted with insulin on an equimolar basis, exhibits linear kinetics in a large range of concentrations, and—differently from insulin—is not extracted by the liver. Therefore, CP plays a key role in quantitative studies of the insulin system; for example, plasma CP concentration data are used to estimate beta cell insulin secretion rate by deconvolution (Chapter 3) or to assess beta cell secretory indexes by the minimal model [2,3]. In the literature, a linear two-compartment model is usually adopted to describe CP kinetics. The model assumes that CP enters the system from the accessible compartment, from which it is cleared from the system and distributes into a peripheral compartment (see Figure 5.1B). In order to provide the parameters of the model in a single individual, an input−output experiment was proposed: after having suppressed spontaneous CP pancreatic secretion by means of a somatostatin infusion, an intravenous bolus of biosynthetic CP is administered and plasma concentration samples are frequently collected (see Chapter 5). If several subjects are available, we could consider a population approach to the modelling of the data. To better understand the abstract structure introduced in the previous subsection, as an exercise, in the following we will try to build a plausible population model. Therefore, we have to define both the first and second stages.

7.4.3.1 First Stage (Observations)

Following Eq. (5.7), we can rewrite the equation of CP kinetics for the jth subject

$$\dot{q}_1^j(t) = -(k_{01}^j + k_{21}^j)q_1^j(t) + k_{12}^j q_2^j(t) + D^j \delta(t) \quad q_1^j(0) = 0$$
$$\dot{q}_2^j(t) = k_{21}^j q_1^j(t) - k_{12}^j q_2^j(t) \quad q_2^j(0) = 0 \tag{7.9}$$
$$y^j(t) = q_1^j(t)/V_1^j$$

with $z_{ij} = y^j(t_{ij}) + v_{ij}$ as the plasma concentration measurements and v_{ij} independent and normally distributed with standard deviation $\sigma_{ij} = CVy_{ij}$.

7.4.3.2   Second Stage (Population)

The simplest model we can imagine does not consider any covariate and assumes that individual parameters are normally distributed around the population value. Then:

$$
\begin{aligned}
k_{01}^{j} &= \theta_1 + \eta_{1j} \\
k_{12}^{j} &= \theta_2 + \eta_{2j} \\
k_{21}^{j} &= \theta_3 + \eta_{3j} \\
V_1^{j} &= \theta_4 + \eta_{3j}
\end{aligned}
\tag{7.10}
$$

with $[\eta_{1j}, \eta_{2j}, \eta_{3j}, \eta_{4j}]^{\mathrm{T}} \sim N_4(0, \Omega)$, where N_4 indicates the four-dimensional multivariate normal distribution. However, in this case study, it is more reasonable to consider parameters log-normally distributed, since they will be forced to be positive. Then:

$$
\begin{aligned}
k_{01}^{j} &= \theta_1 \, e^{\eta_{1j}} \\
k_{12}^{j} &= \theta_2 \, e^{\eta_{2j}} \\
k_{21}^{j} &= \theta_3 \, e^{\eta_{3j}} \\
V_1^{j} &= \theta_4 \, e^{\eta_{4j}}
\end{aligned}
\tag{7.11}
$$

or using a different notation:

$$
\begin{aligned}
\log(k_{01}^{j}) &= \theta_1 + \eta_{1j} \\
\log(k_{12}^{j}) &= \theta_2 + \eta_{2j} \\
\log(k_{21}^{j}) &= \theta_3 + \eta_{3j} \\
\log(V_1^{j}) &= \theta_4 + \eta_{4j}
\end{aligned}
\tag{7.12}
$$

with $[\eta_{1j}, \eta_{2j}, \eta_{3j}, \eta_{4j}]^{\mathrm{T}} \sim N_4(0, \Omega)$ as before. The population parameters θ assume a different meaning in the two formulations (Eqs. (7.11) and (7.12)), since the statistical model is different.

In the literature, a systematic dependence of the CP kinetic parameters from some anthropometric characteristics of the subject, such as the health status, sex, age, and body surface area (BSA), has been reported. These individual (measureable) characteristics can be used to explain part of the observed intersubject variability. Therefore, for example, Eq. (7.11) can be reformulated as follows:[1]

$$
\begin{aligned}
k_{01}^{j} &= \theta_1 \, e^{\eta_{1j}} \\
k_{12}^{j} &= (\theta_2/\mathrm{Age}_j + \theta_3) e^{\eta_{2j}} \\
k_{21}^{j} &= (\theta_4/\mathrm{Age}_j + \theta_5) e^{\eta_{3j}} \\
V_1^{j} &= (\theta_6 \, \mathrm{BSA}_j + \theta_7) e^{\eta_{4j}}
\end{aligned}
\tag{7.13}
$$

where $[\mathrm{Age}_j, \mathrm{BSA}_j] = a_j$ are the covariates.

---

[1]As the one presented here is only a toy example, the actual relationships have been simplified. However, that does not affect the generality of the discussion.

The goal now becomes to find efficient methods to estimate the unknown fixed effects from population time course data and covariate measurements.

7.4.4 Estimation Methods: First Order with Post Hoc, First-Order Conditional Estimation, Laplace, Stochastic Approximation of the Expectation—Maximization Algorithm

Nonlinear mixed-effects modelling simultaneously estimates all the unknown model parameters. It also provides an estimate of the precision of these parameters. Most of the numerous estimation methods available adopt the maximum likelihood approach. The probability of the data should be written as a function of the model parameters and then parameter estimates can be computed maximizing a suitable function of that probability. However, it is difficult to calculate the likelihood of the data for most models of interest to biomedicine, because of the nonlinear dependence of the observations on the random parameters η and possibly v. To deal with this problem, several approximate methods have been proposed. There are three main classes of approaches: (i) linearization methods (e.g., first order (FO), first-order conditional estimation (FOCE)) approximate nonlinear mixed-effects models by a first-order Taylor series expansion to arrive at a pseudomodel that is close to a linear mixed-effects model form; (ii) integral approximation methods (e.g., Laplace) use different approximation techniques to calculate the marginal distribution of data and then maximize the likelihood directly; (iii) EM algorithms approximate the conditional expectation of the log-likelihood in the E-step and then maximize the expected log-likelihood to obtain the estimates in the M-step. Although the mathematical details of the methods are highly complex, in the following only the main ideas will be presented.

7.4.4.1 First Order

Because the main difficulty in population parameter estimation is the nonlinear dependence of the observation on the random parameters η and possible v, the FO method implemented in the NONMEM software is based on a first-order Taylor approximation (linearization) of the relationship between observations and random effect parameters [4]. This linearization is done in each subject, considering as the pivotal point the mean value of the random effects (i.e., zero). Therefore, being $z_j = F_j(x_j, G(\theta, a_j, \eta_j)) + v_j$ then $z_j \approx F_j(x_j, G(\theta, a_j, 0)) + Z_j(\theta, 0)\eta_j + v_j$, where

$$Z_j(\theta, \eta_j) = \frac{\partial F_j(x_j, G(\theta, a_j, \eta_j))}{\partial p_j} \frac{\partial G(\theta, a_j, \eta_j)}{\partial \eta_j}$$

Assuming η and v independent and normally distributed, it has been shown that the z_j after the approximation are normally distributed with

$$E[z_j] = F_j(x_j, G(\theta, a_j, 0))$$

$$\text{Cov}[z_j] \approx Z_j(\theta, 0)\Omega Z_j(\theta, 0)^{\text{T}} + \sum\nolimits_{v_j}(G(\theta, a_j, 0), \xi) = V_j(\theta, 0, \Omega, \xi) \tag{7.14}$$

Therefore, the maximum likelihood estimate of the population parameters can be obtained minimizing the following objective function:

$$O_{FO}(\theta, \Omega, \xi) = \sum_{j=1}^{K} (z_j - F_j(x_j, G(\theta, a_j, 0)))^T V_j(\theta, 0, \Omega, \xi)^{-1}$$

$$(z_j - F_j(x_j, G(\theta, a_j, 0))) + \ln(\det(V_j(\theta, 0, \Omega, \xi)))$$

(7.15)

If the hypothesis of normal distribution of the random effects does not hold, the maximum likelihood estimator cannot be easily computed. In this case, other approaches based on generalized least squares or pseudolikelihood methods have been proposed to exploit the FO linearization [5].

The main advantage of the FO method is the low computational cost. In problems with many (hundreds of) subjects, this can be an important aspect. However, due to the rough linear approximation, it has several drawbacks. It has been shown that it is a biased estimator, it is not asymptotically efficient, and—last but not least—it does not directly provide estimates of the individual parameters p_j. To address the last point, FO is usually coupled with a further step, called *post-hoc* estimation, where maximum *a posteriori* estimation is used to quantify the random effects η_j and then of p_j. In particular, the following objective function has to be minimized to obtain η_j:

$$O_{PH}(\eta_j) = (z_j - F_j(x_j, G(\theta_{FO}, a_j, \eta_j)))^T \sum_{v_j} (G(\theta_{FO}, a_j, \eta_j), \xi_{FO})^{-1}$$

$$(z_j - F_j(x_j, G(\theta_{FO}, a_j, \eta_j))) + \eta_j^T \Omega_{FO}^{-1} \eta_j$$

(7.16)

and then $p_{j_{PH}} = G(\theta_{FO}, a_j, \eta_{j_{PH}})$.

7.4.4.2 FOCE Approximation

The linearization around the pivotal point $\eta_j = 0$ introduced by FO can be rather poor in the presence of large variability among subjects, thus resulting in inconsistent estimates of the fixed effects, mainly when random effects are high. To overcome these problems, a more accurate (but also more computationally expensive) algorithm, called FOCE, has been proposed [4]. The basic idea is to perform the linearization around values that are more representative of individual subjects' characteristics (other than the mean value of ηs). In particular, a proposed choice is to use the maximum *a posteriori* estimate of each individual η_j, given the fixed effects (conditional estimation), as a more appropriate value. Therefore, $z_j \approx F_j(x_j, G(\theta, a_j, \eta_j^*)) + Z_j(\theta, \eta_j^*)(\eta_j - \eta_j^*) + v_j$ (where η_j^* is the pivotal point) and assuming again that η and v are independent and normally distributed, we have that the observations z_j, after the approximation, are normally distributed with

$$E[z_j] = F_j(x_j, G(\theta, a_j, \eta_j^*)) - Z_j(\theta, \eta_j^*)\eta_j^* = M_j(\theta, \eta_j^*)$$

$$\text{cov}[z_j] \approx Z_j(\theta, \eta_j^*)\Omega Z_j(\theta, \eta_j^*)^T + \sum_{v_j} (G(\theta, a_j, \eta_j^*), \xi) = V_j(\theta, \eta_j^*, \Omega, \xi)$$

(7.17)

Then, the maximum likelihood estimate of the population parameters and the maximum *a posteriori* estimate η_j^* can be obtained minimizing the following two objective functions:

$$O_{\text{FOCE}}(\theta, \Omega, \xi) = \sum_{j=1}^{K} (z_j - M_j(\theta, \eta_j^*))^{\text{T}} V_j(\theta, \eta_j^*, \Omega, \xi)^{-1} (z_j - M_j(\theta, \eta_j^*))$$
$$+ \ln(\det(V_j(\theta, \eta_j^*, \Omega, \xi)))$$

$$O_{\text{FOCE}}(\eta_j) = (z_j - F_j(x_j, G(\theta_{\text{FOCE}}, a_j, \eta_j)))^{\text{T}} \sum_{v_j} (G(\theta_{\text{FOCE}}, a_j, \eta_j), \xi_{\text{FOCE}})^{-1}$$
$$(z_j - F_j(x_j, G(\theta_{\text{FOCE}}, a_j, \eta_j))) + \eta_j^{\text{T}} \Omega_{\text{FOCE}}^{-1} \eta_j$$

$$(7.18)$$

This is an iterative procedure that alternates the computation of the population parameters, through the minimization of the first objective function, and of the random effects in each subject, through the minimization of the second objective function.

The main advantage of the FOCE algorithm is that it performs the linearization around a more appropriate point (individual estimate). It simultaneously returns the estimates of the population and of the individual parameters, via the estimation of individual random effects, through $p_{j_{\text{FOCE}}} = G(\theta_{\text{FOCE}}, a_j, \eta_{j_{\text{FOCE}}})$. However, it is more time consuming (due to the iterative process) and it is not asymptotically efficient. It also can suffer in a context of very sparse data, where individual estimates are not reliable.

7.4.4.3 Laplace Method

The basic idea of integral approximation methods, which includes Laplace, is first to approximate the marginal likelihood of the response using a numerical integration routine, then to maximize the approximated likelihood numerically. Integral approximations are in general computationally more demanding than linearization methods. However, integral approximations usually maximize the likelihood of the original data and they can generate more consistent and accurate estimates in parameter estimation compared to linearization methods. Therefore, it is usually a good idea to use linearization methods to provide starting values for the more accurate integral approximation methods. The so-called Laplace method exploits the Laplace approximation, a method for approximating integrals of suitable functions (via a second-order Taylor expansion) by using local information about the integrand at its maximum. Therefore, it is most useful when the integrand is highly concentrated about its maximum value. This approximation can be applied to the mixed-effect model likelihood to marginalize the joint probability of data and random effects over the random effects and then maximizing the marginalized likelihood to derive the population parameters [6]. The Laplace method provides a more accurate approximation when compared to FOCE. However, the computational burden increases, because computation of the second derivative of the likelihood (or at least a good approximation of it) is required.

7.4.4.4   Stochastic Approximation of the Expectation–Maximization Algorithm

EM algorithms are iterative procedures that alternate between performing an expectation step and a maximization step. Their simplicity and stability have made EM methods a popular approach for finding maximum likelihood estimates in statistical models that depend on missing data or unobservable variables (which in our context can be the individual-specific random effects). During each iteration, the E-step generally computes the expectation of the complete data log-likelihood with respect to the conditional distribution of the random effects, given the observations and under the current estimates of the fixed effects; in the M-step, a new estimate for the fixed effects is computed, maximizing the quantity previously computed in the E-step. For nonlinear mixed-effect models, the assessment of the E-step involves the evaluation of a multiple integral that, in general, does not have a close-form expression. Various simulated EM algorithms have been presented to approximate the E-step, such as the Monte Carlo integration (MCEM). A very popular method, available first in the MONOLIX software and now implemented also in other tools such as NONMEM, Matlab, and R, is the SAEM. It is rapidly becoming a reference in the field. It decomposes the E-step into a simulation step and a stochastic approximation step; samples of missing data are either simulated under the conditional distribution of the missed data, given the observed data and the current parameter values, or obtained from a MCMC procedure [7,8]. The SAEM algorithm requires the simulation of only one realization of the missing data for each iteration, thus substantially reducing the computational time compared to MCEM. SAEM (as well as other EM-type algorithms) performs maximum likelihood estimation without any approximation of the statistical model. Near-"optimal" statistical properties (consistency and minimum variance of the estimate) are expected with SAEM. All of the exact EM methods have greater stability in analyzing complex models and can provide accurate results with sparse or rich data. The optimized versions of SAEM can perform more slowly than NONMEM FOCE for simple models, but perform more quickly and stably than NONMEM FOCE for complex models.

7.4.5   Bayesian Approach to Nonlinear Mixed-Effects Models

As discussed in Chapter 5, sometimes it is useful to formulate the identification problem within a Bayesian framework. For this reason, a Bayesian approach to nonlinear mixed-effects models was proposed in the literature and implemented in software tools [9–11]. In this section, the basic ideas will be briefly summarized. The Bayesian setting shares with non-Bayesian approaches the same format for describing the hierarchical models introduced in this section, but involves the addition of a further stage assigned to the specification of the priors. Moreover, all the unknown parameters and data have to be formalized as random variables, for which the distribution has to be defined. The *a posteriori* distribution is the target that has to be computed, in general, through a simulation algorithm such as MCMC methods (see Chapter 5). Then, any desired feature of the posterior density, such as the mode, moments, probabilities, and credible intervals can be derived. In the following, as an example, we will refer to a three-level hierarchy, even if the Bayesian approach can accommodate an arbitrary number of levels associated with random effects terms.

7.4.5.1   Stage 1—Model for the Data

In the first stage, the conditional density of the observations y—given the fixed effects and the random effects—is specified.

$$y_{ij} \sim N(f_i(x_{ij}, p_j), \sigma^2) \tag{7.19}$$

N stays for the normal distribution.

7.4.5.2   Stage 2—Model for Heterogeneity Between Subjects

In the second stage, the conditional density of the random effects given the fixed effects is specified to describe the interindividual variability. For example,

$$p_j \sim N_M(G(\theta, a_j), \Omega) \tag{7.20}$$

N_M stays for the M-dimensional multivariate normal distribution, where M is the number of parameters.

7.4.5.3   Stage 3—Model for the Priors

In the third stage, the prior distributions and the hyperparameters are specified for the residual uncertainty, the population mean parameters and the heterogeneity.

$$\sigma^{-2} \sim \Gamma(a, b)$$
$$\theta \sim N_q\left(\mu, \sum\right) \tag{7.21}$$
$$\Omega^{-1} \sim W_M(\rho\Omega_0, \rho)$$

Γ is the gamma distribution, W_M is the M-dimensional Wishart distribution. a, b, μ, Σ, ρ, and Ω_0 are the fixed parameters of the prior distributions. Equations (7.19)–(7.21) provide a flexible statistical model for the observations.

7.4.6   Evaluation of Modelling Results

7.4.6.1   Individual and Population Parameters—Standard Error of the Estimates

Within the nonlinear mixed-effects modelling estimation framework, different methods are used to obtain the covariance matrix of parameter estimates. One of them is the use of the sandwich matrix or sandwich estimator, often known as the robust covariance matrix estimator or the empirical covariance matrix estimator. The use of the sandwich matrix is based on the seminal results of Huber [12] and White [13], showing that in large samples, the sampling distribution of the maximum likelihood estimator in misspecified models is normal with covariance provided by the sandwich covariance matrix. In other words, it allows an asymptotically consistent estimation of the covariance matrix without making distributional assumptions and even if the assumed model underlying the parameter estimates is incorrect. In nonlinear mixed modelling, the population estimates are of interest and, consequently, the sandwich matrix can be written as:

$$W(\Theta, x) = R^{-1}(\Theta, x) \cdot S(\Theta, x) \cdot R^{-1}(\Theta, x) \tag{7.22}$$

where $\Theta = [\theta, \Omega, \xi]$ and R is the Hessian matrix:

$$R = -\frac{\partial^2 \ln O}{\partial \Theta \, \partial \Theta^{\mathrm{T}}} \tag{7.23}$$

O is the likelihood and S is the cross-gradient product matrix:

$$S = \sum_{N} \frac{\partial \ln O}{\partial \Theta} \times \frac{\partial \ln O}{\partial \Theta^{\mathrm{T}}} \quad N = \text{number of subjects} \tag{7.24}$$

Its name derives from the interpretation of the "ingredients" as:

$$\begin{array}{c} \text{Sandwich}(\Theta, x) = B(\Theta, x)\, M(\Theta, x)\, B(\Theta, x) \\ \text{bread} \\ \text{meat} \\ \text{bread} \end{array} \tag{7.25}$$

Note that it can be proven [14,15] that when K, i.e., the number of individuals, is big and both η and ε are normally distributed, then $S \to R$, and $W(\theta, x) = R^{-1}(\theta, x)$, i.e., the sandwich matrix and the inverse of the Fisher information matrix are both consistent estimators of the covariance matrix of the parameter estimates. However, when the normality assumption of η and ε is violated, the sandwich matrix is a more robust estimator of the covariance matrix. Lastly, for categorical or noncontinuous data, consistency (asymptotically unbiased) and other optimal properties of the estimates are highly dependent on the assumption of normality of the random effects. For these types of data, the R^{-1} matrix is preferable.

As an alternative to the analytical approach, one can obtain the covariance matrix using bootstrapping techniques. Both parametric and nonparametric boot-strap methods have been proposed and used [16,17]. The parametric bootstrap procedure requires specifying the distributions, whose parameters are estimated from the available data (for instance, random effects are generated from a multivariate normal distribution and combined with the fixed parameters values) while the second one is based on the creation of a resampled data vector from the original distribution of the data set [18] or from the resampling of the residuals [19]. Both parametric and nonparametric approaches have advantages and disadvantages. In particular, the nonparametric bootstrap requires fewer hypotheses, assuming only that the simulated data are able to maintain the correlations and relationships in the observed data, but it works really well when the data set is big, while the parametric bootstrap performance is strongly related to the correctness of the stochastic models, but may be advantageous when the data set is small.

In practice, when the R or S matrix is singular, or the R matrix is nonpositive semidefinite, one can resort to bootstrap methods to estimate confidence intervals of the population parameters.

Regarding individual estimates, to investigate their reliability, one can derive the standard error (SE) of η values. A possible method is to obtain this information from the inverse of the Fisher information matrix assuming that the population estimates are the true values and using the formula:

$$\text{cov}(\eta) = 2\left[\frac{\partial^2 O}{\partial \eta_i \, \partial \eta_j}|_{\eta=\hat{\eta}}\right]^{-1} \quad i = 1,\ldots,H, \quad j = 1,\ldots,H \tag{7.26}$$

with H equal to the number of random effects considered in the model [20].

7.4.6.2 Individual and Population Prediction—Weighted Residuals

Assessing nonlinear mixed-effects modelling results means also checking the adequacy of the model predictions to the measured data. This includes the use of two specific plots, i.e., the plot of the individual predicted data versus the measured/observed data of all of the individuals and the plot of the population predicted data versus the measured/observed data of all of the individuals. The individual predicted data are derived by solving the model equations using $[\theta, \eta_j, a_j, \varepsilon]$ estimates of each jth subject. For instance, a classic way to visualize individual predictions versus measured data is shown in Figure 7.2A, where the solid line represents the identity line and circles indicate the data. The closer the data are to the solid line, the better is the adherence of the individual predictions to the observed data. In addition, ideally the data should be randomly distributed along the line of identity. The population predictions are derived by solving the model equations using $[\theta, 0, a_j, \varepsilon]$ estimates, i.e., only fixed effects are considered together with covariates

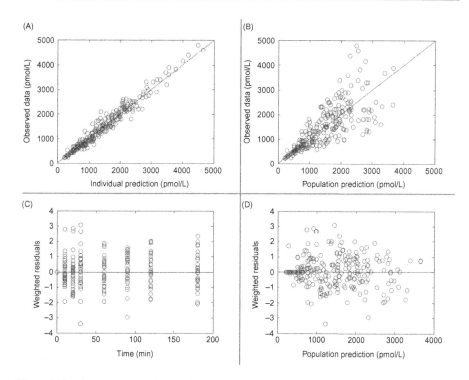

Figure 7.2 Diagnostic plots for nonlinear mixed-effects results. Observed versus predicted plots (A and B), the solid line represents the line of identity. Weighted residual plots (C and D), the mean value is close to zero.

(if present). In Figure 7.2B, it is shown how the data are usually visualized. Even for this plot, the closer the data are to the solid line, the better the adherence of the population prediction to the observations. Also, in this case, ideally the data should be randomly distributed along the line of identity. Considering Figure 7.2A and B, one can conclude that there are not biased predictions in the considered data set, for both individual and population results.

From the individual and population predictions, one can also derive the individual weighted residuals and the population weighted residuals. Population weighted residuals (WRES) are usually plotted in two different ways to check any unwanted heteroscedasticity presence. WRES can be visualized as in Figure 7.2C, i.e., along the time axis and along the population predicted data, Figure 7.2D. In our example, the plots indicate the absence of heteroscedasticity.

7.4.6.3   Population Estimates

Both population and individual parameter estimates have to be checked in the diagnostic phase in terms of accuracy and physiological plausibility of the results.

Table 7.1 Fixed Effect Estimates and Their Precision

	Estimate	SE	SE%
θ_1	0.55	0.039	7
θ_2	1.34	0.079	6

Table 7.2 Between-Subject Variability Estimates and Their Precision

	Estimate	SE	SE%
ω_{11}^2	0.230	0.019	8
ω_{22}^2	0.184	0.022	12

To take a concrete example, we can hypothesize a model characterized by two parameters described for the jth subject by:

$$p_{1j} = \theta_1 \exp(\eta_{1j})$$
$$p_{2j} = \theta_2 + \eta_{2j}$$

(7.27)

with $\eta \sim N(0,\Omega)$ and $\Omega = \begin{bmatrix} \omega_{11}^2 & 0 \\ 0 & \omega_{22}^2 \end{bmatrix}$. In this case, if p_1 is in $[\text{min}^{-1}]$ and p_2 is in [L], θ_1 and θ_2 have the same unit of measurement as p_1 and p_2, respectively, η_1 is unitless, while η_2 has the same unit of measurement as p_2. Regarding Ω, ω_{11} is unitless (i.e., %) and ω_{22} has the same unit of measurement as p_2. To take a concrete example, if the population results are those reported in Table 7.1 and the between-subject variability (BSV) estimates are those in Table 7.2, the square root of 0.230 is 0.48 and, consequently, this means that p_1 has a population (fixed effect) value of 0.55 $[\text{min}^{-1}]$ and its coefficient of variation for BSV is about 48%; p_2 has a population (fixed effect) value of 1.34 [L] and its BSV is about 0.43 [L], i.e., the square root of 0.184. For both parameters, BSV is physiologically reasonable and all of the estimates exhibit a good precision range [7−12%].

7.4.6.4 Individual Estimates

From the estimates of the random effects, one has to verify *a posteriori* if the results are in agreement with the statistical assumption: $\eta_j \sim N(0,\Omega)$. This can be done by using a frequency distribution (histogram) plot as a visual diagnostic tool to confirm the normality assumption for each random effect. Thus, the use of a histogram allows us to understand if the distribution is correctly shaped and to derive insights about presence of outliers or model misspecification. For instance a situation as displayed in Figure 7.3A implies a good agreement with the normality assumption of the random effects (mean and skewness values are close to 0, entropy is close to 3), while in Figure 7.3B, the distribution indicates a potential

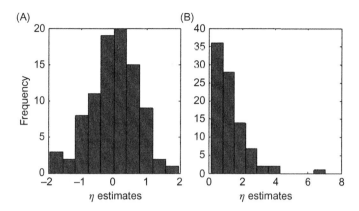

Figure 7.3 Example of possible histograms of the random-effects estimates.

model misspecification. Alternative graphical methods for visual inspection are the stem-and-leaf plot, P–P plot (probability–probability plots display the cumulative probability of a variable versus the cumulative probability of a variable having, in our case, normal distribution), Q–Q plot (quantile–quantile plots display the quartiles of a variable versus those of a variable having, in our case, normal distribution), boxplot, etc. Instead of a visual check, one can use a statistical test for normality, such as the Kolmogorov–Smirnov test.

An additional interesting tool can be used to investigate the setup of the Ω matrix. The Ω matrix can be implemented as a full matrix, meaning that all the parameters are correlated, or as a diagonal matrix, meaning that no correlation between the parameters is assumed, or as a sparse (block) matrix. Plotting the posterior estimates of the random effects as a function of each other allows us to test the suitability of the assumed covariance structure. For instance, suppose we define the two parameters of a model as:

$$p_{1j} = \theta_1^* \exp(\eta_{1j})$$
$$p_{2j} = \theta_2^* \exp(\eta_{2j}) \quad j = 1, \ldots, K \tag{7.28}$$

and

$$\Omega = \begin{bmatrix} \omega_{11}^2 & \omega_{12}^2 \\ \omega_{21}^2 & \omega_{22}^2 \end{bmatrix} \tag{7.29}$$

If the plot of η_1 versus η_2 indicates a nonzero correlation for a model where the corresponding ω_{12} (and consequently ω_{21}) value was fixed to zero, the correlation is generally tested in the model for its significance by allowing the estimator also to derive a value for ω_{12}. If the plot indicates zero correlation for a model where ω_{12} was also zero, the correlation is often not tested for its significance.

7.4.6.5   Visual Predictive Check

Typically in a mixed-effects modelling approach the performance of the models can also be evaluated using the visual predictive check (VPC). The VPC is a diagnostic tool based on a comparison between the statistics obtained from the simulated data using the estimated population parameters and the true observed data. VPC is usually performed using the final θ, Ω, and ζ estimates, while uncertainty in the parameter estimates is not included in this process. The idea of the VPC is to assess by visual inspection whether or not the model is able to describe the variability of the observed data. To do so, multiple simulations are run, keeping the structure of the observed data set. Then, the median and the 5th and the 95th percentiles of the simulated data sets are compared to the corresponding percentiles of the original data. More precisely, the mean and quantiles of the simulated data from the model against the mean and quantiles of the observed data are compared. If the population model adequately describes features in the data, then the simulated and observed means and percentiles should correspond closely. An example of VPC is shown in Figure 7.4.

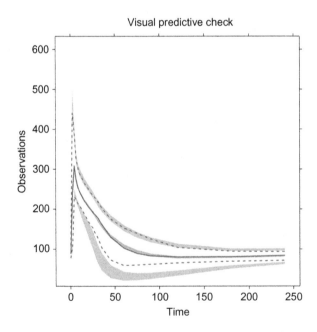

Figure 7.4 Example of VPC. The solid red line represents the median observed data, and the semitransparent red field represents a simulation-based 95% confidence interval for the median. The observed 5% and 95% percentiles are presented with dashed red lines, and the 95% confidence intervals for the corresponding model predicted percentiles are shown as semitransparent blue fields. (For interpretation of the references to color in this figure legend, the reader is referred to the web version of this book.)

The ideal condition is when the 50th, 5th, and 95th quantiles from simulated data superimpose to the corresponding 50th, 5th, and 95th quantiles from observed data. The choice of percentiles for the interval estimates, and the number of simulation replicates are the two factors that can impact the reliability of the VPC plot. A common choice is to perform 1000 simulations but, when a model is very complex or greater precision is desired, then a higher number of simulations may be required. The choice of which quantiles of observed and simulated data to present depends on the number of observed data per sampling time, or, more generally, per binning interval. Thus, if typically 50th, 5th, and 95th quantiles are used, a lower quantile range, such as 50th, 10th, and 90th, may be considered in case the number of observations per bin is small [21]. The VPC is a very powerful tool to assess the performance of population mixed-effects models, and its widespread use is due to the fact that it is simple to interpret even by those without a strong mathematical background and it is easily usable for evaluation of model misspecification and the presence of outliers.

7.4.6.6   Shrinkage

The η- and ε-shrinkage inspection is very useful for investigating the quality of the individual parameter estimates and potential presence of overfitting. In formulas they are respectively defined as:

$$\eta_{\text{sh}} = 1 - \frac{\text{sd}(\eta)}{\omega} \tag{7.30}$$

$$\varepsilon_{\text{sh}} = 1 - \text{sd}(\text{IWRES}) \tag{7.31}$$

where sd indicates standard deviation, ω is the square root value of the Ω matrix diagonal element corresponding to η, and the IWRES are the individual weighted residuals obtained from the difference between the observed data and the individual predicted data divided by the standard deviation of the residual variability. A shrinkage value of zero for both the η- and ε-shrinkage, i.e., no shrinkage, means that the model is reliable and there is enough information in the data to obtain the "true" individual parameter estimates. A shrinkage value of 1 for both the η- and ε-shrinkage, means that the model is misspecified and there is not enough individual information (i.e., few samples per individual), to retrieve the "true" individual parameter estimates and, consequently, they collapse to the population mean. A threshold level of $20-30\%$ is commonly accepted for both η- and ε-shrinkage [22].

Thus, if η-shrinkage is higher than the threshold value, the individual estimates as well as the diagnostic tools listed in Section 7.4.6.4 are not reliable. On the contrary, when η-shrinkage range is between 0 (ideal case) and the threshold value, the individual parameter estimates tend toward the "true" individual parameter or, in other words, the individual estimates are reliable, as well as the diagnostic tools based on them.

A different way to investigate the reliability of the individual estimates is to derive the SE of the η values. A possible method is to derive this information from the inverse of the Fisher information matrix, assuming that the population estimates are the true values (Eq. (7.26)).

Regarding ε-shrinkage, if its value is higher than the threshold, the IWRES distribution exhibits a standard deviation collapsing to zero, and, consequently, the model is excessively parameterized, having too many parameters relative to the number of observations (overfitting). On the contrary, when the ε-shrinkage range is between 0 (ideal case) and the threshold value, we are in the presence of informative data, and the standard deviation of the IWRES approach the ideal value of 1. Thus, in case of model misspecifcation, the presence of significant ε-shrinkage also means that the diagnostic visual tools involving the individual prediction and IWRES/IRES are not reliable.

There is a relation between η- and ε-shrinkage: when there is η-shrinkage, ε-shrinkage is also present. However, as described in Ref. [22], while η-shrinkage may appear substantial even when observations per subject are numerous, ε-shrinkage remains low in these circumstances. Thus, ε-shrinkage is strongly related to the number of measurements compared to the number of parameters of the nonlinear mixed-effects model, resulting in a higher or lower flexibility to describe the observed values.

7.4.7 Model Selection for Nonlinear Mixed-Effects Models

When different model structures are tested on the same data set, the choice of the models can be aided by examining their parsimony. In other words, if the models are equally acceptable in terms of their features, and are also plausible, then the preferred model would be that which had the smaller number of parameters, i.e., the most parsimonious. There are a number of information criteria in the literature. Among them, the Akaike information criterion (AIC) or Bayesian information criterion (BIC) can also be used for nonlinear mixed-effects model selection [23]. Thus, marginal AIC can be calculated as:

$$m\text{AIC} = -2 \cdot \log f(\hat{\theta}) + (p + q) \tag{7.32}$$

with $f(\cdot)$ the (non-log) objective function, p + q the total number of estimated population parameters (i.e., fixed effects and RUV description) or

$$m\text{AIC} = -2 \cdot \log f(\hat{\theta}) + \frac{n - p}{n - p - q - 1}(p + q) \tag{7.33}$$

which has been suggested to adjust [24] for the sample size n. The simplest and most widely used BIC is obtained from the marginal AIC by replacing the constant to obtain:

$$\text{BIC} = -2 \cdot \log f(\hat{\theta}) + \log(n)(p + q) \tag{7.34}$$

Other formulations are also possible. Often a different approach, based on the evaluation of the objective function value, is also used, in particular, but not exclusively, when a decision must be made related to inclusion of covariate relationships in a model [25]. More precisely, the likelihood ratio test is based on the difference in minimum objective function value (O) between models, with and without the covariate relationship. The difference in O between two models is reformulated so as to be compared to an χ^2-distribution with degrees of freedom equal to the number of additional/differing parameters. In this way, one can determine if the difference in O is actually statistically significant, so as to decide whether inclusion of one or more new parameters is justified. Thus, for instance, to achieve a significance level of $p = 0.05$, the addition of one new parameter (1 degree of freedom) to the model has to result in a decrease of the objective function ($O_1 - O_2$) greater than 3.84 points or 6.63 for a significance level of $p = 0.01$. The addition of two new parameters (2 degrees of freedom) to the model has to result in a decrease of the objective function greater than 5.99 points to achieve a significance level of $p = 0.05$. Even though frequently used, the likelihood ratio can induce a wrong model selection, since it can be affected by the use of model linearization techniques. A suggested way to reduce the possibility to detect false covariate relationships is to use a significance level corresponding to a very low p value, often 0.001 [25].

7.5   Covariate Models in Nonlinear Mixed-Effects Models

Variability in biological systems is a widespread trait and poses an extremely challenging problem when trying to inspect its causes. A method to address this issue is to identify whether some independently assessed characteristics of the subjects significantly correlate with the model parameter values. These features, normally referred to in population analysis as covariates, can be integrated into the population model itself to improve its predictive power and, consequently, to explain part of the biological variability. The coefficients driving the relationships between the individual parameter values and the covariates can, in fact, be introduced in the model as additional parameters and therefore optimized together with the remaining population fixed effects. In this way, a part of the population variability, previously accounted for by the individual random effects, is explained in a deterministic fashion.

One of the most used methods for investigating possible inclusion of covariates in a mixed-effects model is stepwise regression. Stepwise regression can be performed with forward selection, backward elimination, or combined modality. In the forward version, firstly, a linear regression analysis of all specified covariate–parameter relationships is made. Subsequently, one has to add the most significant covariate to the model, numerically identify the model parameters, and repeat univariate covariate–parameter relationships analysis with the remaining covariates. This procedure is repeated until no more significant covariates are left. In the backward elimination setup, one starts with the final model in the forward inclusion step and removes covariates one at a time in a stepwise manner. A model selection criterion like the likelihood ratio test (typically with $p < 0.01$ or $p < 0.001$) can be used. Thus, one has to remove the covariate that has the smallest increase in the likelihood ratio test and to continue until all remaining covariates are significant. In the combined modality, one combines the forward selection, i.e., variables are progressively incorporated into larger and larger subsets, with backward elimination, i. e., least promising variables are progressively eliminated. More precisely, one starts with no covariates in the model and adds the best single covariate. After one adds the second and the third covariates, a removing step is accomplished if a statistically worse (based on likelihood ratio test) result is produced before proceeding with an alternation of forward and backward steps.

An additional method is the generalized additive model (GAM) described by Mandema et al. [26]. The GAM can also test nonlinear relationships between the covariate and the parameters through the use of a linear piecewise function. The GAM is built using a stepwise forward/backward method. Once the individual empirical Bayes estimates are obtained, the relationship between these individual estimates and covariates is modelled using a GAM to allow nonlinear covariate–parameter relationships to be discovered, i.e., instead of a linear combination of covariates αs as $p_{ki} = \alpha_{k0} + \sum_{l=1}^{n} \alpha_{kl} X_{ll}$, a more general relation is used:

$$p_{ki} = \alpha_{k0} + \sum_{l=1}^{n} g_{kl}(X_{ll}) \tag{7.35}$$

where, in the original paper, the functions $g_{kl}(\cdot)$ are suggested to be spline functions. At each step, the model is changed by addition (deletion) of the single covariate that results in the largest decrease in the AIC. In this context, the AIC is proportional to the residual sum of squares from the GAM fit, but adds a penalty, proportional to the number of parameters in the model. The search is stopped when the AIC has reached a minimum value.

A general good practice in using both GAM and multivariate stepwise is to center the covariates (with the exception of categorical covariates such as sex) on their mean. This does not affect the covariate selection process or the minimum of the objective function that is reached, but it allows imputing the values of missing covariates to the mean value for several subjects and makes the interpretation of the value of the fixed effects modelling of each parameter's typical values easier, since they can be construed as changes with respect to the population typical values.

References

[1] Sheiner LB, Ludden TM. Population pharmacokinetics/dynamics. Annu Rev Pharmacol Toxicol 1992;32:185−209.
[2] Toffolo G, De Grandi F, Cobelli C. Estimation of beta cell sensitivity from IVGTT C-peptide data. Knowledge of the kinetics avoids errors in modelling the secretion. Diabetes 1995;44:845−54.

[3] Magni P, Sparacino G, Bellazzi R, Toffolo GM, Cobelli C. Insulin minimal model indexes and secretion: proper handling of uncertainty by a Bayesian approach. Ann Biomed Eng 2004;23:1027−37.

[4] Davidian M, Giltinan DM. Nonlinear models for repeated measurement data. New York, NY: Chapman and Hall; 1995.

[5] Vonesh EF. Nonlinear models for the analysis of longitudinal data. Stat Med 1992;11:1929−54.

[6] Wolfinger R. Laplace's approximation for nonlinear mixed models. Biometrika 1993;80:791−5.

[7] Kuhn E, Lavielle M. Coupling a stochastic approximation version of EM with a MCMC procedure. ESAIM P&S 2004;8:115−31.

[8] Lavielle M, Mentré F. Estimation of population pharmacokinetic parameters of saquinavir in HIV patients and covariate analysis with the SAEM algorithm implemented in MONOLIX. J Pharmacokinet Pharmacodyn 2007;34(2):229−49.

[9] Wakefield J, Smith A, Racine-Poon A, Gelfand A. Bayesian analysis of linear and non-linear population models by using the Gibbs sampler. Appl Stat 1994;43:201−21.

[10] Wakefield J. The Bayesian analysis of population pharmacokinetic models. J Am Stat Assoc 1996;91:62−75.

[11] Lunn DJ, Best N, Thomas A, Wakefield J, Spiegelhalter D. Bayesian analysis of population PK/PD models: general concepts and software. J Pharmacokinet Pharmacodyn 2002;29:271−307.

[12] Huber PJ. The behavior of maximum likelihood estimates under nonstandard conditions. Proceedings of the fifth Berkeley symposium on mathematical statistics and probability, 1. Berkeley, CA: University of California Press; 1967 [221−233].

[13] White H. Maximum likelihood estimation of misspecied models. Econometrica 1982;50:1−25.

[14] Kauermann G, Carroll RJ. A note on the efficiency of sandwich covariance matrix estimation. J Am Stat Assoc 2001;96:1387−96.

[15] Beal SL, Sheiner LB. NONMEM users guide—Part II: users supplemental guide. San Francisco, CA: NONMEM Project Group: University of California; 1988.

[16] Baverel PG, Savic RM, Karlsson MO. Two bootstrapping routines for obtaining imprecision estimates for nonparametric parameter distributions in nonlinear mixed effects models. J Pharmacokinet Pharmacodyn 2011;38:63−82.

[17] Das S, Krishen A. Some bootstrap methods in nonlinear mixed-effects models. J Stat Plan Inference 1999;75:237−45.

[18] Parke J, Holford NH, Charles BG. A procedure for generating bootstrap samples for the validation of nonlinear mixed-effects population models. Comp Meth Prog Biomed 1999;59:19−29.

[19] Chernick MR. Bootstrap methods: a guide for practitioners and researchers. New York, NY: John Wiley & Sons; 2008.

[20] Kang D, Bae KS, Houk BE, Savic RM, Karlsson MO. Standard error of empirical Bayes estimate in NONMEM® VI. Korean J Physiol Pharmacol 2012;16:97−106.

[21] Bergstrand M, Hooker AC, Wallin JE, Karlsson MO. Prediction-corrected visual predictive checks for diagnosing nonlinear mixed-effects models. AAPS J 2011;13:143−51.

[22] Savic RM, Karlsson MO. Importance of shrinkage in empirical Bayes estimates for diagnostics: problems and solutions. AAPS J 2009;11:558−69.

[23] Muller S, Scealy JL, Welsh AH. Model selection in linear mixed models. Stat Sci 2013;28:135−67.

[24] Sugiura N. Further analysis of the data by Akaike's information criteria and the finite corrections. Commun Stat A 1978;7:13−26.

[25] Wählby U, Jonsson EN, Karlsson MO. Assessment of actual significance levels for covariate effects in NONMEM. J Pharmacokinet Pharmacodyn 2001;28:231−52.

[26] Mandema JW, Verotta D, Sheiner LB. Building population pharmacokinetic-pharmacodynamic models. I. Models for covariate effects. J Pharmacokinet Biopharm 1992;20:511−28.

8 Systems Biology

Ruoting Yang[a,b], Maria Rodriguez-Fernandez[a,c], Peter C. St. John[c], Francis J Doyle, III[a,c]

[a]Institute for Collaborative Biotechnologies, University of California Santa Barbara, Santa Barbara, CA, [b]Advanced Biomedical Computing Center, SAIC-Frederick National Laboratory, Frederick, MD, [c]Department of Chemical Engineering, University of California Santa Barbara, Santa Barbara, CA

8.1 Introduction

Systems biology aims to understand complex biological systems by integrating high-throughput experimental techniques, data processing, and computational modelling. A major focus of current systems biology is the analysis of biological networks such as gene networks, protein interaction networks, metabolic networks, and signaling networks [1]. Each biological function involves different levels of cellular organization; therefore, individual networks must be integrated into a larger framework. A decade after systems biology was introduced into the language of modern biology, considerable progress has been made in technologies for global cell measurement and in computational analyses of these data to map and model cell function [2]. Computational models are now widely accepted tools for summarizing biological knowledge, guiding the design of new experiments, and serving as test beds for new hypotheses [3]. However, as revealed by the World Technology Evaluation Center (WTEC) survey, the major impediment to further progress of this discipline is the absence of suitable infrastructure for data and software standardization that has yet to be overcome [4].

In Chapter 1, the two fundamental approaches for developing a mathematical model are described; the mechanistic approach (modelling the system) and the data-driven approach (modelling the data). In this chapter, we elaborate on both approaches in the context of biological networks. The use of ordinary differential equations (ODEs) is one of the most common simulation approaches in computational systems biology [5]. Since many biological systems are oscillatory in nature, we focus here on limit cycle models and their application to mammalian circadian rhythms. Moreover, we present an illustrative example of the use of statistical methods for inferring differentially expressed genes (DEGs), pathways, and core modules which can serve as prognostic and diagnostic indicators for posttraumatic stress disorder (PTSD).

Modelling Methodology for Physiology and Medicine. DOI: http://dx.doi.org/10.1016/B978-0-12-411557-6.00008-2

8.2 Modelling the System: ODE Models

The ODE formalism models the concentration of RNAs, proteins, and other species by time-dependent variables governed by a system of rate equations. This type of modelling applies to well-mixed systems in the limit of continuum kinetics (i.e., huge numbers of molecules where reaction rates are proportional to concentration) and relies on the assumption that concentrations vary continuously and deterministically. In contrast to stochastic approaches, which take into account single molecules and their interactions, ODE models typically neglect spatial heterogeneity and randomness in the timing of cellular events.

8.2.1 Model Characterization

A cellular system consists of a network of coupled biochemical reactions, which can consist of: transcription, translation, dimerization, protein or mRNA degradation, enzyme-catalyzed reactions, transport, diffusion, binding or unbinding, DNA or histone methylation, histone acetylation, and phosphorylation [6]. In the following sections, we present basic formulations for individual biochemical reactions that are commonly used to model these processes.

8.2.1.1 Mass Action Law

Molecular networks consist of a series of biochemical reactions, whose kinetics can be described by rate equations according to the mass action law. The mass action law states that the reaction rate is proportional to the probability of a collision of the reactants, which is in turn proportional to the concentration of the reactants. For a reversible reaction of two substrates S_1 and S_2

$$S_1 + S_2 \underset{k_{-1}}{\overset{k_1}{\rightleftharpoons}} P \tag{8.1}$$

the net reaction rate (r) is given by the rate of the forward reaction (r_+) minus the rate of the backward reaction (r_-):

$$r = r_+ - r_- = \frac{d[P]}{dt} = -\frac{d[S_1]}{dt} = -\frac{d[S_2]}{dt} = k_1[S_1][S_2] - k_{-1}[P] \tag{8.2}$$

in which k_1 and k_{-1} are kinetic, or rate, constants.

8.2.1.2 Michaelis–Menten Equation

A commonly used model for enzymatic reactions is the Michaelis–Menten (MM) equation, which approximates the original dynamics under the assumption that the concentration of the enzyme remains constant. The enzyme interacts with the substrate to form an enzyme—substrate complex, which leads to synthesis of the product and the release of the enzyme:

$$E + S \underset{k_{-1}}{\overset{k_1}{\rightleftharpoons}} ES \xrightarrow{\quad k_2 \quad} E + P \tag{8.3}$$

where E, S, and P are enzyme, substrate, and product, respectively. The system of differential equations corresponding to the dynamics of these reactions is

$$\frac{d[S]}{dt} = -k_1[E][S] + k_{-1}[ES] \tag{8.4}$$

$$\frac{d[ES]}{dt} = k_1[E][S] - (k_{-1} + k_2)[ES] \tag{8.5}$$

$$\frac{d[E]}{dt} = -k_1[E][S] + (k_{-1} + k_2)[ES] \tag{8.6}$$

$$\frac{d[P]}{dt} = k_2[ES] \tag{8.7}$$

This ODE system cannot be solved analytically; therefore, some assumptions have been used to simplify the system. The quasi-steady-state assumption for the enzyme—substrate complex ($d[ES]/dt = 0$), under the premise that the conversion of E and S to ES and vice versa is much faster than the decomposition of ES into E and P leads to

$$[ES] = \frac{[E][S]}{K_M} \tag{8.8}$$

with the MM constant

$$K_M = \frac{k_{-1} + k_2}{k_1} \tag{8.9}$$

Combining the quasi-steady-state complex concentration approximation and the conservation law for the enzyme ($[E]_T = [E] + [ES]$, where $[E]_T$ is the total enzyme concentration), results in

$$[ES] = \frac{[E]_T[S]}{K_M + [S]} \tag{8.10}$$

This leads to the well-known MM equation:

$$\frac{d[P]}{dt} = \frac{V_{max}[S]}{K_M + [S]} \tag{8.11}$$

where $V_{max} = k_2[E]_T$ is the maximum reaction rate.

In a molecular network, many cellular processes with no cooperative interactions among molecules (e.g., gene regulation) can be approximated by the MM equation. Gene expression is regulated by transcription factors (TFs) that bind to specific sites in the promoters of the regulated genes. TFs are considered activators if they increase the transcription rate of a gene and repressors if they reduce the transcription rate. Considering the binding of a repressor protein P to an inducer I to form a complex PI:

$$P + I \underset{k_{-1}}{\overset{k_1}{\rightleftharpoons}} PI \tag{8.12}$$

The mass action kinetic equation is

$$\frac{d[PI]}{dt} = k_1[P][I] - k_{-1}[PI] \tag{8.13}$$

At steady state, $d[PI]/dt = 0$, and assuming the conservation of total repressor ($[P_T] = [P] + [PI]$), we arrive at the same MM equation used in the context of enzyme kinetics:

$$[PI] = \frac{[P_T][I]}{K_{eq} + [I]} \tag{8.14}$$

where $K_{eq} = k_{-1}/k_1$ is the dissociation constant.

8.2.1.3   Hill Equation

A more realistic description of inducer binding, able to describe cooperative interactions, is the Hill equation [7]. Many molecules or TFs are composed of several repeated protein subunits and each of them can bind inducer molecules. A repressor P with n active binding sites can be bound by n identical inducer monomers:

$$P + nI \underset{k_{-1}}{\overset{k_1}{\rightleftharpoons}} PI_n \tag{8.15}$$

The total rate of change of the concentration of the complex is given by

$$\frac{d[PI_n]}{dt} = k_1[P][I]^n - k_{-1}[PI_n], \tag{8.16}$$

which usually reaches equilibrium within milliseconds, leading to the steady-state approximation. Assuming conservation of total repressor ($[P_T] = [P] + [PI_n]$), we obtain the binding equation known as the Hill equation:

$$[PI_n] = \frac{P_T[I]^n}{K_{eq}^n + [I]^n} \tag{8.17}$$

where $K_{eq}^n = k_{-1}/k_1$ and n is the Hill coefficient. When $n = 1$, we obtain the MM equation.

8.2.2 Simulation and Parameter Estimation of Oscillating Systems

8.2.2.1 Limit Cycle Models

Many important biological systems contain features that oscillate with time. In studying these systems, it is important that our models capture their oscillatory behavior. For example, in mammalian circadian regulation, gene transcription varies with time of day to meet environmental demands. Other examples of biological processes where oscillations play an important role include embryo development, neuron firing, and cardiac rhythms. Nonlinear ODEs are often used to model these processes, due to their ability to form stable limit cycle solutions—that is, the solution to the equation

$$\frac{dx}{dt} = f(x, p) \tag{8.18}$$

(in which x is the vector of state variables, p is a vector of rate parameters, and f is the vector of model equations) at long periods of time, approaches a periodic trajectory

$$x(t) = x(t + T)$$

in which $T > 0$ is the oscillatory period. The state variables used in ODE modelling typically represent the important time-varying mRNA, protein, and their associated bound and unbound complex species in the considered process. Model equations, which typically contain such mathematical constructs as those described in Section 8.2.1, describe the rates of change of each of the state variables. Kinetic parameters control these rates, typically consisting of rate constants, MM constants, and Hill parameters. We will denote this periodic solution as Γ. Note that since any starting point $x(0)$ close to Γ converges to Γ for long periods of time, the shape of the limit cycle is only a function of the rate parameters in f.

Given a model $f(x, p)$, solving for the time-dependent limit cycle trajectory can typically only be accomplished numerically. The solution $x(t)$ for an arbitrary $x(0)$ can be found using a typical ODE solver. In practice, a limit cycle is often found by integrating the system for long times and checking to see if it reaches a limit cycle solution. However, for precise estimates of the oscillatory period, this method is unsatisfactory. To solve directly for the limit cycle trajectory, a boundary value problem (BVP) formulation is used, where the boundary conditions $x(0) = x(T)$ are enforced. In order to find the correct $[x(0), T]$, such that the boundary conditions are satisfied, the least-squares minimization problem

$$\min_{[x(0),T]} \left[\begin{array}{c} x(0) - x(T) \\ \dot{x}_0(0) \end{array} \right]^2 \tag{8.19}$$

is solved iteratively using Newton minimization. This technique is known as a single shooting method, as in each iteration, the ODE is solved from $t = 0$ to $t = T$ to find suitable values for $\mathbf{x}(T)$. The last entry in Eq. (8.19), $\dot{x}_0(0)$, is the derivative of the first state variable at the initial time. By ensuring this value is set to zero, we specify an exact point on the limit cycle for which to solve, and therefore the minimization problem is identifiable. This type of constraint is known as a phase locking condition. Since any point $\mathbf{x}(t)$ on Γ would satisfy the boundary conditions, this additional constraint forces the target point to be locally identifiable. For numerical tractability, the oscillatory period is often parameterized into the model equations [8]. By defining a characteristic time $\hat{t} = t/T$, the equations become

$$\frac{d\mathbf{x}}{d\hat{t}} = T\mathbf{f}(\mathbf{x}, \mathbf{p})^2 \quad \min_{[x(0),T]} \begin{bmatrix} \mathbf{x}(0) - \mathbf{x}(1) \\ \dot{x}_0(0) \end{bmatrix}^2 \tag{8.20}$$

Many numerical libraries are available for such computations, notably the SUNDIALS suite of nonlinear solvers, which contains the CVODES and KINSOL codes for ODE evaluation and Newton minimization, respectively [9].

8.2.2.2 Optimization Methods

Given a model

$$\frac{d\mathbf{x}}{dt} = \mathbf{f}(\mathbf{x}, \mathbf{p})$$

and cost function $C(\mathbf{p})$ that measures the quality of the model fit, such as χ^2 or the likelihood (L) described in Chapter 5, a common task in modelling biological systems is determining the rate parameters which minimize C:

$$\min_p C(\mathbf{p}) \quad \text{s.t.} \quad \mathbf{p}_{lb} \leq \mathbf{p} \leq \mathbf{p}_{ub}$$

Cost functions of limit cycle models pose a particularly difficult optimization challenge, as a limit cycle Γ does not always exist for any given parameter set $\mathbf{p}$. Thus, the fitness landscape for most parameter space is very flat where $\Gamma(\mathbf{p})$, with isolated regions where limit cycle oscillations are possible. Additional complications arise when constructing Jacobian $dC/d\mathbf{p}$ and Hessian $d^2C/d\mathbf{p}^2$ matrices due to the inherent difficulties in calculating parametric sensitivities for limit cycle systems. For these reasons, stochastic global optimization procedures, which efficiently search high dimensional parameter space without the need for derivative information [10], are often employed. However, analytical bounds on convergence rate—common in deterministic optimization routines—are typically forfeited in stochastic methods.

Evolutionary strategies (sometimes known as genetic algorithms) are one class of methods that have demonstrated success in fitting circadian models and are derived from the biologically inspired process of evolution and mutation. In an evolutionary strategy, a population of many potential solutions is iteratively improved by favoring the proliferation of solutions that have the greatest fitness (or minimal cost) [11]. Evolutionary algorithms are also particularly suited for parallel implementations, as each evaluation routine may be mapped onto an individual processor. Many libraries for such parallel implementations exist, making evolutionary strategies particularly suited for exhaustive minimization of a complex cost function landscape [12].

8.2.3 Sensitivity Analysis for Oscillating Systems

The dynamics of a model are governed by its associated rate parameters. While the constants $\mathbf{p}$ in $\mathbf{f}(\mathbf{x}, \mathbf{p})$ are assumed to not vary with time, it is often the case that external perturbations are manifested by changes to these parameters. A technique known as sensitivity analysis is often used to explore the effects of parameter changes on system dynamics. While it is possible to determine changes using finite difference methods (i.e., resimulating a model under slightly changed parameter values) this approach is often prone to numerical errors. Instead, it is possible to directly differentiate the ODE formulation with respect to its parameter values:

$$\frac{d}{d\mathbf{p}}\left(\frac{d\mathbf{x}}{dt}\right) = \frac{d}{d\mathbf{p}}(\mathbf{f}(\mathbf{x}(\mathbf{p}), \mathbf{p})), \quad \frac{d}{d\mathbf{p}}(\mathbf{x}(0)) = \frac{d}{d\mathbf{p}}(\mathbf{x_0}) \tag{8.21}$$

If we invert the derivative operation in the left-hand side of Eq. (8.21) and take the total derivative of $\mathbf{f}$ with respect to $\mathbf{p}$, we obtain the direct method for solving ODE sensitivities [13]. Here $d\mathbf{x}(t)/d\mathbf{p}$ has been replaced by the matrix symbol $\mathbf{S}(t)$.

$$\frac{d}{dt}\mathbf{S}(t) = \frac{d\mathbf{f}}{d\mathbf{x}}\mathbf{S}(t) + \frac{d\mathbf{f}}{d\mathbf{p}}, \quad \mathbf{S}(0) = 0 \tag{8.22}$$

Here we have made the assumption that the initial conditions of the simulation do not depend on the kinetic parameters. Equation (8.22) can be solved using a standard ODE solver, but performance gains are typically realized by using specialized numerical routines, as the Jacobian matrices are identical to those required in the integration of Eq. (8.18)[14].

8.2.3.1 Period Sensitivity

Many experiments on oscillating systems involve measuring how the period or amplitude of the system changes in response to an increasingly strong external signal. When applied to limit cycle systems, the method described in Eq. (8.22) is often insufficient to explain the complicated parameter dependence of the limit cycle $\Gamma(\mathbf{p})$ (Figure 8.1). Since the initial conditions $\mathbf{x_0}$ found through the method in Eq. (8.20) are parameter dependent, the correct initial conditions for the sensitivities must also be found [8]. Additionally, instead of simply $d\mathbf{x}(t)/d\mathbf{p}$, the

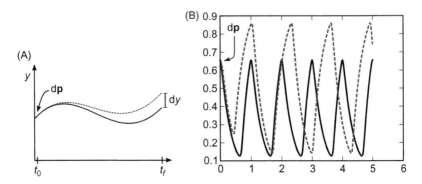

Figure 8.1 Sensitivity metrics for standard (left) and limit cycle (right) models. In limit cycle models, the raw sensitivity solutions are unable to explain characteristic details about the shifted limit cycle.

sensitivities of interest are typically how oscillatory features, such as period or amplitude, change with parameter perturbations. It is possible to calculate these more relevant sensitivities through a decomposition of the time-dependent raw sensitivity matrix [15]. An efficient method for the solution of the period, phase, and amplitude sensitivities is presented in Ref. [8]:

$$\mathbf{S}(t) = \frac{t}{T}\dot{\mathbf{x}}(t)\frac{dT}{d\mathbf{p}} + \mathbf{W}(t) + \dot{\mathbf{x}}(t)\delta(t) \tag{8.23}$$

In Eq. (8.23), $dT/d\mathbf{p}$ is the sensitivity of the period to parameter perturbation, $\mathbf{W}(t)$ is a T-periodic matrix containing information on the sensitivity of the shape, and $\delta(t)$ is a T-periodic matrix containing information on the sensitivity of the phase.

8.2.3.2 Phase Response Curve

Another common experimental measurement with oscillating systems is measuring how the phase of the system changes in response to a *temporary* stimulus signal. Unlike Section 8.2.3.1, where parameter perturbations shift the solution to a new $\Gamma(\mathbf{p}^*)$, temporary parameter perturbations instead advance or delay the system while maintaining an identical limit cycle.

In order to capture the sensitivity of a system to arbitrary temporary perturbations, a mathematical approach known as the velocity response curve (VRC) can be used [16]. In this method, a phase $\phi(\mathbf{x}(t, \mathbf{p}))$ is defined such that $\phi \in [0, T)$ and $d\phi/dt = 1$. To find the sensitivity of the system to an infinitesimal pulse of infinitesimal duration, we must find the derivative [17]:

$$\frac{d}{dt}\frac{d}{d\mathbf{p}}\phi(\mathbf{x}(t, \mathbf{p})) \tag{8.24}$$

The derivative in Eq. (8.24) is most efficiently calculated using an adjoint sensitivity method [14].

8.3 Modelling the Data: Statistical Models

Reductionist approaches have been able to successfully represent biological systems in simplified ODE models, such as models for biological signaling pathways [18] using a few key components in the pathway. The assumptions behind the development of these models usually require homogeneous genotypes and neglect other factors that can also influence the system. However, many diseases, such as cancer, still lack detailed knowledge of the disease mechanisms and the disease development might be the result of a cooperative effort of many genes, proteins, or epigenetic factors. It is therefore practical to collect a large amount of molecular information that is possibly related to the disease. For example, a gene expression microarray quantifies the expression level of a large number of genes simultaneously. In this scenario, statistical methods become an essential tool for analyzing large-scale datasets.

In general, a statistical model is a formalization of relationships between random variables. In contrast to deterministic models, statistical models describe the probability distribution of the dependent variables based on statistical assumptions about the independent variables. The statistical models can not only infer DEGs from a comparative genetic pattern between disease and control biological samples but also highlight differentially expressed signaling pathways, functional clusters, and genetic networks.

8.3.1 Inferential Statistics

Unlike descriptive statistics, which describe the characteristics of a random variable, inferential statistics validate hypotheses about the characteristics of a population based on samples of the population. For example, we can hypothesize that the average height of fifth-grade students is greater than that of fourth-grade students. One may measure the heights of 100 fifth- and fourth-grade students from multiple schools. Here, the 100 students are samples of a much larger population—all fifth-/fourth-grade students. Next, we formulate the following hypothesis test, in which A and B are the average height of all fifth-/fourth-grade students, respectively:

$H_0{:}A = B$ (null hypothesis)
$H_1{:}A \neq B$

We then test whether the null hypothesis H_0 is true: that is, our confidence that A is actually the same as B. Our confidence in the null hypothesis is then represented by the probability of the outcome (called P-value) that A is identical to B. If the P-value is smaller than a preset significant level (commonly 0.05 or 0.01), we say the null hypothesis H_0 is rejected and the difference between A and B is statistically significant (more than 95% chance). Otherwise, the null hypothesis is accepted.

The inference is an estimation based on samples; we have two types of inferential errors: the type I error (the incorrect rejection of a true null hypothesis) and the type II error (the failure to reject a false null hypothesis). All statistics (e.g., Student's t-test and ANOVA) that support the hypothesis test are designed to minimize these errors. It must be noted that an inferential statistic can be applied when its assumptions are satisfied. Incorrect applications can lead to large inferential errors.

8.3.2 Differentially Expressed Genes

Microarray experiments are commonly designed to compare a disease group to a control group. Each group has a number of biological replicates such as mice or human subjects. Gene expression is defined to be the normalized microarray measurement of the messenger RNA concentration. A gene is considered a DEG when the difference between the gene's expression between the two groups is statistically significant.

There are many methods to identify DEGs based on different statistical assumptions. In this chapter, we will introduce the three most commonly used methods, i.e., the Student's t-test, LIMMA (Linear Models for Microarray Data), and the permutation test. A minimum of five samples for each group is necessary for conducting DEG analysis, but using more samples certainly leads to more reliable results.

8.3.2.1 Student's t-Test

The Student's t-test is the most straightforward method of identifying DEGs. Assuming that the gene expression of the disease group, denoted by the variable x_1, has n_1 samples and that of the control group x_2 has n_2 samples, and that both variables follow normal distributions, the t-statistic is calculated to determine whether the average expressions of the two groups are significantly different. In DEG analysis, we will also assume unequal variances of the distributions of the two testing groups. The t-test with this assumption is also called Welch's t-test, which computes the t-score as

$$t = \frac{\bar{x}_1 - \bar{x}_2}{\sqrt{(s_1^2/n_1) + (s_2^2/n_2)}} \tag{8.25}$$

where $\bar{x}$ denotes the mean of a variable and s^2 is the unbiased estimator of the variance of a random variable:

$$s^2 = \frac{\sum_{i=1}^{n} (x_i - \bar{x})^2}{n - 1} \tag{8.26}$$

The P-value is the area under the t-density function from t-score to the nearest tail, with the degrees of freedom given by

$$df = \frac{(s_1^2/n_1 + s_2^2/n_2)^2}{((s_1^2/n_1)^2/(n_1 - 1)) + ((s_2^2/n_2)^2/(n_2 - 1))} \tag{8.27}$$

8.3.2.2 LIMMA (Linear Models for Microarray Data)

LIMMA is an R Bio-conductor package that uses linear models for identifying DEGs [19]. When the gene expression is non-normally distributed, the accuracy of the t-test will be compromised. LIMMA assumes an offset linear model:

$$\begin{bmatrix} x_1 \\ x_2 \end{bmatrix} = \begin{bmatrix} 1 & 1 \\ 1 & 0 \end{bmatrix} \begin{bmatrix} a_0 \\ a_1 \end{bmatrix} + \begin{bmatrix} \omega_0 \\ \omega_1 \end{bmatrix} \tag{8.28}$$

where a_1 represents the offset to the reference basis a_0, and ω_0 and ω_1 are random noises. Thus the hypothesis test is to check whether a_1 is significantly different from 0. A moderated t-statistic is then used to conduct this hypothesis test. Interested readers can find the details of the calculation in Ref. [20].

8.3.2.3 Permutation Test

A permutation test is a nonparametric test that does not assume that the population follows any particular distribution. In contrast to the t-test that assumes normality, the permutation test uses random shuffles of the samples to get the correct distribution. The basic idea is that shuffling the samples will most likely weaken the statistical distinction between two groups. In other words, for a true DEG, the absolute t-statistic of the shuffled samples is often smaller than that of the original samples. Therefore, the simplest permutation test can be formulated by the following four steps:

Step 1. Calculate the t-statistic T_0 based on the original disease and control groups.
Step 2. Randomly switch several disease samples and control samples, and recalculate t-statistic T_1.
Step 3. Repeat step 2 many times resulting in a series of t-statistic T_i.
Step 4. Compute P-value as the ratio of $T_i > T_0$.

It has to be noted that the number of permutations determines the precision of the P-value. With 1000 permutations, the smallest P-value that can be reached is as little as 1/1000. However, since the maximum number of permutations is

$$\binom{n_1 + n_2}{\min(n_1, n_2)} = \frac{(n_1 + n_2)!}{(n_1 + n_2 - \min(n_1, n_2))! \min(n_1, n_2)!},$$

the permutation test yields low precision for data with a small sample size.

After applying the aforementioned DEG analysis methods on the microarray data, one will often get a long list of DEGs. Although the analysis for each gene has controlled type I error to 5%, testing multiple hypotheses simultaneously will increase this type of error to the error of each gene multiplied by the number of tests. This multiple hypothesis testing problem can be corrected by using false discovery rate control algorithms, such as the Bonferroni correction [21], the Benjamini–Hochberg [22], and the Storey approach [23].

8.3.3 Pathway and Gene-Ontology Enrichment

From the DEG analysis, one can identify the significantly changed transcripts. However, genes do not work independently; in fact, genes are highly connected with each other and usually appear coexpressed when they have similar biological functions. The most common way to elucidate the functional clusters is to map the gene lists into pathways and gene-ontology (GO) terms. Pathway datasets are hand-curated databases that cluster the related genes in signaling, regulatory or metabolic functions, while GO terms are expert-curated functional categories. Pathway datasets can be found in several public sources, such as KEGG [24], Reactome [25], Biocarta [26], and many others while GO terms can be downloaded from www. geneontology.org. Enrichment is the method of identifying the significantly changed pathways or GO terms. There are two common ways to conduct pathway and GO enrichment: over-representation analysis (ORA) and gene set enrichment analysis (GSEA) [27].

8.3.3.1 Over-Representation Analysis

ORA computes the probability of having the number of DEGs in a pathway given the total number of DEGs in all genes. Assuming the total number of genes N, DEGs M, and pathway genes J, the probability of having more than K DEGs in a signature follows a hypergeometric distribution (or Fisher exact test):

$$P(\# \text{ of } \mathrm{DEGs} > K) = 1 - \sum_{i=0}^{K} \frac{\binom{J}{i}\binom{N-J}{M-i}}{\binom{N}{M}} \tag{8.29}$$

8.3.3.2 Gene Set Enrichment

In contrast to considering the fraction of the DEGs in a gene set, gene set enrichment summarizes a gene set into an enrichment score (ES) based on the rank of all observed genes:

Step 1. Calculation of an ES. Rank the genes and compute the cumulative sum over the ranked genes; increase the sum when the gene is in the set, decrease it otherwise. The ES is the maximum deviation from zero.

Step 2. Estimation of significance level of ES. Permute the phenotype labels and recompute the ES using the permuted data, so that a null distribution for the ES is generated. The empirical *P*-value of the observed ES is calculated relative to the null distribution.

Step 3. Adjustment for multiple hypothesis testing. Normalize the ES for each gene set based on its size and calculate the false discovery rate corresponding to each normalized ES.

8.3.4 Pathway Inference

Pathway enrichment shows the quantitative importance of a pathway given a DEG list but the *P*-value does not give detailed information about the pathway. For example, assuming that the representative genes in the pathway are unclear, one may want to condense the gene expression information into a pathway expression in order to reduce the feature dimension in classification when identifying the best gene combination for biomarker candidates. In this case, we would conduct pathway inference. Considering a pathway as a large matrix in which the rows are the genes and the columns are the samples, pathway inference aims to collapse all the gene information in a pathway into a row vector called pathway activity (PA). The use of PAs has been proved to improve reproducibility and disease-related functional enrichment for biomarker identification [28,29]. One of the most intuitive ways of creating PAS may be taking the mean or median of the gene expression within the pathway [30]. Whole pathway inference methods include principal component analysis [31] and the log likelihood ratio [32]. The resulting pathway biomarkers usually have better resistance to biological and measurement noise than gene biomarkers. Other studies are more focused on finding the representative genes, including CORG [28] and CMI [29]. In Ref. [29], we compared the classification performance of these six methods as well as individual genes using the statistical framework "COMBINER," which we will discuss in the next section. By cross inference and validation in three breast cancer testing datasets, CORG and CMI performed better than the others, even though they only considered a few genes in the pathway.

The CORG method is a greedy search method that formulates as follows. For a given pathway, we first rank the standardized gene expressions by their *t*-score. If up-regulated genes are dominant, we rank the *t*-score in descending order; otherwise, ascending order is chosen. Next, we aggregate the first two genes using the formula $y_2 = (x_1 + x_2)/\sqrt{2}$; if the expression of this aggregate yields a larger absolute *t*-score than the first gene, this combination is retained as a module with the combined expression becoming the PA. Otherwise, we keep including the third, fourth, ..., *i*th genes by $y_i = \sum_{k=1}^{i} x_k/\sqrt{i}$, until the absolute *t*-score decreases or the pathway limit is reached. The resulting final vector y_i is the PA, and the component genes are called a gene module.

The CMI method is a modification of the CORG method by mixing up-regulated and down-regulated genes. For a pathway, absolute *t*-score is used for ranking the standardized gene expressions in descending order instead of a *t*-score.

Then the PA is the vector with the largest t-score arg max$(t_{score}(P_k))$ in the following candidates:

$$
P_k = \begin{cases} \dfrac{\sum_{k=1}^{i} x_k \, \text{sign}(t_{score}(x_k))}{\sqrt{i}}, & 1 \le k \le \min(|x_k \in \text{DEGs}|, 20), \quad |x_i \in \text{DEGs}| > 0 \\ 0, & |x_i \in \text{DEGs}| = 0 \end{cases}
$$

(8.30)

Here, $|x_i \in \text{DEGs}|$ denotes number of DEGs in the pathway.

8.3.5 Supervised Classification

Supervised classification uses known training data to predict the proper class to which the test data belongs. It has been widely applied to biomarker identification, which tries to find the best gene combination in terms of classification.

For n samples with m features (e.g., gene expressions), the classification problem can be defined as follows. Given a set of features $\{x_1, x_2, \ldots, x_m\}$ with class labels $\{y_1, y_2, \ldots, y_n\} \in \{-1, +1\}$, the task of binary classification is to find a decision function

$$
D(\mathbf{x}) \begin{cases} >0 \Rightarrow \mathbf{x} \in \text{class}(+) \\ <0 \Rightarrow \mathbf{x} \in \text{class}(-) \\ =0 \Rightarrow \mathbf{x} \in \text{decision boundary} \end{cases}
$$

(8.31)

This decision function can be a nonlinear function or a simple separating hyperplane:

$$
D(\mathbf{x}) = \mathbf{w} \cdot \mathbf{x} + b
$$

(8.32)

with $\mathbf{w}$ the weight vector and b the bias value.

There are many classification methods that use differing optimization criteria to estimate the decision function. The linear classification methods include linear discriminant analysis (LDA) (or Fisher's linear discriminant) [33], linear support vector machines (SVMs) [34], naive Bayes classifier [35], nearest shrunken centroid [36], logistic regression [37] and many others; while the nonlinear classification methods include nonlinear SVMs [34], neural networks [38], quadratic discriminant analysis [33], etc.

The classification performance is evaluated by the area under ROC (receiver operating characteristic) curve (AUC) and the error rate. The error rate is the percentage of incorrect predictions in the test data, while the ROC plots the true positive rate versus the false positive rate at various discrimination thresholds.

To reduce variability, the samples must be randomly partitioned into training and test groups. Common cross-validation methods include k-fold, leave-one-out, and leave-k-out cross-validation. k-Fold cross-validation evenly divides the samples into k equal-size groups that contain both classes, assigns $k-1$ groups for training and one for testing, and then rotates to another $k-1$ and 1 training/testing groups. This procedure is repeated k times. The leave-one-out method randomly selects one sample for testing and the rest for training, and rotates for all samples. Similarly, the leave-k-out approach selects k samples for testing. To avoid bias, the cross-validation must be repeated for many random splits to get an average AUC and error rate.

To improve the classification performance, one usually tries to find the best feature combination. Because an exhaustive search for all possible combinations for a large gene set is impractical, removing redundant features backward is much easier to implement. Considering the separating hyperplane (Eq. (8.32)) the weights indicate the importance of the associated features. Guyon et al. proposed recursive feature elimination (RFE), which removes one or multiple features recursively based on their weights [39]. However, classical RFE often lacks stability in feature selection [40] because the weights are totally dependent on the random splits and trying all possible splits is often impossible. Thus, we proposed a consensus feature elimination (CFE) approach to improve the stability of RFE [29]. We first generated 250 groups of 100 alternative 5-fold random splits of samples. Thus, for each group, we constructed 500 classifiers by 500 training and testing cases, that resulted in 500 weights w for each feature and 500 AUCs in testing. The feature was then ranked by the average square weight $\overline{w} = \sum_{j=1}^{500}(w^j)^2/500$. Each group then reported the lowest ranking feature, while the most frequent voting features were removed recursively until the maximum average AUC was achieved.

For some applications, one might need to classify multiple class data, such as multiple types of cancers. Multiple classes classification can be done by one-vs-all strategy [41] that distinguishes one class from all the rest of the classes each time. Prediction is then computed by generalizing all binary classifiers.

8.3.6 COMBINER (Core Module Biomarker Identification with Network Exploration)

In the previous sections, we have described how to identify DEGs and DE pathways from a microarray dataset. Once we have a collection of DEGs and DE pathways for datasets in various conditions (e.g., various tissues, animal strains or human populations), the question that is often raised is: what is the commonality between different conditions? The traditional approach would be to check the overlaps between the DEG lists from different groups. Unfortunately, the common DEGs in three or more datasets are usually very limited due to biological and measurement noise (Figure 8.2). A more biologically motivated approach would be to find the overlapping DE pathways, which often enhance the overlapping rate due to statistical chance. However, an overlapping DE pathway may contain different effective modules that appear differentially expressed (Figure 8.2).

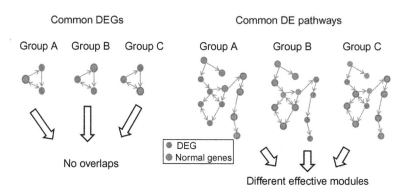

Figure 8.2 The pitfalls of using overlapping DEGs and DE pathways to identify similarity between multiple groups.

COMBINER is a novel modular network biomarker discovery tool that identifies distinct conserved expression modules across multiple conditions [29,42]. The fundamental idea behind COMBINER, depicted in Figure 8.3, is to infer candidate modules, which are the effective modules of pathways, from data of one condition, validate the inferred modules in other conditions using supervised classification, and finally map the resulting modules into the protein−protein interaction (PPI) network. The candidate modules that perform well in classifying samples from multiple conditions are then defined as "core modules," and they are used to construct a modular network. As much experimental and computational work has pointed out, the functional modules are highly conserved across conditions. Therefore the cross-validated module network biomarker is more robust than the putative gene biomarkers extracted from individual datasets. Moreover, the modular network provides a comprehensive interaction map within and between pathways that can be translated to the ODE models discussed in previous sections.

Conventional network analysis approaches utilize an unsupervised approach that constructs a network based on individual gene expression data and identifies functional modules based on network topology [43−48]. For example, the Weighted Gene Coexpression Network Analysis (WCGNA) method [43] presents the high-correlated modules in individual conditions. Our COMBINER approach has adopted a supervised approach that begins with a list of "seed" genes, gradually expands to compact modules in the biological pathways, mutually validates in multiple conditions, and ultimately results in a module network [29,49,50]. There are four major advantages to the COMBINER approach: (i) the resulting modules are compact and thus exclude unrelated downstream signals; (ii) the modules are distinct and well defined with respect to which conditions/tissues/species invoke them; (iii) this method provides multiple robust discriminative biomarkers covalidated in at least two experimental conditions; and (iv) the modular network can be well described by ODE models.

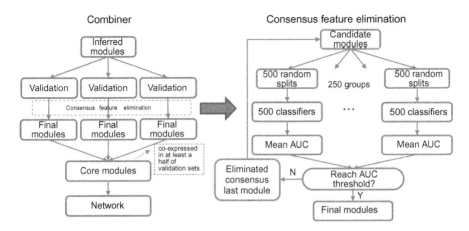

Figure 8.3 Schematic overview of COMBINER. COMBINER first infers candidate modules as activity vectors from each pathway in an inference dataset. It then validates these modules in validation datasets by regenerating activity vectors and performing supervised classification. Finally, the modules present in at least half of the validation sets are considered to be core modules. The resulting core module markers are then projected onto a known PPI network. We generated 250 groups of 500 classifiers in parallel using LDA with RFE. Both the classifier AUC and weight vectors were computed, and each feature was then ranked by its average normalized weight. The most consistently low-ranking feature was then removed recursively until the average AUC threshold was achieved. At this point, the remaining markers were considered to comprise the final modules.

8.4 Applications

8.4.1 Circadian Rhythms

Organisms at all levels of biological complexity demonstrate daily fluctuations in physiology that persist even in the absence of environmental cues. These oscilla- tions, known as circadian rhythms, are an excellent example of a biological limit cycle system. Circadian rhythms in mammals are generated through transcription— translation networks with time-delayed negative feedback, giving rise to sustained transcriptional oscillations. A large percentage of the genome displays some form of circadian rhythm [51], as the molecular clockwork plays a large role in partition- ing competing cellular processes in time.

8.4.1.1 Biological Motivation

Due to the integration of circadian rhythms with key cellular pathways, recent efforts have been made to discover small molecule modulators of circadian rhythms for potential pharmaceutical applications [52,53]. In order to understand how small molecule perturbations result in changes to circadian phenotype, we constructed a model of the core mammalian feedback circuit [54]. In mammalian circadian rhythms, EBOX regulated genes period (*Per1, 2,* and *3*) and Cryptochrome (*Cry1* and *2*) are activated by TFs CLOCK and BMAL1 [55]. PER and CRY protein products, once transcribed, form a heterodimer to reenter the nucleus to repress their transcription. Specifically, our model sought to explain the period-lengthening mechanism of KL001, a small molecule stabilizer of CRY [54].

8.4.1.2   Model Design

The functional roles of the *Cry1* and *Cry2* isoforms in circadian rhythms have long eluded biologists, as although they share a similar structure [56], perturbations to these genes result in opposite period trends [57]. In order to demonstrate how cryptochrome stabilization results in period lengthening, we first developed a mechanistic model to explain the difference between Cry1 and Cry2 period perturbations (Figure 8.4). The differential equations for each state were formulated by using standard Hill-type repression, MM, and mass action kinetics.

A key design component of the mathematical model was the separation of cytoplasmic and nuclear species. Because posttranslational modifications and complex formation are required for the autorepressive genes PER and CRY to enter the nucleus, stabilizing proteins in the cytoplasm or nucleus typically leads to different effects.

8.4.1.3   Parameter Estimation

Parameters were fitted using a genetic algorithm, matching various features of the limit cycle to experimental data. Specifically, protein stoichiometry [58] and key period sensitivities [59] were given the highest weights in the cost function. The model was validated by comparing the simulated dynamics to experimental measurements. First, an unperturbed limit cycle displays reasonable phases and amplitudes, and oscillates with a period of 23.7 h. The knockout periods are 21.7 h for $Cry1^{-/-}$ and 31.5 h for $Cry2^{-/-}$ indicating that the two feedback loops are indeed redundant with different free-running periods.

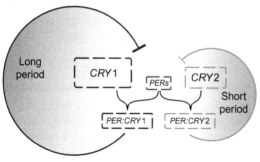

Figure 8.4 Circadian period length is defined by a balance between CRY isoforms. Knockdowns of CRY1 shift the clock to shorter periods and vice versa [54].

8.4.1.4 Model Predictions and Validation

Using the completed model, we investigated possible mechanisms by which KL001 might lengthen the circadian period by analyzing the period sensitivity to different parameters (Figure 8.5). Experimental evidence indicates that stabilization of CRY can either lengthen the period [60] or shorten the period [61]. Assuming equal effect on both isoforms of *Cry*, the model confirms that cytoplasmic stabilization results in period shortening. However, stabilizing nuclear CRY yields the appropriate period lengthening observed experimentally. We therefore predicted that KL001 acts primarily in the nucleus.

After identifying a target mechanism using mathematical modelling, the nuclear stabilization effect was confirmed experimentally by identifying the activity of a nuclear enzyme as the primary target of the small molecule stabilizer [54]. The use of mathematical modelling therefore helped guide future experiments and provided a descriptive mechanism of the activity of a newly identified small molecule. Such studies may lead the way to the development of circadian active pharmaceuticals.

8.4.2 Posttraumatic Stress Disorder

PTSD is an anxiety disorder triggered by exposure to traumatic events, such as war or domestic violence. Typical PTSD symptoms include persistent fear memory (nightmares and flashbacks), emotional numbness, and hypervigilance (often combined with panic attacks). If left untreated, PTSD can be life threatening, as it is often linked to substance abuse and severe depression. PTSD affects about 7.7 million American adults, and members of the military exposed to combat are at high risk for developing PTSD. Currently, PTSD diagnosis depends on psychological evaluation using the DSM IV standard (Diagnostic and Statistical Manual of Mental Disorders) [62].

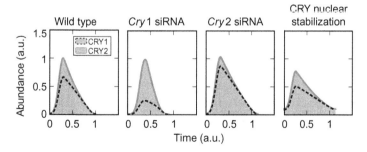

Figure 8.5 Period effects of various clock perturbations. The unperturbed limit cycle (wild type, left) shows oscillations in nuclear protein levels with the slower degrading CRY1 (green) as the majority of the nuclear repressors. Cry1 and Cry2 gene repression (siRNA) show appropriate period lengthening and shortening effects. CRY stabilization in the cytoplasm (CRY2c, second from right) leads to period shortening, while only stabilization in the nucleus (right) matches experimental results. (For interpretation of the references to color in this figure legend, the reader is referred to the web version of this book.)

However, laboratory tests of molecular targets are desirable since they are more accurate and contain less human bias. Unfortunately, PTSD biomarkers and drug targets are currently unavailable for PTSD. In this section, we will illustrate how to use statistical models to identify PTSD biomarkers and drug targets.

8.4.2.1 Biomarkers for PTSD

A biomarker is a measurable and quantifiable biological indicator of a disease, which can be a gene, an mRNA, a protein, any other molecular species, or even an image. Here, we focus on mRNA biomarkers measured by DNA microarray. Although PTSD is a disease that affects the brain, performing brain biopsies and tapping spinal fluid are not practical options. Fortunately, blood has been proven to be a useful surrogate for brain tissue [63]. Specifically, Rollins et al. found over 4100 brain transcripts coexpressed in the blood of healthy human subjects [64]. Recent studies have proposed several candidate brain gene biomarkers that are associated with PTSD [65,66]. Furthermore, it was shown that the mRNA levels of certain transcripts in PTSD patients remain changed with respect to controls even 16 years after the traumatic event [67,68].

Before conducting human studies, animal models have often been used to evaluate human psychological disease; they are amenable to maintaining homogeneous genotypes and collecting brain tissues. We therefore used a mouse model to identify PTSD blood biomarkers and brain drug targets. Recently, Hammemieh et al. [69] used a "social defeat" model that imposes repeated traumatic exposures of mice to a trained aggressor mouse to provoke PTSD-like symptoms. Four different conditions (5-day stress exposure/1 or 10 days recovery, and 10-day stress exposure/1 or 42 days recovery) were chosen to simulate acute and delayed onset PTSD symptoms. After the experiments, blood and tissue samples of seven brain regions were collected, and gene expression levels of these tissues were measured using DNA microarrays [42]. The seven brain regions collected were suspected to be related to PTSD, due to their known roles in fear memory formation, emotion regulation, and decision making—all processes important to the development and pathology of PTSD [66].

We combined the samples from all four conditions, resulting in around 20 treated and 20 control mice for each tissue. After normalizing the microarray data, we computed the DEGs for each tissue using LIMMA (Table 8.1). However, the overlapping DEGs between the tissues are very limited (nearly zero for more than two tissues).

Table 8.1 Numbers of DEGs for each tissue

Tissue	BL	AY	HC	HB	ST	MPFC	SE	VS
Number of DEGs	1612	1789	979	6619	710	700	202	1113

BL, blood; AY, amygdala; HC, hippocampus; HB, hemibrain; ST, stria terminalis; MPFC, medial prefrontal cortex; SE, septal region; and VS, ventral striatum.

8.4.2.2   Core Module Blood Biomarker Network

As mentioned before, blood can be used as a surrogate to the brain for diagnosis of PTSD, because many genes express both in the blood and in the brain. Similarly, in the process of identifying blood biomarkers for PTSD, we would expect the candidate markers also to be expressed in the brain, especially some of the regions related to PTSD mechanisms. As shown in the previous section, the overlapping between the blood and the brain regions is very limited. Thus, we applied the COMBINER to identify the common expression modules active in both blood and multiple brain regions.

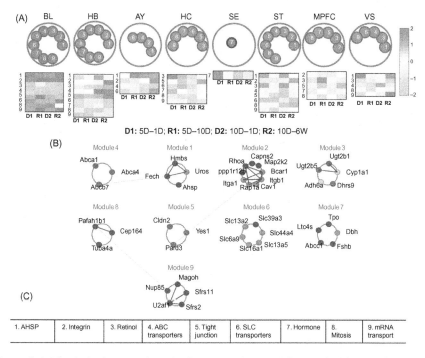

Figure 8.6 Blood–brain network. (A) Nine expression modules resulted from CFE; their brain-specific expression locations are indicated by numbered blue circles. Time-specific blood expression patterns of each module are displayed using average time curves in the form of expression panels. (B) The blood expression level of each gene in the nine modules is indicated with a colored circle. Known PPIs are marked by lines connecting genes—blue lines denote within-module interactions, while gray lines denote between-module interactions. (C) The putative biological functions of the expression modules are listed (as inferred using the KEGG annotation). (5D–1D/10D: 5-day treatment, 1-day/10-day recovery, 10D–1D/6W: 10-day treatment, 10-day/6-week recovery). (For interpretation of the references to color in this figure legend, the reader is referred to the web version of this book.)

We first inferred the blood modules based on biological pathways using the aforementioned CORG method. Starting with the top 100 candidate modules inferred from blood data, we cross-validated the modules using data from each brain region. CFE was used to remove irreproducible modules until the average AUC exceeded 0.75, and the remaining modules were called final modules. If a final module is coexpressed in more than half of the regions, then it was considered a core module, which was mapped into a PPI network.

After this procedure, a total of nine core modules remained. Figure 8.6A presents each module's region-specific brain expression patterns. We used average fold changes to show the time-specific expression pattern of the modules as heat maps in Figure 8.6A. Figure 8.6B further shows the expression of the core modules and the PPIs between their gene products. The color of each gene denotes its expression level in the blood. Blue lines denote known PPIs within modules, while gray lines denote known PPIs between modules. Figure 8.6C lists the putative biological functions of the core modules, i.e., molecular transport, integrin and tight junction function, retinol metabolism, cell cycle, and mRNA transcription. Although initially inferred from blood tissue, most of these processes have been previously implicated in normal and pathological brain function.

As shown in Table 8.2, the gene components of the core modules also exhibit ample evidence of association with blood signatures for PTSD and other PTSD comorbidities such as depression.

8.4.2.3 Core Module Brain Biomarker Network

Another problem of particular interest is to find drug targets for PTSD. Since PTSD is a disease occurring in the brain, the drug targets should be expressed in the brain, but not necessarily in the blood. Thus, we tried to find the common expression modules between brain regions. We inferred 100 modules for each tissue, and validated them using the other tissues data. In total, 37 core modules with 177 genes were identified in the brain—brain network. Figure 8.7A displays the tissue and

Table 8.2 Association between the gene components of the core modules and other related diseases

Genes	Related diseases	References
Abca4, Fech, Magoh, Ppp1r12b, Uros	Human PTSD blood signature	[68]
Ahsp, Dhrs9, Map2k2, Slc13a2, Slc16a1, Slc39a3, U2af1	Human depression blood signature	[70,71]
Hmbs, Pafah1b1, Sfrs2, Yes1	Human bipolar blood signature	[72]
Ugt2b5, Slc6a9	Human blood signature for septic shock	[73]
Dbh, Itgb1, Ltc4s, Rhoa	Mild traumatic brain injury	[74]
Others	Schizophrenia, Alzheimer's disease, sleep disorder	

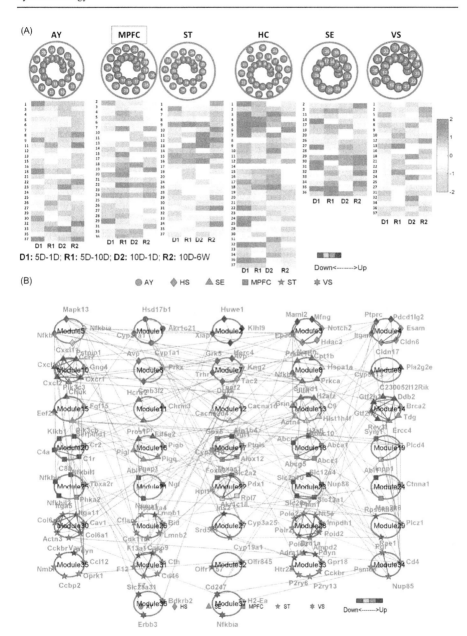

Figure 8.7 Brain—brain network. (A) Application of COMBINER to brain data yields 37 core modules. The tissue- and time-specific expression patterns of each module are presented in the same manner as before. (B) The expression levels and known PPIs of the core module genes are displayed. The shape of a gene represents its inference region, and the color denotes its expression level in that region. Blue lines denote known within-module PPIs, while gray lines denote between-module PPIs. (For interpretation of the references to color in this figure legend, the reader is referred to the web version of this book.)

time-specific expression patterns of each brain–brain core module. Figure 8.7B shows the expression levels of the genes in each module as well as the known PPIs occurring between genes. Unlike the blood–brain network, the shape of the marker used for each gene represents the brain region in which it was inferred.

In the brain–brain core module network, modules 6, 8, 33, and 15 are of particular interest. An active module 6 (*Creb3l2, Prkx, Avp*) in the hippocampus indicates a down-regulated PKA–CREB long-term potentiation pathway, which has been shown to impair memory [75]. In addition, the activity of module 8 (*Prka1b, Hspa1a, Nfkbia, Jun, Cpt1b*) in the septal region shows down regulation of a heat shock protein (HSPA1A). Such activity has previously been found in other PTSD studies [76]. Module 33 depicts an up-regulated dopamine pathway in the ventral striatum. This activity could potentially send excessive dopamine to the amygdala and other brain regions, which has been shown to lead to increased anxiety [77,78]. Finally, module 15 implies an active proinflammatory response in the MPFC that agrees with the study in Ref. [79]. Other validated findings include olfactory impairment in the ST (module 32) [80]; alteration of complement pathways in the MPFC (module 20) [81]; and activated coagulation function in the ST (module 31) [82].

8.5 Conclusions

Biological functions are carried out through interactions of thousands of cellular components (e.g., genes, proteins, and metabolites). Due to the large number of components involved, it is almost impossible to intuitively understand how a molecular network executes various complex cellular functions; therefore, integration of experimental and theoretical research in the form of mathematical modelling is required. This chapter provides an overview of modelling techniques for biological systems. The basic steps for building an ODE model (i.e., structure characterization, model simulation, parameter estimation, and sensitivity analysis) are reviewed; special attention is paid to the case of oscillating systems. The power of these mechanistic approaches to identify target mechanisms that may provide the basis for new therapeutic treatments is illustrated with a model of the core mammalian feedback circuit. Moreover, novel statistical models for inferring DEGs, signaling pathways, functional clusters, and genetic networks are presented. The capabilities of COMBINER for analyzing large-scale datasets and identifying biomarkers and potential drug targets are illustrated with an application to PTSD. Statistical tools such as COMBINER provide a comprehensive interaction map within and between pathways that can be translated to ODE models. Combining statistical and ODE modelling techniques can greatly enhance the predictive capacity.

Acknowledgments

We thank Tsuyoshi Hirota and Steve Kay from UCSD for their collaboration with the circadian rhythms studies and Marti Jett, Rasha Hammamieh, Seid Y Muhie from USACEHR, Uma Mudunuri, Sudhir Chowbina, and Raina Kumar from SAIC-Frederick National Laboratory, and Bernie J. Daigle Jr. and Linda Petzold from UCSB for their collaboration on the PTSD studies. We gratefully acknowledge financial support from U.S. Army Research Office (Grant W911NF-10-2-0111) and NIH (Grant R01GM096873).

References

[1] Klipp E, Herwig R, Kowald A, Wierling C, Lehrach H. Systems biology in practice: concepts, implementation and application. Weinheim: Wiley-VCH Verlag GmbH & Co. KGaA; 2008.

[2] Chuang H-Y, Hofree M, Ideker T. A decade of systems biology. Annu Rev Cell Dev Biol 2010;26:721.

[3] Wolkenhauer O, Mesarovic M. Feedback dynamics and cell function: why systems biology is called systems biology. Mol Biosyst 2005;1(1):14−6.

[4] Cassman M, Arkin A, Katagiri F, Lauffenburger D, Doyle III FJ, Stokes CL. Barriers to progress in systems biology. Nature 2005;438(7071):1079.

[5] Kitano H. Computational systems biology. Nature 2002;420(6912):206−10.

[6] Chen L. Modeling biomolecular networks in cells: structures and dynamics. London: Springer; 2010.

[7] Alon U. An introduction to systems biology: design principles of biological circuits. Boca Raton, FL: Chapman & Hall/CRC; 2007.

[8] Wilkins AK, Tidor B, White J, Barton PI. Sensitivity analysis for oscillating dynamical systems. SIAM J Sci Comput 2009;31(4):2706−32.

[9] Hindmarsh AC, Brown PN, Grant KE, Lee SL, Serban R, Shumaker DE, et al. SUNDIALS: suite of nonlinear and differential/algebraic equation solvers. ACM Trans Math Softw 2005;31(3):363−96.

[10] Holland JH, Reitman JS. Cognitive systems based on adaptive algorithms. ACM SIGART Bull 1977;63:49.

[11] Whitley D. An overview of evolutionary algorithms: practical issues and common pitfalls. Inf Softw Technol 2001;43(14):817−31.

[12] Fortin F-A, De Rainville F-M, Gardner M-A, Parizeau M, Gagne C. DEAP: evolutionary algorithms made easy. J Mach Learn Res 2012;13:2171−5.

[13] Rabitz H, Kramer M, Dacol D. Sensitivity analysis in chemical kinetics. Annu Rev Phys Chem 1983;34(1):419−61.

[14] Serban R, Hindmarsh AC. CVODES, the sensitivity-enabled ODE solver in SUNDIALS. In: Proceedings of the 5th international conference on multibody systems, nonlinear dynamics, and control, Long Beach, CA; 2005.

[15] Kramer MA, Rabitz H, Calo JM. Sensitivity analysis of oscillatory systems. Appl Math Model 1984;8(5):328−40.

[16] Taylor SR, Webb AB, Smith KS, Petzold LR, Doyle FJ. Velocity response curves support the role of continuous entrainment in circadian clocks. J Biol Rhythms 2010;25 (2):138−49.

[17] Taylor SR, Gunawan R, Petzold LR, Doyle III FJ. Sensitivity measures for oscillating systems: application to mammalian circadian gene network. IEEE Trans Autom Control 2008;53:177−88 [Special Issue].

[18] Bhalla US, Iyengar R. Emergent properties of networks of biological signaling pathways. Science 1999;283(5400):381−7.

[19] Smyth G. LIMMA: Linear models for microarray data bioinformatics and computational biology solutions Using R and Bioconductor. In: Gentleman R, Carey VJ, Huber W, Irizarry RA, Dudoit S, editors. New York, NY: Springer; 2005. p. 397−420.

[20] Smyth GK. Linear models and empirical Bayes methods for assessing differential expression in microarray experiments. Stat Appl Genet Mol Biol 2004;3(1):3.

[21] Rice WR. Analyzing tables of statistical tests. Evolution 1989;43(1):223−5.

[22] Benjamini Y, Hochberg Y. Controlling the false discovery rate: a practical and powerful approach to multiple testing. J R Stat Soc Ser B (Method) 1995;289−300.

[23] Storey JD. A direct approach to false discovery rates. J R Stat Soc Ser B (Stat Method) 2002;64(3):479−98.

[24] Kanehisa M, Goto S. KEGG: Kyoto encyclopedia of genes and genomes. Nucleic Acids Res 2000;28(1):27−30.

[25] Joshi-Tope G, Gillespie M, Vastrik I, D'Eustachio P, Schmidt E, de Bono B, et al. Reactome: a knowledgebase of biological pathways. Nucleic Acids Res 2005;33 (Suppl. 1):D428−32.

[26] Nishimura D. BioCarta. Biotech Softw Internet Rep: Comput Softw J Sci 2001;2 (3):117−20.

[27] Subramanian A, Tamayo P, Mootha VK, Mukherjee S, Ebert BL, Gillette MA, et al. Gene set enrichment analysis: a knowledge-based approach for interpreting genome-wide expression profiles. Proc Natl Acad Sci USA 2005;102(43):15545−50.

[28] Lee E, Chuang H-Y, Kim J-W, Ideker T, Lee D. Inferring pathway activity toward precise disease classification. PLoS Comput Biol 2008;4(11):e1000217.

[29] Yang R, Daigle B, Petzold L, Doyle III F. Core module biomarker identification with network exploration for breast cancer metastasis. BMC Bioinformatics 2012; 13(1):12.

[30] Guo Z, Zhang T, Li X, Wang Q, Xu J, Yu H, et al. Towards precise classification of cancers based on robust gene functional expression profiles. BMC Bioinformatics 2005;6(1):58.

[31] Bild AH, Yao G, Chang JT, Wang Q, Potti A, Chasse D, et al. Oncogenic pathway signatures in human cancers as a guide to targeted therapies. Nature 2006;439 (7074):353−7.

[32] Su J, Yoon B-J, Dougherty ER. Accurate and reliable cancer classification based on probabilistic inference of pathway activity. PLoS One 2009;4(12):e8161.

[33] Friedman JH. Regularized discriminant analysis. J Am Stat Assoc 1989;84 (405):165−75.

[34] Vapnik V. Statistical learning theory. New York, NY: Wiley-Interscience; 1998.

[35] Rish I. An empirical study of the naive Bayes classifier. In: IJCAI 2001 workshop on empirical methods in artificial intelligence; 2001.

[36] Tibshirani R, Hastie T, Narasimhan B, Chu G. Diagnosis of multiple cancer types by shrunken centroids of gene expression. Proc Natl Acad Sci USA 2002;99 (10):6567−72.

[37] Hosmer DW, Lemeshow S. Applied logistic regression. New York, NY: Wiley-Interscience; 2004.

[38] Wan EA. Neural network classification: a Bayesian interpretation. IEEE Trans Neural Netw 1990;1(4):303−5.

[39] Guyon I, Weston J, Barnhill S, Vapnik V. Gene selection for cancer classification using support vector machines. Mach Learn 2002;46(1):389−422.

[40] Davis CA, Gerick F, Hintermair V, Friedel CC, Fundel K, Küffner R, et al. Reliable gene signatures for microarray classification: assessment of stability and performance. Bioinformatics 2006;22(19):2356 63.

[41] Rifkin R, Klautau A. In defense of one-vs-all classification. J Mach Learn Res 2004;5:101–41.

[42] Yang R, Daigle JB, Muhie S, Hammamieh R, Jett M, Petzold RL, et al. Core modular blood and brain biomarkers in social defeat mouse model for post-traumatic stress disorder. BMC Syst Biol 2013;7(80).

[43] Zhang B, Horvath S. A general framework for weighted gene co-expression network analysis. Stat Appl Genet Mol Biol 2005;4(1):1128.

[44] Yamaguchi R, Yoshida R, Imoto S, Higuchi T, Miyano S. Finding module-based gene networks with state-space models—mining high-dimensional and short time-course gene expression data. IEEE Signal Proc Mag 2007;24(1):37–46.

[45] Sameith K, Antczak P, Marston E, Turan N, Maier D, Stankovic T, et al. Functional modules integrating essential cellular functions are predictive of the response of leukaemia cells to DNA damage. Bioinformatics 2008;24(22):2602–7.

[46] Keller MP, Choi Y, Wang P, Belt Davis D, Rabaglia ME, Oler AT, et al. A gene expression network model of type 2 diabetes links cell cycle regulation in islets with diabetes susceptibility. Genome Res 2008;18(5):706–16.

[47] De Smet R, Marchal K. Advantages and limitations of current network inference methods. Nat Rev Micro 2010;8(10):717–29.

[48] Xia K, Xue H, Dong D, Zhu S, Wang J, Zhang Q, et al. Identification of the proliferation/differentiation switch in the cellular network of multicellular organisms. PLoS Comput Biol 2006;2(11):e145.

[49] Chuang H-Y, Lee E, Liu Y-T, Lee D, Ideker T. Network-based classification of breast cancer metastasis. Mol Syst Biol 2007;3.

[50] Edelman EJ, Guinney J, Chi J-T, Febbo PG, Mukherjee S. Modeling cancer progression via pathway dependencies. PLoS Comput Biol 2008;4(2):e28.

[51] Pizarro A, Hayer K, Lahens NF, Hogenesch JB. CircaDB: a database of mammalian circadian gene expression profiles. Nucleic Acids Res 2013;41(D1):D1009–13.

[52] Hirota T, Lewis WG, Liu AC, Lee JW, Schultz PG, Kay SA. A chemical biology approach reveals period shortening of the mammalian circadian clock by specific inhibition of GSK-3beta. Proc Natl Acad Sci USA 2008;105(52):20746–51.

[53] Hirota T, Lee JW, Lewis WG, Zhang EE, Breton G, Liu X, et al. High-throughput chemical screen identifies a novel potent modulator of cellular circadian rhythms and reveals CKIα as a clock regulatory kinase. PLoS Biol 2010;8(12):e1000559.

[54] Hirota T, Lee JW, John PCS, Sawa M, Iwaisako K, Noguchi T, et al. Identification of small molecule activators of cryptochrome. Science 2012;337(6098):1094–7.

[55] Ko CH, Takahashi JS. Molecular components of the mammalian circadian clock. Human Mol Genet 2006;15(Suppl. 2):R271–7.

[56] McCarthy EV, Baggs JE, Geskes JM, Hogenesch JB, Green CB. Generation of a novel allelic series of cryptochrome mutants via mutagenesis reveals residues involved in protein–protein interaction and CRY2-specific repression. Mol Cell Biol 2009;29 (20):5465–76.

[57] van der Horst GTJ, Muijtjens M, Kobayashi K, Takano R, Kanno S-I, Takao M, et al. Mammalian Cry1 and Cry2 are essential for maintenance of circadian rhythms. Nature 1999;398(6728):627–30.

[58] Lee C, Etchegaray J-P, Cagampang FRA, Loudon ASI, Reppert SM. Posttranslational mechanisms regulate the mammalian circadian clock. Cell 2001;107(7):855–67.

[59] Zhang EE, Liu AC, Hirota T, Miraglia LJ, Welch G, Pongsawakul PY, et al. A genome-wide RNAi screen for modifiers of the circadian clock in human cells. Cell 2009;139(1):199−210.

[60] Godinho SIH, Maywood ES, Shaw L, Tucci V, Barnard AR, Businò L, et al. The after-hours mutant reveals a role for Fbxl3 in determining mammalian circadian period. Science 2007;316(5826):897−900.

[61] Kurabayashi N, Hirota T, Sakai M, Sanada K, Fukada Y. DYRK1A and glycogen synthase kinase 3beta, a dual-kinase mechanism directing proteasomal degradation of CRY2 for circadian timekeeping. Mol Cell Biol 2010;30(7):1757−68.

[62] APA. Diagnostic and statistical manual of mental disorders. Washington, DC: APA; 2013.

[63] Cai C, Langfelder P, Fuller T, Oldham M, Luo R, van den Berg L, et al. Is human blood a good surrogate for brain tissue in transcriptional studies? BMC Genomics 2010;11(1):589.

[64] Rollins B, Martin MV, Morgan L, Vawter MP. Analysis of whole genome biomarker expression in blood and brain. Am J Med Genet B 2010;153B(4):919−36.

[65] Skelton K, Ressler KJ, Norrholm SD, Jovanovic T, Bradley-Davino B. PTSD and gene variants: new pathways and new thinking. Neuropharmacology 2012;62(2):628−37.

[66] Broekman BFP, Olff M, Boer F. The genetic background to PTSD. Neurosci Biobehav Rev 2007;31(3):348−62.

[67] Zieker J, Zieker D, Jatzko A, Dietzsch J, Nieselt K, Schmitt A, et al. Differential gene expression in peripheral blood of patients suffering from post-traumatic stress disorder. Mol Psychiatr 2007;12(2):116−8.

[68] Segman RH, Shefi N, Goltser-Dubner T, Friedman N, Kaminski N, Shalev AY. Peripheral blood mononuclear cell gene expression profiles identify emergent post-traumatic stress disorder among trauma survivors. Mol Psychiatr 2005;10(5):500−13.

[69] Hammamieh R, Chakraborty N, De Lima TCM, Meyerhoff J, Gautam A, Muhie S, et al. Murine model of repeated exposures to conspecific trained aggressors simulates features of post-traumatic stress disorder. Behav Brain Res 2012;235(1):55−66.

[70] Yi Z, Li Z, Yu S, Yuan C, Hong W, Wang Z, et al. Blood-based gene expression profiles models for classification of subsyndromal symptomatic depression and major depressive disorder. PLoS One 2012;7(2):e31283.

[71] Pajer K, Andrus BM, Gardner W, Lourie A, Strange B, Campo J, et al. Discovery of blood transcriptomic markers for depression in animal models and pilot validation in subjects with early-onset major depression. Transl Psychiatr 2012;2:e101.

[72] Beech RD, Lowthert L, Leffert JJ, Mason PN, Taylor MM, Umlauf S, et al. Increased peripheral blood expression of electron transport chain genes in bipolar depression. Bipolar Disord 2010;12(8):813−24.

[73] Pathan N, Hemingway CA, Alizadeh AA, Stephens AC, Boldrick JC, Oragui EE, et al. Role of interleukin 6 in myocardial dysfunction of meningococcal septic shock. Lancet 2004;363(9404):203−9.

[74] Stevens SL, Leung PY, Vartanian KB, Gopalan B, Yang T, Simon RP, et al. Multiple preconditioning paradigms converge on interferon regulatory factor-dependent signaling to promote tolerance to ischemic brain injury. J Neurosci 2011;31(23):8456−63.

[75] Silva AJ, Kogan JH, Frankland PW, Kida S. Creb and memory. Annu Rev Neurosci 1998;21(1):127−48.

[76] Sriram K, Rodriguez-Fernandez M, Doyle III FJ. A detailed modular analysis of heat-shock protein dynamics under acute and chronic stress and its implication in anxiety disorders. PLoS One 2012;7(8):e42958.

[77] Pezze MA, Feldon J. Mesolimbic dopaminergic pathways in fear conditioning. Prog Neurobiol 2004;74(5):301−20.
[78] Yang R, Sriram K, Doyle III FJ. Control circuitry for fear conditioning associated with post-traumatic stress disorder (PTSD). In: IEEE conference on decision and control (CDC), Atlanta, USA, December 15−17; 2010.
[79] Kubera M, Obuchowicz E, Goehler L, Brzeszcz J, Maes M. In animal models, psychosocial stress-induced (neuro)inflammation, apoptosis and reduced neurogenesis are associated to the onset of depression. Prog Neuropsychopharmacol Biol Psych 2011;35 (3):744−59.
[80] Vasterling JJ, Brailey K, Sutker PB. Olfactory identification in combat-related post-traumatic stress disorder. J Traum Stress 2000;13(2):241−53.
[81] Hovhannisyan L, Mkrtchyan G, Sukiasian S, Boyajyan A. Alterations in the complement cascade in post-traumatic stress disorder. Allergy Asthma Clin Immunol 2010;6 (1):3.
[82] Robicsek O, Makhoul B, Klein E, Brenner B, Sarig G. Hypercoagulation in chronic post-traumatic stress disorder. Isr Med Assoc J 2011;13:548−52.

9 Reverse Engineering of High-Throughput Genomic and Genetic Data

Barbara Di Camillo and Gianna Toffolo

Department of Information Engineering, University of Padova, Padova, Italy

9.1 Introduction

Cellular processes involve millions of molecules playing a coherent role in the exchange of matter, energy, and information, both among themselves and with the environment. These processes are regulated by proteins, whose expression is controlled by a tight network of interactions among genes, proteins, and other molecules. Gene expression and protein ability to interact and exert specific functions might also be influenced by genetic polymorphisms and epigenetic modifications, such as DNA methylation and histone modification, the latter inducing protein binding to specific regions of the DNA, thus regulating or preventing the DNA transcription onto RNA.

Although independent studies on single nucleotide polymorphisms (SNPs), gene and protein expression data and, in part, methylation data have successfully identified a number of significant marker−disease associations, they were able to explain only a fraction of disease heritability and were only partially reproducible in different laboratories and with different patients [1]. One of the reasons for these limitations is that complex pathologies, such as diabetes, cancer, and neurodegenerative disorders, are indeed heterogeneous and multicausal, as a result of the alteration of multiple regulatory pathways and of the interplay between different genes and the environment, rather than imputable to a single dysfunctional gene like monogenic diseases [2].

It is thus of paramount importance to understand this complexity by integrating data on DNA sequences, RNAs, and proteins regulating and controlling gene expression. The term "reverse engineering" indicates the set of methods useful to reconstruct a regulatory network from either dynamic or static multiple stimulus−response experimental data.

Originally, reverse engineering approaches were mainly applied to transcript data since RNA expression data were first available in a quantitative and high-throughput fashion. A number of different methods were proposed to infer

Modelling Methodology for Physiology and Medicine. DOI: http://dx.doi.org/10.1016/B978-0-12-411557-6.00009-4

transcriptional regulatory networks from data collected by microarray technology. Among them were Boolean models, models based on differential equations, Bayesian networks and methods based on pairwise gene expression correlations. However, reverse engineering methods on real or realistically simulated datasets are known to exhibit acceptable performance only if applied in a data rich situation, where the network involves a limited number of genes and their expressions are adequately monitored during multiple experiments, so as to excite different states of the system. Recent applications thus focused on protein data, in particular phosphorylation data, aiming to reconstruct signaling networks, which usually involve some tens of variables, by observing systems that are easy to be perturbed, for example, by using external stimuli such as drug treatments or ligand stimulations. One of the main limitations of these studies arises from the difficulty to quantify protein phosphorylation with acceptable precision, for example, because these measurements depend on protein antibody specificity and are affected by high levels of noise.

In the last few years, the novel paradigm of genome-wide association studies (GWASs) has offered new potentiality to reverse engineering approaches, as a tool to explore the hereditary component of complex multifactorial diseases. A GWAS searches for patterns of genetic variation, in the form of SNPs, between affected and healthy individuals. In some cases, these studies have made available genetic and expression data on the same samples, making it possible to identify the functional consequences of induced and natural genetic variation [3]. In other words, genetic variations can be thought of as a randomized, multifactorial set of perturbations and the gene/protein expression profile of each individual as the system response to a specific set of perturbations. Current systems genetics approaches, known as genetic genomics, try to combine different types of data, such as expression and genetic data, both to improve the performance of reverse engineering application and to get the deepest biological insights.

In this chapter, we will review basic reverse engineering methods referring to transcript data; however, the same concepts can be easily extended to other types of data such as protein data. Finally, we will go into some details of genetic genomics.

9.2   Reverse Engineering Transcriptional Data

Transcription data, either from a microarray or a sequencing experiment, are organized as an $N \times M$ matrix T where T_{ij} gives the expression level of transcript i in sample j. Data can be static or dynamic. Static data consist of steady-state expressions, either collected on cell lines/animals/humans after a specific set of stimuli, or on different classes of subjects not subjected to a particular treatment (e.g., healthy vs. control). Dynamic data consist of samples collected at different times during a perturbation, in general on cell lines, given the ethical and experimental issues related to repeated sampling on animal/human subjects. One important advantage of using dynamic data is the possibility of observing a number of genes and processes changing coordinately in the molecular network; whereas, in the case of static data these processes may be disconnected or loosely correlated. The most informative dynamic experiment has been shown to be a multiple perturbation—response experiment [4]. However, these kinds of experiments are very expensive and time-consuming, whereas reasonable numbers of static data are already available from public databases such as GEO or Array Express [5,6].

In general, the number N of monitored transcripts is of the order of tens of thousands, whereas the number M of samples is of some tens or hundreds. To reduce the number of variables, reverse engineering approaches are usually preceded by a gene selection step to eliminate noninformative profiles. A clustering step is often applied [7,8] to identify similar profiles to be analyzed as a single averaged profile. This step further limits the number of variables and reduces the variability due to noise.

A variety of reverse engineering approaches have been proposed in the last few years to infer gene regulatory networks from gene expression data. The majority of them are quantitative and either are based on pairwise correlation measurements or postulate a model, e.g., Boolean model, model based on differential equations, Bayesian network, to describe the interactions among the biological variables.

9.2.1 Pairwise Methods

Pairwise methods are based on the comparison of pairs of expression profiles, looking for possible cause—effect relationships [9—12]. In general, the analysis pipeline can be summarized in three main steps:

- calculate a score S between each pair of gene to weight the coregulation between them;
- define a confidence threshold on S, corresponding to the desired significance level on a null hypothesis distribution of S (this latter usually obtained by repeated data randomization); the significance level is usually corrected for multiple testing using Bonferroni correction or the false discovery rate [13,14];
- select significant relationships between genes and prune those that are not conditionally independent by the others.

One of the first examples of application of such methods in the literature is given by the relevance networks [9]. In this work, the score S is built on the concepts of entropy (H) and mutual information (MI) [15] to identify regulatory genes able to predict the behavior of the output transcript. Entropy is a measure of information content across different samples, defined as

$$H(T_x) = -\sum_{k=1}^{L} P(T_x \in l_k) \cdot \log_2 P(T_x \in l_k) \tag{9.1}$$

where L is the predefined number of intervals l_k, $k = 1, \ldots, L$, used to discretize gene expression, $P(T_x l_k)$ is the probability that the expression T_x of the transcript x

assumes values within the interval l_k. The entropy value is maximum when T_x is uniformly distributed, that is, $P(T_x l_i)$ is the same in all of the intervals, which corresponds to the state of maximum uncertainty, and diminishes with the bias in gene expression, being 0 if x is expressed in the same interval in all of the samples, which corresponds to a state of non-uncertainty and thus of no information.

The entropy definition can be extended to measure the joint entropy $H(T_x,T_y)$ of two transcripts x and y

$$H(T_x, T_y) = -\sum_{k=1}^{L} \sum_{k'=1}^{L} P(T_x \in l_k, T_y \in l_{k'}) \cdot \log_2 P(T_x \in l_k, T_y \in l_{k'}) \tag{9.2}$$

Finally, mutual information $MI(T_x,T_y)$ is defined as the information of gene x after removing the information not shared with y

$$MI(T_x, T_y) = H(T_x) - H(T_x|T_y) = H(T_x) + H(T_y) - H(T_x, T_y) \tag{9.3}$$

$MI(T_x,T_y)$ is equal to 0 when the joint distribution of T_x and T_y does not give additional information with respect to the distribution of T_x and T_y considered separately, being >0 when the two expression profiles depend on each other. A relationship between T_x and T_y (no direction is specified with this method) is thus identified if the two expression profiles are characterized by a high value of mutual information, that is, exceeding the threshold, identified on randomized data as mentioned above.

More recently, ARACNe (Algorithm for the Reconstruction of Accurate Cellular Networks) was proposed [11,16], in which the definition of mutual information is extended to continuous domain, in order to avoid data discretization:

$$MI(T_x, T_y) = \int_x \int_y f(T_x, T_y) \cdot \log \frac{f(T_x, T_y)}{f(T_x) \cdot f(T_y)} dT_x \, dT_y \tag{9.4}$$

where $f(T_x)$ and $f(T_y)$ are the estimated distributions of variables T_x and T_y, and $f(T_x,T_y)$ is the joint distribution of T_x and T_y [17]. A further innovative feature of the method is an additional pruning step on the inferred network topology, aiming at eliminating those connections between genes x and y that can be explained in terms of their interaction with a third common gene z. The rationale is that, when both x and y interact with z in the gene regulatory network, not only $MI(T_x,T_z)$ and $MI(T_y,T_z)$ assume elevated values but also $MI(T_x,T_y)$ does so, irrespective of x and y being actually coregulated or not (Figure 9.1). The pruning step exploits a well-known property of data transmission theory, the "data processing inequality," stating that if x and y interact with a third gene z but do not interact to each other, then

$$MI(T_x, T_y) < \xi \cdot \max\{MI(T_x, T_y); MI(T_y, T_z)\} \tag{9.5}$$

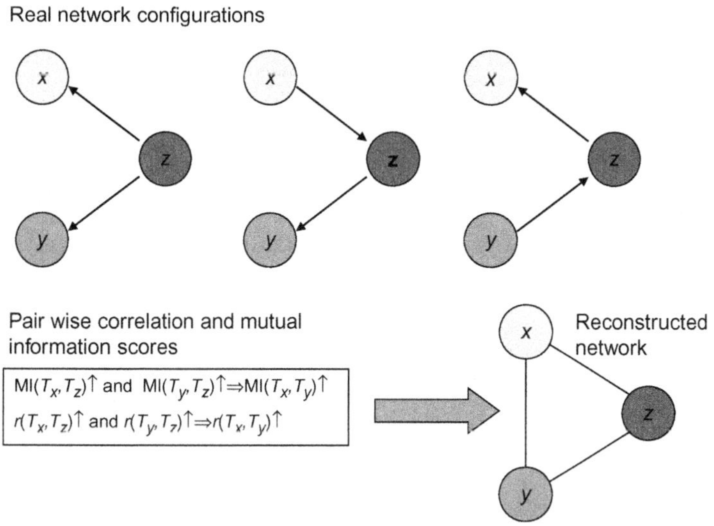

Figure 9.1 If, in the gene regulatory network, both x and y interact with z, but not between them, as shown in the three graphs in the upper panel, then, not only $MI(T_x, T_z)$ and $MI(T_x, T_z)$ but also $MI(T_x, T_y)$ are high, resulting in a false positive edge between x and y, as indicated in the lower panel. The same is true if absolute correlation is used.

where the constant is equal to 1 for noise-free data, whereas a value <1 (and >0) must be set to account for biological and technical variabilities, so as to avoid pruning positive edges. In the ARACNe paper [11], a value equal to 0.85 was estimated to optimally compromise between false positive and false negative pruning of edges.

As an alternative to mutual information, a correlation score is often used. The advantage is that it also provides information on the sign (positive or negative) of the regulation. However, correlation is only able to capture linear relationships between variables. As an example, Schäfer and Strimmer [12] developed an approach based on a graphical Gaussian model to distinguish between directed and undirected interactions based on partial correlation, that is, a measure of correlation between two expression profiles conditioned on the knowledge of all of the other profiles. Partial correlation p_{xy} between two expression profiles T_x and T_y can be expressed from the matrix R of Pearson correlation between all variables, having as elements the correlation r_{xy} between T_x and T_y, since, by denoting as w_{xy} the elements of the inverse matrix $W = R - 1$, the following relation holds

$$p_{xy} = \frac{-w_{xy}}{\sqrt{w_{xx} \cdot w_{yy}}} \qquad (9.6)$$

It is thus possible to reconstruct the pairwise network of interaction between genes based on values of p_{xy} that are significantly different from 0. It is worth noting that if the number of variables is higher than the number of samples, R is not semidefinite positive and thus is not invertible. A pseudo-inverse estimate of R can be used instead [12].

9.2.2   Model-Based Methods

Model-based methods postulate a specific class of models to describe the regulatory network and then select the optimal model configuration and identify model parameters based on experimental data. Methods can be roughly classified into three main groups based on the model used to describe the relationships among variables: Boolean models, models based on differential equations, and Bayesian network models. Since Bayesian networks are described in detail in Chapter 12, we describe here Boolean models and models based on differential equations.

9.2.2.1   Boolean Models

Boolean models require a preliminary quantization of gene expression in two levels. Expression data are continuous in nature, but can be binarized, referring to the state of activation versus inhibition of the promoter site initiating transcription. Different kinds of data such as protein posttranscriptional modification data—such as glycosylation and phosphorylation data—can be easily referred to the state of the protein, e.g., phosphorylated versus unphosphorylated, thus inactive.

Boolean methods aim at identifying, in the space of the logic rules AND−OR−NOT, a rule able to represent gene expression T_x as a logic function of the expression of other genes. For instance, the rule $T_x = T_q$ AND T_y AND T_z means that gene x is expressed if and only if genes q and y are simultaneously expressed and gene z is not expressed. Solving a Boolean model thus implies a search, for each profile T_x, for a rule that explains its state as a logic function of its regulators.

Boolean networks, although simplistic in terms of data representation, are suitable for explaining, at least in part, complex combinations of regulatory actions. In fact, in biological networks, regulators have different possibilities for activating or inhibiting transcription depending on the different combination of regulatory elements that interact in either a competitive or a synergic way. For example, the AND function represents a regulatory effect achieved only if the regulators are simultaneously active. Building on this observation, different functions can be defined based on Boolean fuzzy logic and are able to represent a variety of regulatory interactions and to extend Boolean logic to continuous domain and codomain, since transcription, translation, and gene regulation are intrinsically continuous phenomena in biological networks [18,19].

If dynamic data are available, either an instantaneous or a synchronous regulation model can be adopted: the expression of gene x at time t is modeled in terms of the expression of other genes measured at time t in the former case, at time $t + t$ in the latter. Since instantaneous models do not allow loops in the network, whereas feedback and feed-forward loops are frequent module patterns in biological networks, synchronous models are usually adopted.

A prototype of a Boolean model applied to gene expression data is the algorithm REVEAL (Reverse Engineering ALgorithm) [20], which aims to identify a minimum set of input genes able to predict the behavior of the output gene, building on the concepts of entropy equation (9.1) and mutual information equation (9.3). A pairwise relationship $T_y T_x$, that is, y univocally determines x, is inferred when mutual information between T_x and T_y equals the entropy of T_x:

$$MI(T_x, T_y) = H(T_x) \tag{9.7}$$

Note that, with respect to the pairwise methods considered in the above paragraph, this equality gives information regarding the direction of the relationships between x and y. Equation (9.7) can be extended to describe the relation of gene expression profile T_x with two other profiles T_y and T_z, thus inferring an interaction of order $k = 2$ if $MI(T_x, (T_y, T_z)) = H(T_x)$, analogously for $k = 3$, etc. The REVEAL algorithm searches, for each expression profile T_x, for all of the possible interactions of order $k = 1$, that is, all expression profiles T_y for which Eq. (9.7) is satisfied. If no genes satisfy this condition, the research is extended to $k = 2$ interactions and so forth. In practice, computational cost limits the search to $k = 3$. This also limits the number of false positive regulations, deriving from the low number of available samples as compared to the high number of monitored genes. Once interactions between variables have been identified, Boolean rules can be easily identified from Boolean tables using standard routines.

The most critical step of Boolean model application is the preliminary discretization step, since this produces a simplified data representation and is sensitive to noise. In Di Camillo et al. [21], a quantization strategy based on a model of the experimental error and on a compromise between false positive and false negative classifications (of expression data in 0 and 1) was developed, and is able to optimize downstream reverse engineering performance.

Besides their ability to efficiently describe regulatory interactions, Boolean networks were also used to represent global behavior of large genetic networks and functional differentiation of cells, since attractors—that is, states or sets of states toward which a system evolves over time and that once reached are stable—were interpreted as distinct cell types and states [22,23].

9.2.2.2   Models Based on Differential Equations

Methods based on differential equations are specific to model time series expression data, but with opportune modifications can also be applied to steady-state data. They describe the derivative of gene expression T_x at a generic time t as a function of all other gene expressions and, when present, external inputs. This function can be linear or not and its complexity depends on the model assumed for regulation [24–27]. A simple, yet general, model of regulation assumes the rate of transcription of gene x as a function of a linear combination of the expression of other genes and can be represented, at a generic time t by the following differential equation:

$$\frac{dT_x(t)}{dt} = A_x \cdot f \left(\sum_{j=1}^{N} w_{xj} \cdot T_j(t) + \sum_{u=1}^{U} v_{xu} \cdot c_u(t) + B_x \right) \tag{9.8}$$

where $T_j(t)$ is the observed expression value of gene $j(j = 1, \ldots, N)$ at time t, with N number of genes. A_x is the activation constant for gene x; c_u is the concentration of the external input u ($u = 1, \ldots, U$); B_x is the basal activation level of gene x; w_{xj} and v_{xu} are control parameters, assumed to be time independent and positive, negative, or null depending on the positive, negative, or null control that gene j or input u exert on gene x, respectively. In particular, when $j = x$, w_{xj} represents both the degradation and, if not null, the autoregulation parameters. Finally, f is the activation function, assumed to be linear [24] or sigmoidal [26], depending on model assumptions.

Parameters w_{xj}, v_{xu}, A_x, and B_x are unknown and must be identified from available samples of gene expression collected at different sampling times t_k by optimizing the ability of the model to fit the data. The search space corresponds to the continuous space of system parameters and the function to be optimized (minimized) is usually a global measure of the error between real and predicted temporal profiles. To solve the identification problem, the number of available samples must be at least equal to the number of parameters to be identified for each gene i. This condition is seldom satisfied since the number of samples is usually much lower than the number of analyzed genes. Therefore, even after a gene selection step, the system of model equations (9.8) is usually undetermined and it is necessary to resort to heuristics to solve it. For example, cluster analysis is often applied to diminish the number of profiles to be analyzed, or a nonlinear interpolation followed by a resampling procedure is adopted to augment the number of samples.

Other strategies solve Eq. (9.8) by constraining the weights w_{xj} matrix to be sparse, in line with the scale-free organization observed for metabolic, protein−protein, and transcriptional networks, that is, the probability for each node of having a number of connections K with other nodes follows a power-law distribution:

$$P(k) = \frac{1}{k^\gamma} \tag{9.9}$$

where γ is a parameter characterizing the distribution [28]. This observation suggests that the number of non-null weights w_{ij} for each gene is limited, i.e., $<10-12$. As an example, Yeung et al. [29] used singular value decomposition (SVD) to obtain a set of solutions consistent with data and selected the optimal one based on a criterion of network sparseness. More recently, Sambo et al. [30] separately tackled the discrete component of the problem, i.e., the determination of the biological network topology, and the continuous component of the problem, i.e., the strength of the interactions. This approach allowed both to enforce system sparsity—by globally constraining the number of edges—and to integrate *a priori* information about the structure of the underlying interaction network.

As for Boolean networks, differential equation-based models can be extended to probabilistic models. These models, however, require a higher number of observations and can be considered a special case of Bayesian models.

9.2.3 Performance of Reverse Engineering Algorithms

In order to assess and compare the performance of different reverse engineering algorithms, simulators are helpful to generate *in silico* datasets from a variety of networks of different complexities and observed in a variety of experimental conditions [18,31−33]. Quality of inference depends on the ability of the simulator to resemble the main features of real regulatory networks, in particular to mimic their topology and to describe the regulatory mechanisms.

Different models are available in the literature to represent the topology of biological networks; recently, a model has been developed to simultaneously mimic all of the properties observed in real networks: the power-law distribution of connectivity degree, the independence of the clustering coefficient on the number of genes in the network, and the small-world behavior [34]. To accomplish this goal, the topology is generated as a scale-free network by interconnecting subnetwork structures, which are replicated at different levels of network organization (as in fractals). Regulatory subnetworks are generated by randomly assigning a number of regulators to each gene according to a scale-free structure: the probability for each node of having a number of connections with other nodes follows a power-law distribution. The nodes with the highest number of connections are called hubs. Subnetworks are then connected to each other through nodes randomly selected among the hubs. This strategy is iteratively repeated to generate a network characterized by annidated modules, so as to render the simulated network scale free, but with the clustering coefficient not dependent on the number of nodes in the network.

In regards to the other ingredient of *in silico* networks, e.g., the description of regulatory mechanisms, they are often simulated using differential equations thus providing continuous data, but in general addressing simplistic regulatory logic, e.g., additive or multiplicative effects. In Di Camillo et al. [18], Boolean fuzzy logic, saturative mechanisms, and activation thresholds are integrated with differential equations to generate continuous data, comparable to real data for variety and dynamic complexity.

Reverse engineering methods tested on different simulators result in the ability to correctly infer regulatory interactions among genes, when appropriate perturbation experiments are designed, complying with the algorithm requirements [35,36] and when data are produced with multiple structural perturbations such as gene knockouts [4]. When experiments are not rich enough to observe all of the different states of the system responding to a randomized, multifactorial set of perturbation, the commonly adopted performance measures, such as precision and recall, rarely rise beyond 0.5 in a 0−1 scale. It is, however, important to note that reverse engineering algorithms are superior to classic clustering algorithms for the purpose of finding regulatory interactions among genes [35].

The biologically sound application of reverse engineering algorithms is thus limited in the literature either to small systems or to the few available rich datasets required to obtain good performance. Among them, ARACNe [11] was applied to 340 human B-cell samples under different experimental conditions and was allowed to reconstruct a network of 129,000 interactions among around 6000 genes, with a hierarchical, scale-free organization in which a relatively small number of highly connected genes interacted with most other genes. In particular, the proto-oncogene MYC emerged as one of the largest hubs in the network. Fifty-two percent of the relationships between MYC and other genes were already known in the literature or were successively experimentally validated by the authors.

In Ref. [27], a differential equation model was applied, under steady-state conditions, to a nine-transcript subnetwork of the SOS pathway in *Escherichia coli*. These nine genes were overexpressed in turn and the change in expression relative to unperturbed cells was measured. The algorithm correctly identified the key regulatory connections in the network.

A Boolean approach was developed by Eduati et al. [37] for signaling network modelling and was successfully applied to reconstruct from single-stimulus/inhibitor data a cause—effect network able to predict the protein activity level in multi-stimulus/inhibitor experiments. The method, which was the best performer at the international competition DREAM (Dialogue on Reverse Engineering Assessment and Methods) in 2009, provided reasonable predictions of the IGF1-PI3K and AKT signaling pathway, which is central in cancer studies since it regulates cell proliferation. More recently, the method was integrated with differential equation-based models to extend a given network with links derived from the data using various inference methods [38].

As stated in the introduction, GWAS studies coupled with transcriptomic measurements offer a set of randomized, multifactorial perturbation experiments. The next chapter is dedicated to systems genetics approaches integrating these types of data.

9.3   Reverse Engineering Genetic Genomics Data

Systems genetics refers to the scientific area devoted to the study of complex genetic traits from high-throughput genetic and phenotypic data, obtained using technologies such as gene expression arrays, mass spectrometry, and sequencing. In particular, when polymorphism and expression data are available, the field is referred to as genetic genomics.

Genetic genomics data thus consist of genotype and static gene expression data collected in the same subjects. A dataset is therefore defined by

- an $N \times M$ matrix T where each row represents the expression of a transcript across different subjects and each column represents the expression profile of a subject across different transcripts;

- a $P \times M$ matrix G where each row represents the allelic state of the polymorphism associated with a locus p across different subjects and each column represents the genetic profile of a subject across different loci.

Key assumptions for causal inference are that genetic variations precede phenotypic variation, which provides temporal order, and that mutations in unlinked loci are uncorrelated, which eliminates genetic confounding factors [39]. Under these hypotheses, DNA polymorphisms can directly affect the expression of a gene, for example, because they affect molecular mechanisms involved in transcription, splicing, mRNA decay (cis polymorphisms), or influence the final gene-product affinity with other regulatory complexes that, in turn, regulate other target genes (trans polymorphisms). Usually, cis- and trans-acting polymorphism are distinguished based on their proximity to the target gene. In both cases, the polymorphism leads to an alteration of the expression levels of the downstream genes and can thus be treated as a local perturbation of the gene network. Even without knowing which genetic variant is associated with the active state of the polymorphism and which type of effect, either cis- or trans-, characterizes the polymorphism, it is possible to analyze the alterations in the expression of all genes according to the variations in the genotype of each gene. Usually, the association between the SNP and its target transcripts is performed using standard statistical tests such as ANOVA or its nonparametric variants such as the Kruskal-Wallis test. Classes are defined on the basis of the genotype; for example, in the case of an SNP with two allelic variants A (the most frequent) and a (the less frequent), in diploid organism, as *Homo sapiens*, there are three possible genotypes AA, Aa, and aa. The statistical analysis gives as output a set of regulator-target (polymorphic-transcript) pairs, which constitute the encompassing directed network (EDN), which is called "encompassing" because it contains both directed and undirected effects of the polymorphisms on the targets [40]. In fact, some of the identified cis- and trans polymorphisms might regulate the levels of many different downstream transcripts and it is thus necessary to disentangle the directed (or causal) and undirected (or reactive) relationships between the genotype and the phenotype. For example, given a polymorphism p associated with two target genes x and y, four different causal models are possible between the allelic state G_p and the transcription profiles T_y and T_x (Figure 9.2).

To disentangle the causal and reactive modules, one possibility is to regress T_y on T_x and T_x on T_y and to consider residuals x and y. Computing a t-test for associations between residual x and the SNP, one can test if there is residual variance on T_x that T_y is not sufficient to explain and that is instead explained by G_p. Similarly, computing a t-test for associations between residual y and the SNP, one can test if there is residual variance on T_y that T_x is not sufficient to explain and that is instead explained by G_p. Based on the results of the two t-tests, one can thus choose the best model between independent, causal, and reactive [41]. If the test on x is significant and the one on y is not, the model is causal for the effect of G_p on T_x and reactive for the effect of G_p on T_y; if the test on x is not significant and the one on y is significant, the model is reactive for the effect of G_p on T_x and causal for the effect of G_p on T_y; if the test on x is not significant

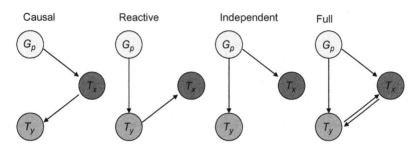

Figure 9.2 Different possible models between genes x and y, whose transcript expression levels, denoted by T_x and T_y, respectively, are statistically associated with the same locus G_p. In the causal model, the transcription level T_x has a causal effect on T_y; in the reactive model T_x is reacting to a causal effect of T_y; in the independent model there is no causal relationship between T_x and T_y; in the full model, there is a causal relationship between T_x and T_y (note that this can be due to an effect of x on y, of y on x, or both).

and the one on y is significant, the model is reactive for the effect of G_p on T_x and causal for the effect of G_p on T_y; if both tests are significant, the model is independent; finally, if both tests are not significant, the model is full. Other pairwise approaches for inferring the causal network are based on the AIC and BIC model selection criteria and on different implementations of the intersection−union test [42−48].

Alternatively, it is possible to solve the problem by applying whole network scoring methods to the EDN [43,49−52]. For example, a simple approach uses regression models to test whether and how a candidate gene target x is affected by the genotype and the gene expression of other genes:

$$T_x = \sum_{p=1}^{P} \alpha_{xp} \cdot G_p + \sum_{j=1}^{M} \beta_{xj} \cdot T_j + \varepsilon_x \qquad (9.10)$$

where α_{xp} are the parameters of the affect of polymorphism G_p on the expression of gene x, x_j are the parameters of the affect of expression profiles T_j on the expression of gene x, and ε_x is a Gaussian residual error term. Given that $M \ll P$, $M \ll N$, and that networks are expected to be sparse, parameters α_{xp} and β_{xj} in Eq. (9.10) can be identified using matrix T and G data using penalized regression such as Lasso regression [53] or elastic nets [54].

While several methods for systems genetics are becoming available, at present not much is known about their strengths and weaknesses. SysGenSIM is a simulator of systems genetics experiments recently developed to generate genetic and kinetic parameters to simulate genotyping, gene expression, and phenotyping data with large gene networks with thousands of nodes [55]. Data produced with SysGenSIM have already been used to assess a number of methods [40,55,56] showing the potentiality of data integration's ability to accurately infer regulatory networks.

Despite the above premises, disentangling the flow of genetic information from genotype to phenotype is still an open issue. Further integration strategies need to

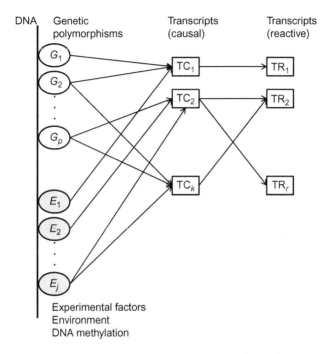

Figure 9.3 Besides genetic polymorphisms, many other variables, such as experimental factors, environment, and DNA methylation, affect transcription. Moreover, some of the affected transcripts might regulate the levels of many different downstream genes, and it is thus necessary to disentangle the directed (or causal) and undirected (or reactive) relationships to reconstruct the regulatory network.

be explored, since many other variables, such as experimental factors, environment, and DNA methylation, affect transcription (Figure 9.3).

Moreover, the *genetic linkage*, that is, the correlation of portions of the genome close to each other, strongly affects the inference of the gene regulatory network. The reason is that the genetic variation is not independent for loci close to each other and thus the correlation between two transcripts targeted by two different but highly correlated polymorphisms is not necessarily to be interpreted as coregulation. Coupling causal inference methods with Bayesian reasoning could be a good strategy to simultaneously consider the effect of multiple genetic and genomic variables plus covariates [41].

9.4 Conclusion

Determining if an alteration in a gene expression is a cause or a consequence of the change in another gene expression and, in turn, if the change in a metabolite level is the cause or the consequence of a disease, is just an example of a burning question in systems biology, which ultimately affects the decision of whether to use the metabolite as a pharmaceutical target or as a prognostic biomarker. In this chapter, reverse engineering methods have been outlined as computational instruments for reconstructing the complex network of interacting DNA sequences, RNAs, and proteins from high-throughput data. As pointed out by international initiatives like the DREAM project (http://www.thedream-project.org/), a comprehensive assessment of different network inference methods on different synthetic and biological network is mandatory [57]. The DREAM initiative as stated by Stolovitzky et al. [58] "is fostering a concerted effort by computational and experimental biologists to understand the limitations and to enhance the strengths of the efforts to reverse engineer cellular networks from high-throughput data." In particular, an important outcome of DREAM has been the observation that no single reverse engineering method performs optimally across different datasets, whereas predictions from multiple inference methods show robust and high performance across diverse datasets [59]. Today, the massive accumulation of high-throughput datasets of different types, plus structural and functional annotations of the genomes, imposes the development of computational frameworks able not only to analyze gene expression profiles *per se* but also to integrate them with other kinds of data and to merge any genomic information. Given these premises, the standard methodologies for the analysis of gene expression profiles should evolve toward approaches that address the integration of high-throughput transcriptional data with a variety of structural and functional information. In this regard, an interesting initiative is Sage Bionetworks (http://sagebase.eu/), which promotes a community-based approach by allowing different scientists to work on the same powerful computational platform where data are stored. This facilitates the solution of the reverse engineering problem by allowing sharing data and methods, otherwise difficult to access and manage.

References

[1] Manolio TA, Collins FS, Cox NJ, Goldstein DB, Hindorff LA, Hunter DJ, et al. Finding the missing heritability of complex diseases. Nature 2009;461:747−53.
[2] Moore JH, Asselbergs FW, Williams SM. Bioinformatics challenges for genome-wide association studies. Bioinformatics 2010;26(4):445−55.
[3] Jansen RC. Studying complex biological systems using multifactorial perturbation. Nat Rev Genet 2003;4:145−51.
[4] Soranzo N, Bianconi G, Altafini C. Comparing association network algorithms for reverse engineering of large-scale gene regulatory networks: synthetic versus real data. Bioinformatics 2007;23:640−7.
[5] Edgar R, Domrachev M, Lash AE. Gene expression Omnibus: NCBI gene expression and hybridization array data repository. Nucleic Acids Res 2002;30(1):207−10.
[6] Rustici G, Kolesnikov N, Brandizi M, Burdett T, Dylag M, Emam I, et al. Functional genomics team, EMBL-EBI, Wellcome trust genome campus. Nucleic Acids Res 2013;41:D987−90.

[7] Bogner-Strauss JG, Prokesch A, Sanchez-Cabo F, Rieder D, Hackl H, Duszka K, et al. Reconstruction of gene association network reveals a transmembrane protein required for adipogenesis and targeted by PPARγ. Cell Mol Life Sci 2010;67 (23):4049–64.

[8] Di Camillo B, Irving BA, Schimke J, Sanavia T, Toffolo G, Cobelli C, et al. Function-based discovery of significant transcriptional temporal patterns in insulin stimulated muscle cells. PLoS One 2012;7(3):e32391.

[9] Butte AJ, Tamayo P, Slonim D, Golub TR, Kohane IS. Discovering functional relationships between RNA expression and chemotherapeutic susceptibility using relevance networks. Proc Natl Acad Sci USA 2000;97(22):12182–6.

[10] Herrero J, Diaz-Uriarte R, Dopazo J. An approach to inferring transcriptional regulation among genes from large-scale expression data. Comp Funct Genom 2003;4:148–54.

[11] Basso K, Margolin AA, Stolovitzky G, Klein U, Dalla-Favera R, Califano A. Reverse engineering of regulatory networks in human B cells. Nat Genet 2005;37(4):382–90.

[12] Schäfer J, Strimmer K. An empirical Bayes approach to inferring large-scale gene association networks. Bioinformatics 2005;21(6):754–64.

[13] Benjamini Y, Hochberg Y. Controlling the false discovery rate: a practical and powerful approach to multiple testing. J R Stat Soc 1995;57(1):289–300.

[14] Storey JD, Tibshirani R. Statistical significance for genome-wide studies. Proc Natl Acad Sci USA 2003;100(16):9440–5.

[15] Shannon CE, Weaver W. The mathematical theory of communication. Chicago, IL: University of Illinois Press; 1963.

[16] Margolin AA, Nemenman I, Basso K, Wiggins C, Stolovitzky G, Dalla Favera R, et al. ARACNe: an algorithm for the reconstruction of gene regulatory networks in a mammalian cellular context. BMC Bioinformatics 2006;7(1):S7.

[17] Steuer R, Kurths J, Daub CO, Weise J, Selbig J. The mutual information: detecting and evaluating dependencies between variables. Bioinformatics 2002;18(2):S231–40.

[18] Di Camillo B, Toffolo G, Cobelli C. A gene network simulator to assess reverse engineering algorithms. Ann N Y Acad Sci 2009;1158:125–42.

[19] Saez-Rodriguez J, Alexopoulos LG, Epperlein J, Samaga R, Lauffenburger DA, Klamt S, et al. Discrete logic modelling as a means to link protein signalling networks with functional analysis of mammalian signal transduction. Mol Syst Biol 2009;5:331.

[20] Liang S, Fuhrman S, Somogyi R. REVEAL, a general reverse engineering algorithm for inference of genetic network architectures. Pac Symp Biocomput 1998;98 (3):18–29.

[21] Di Camillo B, Sanchez-Cabo F, Toffolo G, Nair SK, Trajanoski Z, Cobelli C. A quantization method based on threshold optimization for microarray short time series. BMC Bioinformatics 2005;6(4):S11.

[22] Pal R, Ivanov I, Datta A, Bittner ML, Dougherty ER. Generating Boolean networks with a prescribed attractor structure. Bioinformatics 2005;21:4021–5.

[23] Wuensche A. Genomic regulation modeled as a network with basins of attraction. Pac Symp Biocomput 1998;89–102.

[24] D'haeseleer P, Wen X, Fuhrman S, Somogyi R. Linear modeling of mRNA expression levels during CNS development and injury. Pac Symp Biocomput 1999;4:41–52.

[25] Chen T, Hongyu LH, Church GM. Modelling gene expression with differential equations. Pac Symp Biocomput 1999;4:29–40.

[26] Weaver DC, Workman CT, Stormo GD. Modeling regulatory networks with weight matrices. Pac Symp Biocomput 1999;4:112−23.

[27] Gardner TS, di Bernardo D, Lorenz D, Collins JJ. Inferring genetic networks and identifying compound mode of action via expression profiling. Science 2003;301 (5629):102−5.

[28] Albert R, Barabasi AL. Emergence of scaling in random networks. Science 1999;286:509−12.

[29] Yeung MKS, Tegnér J, Collins J. Reverse engineering gene networks using singular value decomposition and robust regression. Proc Natl Acad Sci USA 2002;99 (9):6163−8.

[30] Sambo F, de Oca MA, Di Camillo B, Toffolo G, Stützle T. MORE: mixed optimization for reverse engineering—an application to modeling biological networks response via sparse systems of nonlinear differential equations. IEEE/ACM Trans Comput Biol Bioinform 2012;9(5):1459−71.

[31] Mendes P, Sha W, Ye K. Artificial gene networks for objective comparison of analysis algorithms. Bioinformatics 2003;19(2):122−9.

[32] Marbach D, Schaffter T, Mattiussi C, Floreano D. Generating realistic *in silico* gene networks for performance assessment of reverse engineering methods. J Comput Biol 2009;16(2):229−39.

[33] Van den Bulcke T, Van Leemput K, Naudts B, van Remortel P, Ma H, Verschoren A, et al. SynTReN: a generator of synthetic gene expression data for design and analysis of structure learning algorithms. BMC Bioinformatics 2006;7:43.

[34] Di Camillo B, Falda M, Toffolo G, Cobelli. C. SimBioNet: a simulator of biological network topology. IEEE/ACM Trans Comput Biol Bioinform 2011;9(2):592−600.

[35] Bansal M, Belcastro V, Ambesi-Impiombato A, di Bernardo D. How to infer gene networks from expression profiles. Mol Syst Biol 2007;3:122.

[36] Corradin A, Di Camillo B, Toffolo G, Cobelli C. *In silico* assessment of four reverse engineering algorithms: role of network complexity and multi-experiment design in network reconstruction and hub detection. In: ENFIN—DREAM conference assessment of computational methods in systems biology, April 28−29, 2008, Madrid.

[37] Eduati F, Corradin A, Di Camillo B, Toffolo GA. Boolean approach to linear prediction for signaling network modeling. PLoS One 2010;5(9).

[38] Eduati F, De Las Rivas J, Di Camillo B, Toffolo G, Saez-Rodriguez J. Integrating literature-constrained and data-driven inference of signalling networks. Bioinformatics 2012;28(18):2311−7.

[39] Neto EC, Broman AT, Keller MP, Attie AD, Zhang B, Zhu J, et al. Modeling causality for pairs of phenotypes in system genetics. Genetics 2013;193(3):1003−13.

[40] Liu B, de la Fuente A, Hoeschele I. Gene network inference via structural equation modeling in genetical genomics experiments. Genetics 2008;178(3):1763−76.

[41] Li Y, Tesson BM, Churchill GA, Jansen RC. Critical reasoning on causal inference in genome-wide linkage and association studies. Trends Genet 2010;26(12):493−8.

[42] Schadt EE, Lamb J, Yang X, Zhu J, Edwards S, Guhathakurta D, et al. An integrative genomics approach to infer causal associations between gene expression and disease. Nat Genet 2005;37(7):710−7.

[43] Li R, Tsaih SW, Shockley K, Stylianou IM, Wergedal J, Paigen B, et al. Structural model analysis of multiple quantitative traits. PLoS Genet 2006;2(7):e114.

[44] Kulp DC, Jagalur M. Causal inference of regulator-target pairs by gene mapping of expression phenotypes. BMC Genomics 2006;7:125.

[45] Chen LS, Emmert-Streib F, Storey JD. Harnessing naturally randomized transcription to infer regulatory relationships among genes. Genome Biol 2007;8:R219.

[46] Aten JE, Fuller TF, Lusis AJ, Horvath S. Using genetic markers to orient the edges in quantitative trait networks: the NEO software. BMC Syst Biol 2008;2:34.

[47] Millstein J, Zhang B, Zhu J, Schadt EE. Disentangling molecular relationships with a causal inference test. BMC Genet 2009;10:23.

[48] Duarte CW, Zeng ZB. High-confidence discovery of genetic network regulators in expression quantitative trait loci data. Genetics 2011;187:955−64.

[49] Zhu J, Wiener MC, Zhang C, Fridman A, Minch E, Lum PY, et al. Increasing the power to detect causal associations by combining genotypic and expression data in segregating populations. PLoS Comput Biol 2007;3(4):e69.

[50] Zhu J, Zhang B, Smith EN, Drees B, Brem RB, Kruglyak L, et al. Integrating large-scale functional genomic data to dissect the complexity of yeast regulatory networks. Nat Genet 2008;40:854−61.

[51] Winrow CJ, Williams DL, Kasarskis A, Millstein J, Laposky AD, Yang HS, et al. Uncovering the genetic landscape for multiple sleep-wake traits. PLoS One 2009;4(4): e5161.

[52] Hageman RS, Leduc MS, Korstanje R, Paigen B, Churchill GAA. Bayesian framework for inference of the genotype-phenotype map for segregating populations. Genetics 2011;187(4):1163−70.

[53] Tibshirani R. Regression shrinkage and selection via the Lasso. J R Stat Soc Ser B 1996;58:267−88.

[54] Zou H, Hastie T. Regularization and variable selection via the elastic net. J R Statist Soc Ser B 2005;67:301−20.

[55] Pinna A, Soranzo N, de la Fuente A. From knockouts to networks: establishing direct cause−effect relationships through graph analysis. PLoS One 2010;5(10):e12912.

[56] Vignes M, Vandel J, Allouche D, Ramadan-Alban N, Cierco-Ayrolles C, Schiex T, et al. Gene regulatory network reconstruction using Bayesian networks, the Dantzig selector, the Lasso and their meta-analysis. PLoS One 2011;6(12):e29165.

[57] Meyer P, Alexopoulos LG, Bonk T, Califano A, Cho CR, de la Fuente A, et al. Verification of systems biology research in the age of collaborative competition. Nat Biotechnol 2011;29(9):811−5.

[58] Stolovitzky G, Monroe D, Califano A. Dialogue on reverse-engineering assessment and methods: the DREAM of high-throughput pathway inference. Ann N Y Acad Sci 2007;1115:1−22.

[59] Marbach D, Costello JC, Küffner R, Vega NM, Prill RJ, Camacho DM, et al. Wisdom of crowds for robust gene network inference. Nat Methods 2012;9(8):796−804.

10 Tracer Experiment Design for Metabolic Fluxes Estimation in Steady and Nonsteady State

Andrea Caumo[a] and Claudio Cobelli[b]

[a]Dipartimento di Scienze Biomediche per la Salute, Università di Milano, Milano, Italy, [b]Department of Information Engineering, University of Padova, Padova, Italy

10.1 Introduction

Understanding the functioning of a metabolic system requires the quantitation of processes that are not directly measurable because they take place in the inaccessible portion of the system. Among these processes, production and utilization of substrates and secretion of hormones are of utmost importance for the investigator. The aim of this chapter is to describe tracer techniques for the quantitation of such fluxes under steady- and nonsteady-state conditions. Although the treatment is fairly general, we will use the glucose system as a prototype.

10.2 Fundamentals

The fundamental concepts underlying the use of tracers to measure production and utilization fluxes of a substrate or a hormone can be better grasped by providing a formal description of the metabolic system under study.

We assume that the metabolic system can be described by a compartmental model having an accessible compartment (usually blood) where the concentration of the substance can be measured (the accessible compartment is denoted by the presence of the dashed line with the bullet) and other inaccessible compartments variously interconnected [1,2]. In Figure 10.1, we can see a compartmental model in which the inaccessible portion is modeled by a five-pool structure. Continuous lines represent fluxes of material from one compartment to another, while dashed lines represent control signals.

Let Figure 10.2 represent the ith compartment of the model, with $Q_i(t)$ denoting the mass of the compartment. The arrows represent fluxes into and out of the

Modelling Methodology for Physiology and Medicine. DOI: http://dx.doi.org/10.1016/B978-0-12-411557-6.00010-0

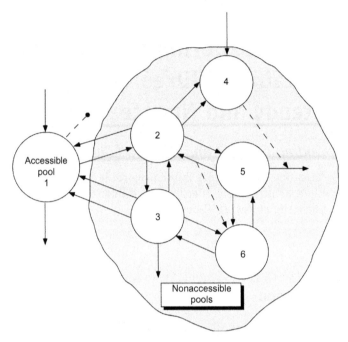

Figure 10.1 A multicompartmental model of a metabolic system. The accessible pool exchanges with the inaccessible portion of the system. Continuous lines represent flux of material and dashed lines represent control actions.

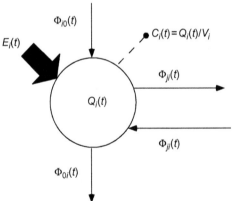

Figure 10.2 The ith compartment of a n-compartment model showing fluxes into and out of the compartment, exogenous inputs, and measurement.

compartment; the input flux into the compartment from outside the system, i.e., the *de novo* synthesis of material, is represented by $\Phi_{i0}(t)$; the flux out of the system (excretion/degradation) by $\Phi_{0i}(t)$; the flux from compartment i to compartment j by $\Phi_{ji}(t)$; finally, $E_i(t)$ denotes the exogenous input. By applying the mass balance principle to each of the n compartments of the system and assuming that compartment 1 represents the accessible pool, we can write:

$$\dot{Q}_i(t) = -\sum_{\substack{j=0 \\ j \neq i}}^{n} \Phi_{ji}(t) + \sum_{\substack{j=0 \\ j \neq i}}^{n} \Phi_{ij}(t) + E_i(t) \quad (i, j = 1, ..., n)$$

$$C(t) = Q_1(t)/V_1$$

$\qquad\qquad\qquad\qquad\qquad\qquad\qquad\qquad\qquad\qquad (10.1)$

where $C(t)$ is the concentration of compound measured in the accessible pool (subscript "1" indicating compartment 1 is omitted) and V_1 is the volume of the accessible pool. Note that, usually, the only input to the system is the one applied to the accessible pool, E_1. In general, the flux of material $\Phi_{ji}(t)$ from compartment i to compartment j is dependent on the mass of the compound, not only in the source compartment i, but also in other compartments of the system:

$$\Phi_{ji}(t) = \Phi_{ji}(Q_1(t), Q_2(t), ..., Q_n(t)) \quad (j = 0, 1, ..., n; I = 1, 2, ..., n; j \neq i)$$

$\qquad\qquad\qquad\qquad\qquad\qquad\qquad\qquad\qquad\qquad (10.2)$

The nature of this functional dependence may be linear or nonlinear and may include threshold/saturation relationships (for instance, Michaelis−Menten kinetics) and control actions (like, for instance, the control exerted by a hormone on the fluxes of a substrate).

It is often more useful to express Eq. (10.1) in terms of the masses of the compound in the compartments. To do so, we render explicit the relationship between the generic flux $\Phi_{ji}(t)$ and the mass in the source compartment, $Q_i(t)$:

$$\Phi_{ji}(t) = k_{ji}(Q_1(t), Q_2(t), ..., Q_n(t)) \cdot Q_i(t) = k_{ji}(t) \cdot Q_i(t) \qquad\qquad (10.3)$$

where k_{ji} (≥ 0) is the fractional transfer coefficient from compartment i to compartment j. If the system is linear and time-invariant, $k_{ji}(t)$ is always constant (i.e., $k_{ji}(t) = k_{ji}$). If the system is linear and time-varying (note that a nonlinear system can be thought of as a linear system with time-varying parameters), we must distinguish between steady- and nonsteady-state conditions. If the system is in steady state, all the fluxes and masses are constant and thus k_{ji} is constant. If the system is in nonsteady state, k_{ji} may vary since saturation and control signals come to play. Using Eq. (10.3), Eq. (10.1) can be written as follows:

$$\dot{Q}_i(t) = -\sum_{\substack{j=0 \\ j \neq i}}^{n} k_{ji}(t)Q_i(t) + \sum_{\substack{j=1 \\ j \neq i}}^{n} k_{ij}(t)Q_j(t) + \Phi_{i0}(t) + E_i(t) \quad (i, j = 1, ..., n)$$

$$C(t) = Q_1(t)/V_1$$

$\qquad\qquad\qquad\qquad\qquad\qquad\qquad\qquad\qquad\qquad (10.4)$

10.3 Accessible Pool and System Fluxes

To put the assessment of production and utilization fluxes on a firm basis, it is useful to make a clear distinction between fluxes pertaining to the accessible pool and
fluxes pertaining to the whole system. Let us begin with the fluxes that refer to the
accessible pool. Ra(t) (rate of appearance) denotes the rate of entry of the compound into the accessible pool (*de novo* entry + exogenous input). Making reference to Eq. (10.1), Ra(t) can be expressed as follows:

$$\text{Ra}(t) = \Phi_{10}(t) + E_1(t) \tag{10.5}$$

Rd(t) (rate of disappearance) is the net outflux of the compound from the accessible pool (resulting from the exchange of material between the accessible and
inaccessible pools). Rd(t) can be expressed in terms of compartmental fluxes as
follows:

$$\text{Rd}(t) = \sum_{\substack{i=0 \\ i \neq 1}}^{n} \Phi_{i1}(t) - \sum_{i=2}^{n} \Phi_{1i}(t) \tag{10.6}$$

Ra(t) and Rd(t) are related by the mass balance equation of the accessible pool:

$$\frac{dQ_1(t)}{dt} = \text{Ra}(t) - \text{Rd}(t) \tag{10.7}$$

As far as the whole system is concerned, $P(t)$ will denote endogenous production
and $U(t)$ will denote whole-body uptake. $P(t)$ equals the sum of all of the *de novo*
fluxes of the compound entering the system:

$$P(t) = \sum_{i=1}^{n} \Phi_{i0}(t) \tag{10.8}$$

$U(t)$ equals the sum of all the fluxes irreversibly leaving the system:

$$U(t) = \sum_{i=1}^{n} k_{0i}(t)Q_i(t) \tag{10.9}$$

$P(t)$ and $U(t)$ are related by the mass balance equation of the whole system:

$$\frac{dQ_T(t)}{dt} = P(t) - U(t) \tag{10.10}$$

where $Q_T = \sum_{i=1}^{n} Q_i(t)$ is the total mass of the compound in the system. In general,
$P(t)$ and $U(t)$ are different from Ra(t) and Rd(t), respectively. If we assume, for the

sake of simplicity, that the accessible pool is the only site of *de novo* entry of the compound, $P(t)$ coincides with Ra(t) (minus any exogenous input of the compound, when present). In contrast, since the compound is usually utilized by tissues in both the accessible and inaccessible pools, $U(t)$ is different from Rd(t), in general. In summary, while Ra(t) and Rd(t) refer to the accessible pool only, $P(t)$ and $U(t)$ refer to the whole system (accessible pool + inaccessible pools).

When the focus of the investigation is the assessment of Ra(t), the development of a compartmental model of the system may turn out to simply not be necessary. For instance, it may happen that one is interested in measuring endogenous glucose production or insulin secretion after a meal, but is not interested in a detailed structural description of glucose or insulin kinetics. The assessment of Ra(t) can be posed as an input-estimation problem, in the sense that one needs to derive the unknown input to the system from its casually related effect on the concentration of the compound in the accessible pool (this issue is discussed in detail in Chapter 3). If the system is linear and time-varying (note, incidentally, that a nonlinear system can be thought of as a linear system with time-varying parameters), the input-estimation problem can be formalized by describing the input−output (I/O) relationship between the unknown input, Ra(t), and the measurable output, $C(t)$, with a Fredholm integral equation of the first kind:

$$C(t) = \int_{-\infty}^{t} h(t, \tau) \mathrm{Ra}(\tau) \mathrm{d}\tau \tag{10.11}$$

where $h(t,\tau)$ describes the I/O behavior the system and is called the kernel of the system. The function $h(t,\tau_0)$ represents the time course of the output when the system is forced by a unitary impulse given at time τ_0. When the system is linear and time-invariant, $h(t,\tau)$ depends only on the difference $t - \tau$, the right-hand side of Eq. (10.11) becomes a convolution integral, and the problem of determining Ra(t) given $h(t)$ and $C(t)$ is called deconvolution. In any case, the assessment of Ra(t) consists in working backward to solve Eq. (10.11), and this requires the preliminary knowledge of the impulse response. How can the impulse response be described? We will distinguish two cases, depending on whether the system is linear and time-invariant or linear and time-varying. If the system is linear and time-invariant, its impulse response can be described by a model of data (no structural, i.e., compartmental model is needed). Usually, a good candidate of $h(t)$ is

$$h(t) = \sum_{i=1}^{n} A_i \, \mathrm{e}^{-\alpha_i t} \tag{10.12}$$

where $\alpha_i > 0$. If the system is linear and time-varying (like the glucose system, for instance), the kernel $h(t,\tau)$ cannot be assessed unless a structural model is formulated that is capable of describing the time-dependency of its parameters.

In summary, in order to quantitate fluxes of appearance/production and disappearance/utilization, one may need to identify a structural model of the system or, at least, determine its impulse response (the two requirements coincide when the system is linear and time-varying). To accomplish these tasks, the ideal probe is a tracer experiment, as we will see in the following section.

10.4 The Tracer Probe

It is easy to realize that the availability of the compound concentration in the accessible pool is not sufficient to quantitate the production and removal processes. For instance, an increased compound plasma concentration might be the consequence of increased production, decreased removal from the circulation, or a combination of both processes. To quantitate these fluxes, we need to perform an I/O experiment in which an adequate data base of dynamic data is generated. The tool of choice is the tracer probe, which is usually a radioactive or a stable isotope. The two tracers differ in some aspects. Radiotracers can be given in negligible amounts (which do not perturb the system) and produce satisfactory signal-to-noise ratios, but can be harmful. Stable-labeled tracers are safe, but are naturally present in the body (natural abundance); in addition, the amount that is given to produce a satisfactory signal-to-noise ratio does not have negligible mass and may perturb the system. This makes the analysis of a stable-isotope tracer experiment more complex than that of a radioactive tracer experiment. However, a kinetic formalism for the analysis of stable-isotope tracer data has been developed, and its link with the radioactive kinetic formalism has been elucidated [3].

An ideal tracer has the following characteristics:

1. it has the same metabolic behavior as the substance being traced (denoted as tracee). This is known as tracer—tracee indistinguishability principle;
2. it is distinguishable from the tracee by the investigator;
3. it does not perturb the system.

Real tracers satisfy such conditions to different extents. Hereafter we will assume, for the sake of simplicity, that the tracer is ideal.

Why do tracers help in enhancing the information that can be gained from an I/O experiment? The reason is that the tracer travels in the system like the tracee, and thus tracer data measured in the accessible pool contain information about the tracee system [4]. By applying the mass conservation law for the tracer to all of the compartments of the model used to describe the behavior of the tracee (Eq. (10.1)), one obtains a system of differential equations:

$$\dot{q}_i(t) = -\sum_{\substack{j=0 \\ j \neq i}}^{n} k_{ji}(t)q_i(t) + \sum_{\substack{j=1 \\ j \neq i}}^{n} k_{ij}(t)q_j(t) + e_i(t) \quad (i,j = 1, ..., n)$$

$$c_1(t) = q_1(t)/V_1$$

(10.13)

Note that we use lower case letters to denote tracer-related variables: $q_i(t)$ is the mass of the tracer in the ith compartment; $e_i(t)$ is the tracer input into the ith compartment (usually the only tracer input to the system is the one applied to the accessible pool, e_1); $c(t)$ is tracer concentration in the accessible pool (subscript "1" indicating compartment 1 is omitted). Note that, thanks to the tracer−tracee indistinguishability principle, the fractional transfer rates of the tracer model are the same as those of the tracee model.

By comparing Eqs. (10.4) and (10.13), one can see that the tracer model has a definite advantage with respect to the tracee model: whereas the endogenous tracee input to the system is unknown (and is often what the investigator wants to determine), the tracer input is known. As a result, a suitably designed tracer experiment allows the investigator to identify the tracer model (using the appropriate parameter estimation techniques). Subsequently, the tracer model can be used, in conjunction with the tracee measurements, to quantitate the tracee model. Analogous considerations apply to the tracer impulse response. In fact, thanks to the tracer−tracee indistinguishability principle, the impulse response of the tracer is the same (apart from the units) as that of the tracee. The tracer impulse response can be determined from tracer I/O data and then used to estimate Ra(t) by working backward to solve Eq. (10.11).

It is useful to point out that, if the tracee is in a constant steady state, all of the tracee fluxes (Φ_{ji}) and masses (Q_i) are constant. As a result, all of the fractional transfer coefficients k_{ji} are constant as well (see Eq. (10.2)). If the k_{ji} is constant, the tracer model described by Eq. (10.13) is linear, irrespective of whether the tracee system is linear or nonlinear. This greatly simplifies tracer data analysis. In particular, the tracer impulse response can be described by a model of the data, such as a multiexponential function. The price to be paid is that the tracer-derived parameters yield a picture of the steady-state operating point, but cannot describe the dynamics of the system in its full nonlinear operation arising, for instance, from saturation kinetics and control signals.

Now that we have a good appreciation of the fundamentals of tracer methodology, we can turn our attention to the issue of tracer experiment design. In the following section, we will discuss the most appropriate tracer administration strategies to measure production and utilization fluxes under steady- and nonsteady-state conditions.

10.5   Estimation of Tracee Fluxes in Steady State

We begin by considering a system which is in steady state with respect to the tracee. In steady state, masses and fluxes in the system are constant. Assuming that there is no exogenous administration of the tracee, the mass balance principle applied to the accessible pool and to the whole system states that:

$$\text{Ra} = \text{Rd} \tag{10.14}$$

$$P = U \tag{10.15}$$

Thus, in steady state, the rate of entry into and exit from the accessible pool (as well as the whole system) is constant. These fluxes are collectively referred to as the turnover rate.

To estimate Ra (and thus the turnover rate) we can exploit the fact that, under steady-state conditions, Eq. (10.11) becomes an algebraic equation (time t in the integral sign goes to infinity), which can be easily solved for the unknown Ra:

$$\text{Ra} = \frac{C}{\int_0^\infty h(\tau)d\tau} \tag{10.16}$$

where C_1 is the steady-state tracee concentration.

In principle, the integral of the impulse response can be estimated from the tracer data generated by any realizable tracer input. In the following, we will examine the three most common formats of tracer administration: the single injection, the constant infusion, and the primed constant infusion (i.e., a combination of the first two).

10.5.1 Single Injection

The single injection technique consists of a tracer bolus rapidly injected into a vein followed by plasma sampling for measurement of the tracer and tracee concentration. This input, at least ideally, is an impulse (making reference to Eq. (10.13): $e_1(t) = d\delta(t)$, where d is the tracer dose), and thus the tracer disappearance curve following the tracer bolus can be interpreted as the impulse response of the system (Figure 10.3A).

Since the system is in steady state, the impulse response of the tracer is that of a linear and time-invariant system and can be described by a sum of decaying exponentials. Thus, the time course of tracer concentration following the tracer injection is given by:

$$c(t) = d \cdot h(t) = d \sum_{i=1}^{n} A_i\, e^{-\alpha_i t} \tag{10.17}$$

where c_1 is tracer concentration in plasma, and A_i, α_i are the coefficients and eigenvalues, respectively, of the multiexponential impulse response. One uses a parameter estimation technique to identify the sum of the exponentials model and employs parsimony criteria, like the Akaike information criterion [1], to select the appropriate number, n, of exponential terms. Once $h(t)$ is known, Ra can be calculated by deconvolution:

$$\text{Ra} = \frac{d \cdot C}{\int_0^\infty c(\tau)d\tau} = \frac{C}{\sum_{i=1}^{n} \frac{A_i}{\alpha_i}} \tag{10.18}$$

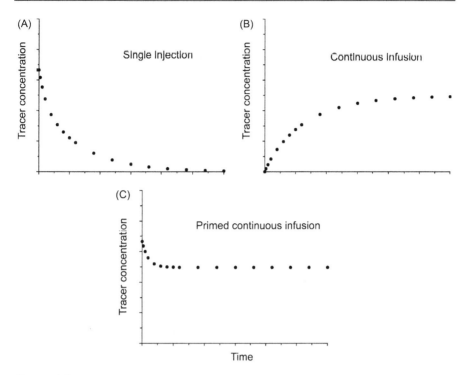

Figure 10.3 Assessment of turnover in steady state: tracer concentration profiles during a bolus injection (A), a constant infusion (B), and a primed continuous infusion (C).

The single injection is the most simple and economical format of tracer administration because it does not require an infusion pump. However, the estimation of turnover from the tracer disappearance curve is not straightforward because it requires a frequent sampling, especially in the initial portion of the study, and a data analysis entailing the use of parameter estimation techniques. In addition, the experiment duration may be rather long since sampling must be continued for at least two to three times the slowest time constant in the system.

10.5.2 Constant Infusion

With the constant infusion technique, the tracer is administered intravenously at a constant rate until the tracer achieves a steady-state level (Figure 10.3B). This input, at least ideally, is a step centered at time 0 (making reference to Eq. (10.13): $e_1(t) = ra\delta_{-1}(t)$, where ra is the constant tracer infusion rate). Since the step is the integral of the impulse, the response to a unit constant infusion will coincide with the integral of the response to a unit dose injection. Thus, by integrating

Eq. (10.17), one obtains the time course of tracer concentration during the constant tracer infusion which is given by:

$$c(t) = \text{ra} \sum_{i=1}^{n} \frac{A_i}{\alpha_i}(1 - e^{-\alpha_i t}) \tag{10.19}$$

Under such conditions of steady state for both the tracer and the tracee, the tracer and tracee concentrations in plasma will be proportional to their respective rates of entry. Calculation of turnover is straightforward since it only requires the measurement of plasma tracer and tracee concentrations at tracer steady state:

$$\text{Ra} = \text{ra}\frac{C}{c} = \frac{\text{ra}}{z} \tag{10.20}$$

where z is the constant tracer-to-tracee ratio [3] (this ratio coincides with the specific activity when a radiotracer is used).

Evaluation of tracee turnover is thus simpler with the constant infusion than with the single injection technique. Another advantage of the constant infusion with respect to the single injection protocol is that one circumvents the need of determining the order of the system and of using a parameter estimation technique.

Achievement of a true tracer steady state is important for a correct calculation of turnover when the constant infusion technique is adopted. The time needed to achieve the steady state depends on the size of the slowest time constant of the system: the smaller its value, the longer the time required to reach the steady state. Thus, it may take a very long time to reach tracer steady state with the constant infusion alone, especially for substances and in subjects having a low turnover. To cope with this problem, the primed continuous infusion strategy has been devised.

10.5.3 Primed Continuous Infusion

When it is of primary importance, a rapid achievement of the tracer steady state (this happens, for instance, when the experimental protocol is long because the steady state is followed by a nonsteady-state period), the tracer administration format of choice is the so-called primed continuous infusion (i.e., an impulse followed by a step input). The reason why this format is preferable to the constant infusion is because the priming dose, when appropriately chosen, considerably speeds up the attainment of the tracer steady state (see Figure 10.3C). To appreciate intuitively why this occurs, we must think that the tracer curve following the primed continuous infusion is, thanks to the linearity of the tracer system in steady state, the superposition of the individual responses to a single injection and to a constant infusion. The final decaying portion of the tracer disappearance curve due to the single injection and the final raising portion of tracer appearance curve are both governed by the slow time constant of the system. Therefore, the time needed to achieve the tracer steady state can be reduced if the amplitudes associated with the slow component in the impulse and in the step responses are identical and, being of opposite signs, cancel each other out. This can be accomplished by selecting the appropriate size of the priming dose with respect to the constant infusion rate. To understand how one can optimally select the priming dose, it is useful to resort to the analytical expression of the system response to a primed continuous infusion, which coincides with the sum of individual responses to an impulse (Eq. (10.17)) and to a step (Eq. (10.19)):

$$c(t) = \mathrm{ra} \sum_{i=1}^{n} \frac{A_i}{\alpha_i} \left[1 + e^{-\alpha_i t} \left(\frac{d}{\mathrm{ra}} \alpha_i - 1 \right) \right] \tag{10.21}$$

From Eq. (10.21), one can see that if the ratio d/ra equals the inverse of the slowest exponential time constant ($1/\alpha_n$), then the contribution of the slowest exponential component, $e^{-\alpha_n t}$, is canceled out and the tracer steady state is achieved more quickly. In particular, if the impulse response contains only a single exponential term, the selection of a priming ratio equal to $1/\alpha_1$ will result in an istantaneous tracer steady state. If the system has a multiexponential impulse response, the achievement of the tracer steady state will require a transient period whose duration will be dictated by the less rapid among the remaining exponential components. For instance, if the system is second order, then the transient period (above the final steady state) will be monoexponential. Thus, the investigator can design the priming ratio provided he/she knows in advance the impulse response of the system. However, in practice, this does not happen so that assumptions about the slowest time constant are to be made on the basis of available physiological knowledge.

When tracer steady state has been obtained, turnover can be calculated as with the constant infusion technique (Eq. (10.20)). It must be emphasized that the calculation is accurate only if a true steady state for the tracer is achieved at the end of the test. However, it may happen (especially with compounds and subjects having a low turnover) that the final portion of the tracer equilibration curve varies so slowly that a false impression of the tracer steady state is given at times when the true plateau level has not been reached (this situation has been extensively investigated for the glucose system in Refs. [5,6]. In this case, the use of Eq. (10.20) would lead to a biased estimate of tracee turnover. If we suspect that the tracer steady state has not been achieved yet at the end of the test, we can still estimate turnover accurately provided we have sampled the tracer curve throughout the experiment, particularly when tracer concentration rapidly decays after the prime. In this case, we will fit all the available tracer data (i.e., the final quasi-steady state, as well as the initial dynamic data) to Eq. (10.21) and we will estimate the impulse response parameters A_i and α_i. These parameters will then be used to calculate the area of the impulse response which, substituted in Eq. (10.16), will yield turnover. What warrants emphasis is that one can resort to this approach to circumvent the lack of a reliable tracer steady only if an adequately frequent sampling schedule has been adopted in the initial part of the test when the fastest components of the system play an important role.

In the study of the glucose system in normal subjects, a priming ratio equal to 100 (corresponding to a slowest exponential component of 0.01 min^{-1}) is commonly used. However, in groups in which the slowest component is altered, the ratio must be changed in order to match the slowest time constant. For instance, the priming modality plays a crucial role in the assessment of basal glucose turnover in non-insulin-dependent diabetes (NIDDM) because, if appropriate adjustment of the priming ratio is not accomplished, the observation period can be insufficient to achieve the tracer steady state and, as a consequence, glucose turnover is overestimated. Fortunately, guidelines have been developed to individualize the priming ratio in NIDDM patients as a function of their fasting glucose concentration [5].

10.6   Estimation of Nonsteady-State Fluxes

Let us suppose that the experimental protocol comprises two phases: a first phase with tracee in steady state and tracer brought to a steady state via a tracer administration—typically a primed continuous infusion—and a second phase in which the investigator imposes a perturbation that pushes the system out of steady state. Such a nonsteady-state transition ends up with a steady state that may be either the initial or a new one. In nonsteady state, Ra, Rd, P, and U are all functions of time and their estimation is much more complex than in steady state. At variance with steady state (see Eqs. (10.14) and (10.15)), in nonsteady state, Ra(t) (which equals the sum of the endogenous and exogenous tracee appearance rates) is different from Rd(t), which is, in turn, different from U(t). We will devote the next sections of this chapter to outline tracer infusion strategies to obtain an accurate estimation of Ra(t), and then an accurate estimation of Rd(t) and U(t).

In nonsteady state, the estimation of Ra(t) on the basis of the convolution integral (Eq. (10.11)) is more complex than in steady state. In principle, Ra(t) estimation is performed in two steps. First, the impulse response $h(t,\tau)$ is identified from tracer data, and then Ra(t) is reconstructed from the impulse response and tracee data by deconvolution. How can $h(t,\tau)$ be assessed? If the system under investigation is linear and time-invariant, the impulse response does not change when the system is pushed out of steady state and can be described by a multiexponential function. The parameters of the multiexponential function can be estimated from the tracer I/O experiment performed in steady state provided that the tracer time course has been adequately sampled (e.g., the rapid decay of tracer concentration immediately following a tracer bolus). In this case, no further tracer administration is required during the nonsteady state. On the other hand, if the system is linear but time-varying (like the glucose system, for instance), the impulse cannot be assessed unless hypotheses are made on the structure of the system. In other words, in order to determine $h(t,\tau)$, it is necessary to formulate and identify from nonsteady-state tracer data a structural model (like the one depicted by Eq. (10.13)) capable of describing the system functioning during the nonsteady state. In particular, one needs to specify which are the time-varying parameters of the model and how they change during the nonsteady state. Different models will yield different estimates of Ra(t). Due to the complexity of metabolic systems, it is difficult to work out a general-purpose model and, usually, models aimed to describe the behavior of the system in each specific experimental situation are developed. In any case, striving for model accuracy has to be balanced against the need for practical identification. It is possible to render the estimation of Ra(t) less dependent on the chosen model if one takes care to design the tracer infusion during the nonsteady state in the most appropriate way. In the next section, we will examine how to do it.

10.6.1   Assessment of Ra: The Tracer-to-Tracee Clamp

Nonsteady-state theory [7,8] suggests that the accuracy of the estimation of Ra(t) can be enhanced if the tracer is infused during the nonsteady state in such a way as to reduce the changes in the tracer-to-tracee ratio, $z(t)$, during the experiment (since this procedure was originally devised for radiotracers, it is known as specific-activity clamp). Ideally, if $z(t)$ is maintained at a perfectly constant rate during the experiment, an accurate estimate of Ra(t) can be obtained irrespective of the model used to interpret the nonsteady state. To understand how this happens, let us suppose that the tracee system is in steady state and a tracer experiment is carried out—typically a primed continuous infusion—until the tracer reaches a steady state throughout the system. In the basal steady state, the tracer-to-tracee ratio, z_b, (subscript "b" denotes "basal") coincides with the ratio between basal tracer infusion and basal production. Let us now suppose that a perturbation, for instance an exogenous administration of tracee, pushes the system out of the steady state. If Ra(t) changes in time with respect to basal P and the tracer is still infused at a constant rate during the nonsteady state, the tracer-to-tracee ratio will change as well. Theory suggests that this change in the tracer-to-tracee ratio should be prevented by infusing the tracer in such a way as to follow the changes of Ra(t). If the tracer infusion rate, ra(t), is adjusted so that ra(t) = Ra(t) $\cdot z_b$, $z(t)$ remains equal to z_b throughout the experiment, and Ra(t) is given by:

$$\text{Ra}(t) = \frac{\text{ra}(t)}{z_b} \qquad (10.22)$$

In other words, ra(t) must have the same shape of Ra(t), with the proportionality factor between the two being the desired target value z_b. Note that Eq. (10.22) is similar to Eq. (10.20) (i.e., the one used to calculate steady-state Ra), but here both ra(t) and Ra(t) change in time. A theoretical proof that the rate of appearance is predicted correctly in nonsteady state by Eq. (10.21) if $z(t)$ remains constant has been given by Norwich [4,7] for a distributed system of the convection–diffusion reaction. For a generic compartmental model describing tracee and tracer dynamics

by Eqs. (10.4) and (10.13), respectively, it has been shown by Cobelli and colleagues [8] that:

$$\text{Ra}(t) = \frac{\text{ra}(t)}{z_1(t)} - \frac{Q_1(t)}{z_1(t)}\frac{dz_1(t)}{dt} - \sum_{i=2}^{n}\left(1 - \frac{z_i(t)}{z_1(t)}\right)k_{1i}Q_i(t) \qquad (10.23)$$

where z_i is the tracer-to-tracee ratio in the ith compartment (z_1 is the tracer-to-
tracee ratio measured in the accessible pool and thus coincides with z). The relevance of this equation lies in the fact that, if the tracer administration is adjusted so
as to induce no changes in $z(t)$ over time, the time derivative of $z(t)$ in the accessible pool is zero and the contributions of the second and third term in Eq. (10.23)
become null. As a result, $\text{Ra}(t)$ coincides with the expression given in Eq. (10.22)
and becomes model-independent because it only hinges on what can be measured
in the accessible pool.

In practice, maintaining plasma $z(t)$ absolutely constant is impossible.
Nevertheless, it is useful for the investigator to try to clamp $z(t)$ at a constant level
by changing the tracer infusion rate in a suitable way because a reduction in the
rate of change of $z(t)$ will provide an estimate of $\text{Ra}(t)$ much less dependent on the
validity of the model used with respect to a "blind" constant tracer infusion.

What are the possible approaches to clamp $z(t)$? The development of the closed-
loop tracer administration scheme requires the availability of measurement methods
of tracer and tracee plasma concentration that are rapid enough. Usually, the tracer
concentration cannot be measured rapidly enough, and an open-loop approach to
clamp $z(t)$ has to be adopted. Therefore, guidelines are needed to design the most
appropriate open-loop scheme of tracer infusion. We have seen additionally that in
order to keep $z(t)$ constant, $\text{ra}(t)$ should have the same shape as the unknown $\text{Ra}(t)$.
In general, there are two sources of tracee contributing to $\text{Ra}(t)$ (Figure 10.4, left
panels): one is the endogenous production, while the other, when present, is the
rate of appearance of the tracee that the investigator administers exogenously
related to the purpose of the study (for instance, during a euglycemic hyperinsulinemic clamp [9], glucose is infused intravenously at a known rate in such a way as to
maintain glucose concentration constant throughout the study). Thus, we can write:

$$\text{Ra}(t) = P(t) + \text{Ra}_{\text{exg}}(t) \qquad (10.24)$$

where $\text{Ra}_{\text{exg}}(t)$ is the rate of appearance of the exogenous tracee. Also, the measured tracee concentration can be thought of as the sum of two components (see
Figure 10.4, right panels): endogenous, $C_{\text{end}}(t)$, and exogenous concentration,
$C_{\text{exg}}(t)$:

$$C(t) = C_{\text{end}}(t) + C_{\text{exg}}(t) \qquad (10.25)$$

The distinction between the endogenous and exogenous tracee sources and concentrations leads naturally to a tracer infusion strategy that has general applicability.

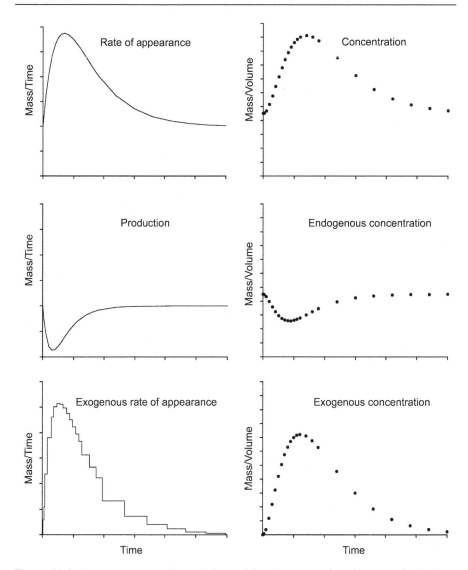

Figure 10.4 Tracee appearance fluxes (left panels) and concentrations (right panels) during the nonsteady state. The rate of appearance of a compound (left, upper panel) can be thought as the sum of two components: endogenous production (right, middle panel) and exogenous appearance rate. Dually, the measured plasma concentration of the compound (left, upper panel) can be thought as the sum of an endogenous (left, middle panel) and an exogenous (left, lower panel) component.

In fact, one viable approach to minimize the changes in $z(t)$ consists in using two distinct tracer infusions, one proportional to the exogenous tracee infusion (Figure 10.5A) and the other proportional to the endogenous tracee production (Figure 10.5B). Implementing the first tracer infusion scheme is simple: one has to

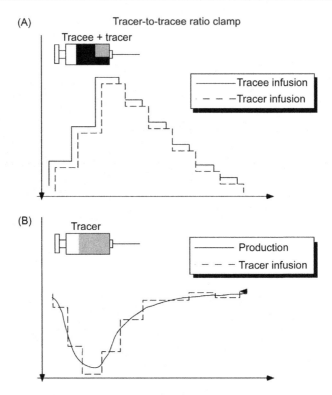

Figure 10.5 A tracer administration strategy to clamp the tracer-to-tracee ratio during the nonsteady state. Two distinct tracer infusions are employed during the experiment: one is proportional to the exogenous tracee infusion (A) and the other to the endogenous tracee production (B). The former tracer infusion can be implemented by adding a proper amount of tracer (see text) to the exogenous tracee infusion, whereas the latter tracer infusion is realized by administering the tracer in such a way as to mimic the expected time course of endogenous production during the experiment.

add some tracer to the exogenous tracee so that the tracer-to-tracee ratio of the labeled infusate, z_{inf}, is equal to z_b. The realization of the second tracer infusion scheme is more difficult because it is necessary to change the basal rate of tracer infusion in such a way as to mimic the expected time course of $P(t)$ during the experiment. This may sound like circular reasoning because adjusting the tracer infusion rate requires the knowledge of $P(t)$, which is exactly what one is trying to determine. However, usually some information about the behavior of $P(t)$ during the nonsteady state is available. This *a priori* knowledge can be used to design a tentative format of tracer administration for the first trial (educated guess). This guess can be later verified by measuring $z(t)$ and, if $z(t)$ is not constant, one learns from the error and refines the format of tracer administration in the subsequent experiment. This procedure can be repeated a few times until a satisfactory format is obtained. In our experience, a few iterations are sufficient to achieve acceptable results.

Usually, the task of changing the tracer infusion in such a way as to mimic the expected time course of $P(t)$ can be automated using a pre-programmable pump. This allows the investigator to change the infusion rate as frequently as he/she wishes without adding complexity to the experiment. On the other hand, it is of no value to change frequently the tracer infusion rate if blood samples are not collected frequently enough. It is sufficient to change the staircase tracer infusion rate immediately after each sample and keep the tracer infusion constant between two consecutive samples (the tracer infusion will be proportional to the average of $P(t)$ in that interval). An important caveat is that the frequency of blood sampling (and thus the number of steps in the tracer infusion rate) should be increased whenever $P(t)$ is expected to change most rapidly, and variability among subjects is elevated.

Using the glucose system as a prototype, we will provide examples of the use of this technique to clamp $z(t)$ under experimental conditions that commonly arise in clinical investigation. Specifically, we will deal with three distinct situations: first, when the exogenous source of the tracee is absent; secondly, when the exogenous source of the tracee is present and known; finally, when the exogenous source of the tracee is present but unknown.

10.6.1.1 No Exogenous Source of Tracee

There are studies in which the only tracee source is endogenous. As a result, the measured tracee concentration coincides with C_{end}. An example of this situation is the study of glucose turnover during physical exercise. During physical exercise, glucose production increases to compensate for the increased glucose utilization by muscles. To clamp $z(t)$ in this situation, one has simply to administer the tracer in such a way as to mimic the expected pattern of increase of $P(t)$. In Ref. [10], Coggan and coworkers approximated the increase of $P(t)$ during exercise by an increasing monoexponential function and achieved an excellent tracer-to-tracee clamp.

10.6.1.2 Known Exogenous Source of Tracee

In many circumstances, the tracee is administered exogenously and its rate of delivery into the systemic circulation is known. Two examples of this situation are the meal-like study and the euglycemic hyperinsulinemic clamp.

The meal-like study consists in an experimental protocol in which glucose and insulin are infused so as to reproduce the plasma glucose and insulin concentration profiles that are typically observed after carbohydrate ingestion in healthy subjects (see, for instance, Ref. [11]). To clamp $z(t)$, a single tracer can be used to clamp the exogenous and endogenous tracee sources: some tracer is added to the glucose infusate, while the basal tracer infusion is changed with a pattern mimicking the expected time course of $P(t)$.

The euglycemic hyperinsulinemic clamp devised by DeFronzo and colleagues [9] is the most widely used approach to investigate the effects of insulin on glucose metabolism. Insulin is infused at a constant rate throughout the study, while glucose is infused at a variable rate so as to maintain its constant level in plasma. Thus, at any time, the exogenous glucose infusion rate, $Ra_{exg}(t)$, equals the difference $Rd(t) - P(t)$, and when a new steady state is attained, one can measure the dose-response effect of insulin on both glucose uptake and production. To clamp $z(t)$ during the euglycemic hyperinsulinemic clamp, one can use the approach described above for the meal-like study (i.e., to add some tracer to the glucose infusate in such a way that $z_{inf} = z_b$, and to change the basal tracer infusion mimicking the expected time course of $P(t)$).

The above-described strategy is the ideal one because it allows one, at least in theory, to achieve a perfect tracer-to-tracee ratio clamp. However, it may turn out to be too complex and labor-intensive, especially if the glucose clamp is to be used in population studies. Finegood and colleagues [12] devised a simpler approach in which one adds some label to the exogenous glucose infusate and maintains the basal tracer infusion unchanged throughout the study. If one interprets Finegood et al.'s approach in the light of the general approach outlined above, it is evident that keeping constant the rate of tracer infusion—instead of mimicking the time course of $P(t)$—will produce an excess of tracer in the system. The idea devised by the authors to compensate for this tracer excess consists in underlabeling the exogenous infusate, i.e., preparing an exogenous infusate having $z_{inf} < z_b$. In order to correctly choose z_{inf}, some *a priori* information about the likely behavior of the glucose system at the end of the clamp is needed. Specifically, assuming that at the end of the clamp both P and Rd have reached a new steady state, the expression for z_{inf} is given by:

$$z_{inf} = z_b \left[1 - \frac{P_b - P_{ss}}{Rd_{ss} - P_{ss}} \right] \tag{10.26}$$

where P_b, P_{ss}, Rd_{ss} are, respectively, basal and final steady-state values of P and Rd.

The advantage of Finegood et al.'s approach with respect to the general approach outlined above is that it keeps to a minimum the experimental effort and does not require one to guess the whole time course of $P(t)$, but only the steady-state values of P and Rd at the end of the study. The disadvantage is that such a technique cannot be refined as more information about the nonsteady-state behavior of $P(t)$ becomes available. In addition, it only ensures that the value of $z(t)$ measured at the end of the clamp is equal to z_b, but does not guarantee that it remains constant throughout the study. We have previously shown [13] that $z(t)$ will remain constant only if the inhibition of production is proportional to the stimulation of glucose disappearance. Since $P(t)$ and $Rd(t)$ may exhibit different time courses or differences between disease states, systematic deviations may arise. Despite these limitations, this approach works reasonably well and, because of its simplicity, is widely used to clamp the tracer-to-tracee ratio during the euglycemic hyperinsulinemic clamp.

In principle, Finegood et al.'s approach is applicable to the hyperglycemic and hypoglycemic clamp as well. The ingredients that allow one to choose z_{inf} under such experimental conditions are always the same from Eq. (10.26) (i.e., the expected values of P and Rd at the end of the experiment). Since during a hyperglycemic and a hypoglycemic clamp, glucose does not remain at the baseline, but is purposefully brought to a new level, systematic deviations of $z(t)$ from constancy are likely to occur in the initial part of the test. For instance, in the initial part of the hyperglycemic clamp, the need of rapidly elevating glucose concentration by means of an exogenous glucose infusion will increase glucose concentration more than tracer concentration, so that $z(t)$ may exhibit a transient undershoot. To prevent this undershoot, one has to add some more tracer in the initial part of the test. This can be done by using, only in the initial part of the test, a labeled infusion having a z_{inf} value higher than the one dictated by Eq. (10.26). Alternatively, one can temporarily increase and then restore the basal tracer infusion. During the hypoglycemic clamp, the exogenous glucose infusion rate initiates when glucose reaches the desired hypoglycemic level. If the basal tracer infusion rate remains unchanged in the initial part of the test when exogenous glucose is not administered, $z(t)$ may exhibit a transient overshoot. To prevent this overshoot, one can temporarily reduce the basal tracer infusion and then restore it in concomitance with the initiation of the exogenous glucose infusion.

10.6.1.3 Unknown Exogenous Source of Tracee

In some experimental situations, the rate of appearance of the exogenously administered tracee is not known. For instance, when glucose is given orally (e.g., during a meal or an oral glucose tolerance test), exogenous glucose is absorbed through the gastrointestinal tract and is delivered into the systemic circulation with a pattern that is unknown and highly variable among subjects. It is intuitive that by adding to the orally administered glucose the same tracer that is infused intravenously would prevent the estimation of $P(t)$ because the contribution of ingested glucose to total Ra(t) could not be determined. Thus, to single out the contribution of glucose production, it is necessary to label the ingested glucose with a tracer that is *different* from the one infused intravenously (see, for instance, [14]) Whereas the tracer given orally is aimed to trace the exogenous glucose source, the tracer infused intravenously at a variable rate mimics the time course of $P(t)$. Since we measure two different tracer concentrations in plasma, $c'(t)$ and $c''(t)$, we will also have to deal with two different z profiles, one referred to the tracer infused intravenously and the other to the tracer given orally. How does one recognize if the tracer-to-tracee clamp is successful? One convenient strategy is to calculate $z(t)$, not with respect to the measured glucose concentration, but with respect to endogenous glucose concentration (i.e., the component of total glucose concentration that is due exclusively to $P(t)$). Endogenous glucose concentration cannot be measured directly, but can be derived in a model-independent way [15] by subtracting from the measured total glucose concentration the exogenous component, where the latter is proportional—thanks to the tracer—tracee indistinguishability principle—to the concentration profile of the tracer mixed to orally administered glucose:

$$C_{end}(t) = C(t) - C_{exg}(t) = C(t) - c'(t)/z_{oral} \tag{10.27}$$

where $c'(t)$ is plasma concentration of the tracer mixed to oral glucose and z_{oral} is the tracer-to-tracee ratio of the oral glucose administration. The endogenous glucose profile, which roughly looks like a delayed version of $P(t)$, can then be compared with the concentration of the tracer infused intravenously, $c'(t)$. If the tracer has been infused in a successful way (i.e., so as to match satisfactorily the profile of $P(t)$), the plasma profiles of $c'(t)$ and $C_{end}(t)$ will change in parallel and their tracer-to-tracee ratio, $z_{end}(t)$, will be constant. In other words, instead of calculating $P(t)$ as the difference between the estimates of $Ra(t)$ and of the appearance rate of orally administered glucose, one first calculates $C_{end}(t)$ and then uses it, in conjunction with $z_{end}(t)$, to estimate $P(t)$. Figure 10.6 shows the results of a study [14] in which glucose production during a meal was estimated with the approach just described.

Two tracers were employed: a stable-isotope glucose tracer, [2-^{2}H]glucose and a radioactive glucose tracer, [3-^{3}H]glucose. The former was added to the meal, while the latter was infused in such a way as to mimic the expected time course of $P(t)$. The format of [3-^{3}H]glucose administration was refined on the basis of the results obtained in the first three subjects who underwent the study and was then kept fixed in the other subjects participating in the study. [3-^{3}H]glucose plasma concentration matched well the endogenous glucose concentration derived from the [2-^{2}H] glucose data (Figure 10.6A), and this provided a relatively stable endogenous tracer-to-tracee ratio (i.e., [3-^{3}H]glucose over endogenous glucose concentration; Figure 10.6B), thus allowing a presumably accurate estimate of $P(t)$ (Figure 10.6C). It is of interest to observe that in the period when $P(t)$ was inhibited (0−200 min), the clamp of endogenous glucose concentration was reasonably good. Clamping $z_{end}(t)$ between 200 and 320 min was more difficult because the time when $P(t)$ began its resumption to the basal level varied markedly among subjects. In addition, in that period both tracer and endogenous glucose concentration were very low so that their ratio was extremely sensitive to changes in either tracer or endogenous glucose concentration. Thus, even a small increase in tracer concentration not accompanied by a concomitant increase in endogenous glucose concentration induced a noticeable increase in $z_{end}(t)$.

It is worth noting that such a dual-tracer protocol allows minimization of the nonsteady-state error associated with the estimate of $P(t)$, but not of the error associated with the estimate of $Ra(t)$. In fact, whereas the estimate of $P(t)$ relies on the tracer-to-tracee ratio of the plasma concentrations of the intravenous tracer and endogenous glucose, the estimate of $Ra(t)$ relies on the tracer-to-tracee ratio of the plasma concentrations of the same tracer and total glucose. If the former $z(t)$ is almost constant, the latter will change considerably, thus making it difficult to accurately estimate $Ra(t)$. In particular, one can expect an overestimation of $Ra(t)$ in the initial part of the test (when $z(t)$ increases) and a subsequent underestimation

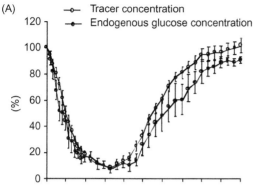

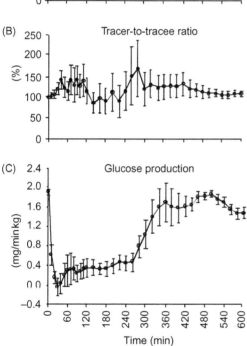

(A) — Tracer concentration — Endogenous glucose concentration

(B) Tracer-to-tracee ratio

(C) Glucose production

Time (min)

Figure 10.6 The tracer-to-tracee-ratio clamp technique for estimating glucose production during a meal [14]. (A) Endogenous glucose concentration and concentration of the tracer infused so as to mimic the expected glucose production profile (percent with respect to basal). (B) Endogenous tracer-to-tracee ratio (percent with respect to basal). (C) Estimated profile of glucose production. Reproduced with permission from [14].

(when $z(t)$ decreases). Of course, the same trend can be expected for the estimate of the rate of appearance of the glucose administered orally, which can be derived by subtracting $P(t)$ from $Ra(t)$.

In order to accurately estimate $Ra(t)$, one has to resort to a third tracer, which will be administered intravenously as a primed continuous infusion in the pre-meal steady state and as a variable infusion during the meal in such a way as to mimic the expected time course of $Ra(t)$. Such triple tracer infusion strategy has been successfully employed in [16]. In that study, a primed continuous infusion of $[6,6-{}^{2}H_2]$ glucose (a stable-isotope glucose tracer) was started two hours prior to the meal, thus allowing the attainment of a constant tracer-to-tracee ratio prior to the initiation of the meal. The task of this first tracer was to trace the endogenous glucose component and thus the format of infusion of this tracer was changed during the meal in such a way as to mimic the expected time course of endogenous glucose production, $P(t)$. The meal ingested by the subjects participating in the study contained a second tracer, $[1-^{13}C]$ glucose, (a stable-isotope glucose tracer). This tracer entered the peripheral circulation in exactly the same manner as the orally administered exogenous glucose and thus allowed the calculation of the exogenous and endogenous components of the measured glucose concentration. The endogenous component of glucose concentration was due to endogenous glucose production, $P(t)$. Thus, by design, the tracer-to-tracee ratio calculated as the ratio between the concentration of the first tracer and the concentration of endogenous glucose concentration (calculated via the second tracer according to Eq. (10.27)) was thus expected to be approximately constant and thus favor a reliable estimation of endogenous glucose production. Up to this point, the tracer infusion strategy is exactly the same as the double tracer infusion strategy outlined before. The novelty here is the adoption of a third tracer. At the time of meal ingestion, a third, radioactive glucose tracer, $[6-^3H]$ glucose, was infused intravenously in a pattern that mimicked the anticipated rate of appearance of orally ingested glucose. Thus, by design, the tracer-to-tracee ratio calculated as the ratio between the concentration of the third tracer and the concentration of exogenous glucose concentration (proportional to the second tracer) was thus expected to be approximately constant and thus favor a reliable estimation of the rate of entry of orally administered glucose into the peripheral circulation.

10.6.2 Assessment of Rd and U

In nonsteady state, $Rd(t)$ and $U(t)$ are no longer equal and, in general, both of them require a model of the system to be estimated [17,18]. Nevertheless, it is easy to recognize that the assessment of $Rd(t)$ is less problematic than that of $U(t)$ because of the different relationship that these two fluxes have with $Ra(t)$:

$$Rd(t) = Ra(t) - \frac{dQ_1(t)}{dt} \tag{10.28}$$

$$U(t) = Ra(t) - \frac{dQ_T(t)}{dt} \tag{10.29}$$

Both $Rd(t)$ and $U(t)$ can be derived from $Ra(t)$ which, as shown above, can be accurately calculated, even with an approximate model, by resorting to the tracer-to-tracee ratio clamp. However, whereas the estimation of $Rd(t)$ only requires measurement of the rate of change of the substance mass in the accessible pool, estimation of $U(t)$ requires the knowledge of the rate of change of the substance mass in the inaccessible compartments as well. Estimation of $Rd(t)$ is thus relatively easier than that of $U(t)$. In fact, if the volume of the accessible pool is available, $Rd(t)$ can be accurately measured—without the need of postulating a model of the system—by keeping $z(t)$ constant. In contrast, the estimation of $U(t)$

always requires a model of the system (accessible and inaccessible pools). Despite the fact Rd(t) pertains to the accessible pool only, its estimation is important because its knowledge allows inferences concerning $U(t)$. In fact, for a system that starts in steady state, goes into a nonsteady-state period, and then returns to the former steady state, the following relationship between Rd(t) and $U(t)$ holds [18]:

$$AUC[Rd(t)] = AUC[U(t)] \tag{10.30}$$

where AUC denotes the area under the curve. A more specific relationship between the time courses of Rd(t) and $U(t)$ during the nonsteady state can be derived if one assumes that the only time-varying parameter is the irreversible loss of the accessible pool [18]. Particularly, for a system that goes into a nonsteady-state period in which substance concentration increases, one has

$$Rd(t) > U(t) \tag{10.31}$$

Conversely, for a system that goes into a period in which substance concentration decreases, one has

$$Rd(t) < U(t) \tag{10.32}$$

10.7 Conclusion

In this chapter, focus has been placed on tracer experiment design strategies needed to quantitate production and utilization fluxes of a substance under both steady- and nonsteady-state conditions. The steady-state problem is easy to tackle and it has been emphasized that, under steady-state conditions, approaches to a rapid attainment of the tracer steady state require attention in selecting the ratio between the priming dose and the constant infusion. The nonsteady-state situation is far more complex and the adoption of "intelligent" infusion strategies becomes a must. In particular, it has been shown that, under nonsteady-state conditions, the tracer infusion should be varied in such a way that the tracer-to-tracee ratio remains as constant as possible to reduce the impact of model error. In this context, the partitioning of the tracee source and concentration into two components, endogenous and exogenous, provides a general framework that helps the investigator to achieve this goal under various experimental conditions.

References

[1] Carson ER, Cobelli C, Finkelstein L. The mathematical modelling of metabolic and endocrine systems. New York, NY: Wiley; 1983.

[2] Cobelli C, Caumo A. Using what is accessible to measure that which is not: necessity of model of system. Metabolism 1998;47(8):1009−35.

[3] Cobelli C, Toffolo G, Foster DM. Tracer-to-tracee ratio for analysis of stable isotope tracer data. Am J Physiol 1992;262(6 Pt 1):E968−75.

[4] Norwich KN. The kinetics of tracers in the intact organism. Oxford: Pergamon; 1977.

[5] Hother-Nielsen O, Beck-Nielsen H. On the determination of basal glucose production rate in patients with type 2 (non-insulin-dependent) diabetes mellitus using primed-continuous 3-3H-glucose infusion. Diabetologia 1990;33(10):603−10.

[6] Hovorka R, Eckland DJ, Halliday D, Lettis S, Robinson CE, Bannister P, et al. Constant infusion and bolus injection of stable-label tracer give reproducible and comparable fasting HGO. Am J Physiol 1997;273(1 Pt 1):E192−201.

[7] Norwich KN. Measuring rates of appearance in systems which are not in steady state. Can J Physiol Pharmacol 1973;51:91−101.

[8] Cobelli C, Mari A, Ferrannini E. Non-steady state: error analysis of Steele's model and development for glucose kinetics. Am J Physiol 1987;252:E679−89.

[9] DeFronzo RA, Tobin JD, Andres R. Glucose clamp technique: a method for quantifying insulin secretion and resistance. Am J Physiol 1979;237(3):E214−23.

[10] Coggan AR, Raguso CA, Williams BD, Sidossis LS, Gastaldelli A. Glucose kinetics during high-intensity exercise in endurance-trained and untrained humans. J Appl Physiol 1995;78(3):1203−7.

[11] Alzaid AA, Dinneen SF, Turk DJ, Caumo A, Cobelli C, Rizza RA. Assessment of insulin action and glucose effectiveness in diabetic and nondiabetic humans. J Clin Invest 1994;94:2341−8.

[12] Finegood DT, Bergman RN, Vranic M. Estimation of endogenous glucose production during hyperinsulinemic-euglycemic glucose clamps: comparison of unlabeled and labeled exogenous glucose infusates. Diabetes 1987;36:914−24.

[13] Butler PC, Caumo A, Zerman A, O'Brien P, Cobelli C, Rizza RA. Methods for assessment of the rate of onset and offset of insulin action during nonsteady state in humans. Am J Physiol 1993;264(4 Pt 1):E548−60.

[14] Taylor R, Magnusson I, Rothman DL, Cline GW, Caumo A, Cobelli C, et al. Direct assessment of liver glycogen storage by ^{13}C nuclear magnetic resonance spectroscopy and regulation of glucose homeostasis after a mixed meal in normal subjects. J Clin Invest 1996;97(1):126−32.

[15] Cobelli C, Toffolo G. Constant specific activity input allows reconstruction of endogenous glucose concentration in non-steady state. Am J Physiol 1990;258(6 Pt 1): E1037−40.

[16] Toffolo G, Basu R, Dalla Man C, Rizza R, Cobelli C. Assessment of postprandial glucose metabolism: conventional dual- vs. triple-tracer method. Am J Physiol 2006;291 (4):E800−6.

[17] Mari A. On the calculation of glucose rate of disappearance in nonsteady state. Am J Physiol 1994;265(5 Pt 1):E825−8.

[18] Caumo A, Homan M, Katz H, Cobelli C, Rizza RA. Glucose turnover in presence of changing glucose concentrations: error analysis for glucose disappearance. Am J Physiol 1995;269(3 Pt 1):E557−67.

11 Stochastic Models of Physiology

Boris Kovatchev[a,b,c] and Stephen Patek[a,c]

[a]Center for Diabetes Technology, University of Virginia, Charlottesville, VA, [b]Department of Psychiatry and Neurobehavioral Sciences, Section Computational Neuroscience, University of Virginia, Charlottesville, VA, [c]Department of Systems and Information Engineering, University of Virginia, Charlottesville, VA

11.1 Introduction

Generally, there are two approaches to the description of probability: a formal-logic approach based on axioms and formally derived properties of abstract mathematical objects, such as random variables and random processes, and a heuristic approach based on one's intuitive understanding of chance, mean, variance, and probability. Many, now classic, books have been written following each of these methods. A most prominent example of the presentation of probability as pure formal logical content is given by William Feller's *An Introduction to Probability and its Applications* which stated: "Axiomatically, mathematics is concerned with relations among undefined things. This aspect is well illustrated by the game of chess. It is impossible to 'define' chess otherwise than by stating a set of rules.... The chessboard and the pieces are helpful, but they can be dispensed with" [1]. An alternative approach to understanding probability is to begin with the intuitive ideas of chance, frequency of events, average tendencies, and the laws of large numbers, and then gradually build theoretical objects, such as random walks, martingales, Markov chains, convergence of probability measures, stochastic processes, and Kalman filters. A classic book adopting this approach is Albert Shiryaev's *Probability* [2]. While both approaches have their merits and have been used in countless mathematics, statistics, and engineering courses, throughout this chapter we will adopt the second, more empirical method, which in our opinion is closer to the applied nature of physiology modelling.

11.2 Randomness and Probability

Typically, a *random (or stochastic) variable* is defined as a variable that can assume more than one value due to chance. The set of values a random variable can assume is called "state space" and, depending on the nature of their state space,

Modelling Methodology for Physiology and Medicine. DOI: http://dx.doi.org/10.1016/B978-0-12-411557-6.00011-2

random variables are classified as discrete (assuming a finite or countable number of values) or continuous, assuming any value from a continuum of possibilities. Some classic examples will clarify this concept: The outcome from tossing a coin is a random variable that can assume two values—heads and tails; hence, the state space of this discrete random variable is finite, comprised of two values $\Omega = \{H, T\}$; throwing dice will result in a discrete random variable that can assume six values $\Omega = \{1,2,3,4,5,6\}$; systolic blood pressure, body temperature, or blood glucose levels are continuous random variables assuming values from a range, for example, of $90-180$ mmHg or $34-42°C$ or $20-600$ mg/dL (in diabetes). Probability is a metric or measure of the chance of a random variable to assume a value of a subset of values in its state space. Mathematically, the probability is a function that maps the state space onto the interval $[0,1]$, which has the following properties: the probability of the entire state space is equal to one, $P(\Omega) = 1$; the probability of an empty subset is zero, $P(\varnothing) = 0$, and the probability is an additive metric, meaning that if $A1$, $A2$, ... are mutually exclusive subsets (events) in Ω, then $P(A1 \cup A2 \cup \ldots) = P(A1) + P(A2) + \cdots$. Further, we frequently use the term "conditional probability," which describes the probability of an event A, given that another related or unrelated event B has occurred:

$$P(A|B) = \frac{P(A \cap B)}{P(B)}$$

It is important to note that the conditional probability is used to define *stochastic independence*: arguably the most important property in probability theory. In it, two events (or random variables) are considered independent if $P(A|B) = P(A)$ or $P(A \cap B) = P(A) \cdot P(B)$. Note a typical mistake: mutually exclusive events, e.g., $A \cap B = \varnothing$, are *not* independent because $P(A \cap B) = 0$. Two important formulas are related to the notion of conditional probability: the total probability formula $P(A) = \sum_{i=1}^{N} P(A \cap Bi) = \sum_{i=1}^{N} P(A|Bi)P(Bi)$ allows the probability of an event to be reconstructed from the conditional probabilities of that event, given conditional probabilities along a set of mutually exclusive events B_1, B_2, ..., B_N; and the Bayes' formula

$$P(Bi|A) = \frac{P(Bi) \cdot P(A|Bi)}{\sum_{i=1}^{N} P(A|Bi)P(Bi)}$$

which relates prior to posterior probabilities for an event and serves as a basis for iterative stochastic estimation of model parameters. Bayesian inference is an iterative estimation process, which proceeds as follows: at iteration zero, a prior distribution of a model parameter is known, which is then refined by accumulation of data into a posterior distribution using the Bayes' formula; at iteration 1, the posterior distribution is considered prior and the procedure continues until a certain degree of convergence is achieved [3]. Bayesian inference is a frequently used method in physiological modelling, with several software modules developed for its use [4,5]. Recently, *Bayesian networks* became a popular approach to deciphering probabilistic relationships in physiological systems. A Bayesian network is defined as a graphical model of the conditional probability relationships among a set of variables [6]. When a time component is included, the network is called a *dynamical Bayesian network* [7]. Bayesian networks have been used to describe the cardiovascular system [7] and glucose−insulin dynamics in intensive care patients [7], have been used in prognostic models of patient management [8], and have been applied to the analysis of gene-expression data [9].

11.3 Probability Distributions and Stochastic Processes

Before continuing further, we need to introduce the notion of *distribution of a random variable* and its time-dependent extension to random (stochastic) processes. Let's review the following example: we observe neuronal behavior (e.g., neuronal spike train) over N consecutive time intervals and assume that within each time interval a neuron fires with a probability p, independently from the previous interval as presented in Figure 11.1.

The simplest random variable related to this experiment is the outcome—spike or no spike—observed within each time interval. This random variable has a state space $\Omega = \{0,1\}$ and a *Bernoulli distribution* $\{1 - p, p\}$ defining the probabilities for no spike or spike in each interval. The process in time of sequential Bernoulli random variables is called *Bernoulli process*, and is equivalent to a sequence of coin tosses with a binary outcome at each attempt. Further, a more complicated question is the probability distribution of the number of spikes over the entire set of N time intervals. This random variable is the sum of N Bernoulli variables, has a state space $\Omega = \{0,1,\ldots,N\}$, and a *binomial distribution*, given by the formula

$$p(k) = \binom{N}{k} \cdot p^k \cdot (1-p)^{N-k}$$

where $k = 0, 1, 2, \ldots, N$ are the possible outcomes of having 0, 1, ... spikes within the N observed time intervals. Both the Bernoulli and the binomial distributions are discrete (i.e., Bernoulli or binomial random variables assume a finite number of

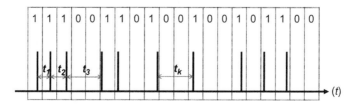

Figure 11.1 Binary presentation of neuronal spike train.

values, 2 and $N+1$, respectively). Another discrete probability distribution is the *Poisson distribution* given by the formula

$$p(k) = \frac{\lambda^k}{k!} \cdot \exp(-\lambda)$$

where $k = 0, 1, 2, \ldots$ are the infinite but countable possible outcomes of our experiment under certain conditions that we will clarify below, and $\lambda > 0$ is a parameter defining the distribution.

Further, let's look at the time intervals $t_1, t_2, \ldots$ between consecutive neuronal spikes. These time intervals are realizations (values) of a random variable representing the time between two consecutive spikes. This random variable has a continuous distribution with a state space $\Omega = [0, \infty]$ (i.e., it can assume any nonnegative value). It is intuitively clear that the probability of such a variable assuming any particular value is zero (i.e., its chance of hitting a particular number is zero). Thus, the distribution of a continuous random variable would have to be defined differently, not by a formula assigning a probability to each value in its state space. Traditionally, the probability distribution of continuous random variable X is given by

$$P(a \leq X \leq b) = \int_a^b f(x) dx$$

where $f(x)$ is the probability density function of X (equivalent to the set of probabilities of discrete distributions). One example of continuous probability distribution is the normal (Gaussian) distribution with a density given by the formula

$$f(x) = \frac{1}{\sigma\sqrt{2\pi}} e^{-(x-\mu)^2/(2\sigma^2)} \quad -\infty < x < \infty$$

where the parameters $-\infty < \mu < \infty$ and $\sigma > 0$ define the shape of the distribution. Another example is the exponential distribution with a density defined by the formula

$$f(x; \lambda) = \begin{cases} \lambda e^{-\lambda x} & x \geq 0 \\ 0 & \text{otherwise} \end{cases}$$

We have clarified earlier the relationship between the neuronal spike train example in Figure 11.1, Bernoulli, and binomial distributions. Continuing further, let us assume that the number of binomial states N is large, such that $N \to \infty$ and that the Bernoulli probability of a spike in any given time interval p is small, such that $p \cdot N \to \lambda$. In this case, the limit of the resulting binomial distribution as $N \to \infty$ is a Poisson distribution with parameter λ and the resulting random process of neuronal spikes is a *Poisson process*. Interestingly, under these conditions, the distribution of the time interval between the sequential points (spikes) in the Poisson process is exponential with that same parameter λ. Thus, the following conceptual relationship emerges: low-intensity discrete physiological processes in which each event occurs independently from the previous one with a certain small probability p can be approximated by a Poisson process, with exponential distribution of the consecutive increments. Moreover, the superposition of many low-intensity processes will be close to a Poisson process regardless of whether the initial processes are of Poisson type or not. Thus, the notion of a discrete Bernoulli process, its continuous counterpart—the Poisson process—and their underlying probability distribution (Bernoulli, binomial, Poisson, and exponential) provide powerful analytical tools for the description of sequences of low-intensity physiological events, such as the spikes of an ensemble (superposition) of neurons, or of the pulses in hormonal secretion. In general, the formal relationship between random variables and random processes can be presented as follows: Let $I(w,t)$ be a random process describing the population of individuals $\{w \in \Omega\}$ across time $\{t\}$. Then, at any point of time t_0, $I(w,t_0)$ is a random variable with a distribution that represents a cross section of the population at time t_0, and for any individual w_0, $I(w_0,t)$ is his/her own trajectory across time.

11.4   The Law of Large Numbers and Limit Theorems

Intuitively, if we toss a (symmetric) coin many times, the overall probability of getting a heads would be ~ 0.5. Similarly, if we throw dice, the probability of getting a 1 would approach $1/6$ with many consecutive attempts. Before mathematically formulating these intuitive properties, we need to introduce mean and variance of a random variable, which are given by the formulas:

$$\mu_X = E(X) = \int_{-\infty}^{\infty} x \cdot f(x)\mathrm{d}x \quad \text{and} \quad \sigma_X^2 = V(x) = \int_{-\infty}^{\infty} (x-\mu)^2 \cdot f(x)\mathrm{d}x$$

where σ is the standard deviation of the distribution.

Now, if X_1, X_2, ... is an infinite sequence of independent and identically distributed (i.i.d) random variables that have mean μ and variance σ^2, then the *law of large numbers* can be formulated as follows:

$$\lim_{N \to \infty} \frac{1}{N}(X_1 + \cdots + X_N) = \mu$$

meaning that the empirical mean of these random variables converges toward their theoretical mean. Further, the limit distribution of the empirical mean is given by the *central limit theorem*, which states that

$$\lim_{N \to \infty} \frac{1}{\sigma\sqrt{N}}(X_1 + \cdots + X_N - N\mu) = \xi \sim \Phi(z) \tag{11.1}$$

where ξ is a random variable with a central (standard) normal distribution with parameters $\mu = 0$ and $\sigma = 1$ and density $f(z; 0, 1) = (1/(\sigma\sqrt{2\pi}))e^{-z^2/2}$. In other words, under very broad conditions the limit distribution of the empirical mean of i.i.d. random variables is normal—a fundamental property that serves as a base for most statistical tests and statistical inference. This is one of the explanations for the widespread applicability of the Gaussian distribution, and the utility of various tests for "normality" or "sphericity" that are intended to guarantee (in a statistical sense) the applicability of standard statistical procedures, such as t-test, regression, and the analysis of variance (ANOVA). Figure 11.2 illustrates one special case of the central limit theorem, known as the Moivre−Laplace theorem, which shows the convergence of a binomial distribution (i.e., of the sum of Bernoulli random variables) to a normal distribution: it is evident that with the increase of N, the binomial distribution gradually assumes the familiar bell-shaped curve of normal probability density.

Another reason for the fundamental role of the normal distribution in nature is its relationship to basic physical and biological processes, such as Brownian motion, Wiener process, and diffusion. The Wiener process W_t has independent increments with $W_t - W_s \sim N(0, t - s)$ for any $0 \leq s < t$, where $N(0, \sigma)$ is a central normal distribution with zero mean and variance equal to the time interval between the consecutive points s and t. The Brownian motion—first observed as the motion of microparticles subjected to hits by water molecules by Robert Brown and formulated as a stochastic process by Albert Einstein [10]—is a common example of a standard Wiener process in which the runs of the microparticles follow normal distribution. Diffusion is one of the basic transport mechanisms in physiology. At a

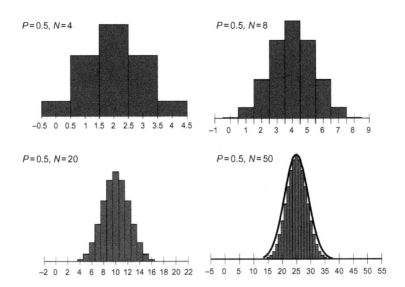

Figure 11.2 Relationship between binomial and central normal distribution.

phenomenological level, the diffusion process is typically described by a differential equation depicting Fick's law, but it can also be derived from the Brownian motion of the diffusing particles. In other words, Fick's law can be considered a macro-result from the individual random walks of the diffusing particles [10,11].

11.5 Analysis of Stochastic Associations: Correlation and Regression

Most modelling efforts are ultimately applied to data for the purpose of analyzing between or within population differences, seeking associations between physiological properties, or finding similarities between phenomena. At the point of data analysis, mathematical modelling meets statistical methods that are exclusively based on the fundamental probability properties discussed above. Here, we provide a brief description of several popular statistical techniques.

The strength of a linear association between two continuous random variables X and Y is typically assessed by the Pearson's correlation coefficient given by the formula:

$$\mathrm{Corr}(X, Y) = \frac{\mathrm{Cov}(X, Y)}{\sigma_X \cdot \sigma_Y}$$

where the covariance between these random variables is given by $\mathrm{Cov}(X, Y) = E[(X - \mu_X)(Y - \mu_Y)]$. Empirically, when series of observations $(x_1, x_2, \ldots, x_n)$ and $(y_1, y_2 \ldots, y_n)$ are available for X and Y, the estimates of their covariance and the standard deviations are given by

$$\hat{\mathrm{Cov}}(X, Y) = \frac{1}{n-1} \sum_{i=1}^{n} (x_i - \bar{x})(y_i - \bar{y}), \quad \hat{\sigma}_X^2 = \frac{1}{n-1} \sum_{i=1}^{n} (x_i - \bar{x})^2, \quad \text{and}$$

$$\hat{\sigma}_Y^2 = \frac{1}{n-1} \sum_{i=1}^{n} (y_i - \bar{y})^2$$

The associations between discrete random variables are typically evaluated by contingency tables and statistical tests, such as Chi-Square and nonparametric correlation, such as Kendall's *tau-b* and *tau-c*, Gamma and Sommer's d coefficients, and so on.

The straight-line fit representing the linear association between the random variables X and Y is typically found using the least-squares criterion depicted in Figure 11.3.

The regression model is analytically described by the conditional expectation of the dependent variable Y along the independent variable X: $E(Y|x) = ax + b$, or as a formula with error term: $Y = ax + b + \varepsilon$, where the error term ε is assumed to have a central normal distribution with zero mean and a certain variance σ^2.

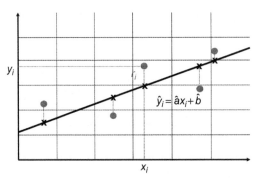

Figure 11.3 Linear regression and least-squares estimation.

The least-squares procedure minimizes the sum of squares of the distances (residuals r_i) between the data points and the straight line. In other words, the procedure finds the coefficients a and b of a straight line that is the best-fit relationship between X and Y presented by the linear formula $\hat{y}_i = \hat{a}x_i + \hat{b}$. Intuitively, the quality of the fit can be evaluated if this linear formula explains Y well, given the values of X. The criteria reflecting this intuitive understanding are based on the notion of "explained" and "residual" variance of the data. Specifically, the entire variance of Y around its mean value can be split into two quantities:

$$\sum_{i=1}^{n} (y_i - \bar{y})^2 = \sum_{i=1}^{n} (y_i - \hat{y}_i)^2 + \sum_{i=1}^{n} (\hat{y}_i - \bar{y})^2 \tag{11.2}$$

one including the residual variance and the other including the variance explained by the linear model. Once the variance is split, the ratio between the variance explained by the model and the total variance of Y is a marker of the quality of model fit. This ratio, given by the formula

$$R^2 = \frac{\sum_{i=1}^{n} (\hat{y}_i - \bar{y})^2}{\sum_{i=1}^{n} (y_i - \bar{y}_i)^2}$$

is called the R^2 (R-square) coefficient, which provides the proportion of variance explained by the model and is usually expressed in percentage. The closer R^2 is to 100%, the better the model fit. In order to come up with a significance level for the goodness-of-fit of the regression model, a similar coefficient is computed:

$$F = \frac{\sum_{i=1}^{n} (\hat{y}_i - \bar{y})^2}{\left(\frac{1}{n-2}\right) \sum_{i=1}^{n} (y_i - \hat{y}_i)^2}$$

which serves as a base for ANOVA comparing the variance explained by the model to the total variance of Y. It has been shown that F has the Fisher F-distribution, which allows testing the hypothesis H_0: the slope of the regression line is zero. The significance level associated with this hypothesis test is the probability for type I error when testing H_0. If this probability is small, we can reject the hypothesis that the regression line is horizontal and conclude that a linear relationship (and nonzero correlation) between X and Y is likely.

Previously we covered the case of a single independent variable (predictor) X and a single dependent variable (response) Y. However, a dependent variable is typically predicted by more than one independent variable. In this case, the linear regression procedure using more than one predictor for Y is called *multiple regression*. In general, the estimation is exactly the same as above, using a least-squares procedure to fit the coefficients of the formula: $y_i = b_0 + b_1 x_{i1} + \cdots + b_p x_{ip} + \varepsilon_i$, where y_i is the value of the ith case of the dependent variable; p is the number of predictors; b_j is the value of the jth coefficient, $j = 0, \ldots, p$ (b_0 is the intercept, the model-predicted value of the dependent variable when the value of every predictor is equal to 0); x_{ij} is the value of the ith case of the jth predictor; and ε_i is the error in the observed value for the ith case. The estimation procedure relies on the assumption that the error term has a normal distribution with a mean of 0.

It is worth noting that least-squares model fit and the associated model goodness-of-fit criteria are applicable to *any* model assessment and are not limited to the simplest case of a linear model and a linear regression. In this case, the stochastic element comes from the error of the fit, which is not necessarily intrinsic randomness associated with the particular physiologic phenomenon.

11.6   Distances, Mean Comparisons, Clustering, and Principal Components

We now review methods that deal with differences or similarities that are hidden within the data. The common element of these methods is that they are based on the mathematical concept of inner product and distance between the data points generally defined as

$$d^2(X, Y)_\Sigma = \sum_{j=1}^{n} \sum_{i=1}^{n} (x_i - y_i) \cdot \sigma^{ij} \cdot (x_j - y_j) \tag{11.3}$$

with respect to a matrix Σ which defines the metric of the data space. For example, if $\Sigma = I$ (the identity matrix), then the inner product is Euclidean:

$$\langle X, Y \rangle_I = (x_1, \ldots, x_n) \cdot \begin{pmatrix} 1 & \cdots & 0 \\ \vdots & 1 & \vdots \\ 0 & \cdots & 1 \end{pmatrix} \begin{pmatrix} y_1 \\ \vdots \\ y_n \end{pmatrix} = \sum_{i=1}^{n} x_i y_i \tag{11.4}$$

Or, if the metric is $\Sigma_n = (1/n - 1)I$, then we have correlation:

$$\langle X, Y \rangle_{\Sigma n} = (x_1, \ldots, x_n) \cdot \begin{pmatrix} \dfrac{1}{n-1} & \cdots & 0 \\ \vdots & \dfrac{1}{n-1} & \vdots \\ 0 & \cdots & \dfrac{1}{n-1} \end{pmatrix} \begin{pmatrix} y_1 \\ \vdots \\ y_n \end{pmatrix} = \dfrac{1}{n-1} \sum_{i=1}^{n} x_i y_i$$

(11.5)

The variety of statistical methods can be classified as depicted in Figures 11.4A−D.

If the goal is to find whether the distance between the means of predefined data subgroups is zero, then the methods of choice fall into the category of *general linear models* (GLMs), which include all *t*-test, ANOVA, and repeated measures analyses (Figure 11.4A); if the goal is to find similarities between cases and the possible subgroups are unknown, then the analytical procedure of choice is *clustering* (Figure 11.4B); if the subgroups of cases are known and we are looking for classification or assignment of new cases to the existing subgroups, then the

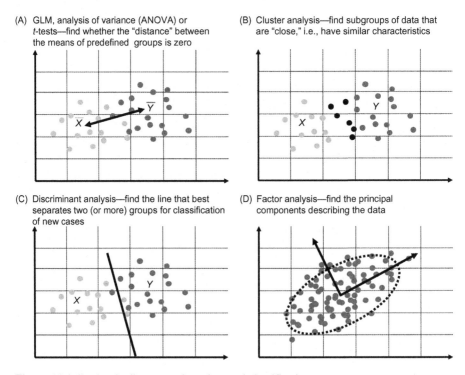

(A) GLM, analysis of variance (ANOVA) or *t*-tests—find whether the "distance" between the means of predefined groups is zero

(B) Cluster analysis—find subgroups of data that are "close," i.e., have similar characteristics

(C) Discriminant analysis—find the line that best separates two (or more) groups for classification of new cases

(D) Factor analysis—find the principal components describing the data

Figure 11.4 Stochastic distances, clustering, and classification.

recommended procedure would be *discriminant analysis* (Figure 11.4C), and if the goal is to find hidden factors underlying the data, then the procedure would be *factor analysis* (Figure 11.4D).

11.6.1 General Linear Models

GLMs present a common framework for statistical comparisons of averages, which includes several methods that can be classified as follows:

- Predictors: *If discrete*, the predictors are generally called factors; in this case, typical analytical procedures include *t*-test and ANOVA; *if continuous*, the predictors are called independent variables that are typically used in regressions, or as covariates in ANOVA.
- Factors in factorial models (below) can be *between-subject* if defined by a grouping variable identifying two or more groups (typical for *t*-test and ANOVA), or *within-subject* if defined by variables that are measured on the same subjects on repeated occasions; typical procedures including within-subject factors are paired *t*-test and repeated measures ANOVA.
- Fixed factorial design is generally presented as $(k_1 \times \cdots \times k_s \times m_1 \times \cdots \times m_t)$, where k_i are the levels of between-subject factors i, $i = 1, 2, \ldots, s$, and m_j are the levels of within-subject factors j, $j = 1, 2, \ldots, t$. For example, a 2×2 ANOVA would include 1 two-level between-subject factor and 1 two-level within-subject factor. Depending on the number and the type of factors, the analytical procedures can be (i) if $s = 1$, $t = 0$, and $k_1 = 2 \rightarrow$ independent samples *t*-test (two groups, one measure); (ii) if $s = 0$, $t = 1$, and $m_1 = 2 \rightarrow$ paired *t*-test (one group, two repeated measures); (iii) any other s and $t \rightarrow$ ANOVA, repeated measures, or GLM procedures that could include contrasts and covariates.

11.6.2 Cluster Analysis

Clustering is performed to identify hidden subgroups of data that coalesce (or cluster) around certain, initially unknown, cluster centers. This is a basic statistical tool used in many classification studies, in particular in gene-expression investigations. Its theoretical base is provided by the concept of distance between "objects" within a certain "space." Depending on the distance that is adopted, the cluster analysis has different names, but the general concept of distance and proximity between data points always remains the same. An incomplete list of various clustering techniques is follows: (i) nearest neighbor—the distance between two clusters is the distance between their closest neighboring points; (ii) farthest neighbor—the distance between two clusters is the distance between their two furthest member points; (iii) unweighted pair-group method using averages—the distance between two clusters is the average distance between all intercluster pairs; (iii) average linkage within groups is the mean distance between all possible inter- or intracluster pairs; (iv) Ward's method calculates the sum of squared Euclidean distances from each case in a cluster to the mean of all variables; (v) centroid method—the cluster to be merged is the one with the smallest sum of Euclidean distances between cluster means for all variables; (vi) median method—clusters are weighted equally regardless of group size when computing centroids of two clusters being combined.

11.6.3 Discriminant Analysis

In Figure 11.4C, there are two known groups of data X and Y, and the objective is to identify a line that best separates the two groups, given certain independent variables (predictors). In other words, we are trying to predict group membership using some independent variables. In some sense, this is similar to a linear regression: while the regression model finds a line that best fits all data points (i.e., minimizes the distance between the data and a straight line), the discriminant model finds a line that best separates predefined groups of data points (i.e., maximizes the distance between the groups defined by the dependent variable). In linear regression the dependent variable is continuous, in discriminate analysis, it is categorical (also called criterion variable). Thus, discriminant analysis identifies principal characteristics (one or more linear functions) that best separate the observed groups. These principal characteristics are derived from a set of predictors and can therefore be used later for the classification of new unknown cases. For example, if a certain combination of age, body mass index, and lipid parameters is favorable for a metabolic syndrome, a person with that combination would be classified as having metabolic syndrome. If a discriminant model was built on two groups, a single line is identified as a discriminant function. If there are three groups, then two lines would be needed. In general, N groups are separated by $N - 1$ discriminant lines if $N - 1$ is greater than the number of predictors. If $N - 1$ is less than the number of predictors, then the number of discriminant lines is equal to the number of predictors. The significance of the discriminant model (as well as the significance of each of the predictors included in the model) can be judged by the values of a coefficient called Wilks' lambda. This coefficient varies from 0 to 1, with 1 meaning that all group means are the same. Thus, a smaller value of Wilks' lambda would indicate a better separation between the groups. An F-test associated with Wilks' lambda is used to reject the null hypothesis H_0: Wilks' lambda $= 1$, and thereby assign a significance level to the discriminant model. Similarly, for each predictor in the model, F-test values show which variables contribute significantly to the group classification.

11.6.4 Factor Analysis

This type of analysis is used to identify hidden (latent) variables that are responsible for the variance in the observed population. These hidden variables are called *principal components* or *factors* and are graphically represented in Figure 11.4D. Geometrically, the data in Figure 11.4D form a cloud that has an ellipsoid shape. The main axes of that ellipsoid represent the principal directions of spread of the data (i.e., the principal components of data variance). Thus, factor analysis seeks to identify these principal components and thereby explain the variance in the data. The strength of the factor analysis is in situations where the data cloud includes multiple variables (i.e., has multiple dimensions) but the real hidden dimensionality of the data is lower. In particular, if two variables are highly correlated, then the variance of these two variables can be explained by a single underlying factor (principal component) that is responsible for most of the spread of the data ellipsoid. In this case, the two-dimensional data cloud can be reduced to one-dimensional representation of its variance. In other words, factor analysis is a data-reduction technique used to identify principal components in the data. The analytical procedure uses the correlation matrix of the observed variables to identify which variables are close (in terms of a correlation distance). The most commonly used analytical procedure is called *principal components analysis*, which identifies several linear combinations (factors) from the variables in hand, such that these factors extract maximum variance from the data. This is done sequentially—one linear combination (Factor 1) is extracted first, then the procedure removes the variance explained by Factor 1 and seeks a second linear combination—Factor 2—which explains the maximum proportion of the remaining variance, and so on. This procedure results in uncorrelated (orthogonal) factors, which explain the variance with a minimum number of principal components (dimensions) (i.e., reveal the true dimensionality of the data).

11.7 Markov Chains

For a given probability model, any indexed collection of random variables $\{X(t)\}_{t \in T}$ can be regarded as a stochastic process. There are many different types of stochastic process models, all distinguished by (i) the "range" space of $X(t)$ (i.e., whether it is continuous or discrete), (ii) the nature of the index set T (i.e., whether it is continuous or discrete), and (iii) joint distributions of collections of random variables $\{X(t_1), X(t_2), \ldots, X(t_n)\}$ within the sequence (i.e., whether those random variables are independent or correlated in some specific way). When the index t is scalar-valued, the index is often referred to as "time," and when index set T is the set of integers $\{\ldots, -2, -1, 0, 1, 2, \ldots\}$, the process is thought as a "discrete-time" process. Similarly, if T is the set of real numbers, then the process is referred to as a process in "continuous time." Inferences about stochastic processes typically derive from some specific regularity structure that ultimately specifies the distribution of $X(t)$ and the relationship of $X(t)$ to other random variables in the sequence. The general theory of stochastic processes has been developed over many years, and is well represented in Refs. [1,12,13–18].

A *Markov chain* is a discrete-time stochastic process satisfying the *Markov property*, requiring that the conditional distribution for $X(t + 1)$ given the entire history of the process up to time t depends explicitly only on the value of the process at time t (i.e., given that $X(0) = x_0, \ldots, X(t) = x_t$). That is, $P(X(t + 1) = x' | X(0) = x_0, \ldots, X(t) = x_t) = P(X(t + 1) = x' | X(t) = x_t)$. In this sense, $X(t)$ for a Markov chain can be regarded as the "state" of the process at time t because the knowledge that $X(t) = x_t$ is all that is needed to fully characterize the conditional distribution of $X(t + 1)$, and the rest of the history of the process up to time t is irrelevant. If the conditional distribution for $X(t + 1)$ given $X(t) = x_t$ does not explicitly depend on time t, then the Markov chain is said to be *homogeneous* (or *stationary*). We refer to the set of values S that $X(t)$ can take as the *state space* of the Markov chain. If the state space is countable, then the Markov chain is said to be *countable* (or *denumerable*). It is a *finite* Markov chain if S is a finite set.

Markov chain models are widely used in modelling physiological processes in which discrete state transitions occur randomly. In reviewing recently published works, Markov chain models have been used in neuroscience, e.g., in modelling brain activity [19], modelling the behavior of neurons [20−23], and modelling spontaneous seizure generation [24]. Markov chain models have been used in modelling abnormal electrocardiographic rhythms [25] and the action of ventricular muscle cells [26,27]. Markov chain models have been proposed for postural control [28] and in modelling human physical activities [29]. Markov chain models have recently been proposed for modelling gene regulatory networks [30−32] and tumor angiogenesis [33], as well as modelling the changing structure of bones [34].

The focus of this section is on the theory of stationary finite Markov chains in discrete time, and the discussion here largely follows the treatment of Ref. [35]. Other references including [18,36] cover the general case of denumerable Markov chains and Markov chains in continuous time.

Let $S = \{1, 2, \ldots, n\}$ denote the state space of a stationary Markov chain. The evolution of the Markov chain can be described in terms of *transition probabilities*: $p_{ij} = P(X(t + 1) = j | X(t) = i)$, with $i, j \in S$. The *transition probability matrix* of the Markov chain is the matrix $P = \{p_{ij}\}_{i,j \in S}$. Both the transient and steady-state characteristics of the Markov chain are ultimately governed by P. To begin with, looking at the nonzero entries of P, it is possible to illustrate all possible state transitions of the Markov chain within a single diagram, as shown in Figure 11.5.

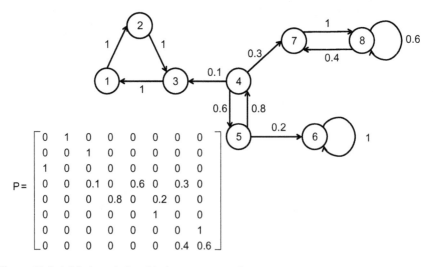

$$P = \begin{bmatrix} 0 & 1 & 0 & 0 & 0 & 0 & 0 & 0 \\ 0 & 0 & 1 & 0 & 0 & 0 & 0 & 0 \\ 1 & 0 & 0 & 0 & 0 & 0 & 0 & 0 \\ 0 & 0 & 0.1 & 0 & 0.6 & 0 & 0.3 & 0 \\ 0 & 0 & 0 & 0.8 & 0 & 0.2 & 0 & 0 \\ 0 & 0 & 0 & 0 & 0 & 1 & 0 & 0 \\ 0 & 0 & 0 & 0 & 0 & 0 & 0 & 1 \\ 0 & 0 & 0 & 0 & 0 & 0 & 0.4 & 0.6 \end{bmatrix}$$

Figure 11.5 A Markov chain with three recurrent classes.

11.7.1 Chapman–Kolmogorov Equation and k-Step Transition Probabilities

Clearly, the state transition probabilities p_{ij} describe the conditional probability that, given that the current state of the chain is i, the state of the Markov chain will be j after one transition. Note that the Markov property also makes it easy to compute the probability that the chain will be in state j after two state transitions: $p_{ij}(2) \equiv P(X(s+2) = j|X(s) = i) = \sum_{k \in S} p_{ik} p_{kj}$ for any time $s = 0, 1, 2, \ldots$, where the formula accounts for (i) the possibility of transitioning from i to any state k in the first step and (ii) making the necessary final transition to j. We refer to $p_{ij}(2)$ as a two-step transition probability. Note that the two-step transition matrix $P(2) = \{p_{ij}(2)\}_{i,j \in S}$ can be computed as the square of the transition probability matrix (i.e., $P(2) = P^2$). From here, any t-step transition probability can be computed recursively as $p_{ij}(t) \equiv P(X(s+t) = j|X(s) = i) = \sum_{k \in S} p_{ik}(t-1) p_{kj}$, the result of which is known as the *Chapman–Kolmogorov equation* for the t-step transition probabilities. Note that the t-step transition probability matrix can be computed as $P(t) = P^t$.

Knowing the probability distribution (mass function) for the state of the Markov chain at some stage s, it is easy to compute the distribution for the state of the system at stage $s + t$. Specifically, representing the distribution at time s as the row vector $\pi^s = (P(X(s)) = 1, P(X(s)) = 2, \ldots, P(X(s)) = n)$, the distribution of the state at stage $s + t$ is $\pi^{s+t} = \pi^s P^t$. (This is because $P(X(s+t) = j) = \sum_{i \in S} P(X(s) = i) p_{ij}(t)$ for all $j \in S$.)

Generally, the analytical questions that one may ask about Markov chains relate to the behavior of $\pi^t = \pi^0 P^t$ as $t \to \infty$, including the following: (i) Does the chain get trapped within a particular subset of states? (ii) What happens along the way to getting trapped? (iii) Does the chain reach some kind of probabilistic steady state? It shall be clear in the following section that the answers to such questions are intimately connected to and governed by the transition probability matrix P.

11.7.2 Classification of States

The state $j \in S$ is said to be *accessible* from $i \in S$ if the conditional probability of being in state j is positive after some number t of state transitions starting from i (i.e., if $p_{ij}(t) > 0$ for some $t = 1, 2, \ldots$). In addition, let $A(i) \subseteq S$ be the subset of states that are accessible from i. The state i is said to be *recurrent* if it itself is accessible from all of the states in $A(i)$. It is not hard to show that if i is a recurrent state and j is accessible from i, then $A(i) = A(j)$, and recurrent states of the chain may be partitioned into a set of one or more *recurrent classes*. Starting within a given recurrent class, the Markov chain may only transition between states within that recurrent class; there is no opportunity to escape. If a state i is recurrent and is the only member of its recurrent class, then i is said to be an *absorbing state*. States i that are not recurrent are said to be *transient*. It is not difficult to show that, starting from a transient state, the process will eventually (with probability 1) transition into one of the recurrent classes of the chain, so that the number of transitions back to the recurrent state is finite (with probability 1). If the entire state space S comprises a single recurrent class, then the chain is said to be *irreducible*. Note that for *finite* Markov chains, there must exist at least one recurrent class. For the Markov chain of Figure 11.5 there are three recurrent classes $\{1, 2, 3\}$, $\{6\}$, and $\{7, 8\}$; the states 4 and 5 are both transient, and state 6 is absorbing.

A given recurrent class is said to be periodic if for some integer $d > 1$, the class can be partitioned into subsets $S_1, \ldots, S_d$ such that $P(X(t+1) \in S_2 | X(t) \in S_1) = 1$, $P(X(t+1) \in S_3 | X(t) \in S_2) = 1$, and so on, cycling eventually back to S_1, i.e., $P(X(t+1) \in S_1 | X(t) \in S_d) = 1$. Note that for such a recurrent class, $p_{ii}(1) = 0$, $p_{ii}(2) = 0, \ldots, p_{ii}(d-1) = 0$; however, it may be that $p_{ii}(d) > 0$. If no such $d > 1$ exists, then the recurrent class is said to be aperiodic. For the Markov chain of Figure 11.5 the recurrent classes $\{6\}$ and $\{7, 8\}$ are aperiodic, and the recurrent class $\{1, 2, 3\}$ is periodic with $d = 3$.

11.7.3   Transient Behavior of Markov Chains

In many applications, it is important to be able to characterize the behavior of the Markov chain in the transitory period prior to reaching one of its recurrent classes.

11.7.3.1   Computing Absorption Probabilities

For Markov chains with two or more recurrent classes, it is of interest to know the probability with which the chain will be absorbed by any one of the recurrent classes. Without loss of generality, one can consider the case where every recurrent state of the chain is an absorbing state. Using $A \subseteq S$ to represent the set of absorbing states, let α_{ia} denote the probability of eventually reaching the absorbing state $a \in A$ starting from the state $i \in S$. Clearly, $\alpha_{ia} = 1$ if $i = a$, and $\alpha_{ia} = 0$ if i is any other absorbing state. For all other starting states i, it can be shown that $\alpha_{ia} = \sum_{j \in S} p_{ij} \alpha_{ja}$, yielding an independent linear equation for each transient state. For the Markov chain of Figure 11.5, the probability of eventually reaching the recurrent class $\{1, 2, 3\}$ from the initial transient state 5 is 4/26, and the corresponding probabilities of reaching $\{6\}$ and $\{7,8\}$ from state 5 are 12/26 and 10/26, respectively.

11.7.3.2   Computing the Expected Time to Reach an Absorbing State

In Markov chains with a single recurrent class, it is of interest to know the expected number of state transitions required to eventually reach the recurrent class from any initial transient state. Without loss of generality, one can consider the case in which the single recurrent class consists of a single absorbing state $a \in S$. Let c_{ij} denote a cost associated with single-step transitions from i to j. In addition, let μ_i denote the expected accumulated cost following a trajectory that eventually reaches a starting from i. Clearly, $\mu_a = 0$, since no state transitions are required to reach a starting from a. For all other initial states $i \neq a$, it can be shown that $\mu_i = \sum_{j \in S} p_{ij}(c_{ij} + \mu_j)$, again yielding an independent linear equation for each transient state. To compute the expected number of transitions required to reach a, it is a simple matter of choosing costs $c_{ij} = 1$ for all transient states i and all $j \in S$. For the Markov chain of Figure 11.5, the expected number of state transitions required to reach any one of the recurrent classes starting from state 5 is 90/26.

11.7.4 Steady-State Behavior

In some applications, it is important to have a simple characterization of the "limiting" probability of the process being in any particular state a long time from now. In this vein, it is of interest to know whether the Markov chain has an associated *steady-state distribution* π^* such that $\pi^* = \lim_{t \to \infty} \pi^0 P^t$ for all initial π^0. (Thus, in order for the process to have a steady-state distribution π^*, it must be true that the limit exists for all π^0 *and* that the value of the limit actually does not depend on π^0.)

The Markov chain of Figure 11.5 demonstrates that a chain need not have a steady-state distribution. For example, for the degenerate initial distribution in which $P(X(0) = 1) = 1$, it is clear that π^t will forever cycle through the sequence of distributions $(1, 0, 0, 0, 0, 0, 0, 0)$, $(0, 1, 0, 0, 0, 0, 0, 0)$, and $(0, 0, 1, 0, 0, 0, 0, 0)$, never converging. Moreover, for the initial distribution in which $P(X(0) = 6) = 1$, it follows that $\pi^t = (0, 0, 0, 0, 0, 1, 0, 0)$ for all t. Thus, it is clear that the limiting behavior of π^t (including its failure to converge) can depend strongly upon the initial state of the process. In particular, periodicity (as in the periodicity of the recurrent class $\{1, 2, 3\}$) contributes to π^t not converging for at least some initial distributions. Similarly, the existence of multiple recurrent classes jeopardizes the uniqueness of the limit if it exists.

Not surprisingly, it can be shown that (i) if the Markov chain has a single class of recurrent states *and* (ii) if the single class of recurrent states is aperiodic, then the chain exhibits a unique steady-state distribution π^*. Moreover, the steady-state distribution is the unique nonnegative solution to the equation $\pi = \pi P$ such that $\pi e = 1$, where $e = (1, 1, \ldots, 1)'$. (The fact that a nonnegative solution exists for a case of finite Markov chains is a result of the celebrated Perron−Frobenius theorem, which relates to the eigenstructure of P. Probabilistic methods are required to derive similar results for Markov chains with countable state spaces.)

Note that the results above can be adapted to analyze individual (aperiodic) recurrent classes within a larger Markov chain. For example, for the Markov chain of Figure 11.5, for any initial distribution π^0 such that $P(X(0) = 7) + P(X(0) = 8) = 1$, then $\pi^t \to (0, 0, 0, 0, 0, 0, 2/7, 5/7)$.

11.7.5 Imperfect Information: Hidden Markov Chains

A *hidden Markov chain* is a Markov chain with transition probability matrix P for which state transitions are only partially revealed as follows: when a state transition takes place, say from i to j, an observation value z is produced that is drawn from a set of observation symbols Z according to a given conditional probability distribution $r(z; i, j) = P(Z = z | X(t) = i, X(t + 1) = j)$. Of course, the underlying (hidden)

Markov chain will have transient and long-term behavior according to P as discussed in Sections 11.7.3 and 11.7.4. However, what is interesting is the prospect of inferring the underlying state of the chain from histories of observations $H_t = \{Z_1 = z_1, Z_2 = z_2, \ldots, Z_t = z_t\}$.

Computing the posterior distribution of X(t): With a prior probability distribution on the initial state of the Markov chain, the posterior distribution can be computed recursively using Bayes' rule:

$$P(X(t+1) = j|H_{t+1}) = \frac{\sum_{i \in S} r(z_{t+1}; i, j) p_{ij} P(X(t) = i|H_t)}{\sum_{l \in S} \sum_{i \in S} r(z_{t+1}; i, l) p_{il} P(X(t) = i|H_t)} \quad (11.6)$$

for each $j \in S$.

Computing the maximum likelihood estimate: In applications where there is no prior distribution on the initial state of the system, it is of interest to be able to compute from H_N the maximum likelihood estimate $(\hat{x}_0, \hat{x}_1, \hat{x}_2, \ldots, \hat{x}_N)$ of the hidden sequence of state transitions $\{X_0 = x_0, X_1 = x_1, X_2 = x_2, \ldots, X_N = x_N\}$. In this case, the maximum likelihood estimate can be computed as the solution to a shortest path problem, as discussed in Ref. [37].

11.7.6 Markov Chains in Continuous Time

Many of the physiological applications identified in the introduction to this section relate to systems in which state transitions occur randomly in time, but where an "embedded" Markov chain describes the specific sequence of state transitions. A *continuous-time Markov chain* is one in which the holding times between states are exponentially distributed with state-dependent transition rates. Figure 11.6 illustrates a hypothetical model of glucose—insulin interactions using a continuous-time Markov chain, where the state of the chains is multivariate accounting for the numbers of glucose (ngut, ng), where ng is proportional to blood glucose concentration BG, and insulin (nsc, ni) "customers" in parallel compartmental systems. Interestingly, the theoretical results that characterize the steady-state behavior of continuous-time Markov chains can be derived from the theory of discrete-time Markov chains via a sampling approach; see for example Ref. [35]. More generally, a semi-Markov chain is one in which holding times are i.i.d for each return to a given state.

11.8 State Estimation for Discrete-Time Linear Systems: Kalman Filtering

Many physiological processes can be described as dynamic systems, and, to the extent that (i) only some of the relevant variables can be measured, (ii) measurements are imperfect, often with significant measurement noise, and (iii) the underlying processes are also subject to unobservable and random external influences (disturbances or "process noise") sophisticated statistical tools are required to

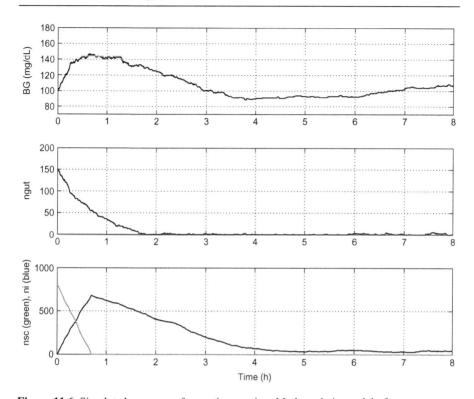

Figure 11.6 Simulated response of a continuous-time Markov chain model of glucose−insulin dynamics.

achieve the best possible understanding of the state of the system from the available data. A popular objective in analyzing the data is to compute the estimate that minimizes the expected square-error between the estimate and the true state of the system given all of the available data. Closed-form solutions exist for the least-square problem when the system model is linear and disturbances and measurement noise are white, zero-mean, Gaussian processes. Kalman filtering provides an efficient technique for recursively computing the least-squares estimate as each new measurement value becomes available. The following paragraphs briefly lay out the basic structure of the Kalman filter for both the time-varying and time-invariant (steady-state) cases. Statistical methods for state estimation are a standard topic in engineering systems, and the material from this section can be found in many places, including Refs. [37−43].

Kalman filtering is ubiquitous in biomedical applications. Recent applications in neuroscience include estimating brain activity [44−46] and fiber tracking in tractography (neuroscience) [47]. Kalman filters have been used in guiding high-intensity focused ultrasound interventions [48,49]. Kalman filters have been used in classifying cardiac arrhythmia [50] and in estimating abnormal motions of the left ventricle [51]. In tracking vital signs, Kalman filters have been used to estimate respiration rate [52] and to estimate blood pressure and heart rate [53]. Recently,

Kalman filters have found application in estimation, control, and safety of automated treatment of diabetes [54–59].

11.8.1 The Transient Case

In its most basic form, Kalman filtering applies to the time-varying Gauss–Markov model below:

$$x(k + 1) = A_k x(k) + B_k u(k) + \omega(k), \quad k = 0, 1, \ldots$$

where

- A_k and B_k are appropriately dimensioned state space matrices,
- $\{\omega(k)\}_{k=0,1,\ldots}$ is a zero-mean white Gaussian process with covariance W_k (positive definite),
- $\{u(k)\}_{k=0,1,\ldots}$ is a systematic input (a known input signal).

Let $y(k) = C_k x(k) + v(k)$ represent a noisy measurement of the system ($k = 1, 2, \ldots$), where C_k is an appropriately dimensioned matrix and $\{v(k)\}_{k=1,2,\ldots}$ is a zero-mean white Gaussian process with covariance V_k (positive definite).

The optimal least-squares estimate of $x(k)$ at any stage is given by the conditional expectation $\hat{x}(k) = E\{x(k)|I(k)\}$, where $I(k)$ is the history of all available measurements and previous control actions:

$$I(k) \equiv \{y(k), y(k-1), \ldots, y(1); u(k-1), \ldots, u(0)\} = \{y(k), u(k-1); I(k-1)\}$$

The Kalman filter computes $\hat{x}(k)$ recursively, as follows. Given $\hat{x}(k-1)$, its error covariance P_{k-1}, and the stage k measurement $y(k)$:

    Step 1. Predict from last estimate: $\breve{x}(k) = A_{k-1}\hat{x}(k-1) + B_{(k-1)}u(k-1)$
    Step 2. Propagate covariance: $\breve{P}_k = A_{k-1}P_{k-1}A'_{k-1} + W_{k-1}$
    Step 3. Compute Kalman gain: $L_k = \breve{P}_k C'_k [C_k \breve{P}_k C'_k + V_k]^{-1}$
    Step 4. Update estimate: $\hat{x}(k) = \breve{x}(k) + L_k[y(k) - C_k \breve{x}(k)]$
    Step 5. Update error covariance: $P_k = [I - L_k C_k]\breve{P}_k$

with the process being initialized by an unbiased initial estimate $\hat{x}(0)$ of $x(0)$ such that $(x(0) - \hat{x}(0))$ is normally distributed with covariance P_0. It is well known [38] that $x(k)$ is conditionally normal with mean $\hat{x}(k)$ and covariance P_k as computed above. Likewise, the anticipated measurement is conditionally normal with mean $\breve{y}(k) = C_k \breve{x}(k)$ and covariance $Y_k = C_k \breve{P}_k C'_k + V_k$.

It is important to note that, even if the linear dynamic system is time invariant and/or the disturbance and measurement noise processes are stationary, it is appropriate to use this time-varying form of the Kalman filter to account for the transient effect of the initial estimate $\hat{x}(0) = \bar{x}(0)$ and covariance $P_0 = X_0$.

11.8.2 Time-Invariant Steady-State Estimation

In the case where the linear system is time invariant and the disturbance and noise processes are stationary, i.e., where $A_k = A, B_k = B, H_k = H, C_k = C, V_k = V$ for all

$k = 0, 1, \ldots$, the covariance matrix of the error of the state estimate may converge. For example, if both (A, C) is observable and (A, D) is controllable (where D is the "square root" of the disturbance process covariance W, i.e., $W = DD'$), then it can be shown (see, e.g., Ref. [37]) that $\check{P}_k$ converges to the unique positive definite solution P to the algebraic Riccati equation:

$$\check{P} = A(\check{P} - \check{P}C'(C\check{P}C' + V)^{-1}C\check{P})A' + W \tag{11.7}$$

and the corresponding steady-state error covariance (the limit of P_k) is

$$P = \check{P} - \check{P}C'(C\check{P}C' + V)^{-1}C\check{P} \tag{11.8}$$

In addition, the steady-state operation of the Kalman filter can be expressed in reduced form as

$$\hat{x}(k + 1) = A\hat{x}(k) + Bu(k) + PC'V^{-1}(y - C(A\hat{x}(k) + Bu(k))), \tag{11.9}$$

which, under the same assumption of observability and controllability, is stable in the sense that the error covariance of the estimate produced this way will asymptotically converge to P, regardless of the actual initial error covariance. (These results can be generalized to the case where (A, C) is detectable and (A, D) is stabilizable.)

11.8.3   Example

To illustrate the operation of the time-varying form of the Kalman filter, and the transition into steady-state operation, consider the two-state example below:

$$A = \begin{bmatrix} 0.6 & 0 \\ 0.3 & 0.9 \end{bmatrix}, \quad B = \begin{bmatrix} 1 \\ 0 \end{bmatrix}, \quad W = \begin{bmatrix} 0.04 & 0 \\ 0 & 0.04 \end{bmatrix}$$
$$C = \begin{bmatrix} 0 & 1 \end{bmatrix}, \quad V = 1$$

and suppose that true initial state of the process is $x(0) = (1 \quad 3)'$, while the estimate of the initial state is $\hat{x}(0) = (2 \quad 6)'$, which is assumed to have zero error

$$\left(\text{i.e., } P_0 = \begin{bmatrix} 0 & 0 \\ 0 & 0 \end{bmatrix} \right)$$

Figure 11.7 illustrates the response of the system to a constant systematic input $u(k) = 0.5$ for $k = 0, 1, \ldots, 100$.

Note that while the time-varying and steady-state estimates both start at $\hat{x}(0) = (2 \quad 6)'$,

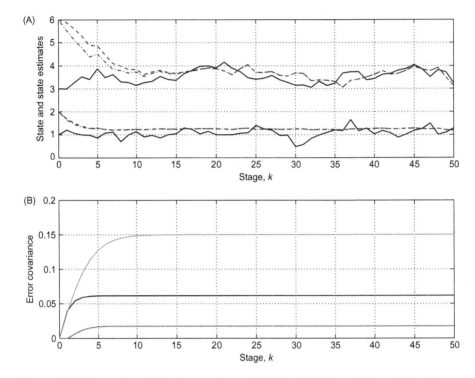

Figure 11.7 Comparison of the time-varying and steady-state Kalman filter estimates for a two-state system. (A) The solid (black in web version) traces correspond to the true state of the system, and the dashed (blue in web version) and dash−dot (blue in web version) traces correspond to the time-varying and steady-state Kalman filter estimates, respectively. (B) The convergence of P_k for the time-varying Kalman filter.

- the two estimates diverge initially corresponding to the fact that the steady-state Kalman filter essentially always assumes a constant error covariance P while the time-varying Kalman filter acknowledges and modifies over time the initial error covariance P_0 and
- the estimates eventually coalesce as $P_k \rightarrow P$.

11.8.4 Extensions

The Kalman filter has been extended in various ways to accommodate different modelling assumptions. For example, the Kalman−Bucy filter [60] provides the least-squares state estimate for linear systems in continuous time, where the filter now takes the form of a set of differential equations. Nonwhite disturbance and measurement noise processes can be addressed through state augmentation [61]. The extended Kalman filter (see Ref. [38] for example) is a standard method of estimating the state of a differentiable nonlinear system via linearization of the model about the current estimate of the process. The unscented Kalman filter [62] is a popular method for nonlinear estimation using a deterministic sampling technique to avoid the difficulties of linearization.

11.9 Conclusion

Clinical criteria for diagnosis and treatment, and now algorithmic and closed-loop treatments of disease all depend on accurate models of physiological phenomena. Whereas inter- and intrapatient/subject variability are important issues in characterizing these phenomena, the available models are often fundamentally deterministic. While intersubject variability can be addressed to some extent through the development of a representative population of deterministic model instances (with each member of the population having a distinct set of parameters), intrasubject variability is difficult to capture without resorting to stochastic models. In this chapter, we have reviewed some of the key concepts from probability theory, stochastic processes, and statistical inference as they relate to physiological modelling. We have also reviewed basic probability and statistical methods that are pertinent to the analysis and interpretation of physiological data. Stochastic models of the type presented here will become increasingly important as medical interventions become more and more personalized.

References

[1] Feller W. An introduction to probability and its applications. New York, NY: John Wiley & Sons; 1950, 1957, 1968.

[2] Shiryaev A. Probability. New York, NY: Springer Graduate Texts in Mathematics; 1995.

[3] George EP. Box. Bayesian inference in statistical analysis. New York, NY: Wiley Classics Library; 1992.

[4] Vyshemirsky V, Girolami M. BioBayes: a software package for Bayesian inference in systems biology. Bioinformatics 2008;24(17):1933−4.

[5] Chowdhary R, Zhang J, Liu JS. Bayesian inference of protein−protein interactions from biological literature. Bioinformatics 2009;25(12):1536−42.

[6] Ben-Gal I. Bayesian networks. In: Ruggeri F, Faltin F, Kenett R, editors. Encyclopedia of statistics in quality & reliability. New York, NY: John Wiley & Sons; 2007.

[7] Hulst J. Modeling physiological processes with dynamic Bayesian networks. Doctoral thesis. Faculty of Electrical Engineering, Mathematics and Computer Science, Delft University of Technology; 2006.

[8] Van Gerven MA, Taal BG, Lucas PJ. Dynamic Bayesian networks as prognostic models for clinical patient management. J Biomed Inf 2008;41(4):515−29.

[9] Friedman N, Linial M, Nachman I, Pe'er D. Using Bayesian networks to analyze expression data. J Comput Biol 2000;7(3-4):601−20.

[10] Einstein A. On the motion of small particles suspended in a stationary liquid, as required by the Molecular Kinetic Theory of Heat 1905 [English translation]

Investigations on the theory of the Brownian motion. Mineola, NY: Dover Publications; 1926

[11] Philibert J. One and a half century of diffusion: Fick, Einstein, before and beyond. Diffus Fundam 2005;2(6):1–19.

[12] Cinlar E. Introduction to stochastic processes. Englewood Cliffs, NJ: Prentice-Hall; 1975.

[13] Grimmett G, Stirzakir D. Probability and random processes. 3rd ed. Oxford: Oxford University Press; 2001.

[14] Jacod J, Shiryaev AN. Limit theorems for stochastic processes. Berlin: Springer-Verlag; 1987.

[15] Lipster RS, Shiryayev AN. Statistics of random processes I: general theory. New York, NY: Springer-Verlag; 1977.

[16] Lipster RS, Shiryayev AN. Statistics of random processes II: applications. 2nd ed. Berlin: Springer; 2001.

[17] Papoulis A. Probability, random variables, and stochastic processes. 3rd ed. New York, NY: McGraw-Hill; 1991.

[18] Ross SM. Stochastic process. 2nd ed. New York, NY: John Wiley & Sons; 1996.

[19] Chaari L, Vincent T, Forbes F, Dojat M, Ciuciu P. Fast joint detection-estimation of evoked brain activity in event-related fMRI using a variational approach. IEEE Trans Med Imaging 2013;32(5):821–37.

[20] Groff J, DeRemigio H, Smith G. Markov chain models of ion channels and calcium release sites. In: Laing C, Lord G, editors. Stochastic methods in neuroscience. Oxford: Oxford University Press; 2009.

[21] Ivanov AV, Petrovsky AA. Markov coding strategy of the simple spiking of auditory neuron. Proceedings of the international joint conference on neural networks; 2006. p. 4235–42.

[22] Kostal L, Lansky P. Classification of stationary neuronal activity according to its information rate. Network: Comp Neural Syst 2006;17(2):193–210.

[23] Roberts WJJ, Ephraim Y. An EM algorithm for ion-channel current estimation. IEEE Trans Signal Process 2008;56(1):26–33.

[24] Shayegh F, Sadri S, AmirFattahi R. A theoretical model for spontaneous seizure generation based on Markov chain process. Proceedings of the IEEE/EMBS conference on neural engineering; 2009. p. 637–40.

[25] Clifford GD, Nemati S, Sameni R. An artificial vector model for generating abnormal electrocardiographic rhythms. Physiol Meas 2010;31(5):595–609.

[26] Iyer V, Mazhari R, Winslow RL. A computational model of the human left-ventricular epicardial myocyte. Biophys J 2004;87(3):1507–25.

[27] Passini E, Severi S. Computational analysis of extracellular effects on an improved human ventricular action potential model. Proceedings of the 2012 computing in cardiology conference; 2012. p. 873–76.

[28] Hur P, Shorter KA, Mehta PG, Hsiao-Wecksler ET. Invariant density analysis: modeling and analysis of the postural control system using Markov chains. IEEE Trans Biomed Eng 2012;59(4):1094–100.

[29] Mannini A, Sabatini AM. Classification of human physical activities from on-body accelerometers: a Markov modeling approach. Proceedings of the international conference on bio-inspired systems and signal processing; 2011. p. 201–8.

[30] El Samad H, Khammash M, Petzold L, Gillespie D. Stochastic modeling of gene regulatory networks. Int J Robust Nonlinear Control 2005;15:691–711.

[31] Pal R, Bhattacharya S. Transient dynamics of reduced-order models of genetic regularity networks. IEEE/ACM Trans Comput Biol Bioinform 2012;9(4):1230−44.

[32] Zhang S-Q, Ching W-K, Jiao Y, Wu L-W, Chan RH. Construction and control of genetic regulatory networks: a multivariate Markov chain approach. J Biomed Sci Eng 2008;1(1):15−21.

[33] Mohammadi B, Haghpanah V, Larijani B. A stochastic model of tumor angiogenesis. Comput Biol Med 2008;38(9):1007−11.

[34] Rusconi M, Valleriani A, Dunlop JWC, Kurths J, Weinkamer R. Quantitative approach to the stochastics of bone remodelling. Europhys Lett 2012;97(2):28009.

[35] Bertsekas DP, Tsitsiklis JN. Introduction to probability. 2nd ed. Belmont, MA: Athena Scientific; 2008.

[36] Kemeny JG, Snell JL, Knapp AW. Denumerable Markov chains. New York, NY: Springer-Verlag; 1976.

[37] Bertsekas DP. 3rd ed. Dynamic programming and optimal control, vol. 1. Belmont, MA: Athena Scientific; 2005.

[38] Anderson BDO, Moore JB. Optimal filtering. Englewood Cliffs, NJ: Prentice-Hall; 1979.

[39] Gelb A, editor. Applied optimal estimation. Cambridge, MA: MIT Press; 1974.

[40] Goodwin GC, Sin KS. Adaptive filtering prediction and control. Englewood Cliffs, NJ: Prentice-Hall; 1984.

[41] Kalman RE. A new approach to linear filtering and prediction problems. Trans ASME J Basic Eng 1960;82(Series D):35−45.

[42] Krishnan V. Nonlinear filtering and smoothing. New York, NY: John Wiley & Sons; 1984.

[43] Maybeck P. Stochastic models, estimation, and control. New York, San Francisco, London: Academic Press; 1982.

[44] Deneux T, Faugeras O. EEG-fMRI fusion of paradigm-free activity using Kalman filtering. Neural Comput 2010;22(4):906−48.

[45] Hu X-S, Hong K-S, Ge SS, Jeong MY. Kalman estimator- and general linear model-based on-line brain activation mapping by near-infrared spectroscopy. Biomed Eng Online 2010;9:82.

[46] Lenz M, Musso M, Linke Y, Tuscher O, Timmer J, Weiller C, Schelter B. Joint EEG/fMRI state space model for the detection of directed interactions in human brains—a simulation study. Physiol Meas 2011;32(11):1725−36.

[47] Malcolm JG, Shenton ME, Rathi Y. Filtered multisensor tractography. IEEE Trans Med Imaging 2010;29(9):1664−75.

[48] de Senneville BD, Roujol S, Hey S, Moonen C, Ries M. Extended Kalman filtering for continuous volumetric MR-temperature imaging. IEEE Trans Med Imaging 2013;32 (4):711−8.

[49] Roujol S, de Senneville BD, Hey S, Moonen C, Ries M. Robust adaptive extended Kalman filtering for real time MR-thermometry guided HIFU interventions. IEEE Trans Med Imaging 2012;31(3):533−42.

[50] Sayadi O, Shamsollahi MB. Life-threatening arrhythmia verification in ICU patients using the joint cardiovascular dynamical model and a Bayesian filter. IEEE Trans Biomed Eng 2011;58(10):2748−57.

[51] Punithakumar K, Ben Ayed I, Ross IG, Islam A, Chong J, Li S. Detection of left ventricular motion abnormality via information measures and Bayesian filtering. IEEE Trans Inf Technol Biomed 2010;14(4):1106−13.

[52] Nemati S, Malhotra A, Clifford GD. Data fusion for improved respiration rate estimation. EURASIP J Adv Signal Process 2010; 926305.

[53] Li Q, Mark RG, Clifford GD. Artificial arterial blood pressure artifact models and an evaluation of a robust blood pressure and heart rate monitor. Biomed Eng Online 2009;8:13.

[54] Hughes-Karvetski C, Patek SD, Breton MD, Kovatchev BP. Historical data enhances safety supervision system performance in T1DM insulin therapy. Comput Methods Programs Biomed 2013;109(2):220−5.

[55] Facchinetti A, Sparacino G, Cobelli C. An online self-tunable method to denoise CGM sensor data. IEEE Trans Biomed Eng 2010;57(3):634−41.

[56] Facchinetti A, Sparacino G, Cobelli C. Online denoising method to handle intraindividual variability of signal-to-noise ratio in continuous glucose monitoring. IEEE Trans Biomed Eng 2011;58(9):2664−71.

[57] Facchinetti A, Favero S, Sparacino G, Cobelli C. An online failure detection method of the glucose sensor-insulin pump system: improved overnight safety of type-1 diabetic subjects. IEEE Trans Biomed Eng 2013;60(2):406−16.

[58] Parker RS, Doyle III FJ, Peppas NA. A model-based algorithm for blood glucose control in type I diabetic patients. IEEE Trans Biomed Eng 1999;46(2):148−57.

[59] Cameron F, Niemeyer G, Bequette BW. Extended multiple model prediction with application to blood glucose regulation. J Process Control 2012;22(8):1422−32.

[60] Kalman RE, Bucy RS. New results in linear filtering and prediction theory. Trans ASME, J Basic Eng 1961;83(1):95−107.

[61] Bryson AE, Henrikson LJ. Estimation using sampled data containing sequentially correlated noise. J Spacecr Rockets 1968;5(6):662−5.

[62] Julier SJ, Uhlmann JK. Unscented filtering and nonlinear estimation. Proc IEEE 2004;92(3):401−22.

12 Probabilistic Modelling with Bayesian Networks

Francesco Sambo[a], Fulvia Ferrazzi[b] and Riccardo Bellazzi[c]

[a]Department of Information Engineering, University of Padova, Padova, Italy, [b]Institute of Human Genetics, Friedrich-Alexander-Universität Erlangen-Nürnberg, Erlangen, Germany, [c]Department of Electrical, Computer and Biomedical Engineering, University of Pavia, Pavia, Italy

12.1 Introduction

Probabilistic graphical models have been used by statisticians since the early 1920s as a modelling framework useful for describing the dependencies among a set of variables in multivariate problems [1,2]. Within this framework, each variable is represented as a node of a graph, and the edges between two nodes express a probabilistic relationship between such nodes. Probabilistic graphical models possess two important characteristics: (i) they clearly express the conditional independence between the variables, thus allowing an intuitive but sound way to describe the assumptions underlying the modelling process and (ii) they associate to the graph a probabilistic model that can be used for performing inference, and thus, estimation, simulation, and prediction [3]. Their nature has made them progressively more attractive as soon as computers have been made available. A decisive step has been represented by the work carried on by artificial intelligence researchers in the 1980s, since such models have been considered as a powerful means for implementing probabilistic reasoning strategies, in which the goal was to reason about hypotheses given the available data, such as the ranking of plausible diagnoses given the patient's symptoms [4]. In particular, researchers further clarified probabilistic model semantics by introducing Bayesian networks (BNs), directed acyclic graphs (DAGs) in which nodes represent the problem variables, and the probabilistic relationships between variables are expressed by the conditional probability of each node given its parents in the graph [3,5]. Some authors have then restricted the definition of BNs to discrete variables only, so that the conditional probability is expressed by means of probability tables [3]. Rather interestingly, such a model allows expressing the joint probability distribution of all problem variables through

Modelling Methodology for Physiology and Medicine. DOI: http://dx.doi.org/10.1016/B978-0-12-411557-6.00012-4

the multiplication of the conditional probabilities attached to the graph, thanks to a rule called the *chain rule*, described in detail in the following sections [3,5]. This shows that BNs are systems that express a joint probability distribution in a compact way, since they require specifying local models only (i.e., the conditional probability tables, CPTs). Moreover, relying on the clear semantic of DAGs, they allow expressing conditional independence relationships in a clear way, at the same time understandable by humans and computable by machines [6]. Since their first introduction, BNs have further stimulated research in several areas.

First of all, they have been studied as instruments for computing posterior probability distributions (i.e., the posterior probability of any of the problem variables given the knowledge about any of the other variables of the problem). This theme is called *inference*. Given that the general inference problem of BNs has been proven to be NP-hard, a variety of methods have been defined, ranging from exact algorithms to approximate solutions of the exact ones, to stochastic simulation algorithms [3,5,7].

Noteworthy research efforts have been carried on to extend BNs to continuous variables, although in this case general results can be obtained under specific assumptions on the conditional probabilities, such as in the presence of conditionally Gaussian models [8].

Another important research direction has been related to the automated construction of BNs from data. Also in this case, a large number of algorithms have been designed and tested, including methods to learn CPTs from data, methods that induce the graph and learn the conditional probability models, and, finally, approaches that allow learning the model with hidden variables [7,9−11].

A final crucial research topic dealt with representing time and dynamical systems. In this case a general discrete-time extension of BNs have been introduced called dynamic Bayesian networks (DBNs) [12]. DBNs are in fact a very general model, in which the variables are related to each other over adjacent time steps, called time slices. If the problem variables are the input, state, and output variables, a DBN may represent any probabilistic discrete-time dynamical system. For this reason, DBNs can easily represent hidden Markov models and Kalman filters [7].

The interest in general instruments able to compute posterior probability distributions has been quite high in the Bioengineering and Biomedical Informatics community. As a matter of fact, BNs and DBNs allow the dealing with a variety of crucial problems in biomedicine, ranging from classification to prediction, and from simulation to parameter estimation. There are some experiences that have been landmark points in BN applications in medicine and biology. A noteworthy example was represented by the re-engineering of an expert system called quick medical reference (QMR), an information resource to help physicians in diagnosing adult diseases, as an internist [13]. QMR-DT was a probabilistic version based on a two-level BN [14], representing disease at the top level and findings at the bottom level. Once the findings were observed, the posterior probability of all diseases was computed. Another large system that exploited BNs for diagnosis and interpretation was MUNIN [15−17]. MUNIN was aimed to be a "full expert system" for neuromuscular diagnosis, exploiting a probabilistic representation of the nervous system to interpret electromyography (EMG). The overall goal was to include many aspects of EMG assessment, comprising test planning, test guidance, test setup, signal processing of test results, diagnosis, and treatment recommendation. Leveraging the experience carried on in MUNIN, the same researchers started a project on the definition of a model-based probabilistic expert system, called DIAS [18–21]. DIAS used a DBN to model glucose metabolism and a simulation strategy to support patients and physicians in defining the best insulin regimen in type 1 and type 2 diabetes.

BNs have been largely used also as a conceptual framework for parameter estimation in complex probabilistic modelling [22] and as an instrument for state reconstruction in smoothing and deconvolution [23].

Later, BNs have been largely used in bioinformatics. In this area, they have been exploited in different contexts. As a classification algorithm, they showed the ability to integrate genotype and phenotype data in a coherent diagnostic framework. An example is represented by the work of Sebastiani et al., who learned a BN model from data coming from over 1300 individuals with sickle cell anemia (SCA) who have been genotyped, in order to estimate their probability of stroke given the genetic profile [24]. Moreover, they have been used to represent the complex intertwined relationships between genes by learning a DBN from gene expression time series in a series of papers describing different modelling domains, including cell cycle [25], influenza A (H1N1) infection [26], and SOS DNA repair network [27].

Rather interestingly, a number of tools are available to design and learn BNs, as well as to perform inference and simulation [28,29]. For these reasons, BNs represent nowadays a very important modelling tool in biomedicine and this chapter is conceived to provide the basic theoretical and practical elements to start working with this instrument.

The chapter develops as follows. First, we will describe the theoretical foundations of BNs, including a brief review of probability and graph theory, a formal definition of BNs and details on discrete, continuous, and dynamic BNs. Then, we will report some of the algorithms for inference, conditional probability learning, and structure learning. Finally, we will describe a number of examples of BN applications in biomedicine.

12.2   Theoretical Foundations

This section introduces some key concepts from probability theory and graph theory, which are then exploited to formally define the BN model. Whenever possible, we will follow the same notation used in Ref. [5], to which we refer for a more detailed and extensive presentation of the topic.

12.2.1   Probability and Graph Theory

We begin by defining a *sample space* Ω as a set of *outcomes*, i.e., $\Omega = \{\omega_1, \omega_2, \ldots, \omega_n\}$. In addition, we assume that there is a set of *measurable events*

S to which we are willing to assign probabilities: formally, each *event* $\alpha \in S$ is a subset of Ω. Events can be considered as the fundamental units of probability theory.

A *probability distribution P* over (Ω, S) is a *function* from events in S to real values between 0 and 1. Since events are sets, we can perform set operations on them: for example, the probability of two events, α and β, simultaneously occurring can be specified by $P(\alpha \cap \beta)$.

To model how the evidence of an event α affects the probability of another event β, we introduce the concept of *conditional probability*: formally, the conditional probability of β given α is defined as

$$P(\beta | \alpha) = \frac{P(\alpha \cap \beta)}{P(\alpha)}$$

From the formula above one can note that $P(\alpha \cap \beta) = P(\alpha)P(\beta|\alpha) = P(\beta)P(\alpha|\beta)$, which implies what is known as the *Bayes' rule*

$$P(\alpha | \beta) = \frac{P(\beta | \alpha)P(\alpha)}{P(\beta)}$$

Bayes' rule has a central role in probabilistic modelling, in that it allows the computing of the conditional probability $P(\alpha|\beta)$, also known as the *posterior probability* of α given β, from the *prior* or *marginal* probability $P(\alpha)$ and from the *likelihood* of α given β, $P(\beta|\alpha)$.

Two events α and β are said to be *independent* if and only if $P(\alpha \cap \beta) = P(\alpha) \cdot P(\beta)$. This directly implies that event α is independent of event β, denoted $(\alpha \perp \beta)$, if $P(\alpha|\beta) = P(\alpha)$ or if $P(\beta) = 0$. In addition, independence is a symmetric notion, that is, $(\alpha \perp \beta)$ implies $(\beta \perp \alpha)$.

Independent events, however, are not often encountered in physiology and medicine: a common situation is when two events are independent given an additional event. We say that an event α is *conditionally independent* of event β given event γ, denoted $(\alpha \perp \beta | \gamma)$, if $P(\alpha|\beta \cap \gamma) = P(\alpha|\gamma)$ or if $P(\beta \cap \gamma) = 0$.

Another major building block of probability theory is the concept of *random variable*: a random variable is defined by a function that associates a value to each outcome in Ω. Random variables can take different sets of values: *categorical* (or *discrete*) random variables can take one of a few values, while *continuous* random variables can take infinitely many numbers. We will use upper case roman letters X, Y, Z to denote random variables, lowercase letters x, y, z to refer to a value of a random variable and Val(X) to denote the set of values that X can take. Moreover, we will use bold face letters $\mathbf{X}$, $\mathbf{Y}$, $\mathbf{Z}$ to denote sets of random variables, while $\mathbf{x}$, $\mathbf{y}$, $\mathbf{z}$ will denote assignments of values to the variables in these sets.

Once a random variable X is defined, we can consider the *marginal* distribution over events that can be described using X and denote it by $P(X)$. As an example, consider the random variable *Obesity* (O). The *marginal* distribution over *Obesity* assigns probability to specific events such as $P(O = true)$ and $P(O = false)$; suppose $P(O = true) = 0.3$ and $P(O = false) = 0.7$. If we consider the random variable *GlucoseTolerance* (*GT*), with possible values *normal* (*ngt*), *impaired* (*igt*), and *type 2 diabetes* (*t2d*), we can also define a distribution such as $P(GT = ngt) = 0.65$, $P(GT = igt) = 0.2$, and $P(GT = t2d) = 0.15$.

In what follows, when the identity of a random variable is clear from the context, we will often write $P(x)$ as a shorthand for $P(X = x)$.

In many situations, we are interested in questions that involve the values of several random variables: for example, we might be interested in the event "*Obesity = true* and *GlucoseTolerance = impaired.*" To model such events, we need to consider the *joint distribution* over the two random variables. In general, the joint distribution over a set $X = \{X_1, \ldots, X_n\}$ of random variables is denoted $P(X_1, \ldots, X_n)$ and is the distribution that assigns probabilities to events that are specified in terms of these random variables. An example of joint distribution of the two random variables *Obesity* and *GlucoseTolerance* can be seen in Table 12.1.

The joint distribution of two random variables has to be consistent with the marginal distributions, in that $P(x) = \sum_{y \in \text{Val}(Y)} P(x, y)$. As shown in Table 12.1, the marginal distribution of *Obesity* can in fact be computed by summing the probabilities across each row, while the marginal distribution of *GlucoseTolerance* is computed by summing across columns.

The concept of conditional independence can be extended from events to random variables.

Let X, Y, and Z be sets of random variables. We say that X is *conditionally independent* of Y *given* Z if $(X = x \perp Y = y | Z = z)$ for all values $x \in \text{Val}(X)$, $y \in \text{Val}(Y)$, and $z \in \text{Val}(Z)$.

We conclude this section by providing some definitions from graph theory.

A *graph* $\mathcal{G}$ is a data structure consisting of a set of nodes $\mathcal{X} = \{X_1, \ldots, X_n\}$ and a set of edges ε. A pair of nodes X_i, X_j can be connected by a *directed edge* $X_i \rightarrow X_j$ or an *undirected edge* $X_i - X_j$: the set of edges ε is thus a set of pairs with or without direction. We say a graph is *directed* if all of its edges are directed, *undirected* if all edges are undirected, and *partially directed* otherwise. Moreover, the graph $\mathcal{G}'$, obtained from another graph $\mathcal{G}$ by replacing all directed edges with undirected edges, is called the *skeleton* of $\mathcal{G}$.

Table 12.1 Example of a Joint Distribution $P(Obesity, GlucoseTolerance)$

		Obesity		
		FALSE	TRUE	
Glucose Tolerance	ngt	0.45	0.2	0.65
	igt	0.1	0.1	0.2
	t2d	0.05	0.1	0.15
		0.6	0.4	1

A probability is specified for each combination of values of the two random variables. Summing by rows or columns, it is possible to obtain the marginal distribution of each random variable. The sum of all the entries in the table is 1.

Whenever we have an edge $X_i \to X_j$, we say that X_j is a *child* of X_i and that X_i is a *parent* of X_j. We use Pa_X to denote the set of parents of X. We call a *path* in a graph an oriented succession of directed or undirected edges and a *cycle* a path which begins and ends at the same node. If a node X_i precedes a node X_j in a path, we say that X_i is an *ancestor* of X_j and that X_j is a *descendant* of X_i. Finally, a graph is *acyclic* if it does not contain any cycle.

12.2.2 Bayesian Networks

A BN is a probabilistic graphical model that represents conditional dependence over a set of random variables in a compact and human-readable form.

A BN is completely determined by a DAG, known also as the *network structure*, and by a set of conditional probability distributions: each node of the network corresponds to a random variable and each edge corresponds to a probabilistic dependence between the two nodes. In what follows, thus, the terms *node* and *random variable* will be used as synonyms.

More precisely, a BN represents a joint probability distribution between its nodes for which the *Markov condition* holds: any node in a BN is conditionally independent of its nondescendants, given its parents.

The Markov condition implies that the joint probability distribution of the nodes can be decomposed as

$$P(X_1, \ldots, X_n) = \prod_{i=1}^{n} P(X_i | Pa_{X_i}) \tag{12.1}$$

The decomposition of Eq. (12.1) is called *chain rule for* BNs and allows a more compact representation of the full joint probability distribution, requiring fewer parameters to be completely specified: the probability distribution of each node can in fact be expressed simply as a function of the states of its parent nodes.

Figure 12.1A reports an example of discrete BN with four nodes, modelling the hypothetical probabilistic relations between the variables *HighFatDiet* (*HFD*), *GlucoseTolerance* (*GT*), *Obesity* (*OB*), and *RiskOfCardiovascularDisease* (*CVD*). As it is clear from the figure, the probability distribution of each node is expressed as a function of all possible combinations of values of its parents, in the form of a *CPTs*. Such a network can be used to answer queries like: "What is the probability of being obese, if on a high fat diet?"; "What is the probability of having impaired glucose tolerance, if at risk of cardiovascular disease?"; "If obese and on a high fat diet, what is the probability of being at risk of cardiovascular disease?" Moreover, the Markov condition can be used to infer conditional independence relations from the network. For example, we can infer from the network structure that, once the values of *GlucoseTolerance* and *Obesity* are known, *RiskOfCardiovascularDisease* becomes independent of *HighFatDiet*.

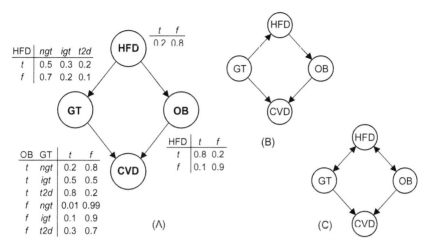

Figure 12.1 (A) Example of a 4-node discrete BN. (B) BN equivalent to network (A). (C) Equivalence class of networks (A) and (B).

Two network structures $\mathscr{G}$ and $\mathscr{G}'$ are said to be *equivalent* if the set of distributions that can be represented using $\mathscr{G}$ is identical to the set of distributions that can be represented using $\mathscr{G}'$. Two equivalent networks always share the same set of nodes, the same skeleton and the same set of *immoralities* [30]: an immorality is a structural pattern in the form $X \rightarrow Z \leftarrow Y$ with no direct edge between X and Y. Figure 12.1B represents a BN equivalent to the one of Figure 12.1A.

Equivalent networks can be grouped in an *equivalence class*, which can be represented by a completed partially directed acyclic graph (CPDAG). In a CPDAG, the edges compelled to a specific direction are directed, while the edges that can take both directions in the equivalence class are left undirected. The CPDAG of an equivalence class is unique and can be computed from any of the DAGs in the class in polynomial time [31]. Figure 12.1C shows the CPDAG representing the equivalence class of the networks in Figure 12.1A and B: the direction of the two edges incoming to *CVD* is compelled, since they form an immorality, while the other two edges can be freely reversed.

12.2.2.1 Continuous Bayesian Networks

The example in Figure 12.1 is a BN with discrete variables (i.e., variables with a finite number of possible values). Conditional distributions in discrete variable BNs can be conveniently represented with probability tables and are able to model dependencies between variables without making any assumption on the underlying relationship (e.g., linearity). Many real-world variables are of a continuous nature (e.g., blood glucose concentration or gene expression levels). In these cases, a possible solution is to discretize these variables and resort to discrete BNs. In some cases, though, discretization would lead to a major loss of information, unless a high number of discrete states is employed, which would significantly increase model complexity. The other solution is to employ continuous-variable BNs. The general BN framework so far presented holds for both discrete and continuous variables, as long as the conditional distribution $P(X|Pa_X)$ assigned to each node represents for each possible value pa_X of Pa_X a distribution over X [5]. When all variables in the network are continuous, the most commonly employed distribution is a linear Gaussian distribution model. Given the continuous variable Y with continuous parents $X_1, \ldots, X_k$, the probability density of Y as a function of its parents is

$$P(Y|x_1, \ldots, x_k) = N(\beta_0 + \beta_1 x_1 + \cdots + \beta_k x_k; \sigma^2)$$

This simple model can be extended to cases in which the mean of Y depends on its parents in a nonlinear way, or in which the variance also depends on the parent values. Hybrid models are also possible, which incorporate both discrete and continuous variables.

In the following section, we will focus on inference and learning algorithms for discrete variable BNs. Some of these algorithms can be directly extended to the linear Gaussian case, as we will point out, while we refer the reader to the literature for other cases [5].

12.2.2.2 Dynamic Bayesian Networks

DBNs are an extension of BNs that represent the temporal evolution of variables over time. Nodes in the DAG associated with a DBN continue to represent random variables, while edges represent temporal dependencies. The key assumption is that the probability distributions describing the temporal dependencies are time invariant so that the overall temporal evolution of the analyzed process can be entirely reconstructed by knowing the temporal dependencies represented in the DBN graph [32]. Figure 12.2 shows an example of DBN describing the evolution of the expression values of three genes $G1$, $G2$, and $G3$. The graph shows that the expression value of each gene at time $(t + 1)$ is assumed to depend on the gene's expression at time t as well as on the expression of one or two of the other genes. Furthermore, the example shows that the temporal dimension of DBNs allows encoding feedback regulation such as the one occurring between $G1$ and $G2$, which is not possible in

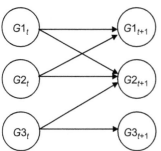

Figure 12.2 Example of a DBN describing the temporal evolution of three genes' expression values.

static BNs because of the required acyclicity of the graph. The example in the figure is a DBN of order 1, as all temporal dependencies occur between consecutive time points; yet DBNs are not restricted to dependencies of order 1 but can also represent higher order dependencies.

12.3   Algorithms

12.3.1   Inference

A BN provides a compact model to completely describe the joint probability distribution of a set of variables. For this reason, it can be used to compute the marginal probability distribution of any of the variables as well as for computing the posterior probability of a subset of variables given the observations gathered on other variables. This step is often referred to as *inference* [3,5]. A plethora of methods have been studied over the last 20 years in order to improve the computational efficiency of the inference process, also because the general problem of deriving posterior probabilities in BNs has been proven to be NP-hard [33,34].

The approaches reported in the literature can be divided into *exact* or *approximate* methods. Among exact inference methods we can distinguish the following strategies [3,5]: variable elimination, which progressively eliminates (by marginalization) the variables not observed and not interesting [3]; junction tree propagation, which allows fast computations by message passing algorithm on an auxiliary tree [35,36]; and recursive conditioning, which builds inference model compromising between time and size of the probability tables needed for calculations of intermediate results [37]. Among approximate inference algorithms are the stochastic simulation strategies, now referred to as particle filters, including logic sampling [38,39] and likelihood weighting [40], Gibbs sampling [41], optimization strategies such as loopy belief propagation [4,42], and variational methods [43]. Interestingly, approximate methods are also NP-hard [34]. As a note, while exact methods are particularly suited for discrete distributions in networks with relatively low graph connectivity, since they provide very fast updating of the posterior probabilities given the evidence, sampling strategies allow dealing with more general problems, including continuous probability distributions or hybrid networks, while paying the price of slower performances.

In this section, we present the basic ideas of the variable elimination algorithm of the junction tree propagation and of stochastic simulation approaches.

12.3.1.1   Variable Elimination

In order to illustrate the basic idea of the variable elimination algorithms, let us consider the example reported in Figure 12.3, a simplified problem taken from Ref. [35]. We assume that all variables are discrete, and that we have observed the values of the two nodes D and F, so that our observed evidence $Obs = \{D = d, F = f\}$. We want to compute the posterior probability $P(A|Obs)$. If we apply the

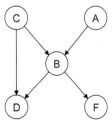

Figure 12.3 A simple Bayesian network.

chain rule, the joint probability distribution of all variables, denoting with U the unobserved variables, can be written as

$$P(U, Obs) = P(A, B, C, d, f) = P(A)P(C)P(B|A, C)P(d|B, C)P(f|B)$$

where the lowercase letters mean that the corresponding variables are known, so that their CPTs refer only to the corresponding entry. To compute $P(A|Obs)$ we *eliminate* (i.e., we marginalize out) the variables that are not known and the ones not of interest, namely B and C. The order of marginalization does not affect the final results. We can start with C:

$$\sum_C P(A, B, d, f, C) = \sum_C P(A)P(C)P(B|A, C)P(d|B, C)P(f|B)$$
$$= P(A)P(f|B) \sum_C P(C)P(B|A, C)P(d|B, C)$$

In order to compute the marginalization, we need to multiply the three probability tables in the summation and to sum over C. The final result will be the table $T(A, B, d)$, so that

$$P(A, B, d, f) = P(A)P(f|B)T(A, B, d)$$

The next step consists in marginalizing B out. Following the same steps:

$$\sum_B P(A, B, d, f) = \sum_B P(A)P(f|B)T(A, B, d) = P(A) \sum_B P(f|B)T(A, B, d)$$

Again, the multiplication and the summation will give us a new table $T(A,f,d)$, so that

$$P(A, Obs) = P(A, d, f) = P(A)T(A, d, f)$$

Once $P(A,Obs)$ is known, it is straightforward to compute the posterior probability of A given Obs as

$$P(A|Obs) = \frac{P(A, Obs)}{P(Obs)} = \frac{P(A, d, f)}{\sum_A P(A, d, f)}$$

The variable elimination algorithm clearly gives the idea of the advantages that BNs provide to computing posterior probabilities of complex models, since, thanks to the chain rule, it is possible to perform "local" inference steps that involve a reduced number of variables at a time. However, the basic operations of product of CPTs and summation may quickly turn out to be intractable, since the joint probability distribution (including the local one) increases exponentially with the number of variables. In the following section, we will show an example of one of the most efficient algorithms based on junction tree propagation.

12.3.1.2 Junction Tree Propagation

Given a certain BN, it is possible to derive an ancillary tree called *junction tree*. First of all, the BN graph is transformed into its *moral graph*: the parents of each node are connected with undirected edges and then all directed edges are transformed into indirect ones. In the second step, the graph is triangulated by adding edges to the moral graph. A graph is triangulated if each loop of length more than three has an edge that connects two nodes that are not adjacent in the loop. The third step identifies clusters of nodes, called "cliques." A clique of an undirected graph is a subgraph *complete* (i.e., each couple of nodes of the subgraph are connected) and *maximal* (i.e., there is no other complete subgraph containing the clique).[1] The fourth step consists in building the junction tree with cliques as nodes. Special nodes, called separators S, are finally added to the tree, containing the variables shared between two adjacent cliques.

Figure 12.4 shows a BN and the steps to build the corresponding junction tree (example from Ref. [44]).

Once the junction tree has been derived from the BN, it is necessary to quantify it. This is done by resorting to *potential functions*. A potential function φ_X over a set of variables X is a function that maps each assignment of values x of X to a positive real number. Given a set of variables Y and its potential function φ_Y and a set of variables $X \subseteq Y$, the potential function φ_Y can be computed by marginalization, $\varphi_X = \sum_{Y/X} \varphi_{XY}$. Each node (X) and each separator (S) of the junction tree is assigned a potential function. Some conditions must hold in order to perform a correct quantification. First, for each node X and its neighbor separators S, it must hold that $\varphi_S = \sum_{X/S} \varphi_X$. This means that the two potentials are consistent. Second, the joint probability distribution of all variables V of the BN is obtained by the equation:

$$P(V) = \frac{\prod_i \varphi_{X_i}}{\prod_j \varphi_{S_j}}$$

---

[1]Finding cliques over a general graph is a complex problem that turns out to be intractable for highly connected graphs.

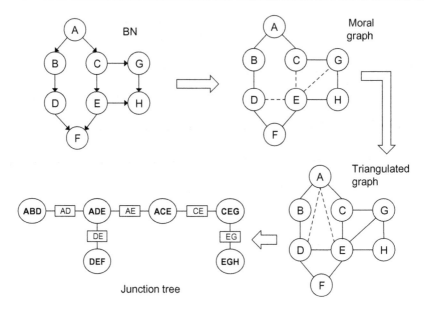

Figure 12.4 Building a junction tree.

The junction tree must thus ensure that, given a node X, $\varphi_X = P(X)$, so that the marginal probability distribution of a BN variable V can be computed starting from any node X containing V by marginalization: $P(V) = \sum_{X/V} \varphi_X$.

In order to ensure that the junction tree is consistent and that it is possible to correctly compute the marginal probability distribution of any node V of the BN, the inference algorithm relies on two steps:

1. A starting set of potentials is assigned to each node, thus leading to an inconsistent tree; a suitable strategy consists in setting all the potentials equal to 1. Then, each variable V is assigned to a node X that contains V and its parents, and the corresponding potential is updated as $\varphi_X^{new} = \varphi_X^{old} P(V|pa(V))$. In this way, after initialization, each CPT has been multiplied to the potential of one node.

2. A message passing strategy is performed to derive a consistent tree. Once this is done it is possible to obtain the probability distributions of any node by marginalization. Given two adjacent clusters X and Y with a separator S, a single message pass is made by a projection step (i.e., $\varphi_S^{old} = \varphi_S$ and $\varphi_S^{new} = \sum_{X/S} \varphi_X$) and by an adsorption step: $\varphi_Y^{new} = \varphi_Y^{old}(\varphi_S/\varphi_S^{old})$. The global propagation of the messages is then performed following a strategy that ensures it will reach all nodes. An arbitrary node X is selected. Then, each cluster passes a message to its neighbor node in the direction of X, starting from the furthest node from X (collect evidence phase). Finally, starting from X and moving from X, each cluster passes a message to its neighbor node (distribute evidence phase).

Figure 12.5 describes the basic mechanism.

By using the message passing algorithm, it is possible to compute any posterior probability given a set of observations Obs. In this case, the first step is to define a likelihood function Λ for every node of the BN. Λ has values corresponding to

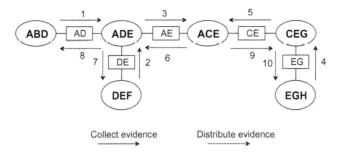

Figure 12.5 Reaching a consistent distribution of potentials having selected the node ACE. The information flows following the order of the arrows.

each state of the variables: it is set to 1 in correspondence to the observed state and 0 otherwise. For example, if C is a binary variable with states on and off, if we observe $C =$ on, $\Lambda_C(\text{on}) = 1$ and $\Lambda_C(\text{off}) = 0$. In case a node has no observed states, all values of Λ are set to 1. Afterwards, the message passing algorithm is reapplied, with two main changes: (i) observations are handled by propagating the likelihood: if given a node V, the observation is $V = v$, a node X that contains V is selected, and $\varphi_X^{\text{new}} = \varphi_X^{\text{old}} \Lambda_V^{\text{new}}$; (ii) once the junction tree has been made consistent with the message passing algorithm, it is possible to derive the required posterior probabilities: for example, if one wants to compute the posterior probability of a BN variable W given Obs, given a junction tree node Y containing W we have $P(W, Obs) = \sum_{Y/W} \varphi_W$ and

$$P(W|Obs) = \frac{P(W, Obs)}{P(Obs)} = \frac{P(W, Obs)}{\sum_W P(W, Obs)}$$

12.3.1.3 Stochastic Simulation

Another class of very important methods that is less sensitive to the graph connectivity than junction tree propagation and is more easily adaptable to continuous distribution is represented by approximate inference performed by stochastic simulation. In a nutshell, stochastic simulation aims to simulate a database of cases extracted from the joint probability distribution described by the BN. The marginal probability distribution of any configuration can be derived simply by counting the occurrences in the database of that configuration divided by the total number of cases. In other words, being interested in the probability of $P(A = a)$, what is needed is the computation of

$$P(A = a) \approx \frac{N(A = a)}{T}$$

where $N(A = a)$ is the number of times A is equal to a in the database and T is the total number of samples in the database.

There are several stochastic simulation algorithms published in the literature, all of them able to exploit the conditional probability distribution of the BN to generate samples from the posterior probability of the variables given the available evidence.

For example, the main idea of *logic sampling* is to exploit the topological ordering of the variables to generate scenarios made of a set of samples. Given a set of observations, *Obs*, the algorithm works as follows: let $(V_1, \ldots, V_n)$ be the topological order of the *n* variables of the BN. Starting from 1 to *n*, a scenario is generated by sampling a state v_i from $P(V_i|Pa(V_i))$, given the states of $Pa(V_i)$ already sampled in the same scenario. If the scenario is compatible with *Obs* (i.e., all samples of the observed variables have states equal to the observations) retain the scenario, otherwise discard it. Repeat the sampling process *M* times, and suppose that *T* scenarios have been retained. The final probability estimate is

$$P(V_i = v_i|Obs) \approx \frac{N(V_i = v_i)}{T}$$

Referring to the network reported in Figure 12.3, a topological order is *A*, *C*, *B*, *D*, *F*. For each scenario, the logic sampling algorithm samples *A* from the distribution $P(A)$, thus obtaining a value, say *a*. Then, a sample is extracted from $P(C)$, thus obtaining a value *c*. In the next step, a sample is drawn from $P(B|a,c)$, say *b*. Then, another sample *d* is taken from $P(D|b,c)$ and finally the simulation is completed with a sample *f* from $P(F|b)$. The values obtained, *a*, *b*, *c*, *d*, *f*, make the *i*th scenario that is stored in the database, in case no observation is available or in case all sampled values are compatible with the observation.

The logic sampling algorithm is quite inefficient, in particular when the available evidence is related to few variables with low marginal probability. For this reason, more efficient strategies have been defined. One of them is the likelihood weighting algorithm. The main idea is to avoid sampling the variables in *Obs*, while fixing them to the observed states in each scenario, and to weight the scenario with the likelihood of the sampled configuration.

The "forward simulation" strategies, such as logic sampling and likelihood weighting, require a large number of scenarios in order to extract reliable posterior probabilities. Moreover, they may turn out to be even more inefficient (or, in the case of logic sampling, inapplicable) in the case of BNs with continuous variables. An interesting alternative is represented by Gibbs sampling, or, more generally, by Monte Carlo Markov chain simulations. The basic idea, in this case, is to run a simulation process that, starting from a scenario that is compatible with the observations and then following a topological order, samples a new value of each unobserved variable given the values of all the other variables sampled so far. In this way a chain of values is generated. This type of chain is known to converge with the posterior distribution of the variables given the observations. In Gibbs sampling, the values of each unobserved variable are simulated from a probability distribution known as *full conditional*, which can be obtained multiplying the probabilities (or the probability density functions) of the *Markov blanket* of a node.

The Markov blanket of a node can be easily found in the graphical structure of the BN, since it comprises the parents of the node, its children, and the parents of its children. The advantage of Gibbs sampling stands in its capability of generating scenarios that are extracted from the posterior probability distribution; moreover, it can be easily used with continuous variables to compute posterior moments. The disadvantages are related to the need of running a number of simulations before the chains converge to the posterior distribution, a step known as *burn-in*; moreover, the speed of convergence heavily depends on the initial configuration. Finally, sometimes exploring the posterior distribution may require a considerable number of simulations or *ad hoc* strategies to avoid local minima.

12.3.2 Parameter Learning

Given a BN structure $\mathscr{G}$ and a dataset $\mathscr{D}$, containing multiple samples of all of the random variables in $\mathscr{G}$, the problem of learning the conditional distributions implied by $\mathscr{G}$ consists in estimating, for each variable X, a set of parameters $\boldsymbol{\theta}_{X|Pa_X}$ describing the dependency of X over its parents Pa_X.

For BNs with discrete variables, such as the network in Figure 12.1A, the parameters to be estimated are all of the entries of the CPT of each variable (i.e., the values $\theta_{x|pa_x} = P(X = x|Pa_X = pa_X, \mathscr{D})$ for each of the possible values of X and of its parents Pa_X).

To this aim, one can exploit either *maximum likelihood* estimates, based on calculating the relative frequencies of the different events in the data, or Bayesian maximum *a posteriori* (MAP) estimates, augmenting this observed data with prior distributions over the values of these parameters.

It can be shown that the maximum likelihood estimate of $\theta_{x|pa_x}$ is given by

$$\hat{\theta}_{x|pa_x} = \frac{N(X = x \wedge Pa_X = pa_X)}{N(Pa_X = pa_X)}$$

where $N(c)$ counts the number of observations in $\mathscr{D}$ satisfying condition c.

One of the risks of maximum likelihood is that it can sometimes return estimates equal to zero, in case no example satisfying the condition at the numerator is observed in $\mathscr{D}$. To avoid this situation, it is often preferable to smooth the estimate with a coefficient α known as *equivalent sample size* (ESS). The smoothed estimate is given by

$$\hat{\theta}_{x|pa_x} = \frac{N(X = x \wedge Pa_X = pa_X) + \alpha}{N(Pa_X = pa_X) + |\text{Val}(Pa_X)| \cdot \alpha}$$

where $|\text{Val}(Pa_X)|$ is the number of distinct values Pa_X can take.

Formally, this expression corresponds to a MAP estimate of $\theta_{x|pa_x}$, assuming a Dirichlet prior distribution with equal-valued hyperparameters α. An intuitive interpretation of α is the number of imaginary samples, for each combination of values of X and Pa_X, assumed to have been observed before estimating $\theta_{x|pa_x}$ from $\mathscr{D}$.

In the case of continuous-variable networks with linear Gaussian distributions, a closed-form solution exists for the estimate of the parameters β and $\tau = 1/\sigma^2$ when certain prior distributions are employed, namely Gamma distributions or their multivariate generalizations, Wishart distributions [32].

12.3.3 Structure Learning

Structure learning of BNs has been an active research topic in the last 20 years, because the problem is NP-complete in the general case: given a dataset $\mathcal{D}$, containing multiple samples of a set of random variables, the objective is to find the best, or the most probable, BN structure in the exponential space of all possible structures. Several scoring functions have been proposed to assess the quality of a BN structure: some of the most notable are the Akaike Information Criterion (AIC), the Bayesian Information Criterion (BIC), and the Bayesian Dirichlet equivalent (BDe) [10].

From the vast literature, we can identify three main approaches to BN structure learning: greedy search, complete search, and search based on independence tests. We will provide an example algorithm for each of them and refer the reader to Ref. [5] for a more extensive presentation of this topic.

Historically one of the first approaches to structure learning, *greedy search* attempts to construct a BN structure starting with a network without any edge and iteratively adding the "best" set of parents to each node, according to a local score. The most famous example of the greedy approach is the K2 algorithm of Cooper and Herskovits [9]: the authors formulate a local scoring function, which can be separately computed for each node and its parent set, and a polynomial time greedy algorithm. Polynomial complexity is achieved by requiring in input an upper bound on the number of parents for each node and an ordering of the nodes, with the additional constraint that the parents of a node can be searched for only among the nodes that precede it in the ordering. Clearly, the K2 algorithm offers no guarantees of finding the optimal BN structure, but its low complexity makes it a powerful tool for datasets with a large number of variables.

Complete search, conversely, explores the entire space of possible networks and is guaranteed to return an optimal network; the huge memory and time requirements, however, limit its application to small sized networks (up to 30−35 nodes). A representative of this approach is the recent branch-and-bound algorithm of De Campos and Ji [45]. The algorithm consists of two steps: first a cache is built for storing the local scores of all the possible parent sets for each node; then, a branch-and-bound algorithm exploits the cache to explore the search space. Several theorems, formulated by the authors, are exploited to avoid the computation of many combinations of parents for each node, in order to speed up the process and limit the amount of memory required by the cache.

Finally, approaches based on *independence tests* start with a complete network and aim at forbidding as many edges as possible, by assessing conditional independence between variables with statistical tests. We report as an example the max−min hill climbing algorithm of Tsamardinos et al. [46], where the G^2 statistic for conditional independence is used to identify, for each node, a set of candidate parents or children (CPC). The CPC sets are then used to restrict the search space of a hill climbing algorithm, which starts from an arbitrary structure, consistent with the CPC sets, and iteratively tries to increase its score with local search operators such as addition, removal, or inversion of an edge.

Hybrid approaches, combining the speedup gained by independence tests with the power of complete search, have also been explored in the literature [47,48].

Analogously to the parameter learning case, for linear Gaussian networks with appropriate prior distribution choices, the scoring functions can be calculated in closed form, thus allowing the use of the same approaches for structure learning as for discrete BNs.

12.4   Examples

In this section, we present several example applications of the BN model for physiological and biomedical data.

12.4.1   The Diabetes Advisory System

The Diabetes Advisory System (DIAS, [49]) was proposed in the mid-1990s as a decision support system to provide advice on the insulin therapy for type 1 diabetic patients. A few years later, the system has been further extended to type 2 diabetes (DIAS-NIDDM, [50]).

The system is based on a DBN, modelling the evolution of blood glucose over 24 h. The model is obtained by cascading 24 identical subnetworks. Within each subnetwork, nodes represent state variables, such as blood glucose and active insulin, their interaction with the inputs, such as the increase in blood glucose due to an ingested meal or the increase in active insulin due to an injection, and their interaction with blood glucose in the previous and subsequent subnetwork, for example, by renal clearance or by insulin-dependent utilization.

The discrete set of possible values for each node is the result of a quantization of the corresponding continuous metabolic measure. Each node also has an associated probability distribution over its values, which depends on the values of the node's parents. All of the parameters of the distributions have been manually set by the authors, based on the literature on glucose metabolism, apart from two parameters for type 1 diabetes, namely insulin sensitivity and time from insulin injection to maximum insulin absorption, and one additional parameter for type 2 diabetes, pancreatic sensitivity.

The three latter parameters are exploited to tune the network to a specific individual and are set to allow the best fit to observed blood glucose data over one or more days. Once tuned to an individual, the system can be used to forecast the blood glucose evolution over a period of 24 h, given the amount and timing of carbohydrate intakes and insulin injections.

Finally, the authors assembled a penalty-score function of blood glucose values, starting from the advice of a pool of diabetologists. The score is used to provide recommendations on the diet and on the insulin dosage and schedule, searching for the therapy that minimizes the predicted penalty across the 24 h.

The system has been extensively tested and is shown to be able to reproduce evolution of blood glucose over the day and its probabilistic dependence on carbohydrate intakes and insulin injections.

12.4.2   Identifying Feedback Control in Cell Cycle

BNs offer a flexible and powerful framework for modelling the interactions between cellular system components, genes, proteins, or other small molecules. The probabilistic framework allows taking into account the intrinsic variability of cellular systems and the presence of noise in the data, at the same time offering the opportunity to model unobserved system components. For these reasons BNs have become increasingly popular for the "reverse engineering" of cellular systems from high-throughput data (i.e., the inference of the regulatory network most likely to have generated the observed data [51,52]).

DBNs have been employed to discover key regulatory genes in cell cycle control [25]. The hypothesis was that feedback control structures highlight central genes in cell cycle regulation, which may constitute novel pharmacological targets. To this aim continuous DBNs [53] have been exploited to reverse engineer gene expression microarray data measured during cell cycle in HeLa cells, a human cancer cell line [54]. We concentrated our analysis on about a 1000 microarray probes identified by the authors as having periodic expression profiles. The inferred network contained a complex feedback structure involving 12 probes, mapping to 10 different genes. Five of these genes were well-known cell cycle regulators and *in silico* analyses showed that feedback genes hold important roles in system dynamics. Furthermore, results of a high-throughput gene silencing study in HeLa cells [55] allowed us to assess that feedback genes are highly enriched in cell cycle regulators with respect to the starting gene list of periodically expressed genes (P value = 0.0043 calculated with hypergeometric distribution).

Gene networks inferred with DBN from expression data represent probabilistic dependencies between genes, which may or may not correspond to a biological regulatory relationship. As the study shows, these network models are able to summarize the gene expression dynamics of the analyzed system and can be effectively employed to prioritize genes for successive biological investigations.

12.4.3   Stroke Prediction in SCA Patients

The ability of BNs to model multivariate dependencies makes them a very interesting framework to describe interactions between genes, environment, and phenotype(s). In addition, inference algorithms make it possible to use BNs as classifiers to predict probability of occurrence of the phenotype of interest (a node in the BN) given a particular genotype. The first significant example of use of BNs for the genetic dissection of complex traits is the study by Sebastiani et al. on modelling of stroke in SCA patients [24]. In order to study the genetic determinants of stroke in SCA, 108 SNPs in 80 candidate genes have been analyzed in 1398 African Americans with SCA, 92 of them with reported overt stroke. The authors focused on "diagnostic type" networks (i.e., networks describing the dependencies of genotypes on phenotype), as this type of analysis can reduce search complexity as well as identify larger sets of SNP-phenotype associations. The overall identified dependency network links 69 SNPs in 20 genes as well as additional phenotypic variables (such as hemoglobin levels) to stroke. In particular, 25 SNPs in 11 genes were found to be directly associated with stroke (i.e., they had the largest independent affect on predicting risk of stroke). The inferred BN classifier achieved 98.5% accuracy in fivefold cross-validation. Furthermore, the model predicted the occurrence of stroke in an independent population of 114 individuals (7 with reported stroke) with 98.2% accuracy and a 100% true positive rate. A logistic regression model built for comparison achieved 88% accuracy on the same validation set. The BN model also offered the possibility of examining dependencies between SNPs and showed that dependencies also exist across different genes, supporting the hypothesis that SCA is a complex trait caused by the interaction of multiple genes.

In recent years BNs have been increasingly exploited for genetic association studies and variants of the BN modelling framework have been proposed [56−58]. For example, an approach that combines BNs and classification trees to build classifiers on metavariables derived from SNPs mapping to the same gene has been proposed [59]. The learned BN model results were more parsimonious than the one based on single SNPs and had higher prediction accuracy. The analysis of genome-wide data with hundreds of thousands of SNPs poses significant challenges in terms of tractability. Research efforts have thus also been devoted to extending Bayesian learning to a genome-wide scale [60,61].

12.5 Conclusions and Future Perspectives

BNs are nowadays a robust and well-studied methodological framework, suitable for modelling a variety of problems where it is needed for representing the probabilistic relationships between problems' variables. Their capability of coupling clear graphical semantics with probabilistic inference allows their full exploitation in biomedical modelling, in relation to classification, simulation, and prediction tasks. Moreover, thanks to their extension to time-dependent domain (i.e., the DBNs) it is possible to use them to easily represent complex dynamical systems. This chapter was aimed at providing the basic ideas underlying BNs and DBNs and their use in biomedical modelling.

Rather interestingly, further research is now ongoing in several directions. One of them is related to overcoming some BN limitations, in particular the need for a discrete timescale. In particular, following the work of Nodelman et al. [62,63], the continuous-time Bayesian networks (CTBNs) have recently gained some popularity

[64]. CTBNs have been used to model social networks [65], cardiogenic heart failure [66], and stroke rehabilitation [67].

A CTBN is a probabilistic graphical model in which nodes are discrete random variables, where the state evolves continuously over time. The probability law that governs the state transitions depends on the state of the node parents in the graph.

As for BNs, as well as CTBNs, exact inference is NP-hard. In this case, a variety of approximate algorithms have been proposed, including expectation propagation [63] and stochastic simulation strategies [64,68].

The problem of learning CTBNs structure has been studied too [62,63]. In this case, the solution is somehow simpler than in the BN and DBN cases, since the network may have cycles and it is thus possible to learn the local structure of each CTBN node independently.

A specific characteristic of CTBNs resides in their capability of managing not only observations at a given time stamp but also evidence streams (i.e., values observed along a time interval). This aspect seems particularly interesting, given the current emphasis on the analysis of data streams: continuous, rapid data records increasingly collected in big data scenarios.

BNs and DBNs have also become increasingly popular thanks to the availability of a plethora of software tools: commercial, free, and open source. A nice updated summary is maintained by Murphy ([69]http://www.cs.ubc.ca/~murphyk/Software/bnsoft.html). Many of the software tools include inference and learning of BNs and DBNs. Some of them are provided with a graphical user interface, which can be used also for teaching and for interactions with nonexpert users. As examples, Hugin is one of the most well-known types of commercial software [70], Genie (and its library Smile) is a free system with an easy graphical interface [29], and, finally, the Bayes Network Toolbox in Matlab is a set of widely used software programs [28]. Rather interestingly, BNs have been also used in data mining tools, like Weka, to learn classifiers from data [71].

References

[1] Wright S. Correlation and causation. Part I. Method of path coefficients. J Agric Res 1920;20:557—85.
[2] Wright S. The method of path coefficients. Ann Math Stat 1934;5:161—215.
[3] Jensen FV, Nielsen TD. Bayesian networks and decision graphs. New York, NY: Springer; 2007.
[4] Pearl J. Probabilistic reasoning in intelligent systems: networks of plausible inference. San Francisco, CA: Morgan Kaufmann Publishers Inc.; 1988.
[5] Koller D, Friedman N. Probabilistic graphical models: principles and techniques. Cambridge, MA: MIT Press; 2009.
[6] Charniak E. Bayesian networks without Tears. AI Mag 1991;12(4):50—63.
[7] Russell SJ, Norvig P. Artificial intelligence: a modern approach. Upper Saddle River, NJ: Prentice-Hall; 2003 [Chapters 13—15].
[8] Cowell RG. Local propagation in conditional Gaussian Bayesian networks. J Mach Learn Res 2005;6:1517—50.

[9] Cooper GF, Herskovits EA. Bayesian method for the induction of probabilistic net-
     works from data. Mach Learn 1992;9(4):309−47.
[10] Heckerman D, Geiger D, Chickering DM. Learning Bayesian networks—the combina-
     tion of knowledge and statistical data. Mach Learn 1995;20(3):197−243.
[11] Buntine W. Operations for learning with graphical models. J Artif Intell Res
     1994;2:159−225.
[12] Dean T, Kanazawa K. A model for reasoning about persistence and causation.
     J Comput Intell 1990;5(3):142−50.
[13] Miller RAA. History of the INTERNIST-1 and Quick Medical Reference (QMR)
     computer-assisted diagnosis projects, with lessons learned. Yearb Med Inform
     2010;121−36.
[14] Shwe M, Cooper G. An empirical analysis of likelihood-weighting simulation on a
     large, multiply connected medical belief network. Comput Biomed Res 1991;24
     (5):453−75.
[15] Andreassen S, Falck B, Olesen KG. Diagnostic function of the microhuman prototype
     of the expert system—Munin. Electroen Clin Neuro 1992;85(2):143−57.
[16] Andreassen S, Rosenfalck A, Falck B, Olesen KG, Andersen SK. Evaluation of the
     diagnostic performance of the expert EMG assistant MUNIN. Electromyogr Motor C
     1996;101(2):129−44.
[17] Suojanen M, Andreassen S, Olesen KG. The EMG diagnosis—an interpretation based
     on partial information. Med Eng Phys 1999;21(6−7):517−23.
[18] Cavan DA, Hovorka R, Hejlesen OK, Andreassen S, Sonksen PH. Use of the DIAS
     model to predict unrecognised hypoglycaemia inpatients with insulin-dependent diabe-
     tes. Comput Meth Prog Bio 1996;50(3):241−6.
[19] Hovorka R, Tudor RS, Southerden D, Meeking DR, Andreassen S, Hejlesen OK, et al.
     Dynamic updating in DIAS-NIDDM and DIAS causal probabilistic networks. IEEE
     Trans Biomed Eng 1999;46(2):158−68.
[20] Turner BC, Hejlesen OK, Kerr D, Cavan DA. Impaired absorption and omission of
     insulin: a novel method of detection using the diabetes advisory system computer
     model. Diabetes Technol Ther 2001;3(1):99−109.
[21] Plougmann S, Hejlesen OK, Cavan DA. DiasNet—a diabetes advisory system for com-
     munication and education via the internet. Int J Med Inform 2001;64(2−3):319−30.
[22] Magni P, Bellazzi R, Sparacino G, Cobelli C. Bayesian identification of a population
     compartmental model of C-peptide kinetics. Ann Biomed Eng 2000;28(7):812−23.
[23] Magni P, Bellazzi R, De Nicolao G. Bayesian function learning using MCMC methods.
     IEEE Trans Pattern Anal 1998;20(12):1319−31.
[24] Sebastiani P, Ramoni MF, Nolan V, Baldwin CT, Steinberg MH. Genetic dissection
     and prognostic modeling of overt stroke in sickle cell anemia. Nat Genet 2005;37
     (4):435−40.
[25] Ferrazzi F, Engel FB, Wu E, Moseman AP, Kohane IS, Bellazzi R, et al. Inferring cell
     cycle feedback regulation from gene expression data. J Biomed Inform 2011;44
     (4):565−75.
[26] Dimitrakopoulou K, Tsimpouris C, Papadopoulos G, Pommerenke C, Wilk E, Sgarbas
     KN, et al. Dynamic gene network reconstruction from gene expression data in mice
     after influenza A (H1N1) infection. J Clin Bioinform 2011;1:27.
[27] Perrin BE, Ralaivola L, Mazurie A, Bottani S, Mallet J, D'Alche-Buc F. Gene net-
     works inference using dynamic Bayesian networks. Bioinformatics 2003;19(Suppl. 2):
     II138−48.
[28] Murphy KP. The Bayes net toolbox for MATLAB. Comput Sci Stat 2001;33:2001.

[29] Druzdzel MJ. SMILE: Structural Modeling, Inference, and Learning Engine and GeNIe: a development environment for graphical decision-theoretic models. In: Proceedings of the 16th national conference on artificial intelligence and the 11th innovative applications of artificial intelligence conference innovative applications of artificial intelligence; 1999. p. 902–903.

[30] Verma T, Pearl J. Equivalence and synthesis of causal models. In: Uncertainty in artificial intelligence 6 annual conference (UAI'90); 1990. p. 255–268.

[31] Chickering DA. Transformational characterization of Bayesian networks structures. In: Proceedings of the 11th conference on uncertainty in artificial intelligence (UAI'95); 1995. p. 87–98.

[32] Sebastiani P, Abad M, Ramoni MF. Bayesian networks for genomic analysis. In: Dougherty ER, Shmulevich I, Chen J, Wang ZJ, editors. Genomic signal processing and statistics. EURASIP book series on signal processing and communications. New York, NY: Hindawi; 2005. p. 281–320.

[33] Cooper GF. The computational-complexity of probabilistic inference using Bayesian belief networks. Artif Intell 1990;42(2–3):393–405.

[34] Dagum P, Luby M. Approximating probabilistic inference in Bayesian belief networks is NP-hard. Artif Intell 1993;60(1):141–53.

[35] Jensen FV, Lauritzen SL, Olesen KG. Bayesian updating in causal probabilistic networks by local computations. Comput Stat Q 1990;4:269–82.

[36] Lauritzen S, Spiegelhalter DJ. Local computation and probabilities on graphical structures and their applications to expert systems. J R Stat Soc 1988;50(2):157–224.

[37] Darwiche A. Recursive conditioning. Artif Intell 2001;126(1–2):5–41.

[38] Henrion M. Propagating uncertainty in Bayesian networks by probabilistic logic sampling. In: Uncertainty in artificial intelligence 2 annual conference (UAI'86); 1986. p. 149–163.

[39] Fung R, Chang KC. Weighing and integrating evidence for stochastic simulation in Bayesian networks. In: Proceedings of the fifth conference on uncertainty in artificial intelligence (UAI'89); 1989. p. 112–117.

[40] Shachter RD, Peot MA. Simulation approaches to general probabilistic inference on Bayesian networks. Proceedings of the fifth conference on uncertainty in artificial intelligence (UAI'89); 1989; p. 311–318.

[41] Gelfand AE. Smith AFM. Sampling-based approaches to calculating marginal densities. J Am Stat Assoc 1990;85(410):398–409.

[42] McEliece R, Mackay D, Cheng J. Turbo decoding as an instance of Pearl's belief propagation algorithm. IEEE J Sel Areas Commun 1998;16:140–52.

[43] Jaakkola TS, Jordan MI. Variational probabilistic inference and the QMR-DT network. J Artif Intell Res 1999;10:291–322.

[44] Huang C, Darwiche A. Inference in belief networks: a procedural guide. Int J Approx Reason 1996;15(3):225–63.

[45] de Campos CP, Ji Q. Efficient structure learning of Bayesian networks using constraints. J Mach Learn Res 2011;12:663–89.

[46] Tsamardinos I, Brown LE, Aliferis CF. The max–min hill-climbing Bayesian network structure learning algorithm. Mach Learn 2006;65(1):31–78.

[47] Badaloni S, Sambo F, Venco F. Bayesian networks structure learning: hybridizing complete search with independence tests. AI Commun, 2014.

[48] Perrier E, Imoto S, Miyano S. Finding optimal Bayesian network given a superstructure. J Mach Learn Res 2008;9:2251–86.

[49] Andreassen S, Benn JJ, Hovorka R, Olesen KG, Carson ERA. Probabilistic approach to glucose prediction and insulin dose adjustment—description of metabolic model and pilot evaluation study. Comput Meth Prog Bio 1994;41(3−4):153−65.

[50] Tudor RS, Hovorka R, Cavan DA, Meeking D, Hejlesen OK, Andreassen S. DIAS-NIDDM—a model-based decision support system for insulin dose adjustment in insulin-treated subjects with NIDDM. Comput Meth Prog Bio 1998;56(2):175−91.

[51] Friedman N. Inferring cellular networks using probabilistic graphical models. Science 2004;303(5659):799−805.

[52] Markowetz F, Spang R. Inferring cellular networks—a review. BMC Bioinformatics 2007;8(Suppl. 6):S5.

[53] Ferrazzi F, Sebastiani P, Ramoni MF, Bellazzi R. Bayesian approaches to reverse engineer cellular systems: a simulation study on nonlinear Gaussian networks. BMC Bioinformatics 2007;8(Suppl. 5):S2.

[54] Whitfield ML, Sherlock G, Saldanha AJ, Murray JI, Ball CA, Alexander KE, et al. Identification of genes periodically expressed in the human cell cycle and their expression in tumors. Mol Biol Cell 2002;13(6):1977−2000.

[55] Kittler R, Pelletier L, Heninger AK, Slabicki M, Theis M, Miroslaw L, et al. Genome-scale RNAi profiling of cell division in human tissue culture cells. Nat Cell Biol 2007;9(12):1401−12.

[56] Mourad R, Sinoquet C, Leray P. Probabilistic graphical models for genetic association studies. Brief Bioinform 2012;13(1):20−33.

[57] Su C, Andrew A, Karagas MR, Borsuk ME. Using Bayesian networks to discover relations between genes, environment, and disease. BioData Min 2013;6(1):6.

[58] Sebastiani P, Abad-Grau MM. Bayesian networks for genetic analysis. In: Alterovitz G, Ramoni MF, editors. Systems bioinformatics: an engineering case-based approach. Norwood, MA: Artech House; 2007. p. 205−27.

[59] Malovini A, Nuzzo A, Ferrazzi F, Puca AA, Bellazzi R. Phenotype forecasting with SNPs data through gene-based Bayesian networks. BMC Bioinformatics 2009;10 (Suppl. 2):S7.

[60] Zhang Y. A novel Bayesian graphical model for genome-wide multi-SNP association mapping. Genet Epidemiol 2012;36(1):36−47.

[61] Sambo F, Trifoglio E, Di Camillo B, Toffolo GM, Cobelli C. Bag of naive Bayes: biomarker selection and classification from genome-wide SNP data. BMC Bioinformatics 2012;13(Suppl. 14):S2.

[62] Nodelman U, Koller D, Shelton C. Expectation propagation for continuous time Bayesian networks. In: Proceedings of the 21st conference on uncertainty in artificial intelligence (UAI'05); 2005. p. 431−440.

[63] Nodelman U, Shelton C, Koller D. Continuous time Bayesian networks. In: Proceedings of the 18th conference on uncertainty in artificial intelligence (UAI'02); 2002. p. 378−387.

[64] Fan Y, Xu J, Shelton CR. Importance sampling for continuous time Bayesian networks. J Mach Learn Res 2010;11:2115−40.

[65] Fan Y, Shelton C. Learning continuous-time social network dynamics. In: Proceedings of the 25th conference on uncertainty in artificial intelligence (UAI'09); 2009. p. 161−168.

[66] Gatti E, Luciani D, Stella F. A continuous time Bayesian network model for cardiogenic heart failure. Flex Serv Manuf J 2012;24(4):496−515.

[67] Stella F, Amer Y. Continuous time Bayesian network classifiers. J Biomed Inform 2012;45(6):1108−19.

[68] El-Hay T, Friedman N, Kupferman R. Gibbs sampling in factorized continuous-time Markov processes. In: Proceedings of the 24th conference on uncertainty in artificial intelligence (UAI'08); 2008. p. 169−178.

[69] Murphy K. Software for graphical models: a review. ISBA (International Society for Bayesian Analysis) Bulletin 2007;4(4):13−5.

[70] Madsen AL, Michael L, Kjærulff UB, Jensen F. The Hugin tool for learning Bayesian networks. In: Nielsen TD, Zhang NL, editors. Symbolic and quantitative approaches to reasoning with uncertainty. Berlin: Springer-Verlag; 2003. p. 594−605.

[71] Hall M, Frank E, Holmes G, Pfahringer B, Reutemann P, Witten IH. The WEKA data mining software: an update. SIGKDD Explor Newsl 2009;11(1):10−8.

13 Mathematical Modelling of Pulmonary Gas Exchange

Dan S. Karbing[a], Søren Kjærgaard[b], Steen Andreassen[c] and Stephen E. Rees[a]

[a]Respiratory and Critical Care Group (rcare), MMDS, Aalborg University, Denmark, [b]Anaesthesia and Intensive Care, Aalborg University Hospital, Denmark, [c]Center for Model-Based Medical Decision Support (MMDS), Aalborg University, Denmark

The purpose of models described in this chapter is to quantify pulmonary gas exchange from clinical or experimental measurements. The application of these models includes the description of pathophysiology of pulmonary disorders and effects of therapeutic interventions. Such applications require models with a limited number of parameters that can be tuned to fit measurement data. In contrast, the lungs contain approximately 33–50,000 effective gas exchanging units [1], each of which may have individual gas exchange properties. While it is not possible to characterize gas exchange in all of these individual lung units, models presented in this chapter are based on the assumption that gas exchange properties, like any distributed biological variable, have distributions that may be represented by fewer parameters, and as such the lungs can be represented as consisting of relatively few gas exchanging compartments.

13.1 Standard Equations Used to Describe Gas Transport in the Lungs

Figure 13.1 illustrates a conceptual model of the lungs including three compartments representing alveolar dead space (i.e., alveoli that are ventilated but not perfused), pulmonary shunt (i.e., alveoli that are perfused but not ventilated), and effective alveoli (i.e., those that are involved in gas exchange and are both ventilated and perfused). For alveoli involved in gas exchange, Figure 13.1 includes a resistance to gas diffusion across the alveoli/lung capillary membrane.

The conceptual model illustrated in Figure 13.1 is simple and does not include a large number of compartments describing heterogeneity of ventilation, diffusion, or perfusion within the lungs. As such, the mathematical formulation of this model

Modelling Methodology for Physiology and Medicine. DOI: http://dx.doi.org/10.1016/B978-0-12-411557-6.00013-6

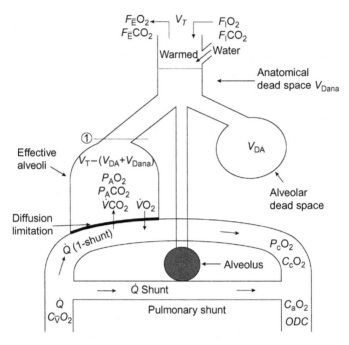

Figure 13.1 A three-compartment model of the lungs, including alveolar dead space, pulmonary shunt, and resistance to gas diffusion across the alveoli−lung capillary membrane.

includes equations which are well-recognized within the field of respiratory physiology, equations which can be seen as the building blocks of more complex models of gas exchange. These equations are the alveolar air equation, the venous admixture equation, Fick's law of diffusion, the Fick principle of blood flow, and the Bohr equation for the estimation of dead space. The derivation and description of these equations now follow; all relevant nomenclature is included in the glossary (Appendix A).

13.1.1 The Alveolar Air Equation

The alveolar air equation can be derived by considering the volume of gas flowing into and out of the effective alveoli. The volume of gas flowing into the effective alveoli (across line 1 Figure 13.1) is the product of the fraction of gas in the inspired air (F_I) warmed and humidified ($F_{I,BTPS}$) and the ventilation of the effective alveoli ($\dot{V}_{A,BTPS}$). Similarly, the volume of gas flowing out of the effective alveoli is the fraction of gas in the effective alveoli on expiration ($F_{A,BTPS}$) multiplied by $\dot{V}_{A,BTPS}$, plus the gas diffusing into blood during gas exchange ($\dot{V}$). Assuming all gases are expressed at body temperature and pressure and saturated with water, this gives

$$F_A \cdot \dot{V}_A + \dot{V} = F_I \cdot \dot{V}_A \tag{13.1}$$

Using barometric pressure (P_B) to convert alveolar fraction to pressure, ($P_A = F_A \cdot P_B$), Eq. (13.1) can be written as:

$$\frac{P_A}{P_B} \dot{V}_A + \dot{V} = F_I \cdot \dot{V}_A \tag{13.2}$$

Rearranging Eq. (13.2) gives the most general form of the alveolar air equation:

$$P_A = F_I \cdot P_B - \frac{\dot{V} \cdot P_B}{\dot{V}_A} \tag{13.3}$$

For oxygen and carbon dioxide this equation may be written as:

$$P_A O_2 = F_I O_2 \cdot P_B - \frac{\dot{V} O_2 \cdot P_B}{\dot{V}_A} \tag{13.4}$$

$$P_A CO_2 = F_I CO_2 \cdot P_B - \frac{\dot{V} CO_2 \cdot P_B}{\dot{V}_A} \tag{13.5}$$

where $\dot{V} O_2$ is the oxygen consumption and $\dot{V} CO_2$ is the carbon dioxide production. $\dot{V}_A$ is calculated as the respiratory frequency, f, multiplied by the effective tidal volume ($V_{T,BTPS} - (V_{Dana} + V_{Dalv})$), which is:

$$\dot{V}_A = f(V_{T,BTPS} - (V_{Dana} + V_{Dalv})) \tag{13.6}$$

In Eqs. (13.1)–(13.6), ventilation of effective alveoli ($\dot{V}_A$), the fraction of gas flowing into the alveoli (F_I), and the tidal volume (V_T) are described at body temperature and pressure and fully saturated with water or BTPS. Inspired gas is usually dry and at ambient temperature and pressure (ATPD), as it is warmed and humidified during inspiration. Inspired volumes and fractions are usually measured at ATPD, but may be converted into the equivalent values at BTPS, as described in Appendix B.

Equations (13.4)–(13.6), and the corrections for humidification and warming of the inspiration (Eqs. (B.1)–(B.10)), are often used to estimate alveolar oxygen partial pressure ($P_A O_2$) which cannot be directly measured. When measurements of mixed venous blood ($C_{\bar{v}} O_2$) and cardiac output ($\dot{Q}$) are possible, $\dot{V} O_2$ can be calculated from the Fick principle of blood flow (see Eq. (13.13)), and used with measurements of ventilation ($F_I O_2$, f, V_T) and estimates of dead space (V_{Dana}, V_{Dalv}) to calculate $\dot{V}_A$ from Eq. (13.6), and therefore $P_A O_2$ from Eq. (13.4). In the absence of pulmonary artery catheter measurements of $C_{\bar{v}} O_2$ or $\dot{Q}$, an indirect estimate of $P_A O_2$ is possible using information from the carbon dioxide (CO_2) system. First, Eq. (13.5) is solved to obtain an estimate for $\dot{V} CO_2$. This is done by making two assumptions: the alveolar and arterial PCO_2 are equivalent ($P_ACO_2 \approx P_aCO_2$), when P_aCO_2 is measured from an arterial blood sample; the inspired CO_2 concentration is negligible ($F_1CO_2 = 0$). Calculated $\dot{V}CO_2$ can then be used to estimate $\dot{V}O_2$, assuming a value of the respiratory quotient (R, normal value 0.8) and using the equation:

$$\dot{V}CO_2 = R \cdot \dot{V}O_2 \tag{13.7}$$

which links oxygen consumption ($\dot{V}O_2$) and carbon dioxide production ($\dot{V}CO_2$) under steady-state conditions of both O_2 and CO_2. The estimated value of $\dot{V}O_2$ can then be inserted into Eq. (13.4) and P_AO_2 calculated. The alveolar air equation can be written in many forms, with the appropriate form chosen based upon the available measurements. A common form of this equation, used when mixed expired gases can be measured, is given below:

$$P_AO_2 = P_1O_2 - P_aO_2 \frac{F_1O_2 - F_{\overline{E}}O_2}{F_{\overline{E}}CO_2} \tag{13.8}$$

where $F_{\overline{E}}O_2$ and $F_{\overline{E}}CO_2$ are mixed expired fractions of O_2 and CO_2, respectively.

13.1.2 Venous Admixture

The venous admixture or "shunt fraction (shunt)" is the fraction of total cardiac output ($\dot{Q}$) that is not involved in gas exchange. As illustrated in Figure 13.1, total blood flow ($\dot{Q}$) is the sum of that flowing through lung capillaries ($\dot{Q}(1\text{-shunt})$) and that which is not involved in gas exchange ($\dot{Q}$ shunt):

$$\dot{Q} = \dot{Q} \cdot \text{shunt} + \dot{Q}(1 - \text{shunt}) \tag{13.9}$$

In the same way, the flow of oxygen into the arteries ($\dot{Q} \cdot C_aO_2$) can be described as the sum of that coming from blood leaving the lung capillaries ($\dot{Q}(1 - \text{shunt}) \cdot C_cO_2$) and that coming from shunted mixed venous blood ($\dot{Q} \cdot \text{shunt} \cdot C_{\overline{V}}O_2$):

$$\dot{Q} \cdot C_aO_2 = \dot{Q} \cdot \text{shunt} \cdot C_{\overline{V}}O_2 + \dot{Q}(\text{shunt}) \cdot C_cO_2 \tag{13.10}$$

By rearranging Eq. (13.10), we obtain the shunt equation

$$\text{shunt} = \frac{C_cO_2 - C_aO_2}{C_cO_2 - C_{\overline{V}}O_2} \tag{13.11}$$

The oxygen content of blood leaving the lung capillaries (C_cO_2) cannot be measured. To calculate shunt, it is usually assumed that alveoli and blood leaving the lung capillaries are at equilibrium for PO_2, or $P_cO_2 = P_AO_2 \cdot P_AO_2$, estimated from the alveolar air equation, which can then be used to calculate C_cO_2.

13.1.3 Fick's First Law of Diffusion

This law states that the rate of diffusion ($\dot{V}$) of a gas across a tissue barrier is proportional to the partial pressure difference across that barrier:

$$\dot{V} = D_L(P_A - P_c) \tag{13.12}$$

where P_c is the partial pressure of gas in the lung capillary gas and D_L is the diffusion capacity of the gas.

13.1.4 The Fick Principle of Blood Flow

The Fick principle describes the oxygen consumption ($\dot{V}O_2$) as a function of the arteriovenous oxygen concentration difference and the cardiac output ($\dot{Q}$), and can be used to calculate either $\dot{Q}$ or $\dot{V}O_2$. The equation is written as:

$$\dot{V}O_2 = \dot{Q}(C_aO_2 - C_{\bar{v}}O_2) \tag{13.13}$$

13.1.5 Estimation of Respiratory Dead Space

Respiratory dead space is that part of the tidal volume (V_T) that does not take part in gas exchange. This includes all nonalveolar ventilation (anatomical dead space V_{Dana}), and ventilation of alveoli that are not perfused (alveolar dead space V_{DA}). The sum of V_{Dana} and V_{DA}, is known as physiological dead space (V_{Dp}).

The Bohr equation, used to describe dead space, can be derived by describing CO_2 expiration in three ways. At the mouth, the total CO_2 expired can be described either as the product of mixed expired CO_2 fraction ($F_{\bar{E}}CO_{2,BTPS}$) and tidal volume ($V_{T,BTPS}$), or as the product of end tidal CO_2 fraction ($F_{E'}CO_2$) and the volume of end tidal gases ($V_{T,BTPS} - V_{Dana}$), where end tidal gases are those expired after anatomical dead space has been washed out. Equating these and converting fractions to pressure (i.e., $P_{\bar{E}}CO_2 = F_{\bar{E}}CO_2 \cdot P_B$, $P_{E'}CO_2 = F_{E'}CO_2 \cdot P_B$) gives:

$$P_{E'}CO_2(V_{T,BTPS} - V_{Dana}) = P_{\bar{E}}CO_2 \cdot V_{T,BTPS} \tag{13.14}$$

Equation (13.14) can be rearranged to give the Bohr equation for the estimation of anatomical dead space:

$$V_{Dana} = V_T \frac{(P_{E'}CO_2 - P_{\bar{E}}CO_2)}{P_{E'}CO_2} \tag{13.15}$$

where all gases are represented at BTPS.

Total CO_2 expiration can also be described at the effective alveoli, as the product of alveolar CO_2 fraction ($F_ACO_{2,BTPS}$) and the volume of effective alveolar ventilation

$(V_{T,BTPS} - (V_{Dana} + V_{DA}))$. Converting fraction to pressure $(P_A CO_2 = F_A CO_2 \cdot P_B)$ and equating CO_2 expiration at the effective alveoli and the mouth gives:

$$P_A CO_2(V_{T,BTPS} - (V_{Dana} + V_{DA})) = P_{E'} CO_2 \cdot V_{T,BTPS} \tag{13.16}$$

which can be rearranged to give the Bohr equation for the estimation of physiological dead space:

$$V_{Dp} = V_{Dana} + V_{DA} = V_T \frac{(P_A CO_2 - P_{\bar{E}} CO_2)}{P_A CO_2} \tag{13.17}$$

where all gases are expressed in BTPS.

13.2 Models of Diffusion Limitation

Figure 13.2 illustrates the diffusion of oxygen from alveoli to lung capillary blood. During the transport of blood through the lung capillaries, the partial pressure of oxygen in the blood equilibrates with the partial pressure of oxygen in the alveoli, with equilibrium usually being reached during the blood transport through the capillaries [2]. When diffusion is impaired, equilibrium may not be reached and an alveolar–arterial oxygen pressure difference may occur. Figure 13.2 illustrates some of the possible causes of this difference due to "diffusion" limitation. These are a short transit time of the blood through the lung capillary, and a reduced diffusion capacity from alveoli to lung capillaries $(D_L O_2)$, which can be due to either

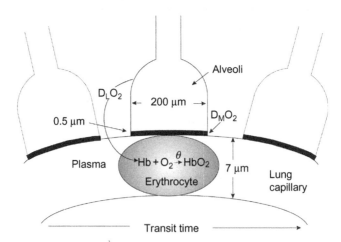

Figure 13.2 Diffusion of oxygen from alveoli to lung capillary blood. $D_L O_2$ is the total diffusion capacity of O_2 from alveoli to the blood, $D_M O_2$ is the diffusion capacity of the lung capillary membrane, and θ is the rate of oxygen binding in the blood.

reduced diffusion capacity across the lung capillary membrane ($D_M O_2$) or to slow binding of oxygen to Hb in the erythrocyte.

Much of the theory of mathematical models constructed to explain diffusion abnormalities within the lungs can be understood by the application of Fick's first law of diffusion. This law was first applied to study the diffusion properties of oxygen by Bohr [3], and states that for blood transversing the lung capillary, the total oxygen transport ($\dot{V}O_2$) is proportional to the alveolar–lung capillary PO_2 difference:

$$\dot{V}O_2 = D_L O_2 (P_A O_2 - P_c O_2) \tag{13.18}$$

An expression for the change in $P_c O_2$ as blood flows through the lung capillaries can be derived by considering the flow of oxygen into a single "slice" of blood flowing through the capillary. Assuming that blood flow through the lungs can be represented as a single capillary length X and cross-sectional area A, then for a single slice of blood, width dx, total $\dot{V}O_2$ is the sum of $\dot{V}O_2$ values as the slice flows through the capillary:

$$\dot{V}O_2 = \int_0^X \dot{V}O_2(x)dx \tag{13.19}$$

If the slice of blood flows through the capillary from time t to time $t + dt$, oxygen transport into the slice ($\dot{V}O_2(x)dx\,dt$) is equal to the change in O_2 mass in the slice ($[C_c O_2(t + dt) - C_c O_2(t)]V_s$) during this time:

$$\dot{V}O_2 dx\,dt = [C_c O_2(t + dt) - C_c O_2(t)]V_s \tag{13.20}$$

where V_s is the volume of the slice of blood and $[C_c O_2(t + dt) - C_c O_2(t)]$ is the change in oxygen concentration ($dC_c O_2(t)$). V_s is the product of the cross-sectional area of the capillary (A) and dx. Substituting for V_s and $dC_c O_2(t)$ in Eq. (13.20) gives:

$$\dot{V}O_2(x) = A \frac{dC_c O_2(t)}{dt} \tag{13.21}$$

The flow of oxygen from the alveoli into the slice of blood ($\dot{V}O_2(x)dx$) can also be described using Fick's first law of diffusion, which means as the product of the diffusion capacity over the slice ($D_L O_2\,dx/X$), and the partial pressure difference at time t ($P_A O_2 - P_c O_2(t)$):

$$\dot{V}O_2(x)dx = \frac{D_L O_2\,dx}{X}(P_A O_2 - P_c O_2(t)) \tag{13.22}$$

Eliminating $\dot{V}O_2(x)$ from Eqs. (13.21) and (13.22) gives:

$$A \frac{dC_c O_2(t)}{dt} = \frac{D_L O_2\,dx}{X}(P_A O_2 - P_c O_2(t)) \tag{13.23}$$

The volume of lung capillary blood V_c equals the product of the cross-sectional area (A) and the capillary length (X). Replacing for A and X in Eq. (13.23) gives:

$$\frac{dC_cO_2(t)}{dt} = \frac{D_LO_2 \, dx}{V_c}(P_AO_2 - P_cO_2(t)) \tag{13.24}$$

which can be rearranged to give:

$$\int \frac{1}{P_AO_2 - P_cO_2(t)} dC_cO_2 = \int \frac{D_LO_2}{V_c} dt \tag{13.25}$$

This equation cannot be solved algebraically to obtain an expression describing the change in P_cO_2 as blood passes through the lung capillary because of the complicated relation between oxygen partial pressure (P_cO_2) and concentration (C_cO_2). Binding of O_2 to hemoglobin is described by the oxygen dissociation curve (ODC), a nonlinear function relating PO_2 to the oxygen saturation of hemoglobin in the blood (S_cO_2), which means $S_cO_2 = ODC(P_cO_2)$. Total O_2 concentration is therefore:

$$C_cO_2 = P_cO_2 \cdot \alpha_{O_2} + Hb \cdot ODC(P_cO_2) \tag{13.26}$$

where α_{O_2} is the solubility of oxygen in the blood (normal value 0.01 (mmol/(L kPa)), 0.0014 (mmol/(L mmHg)) [4]).

With knowledge of the concentration and pressure of oxygen in the venous blood and in blood coming from the lung capillaries, assuming a constant alveolar oxygen partial pressure (P_AO_2) and pulmonary capillary blood volume (V_c), and using the oxygen dissociation curve, Bohr was able to use graphical/numerical methods, known as *Bohr integration*, to solve Eq. (13.25) and estimate the diffusing capacity for oxygen (D_LO_2).

For an inert gas, Henry's law applies such that the partial pressure is proportional to concentration:

$$C = \beta \cdot P \tag{13.27}$$

where β is the solubility of the gas in blood. As shown by Wagner [5], for inert gases it is possible to obtain an expression for the partial pressure of gas in blood as it passes through the lung capillary that is algebraically solvable, as follows. Equation (13.24) can be written for an inert gas as:

$$\frac{dP_c(t)}{dt} = \frac{D_L}{V_c \cdot \beta}(P_A - P_c(t)) \tag{13.28}$$

By integrating Eq. (13.28), an algebraic expression can be obtained for the partial pressure of the gas in blood as it passes through the lung capillary:

$$P_c(t) = P_A - (P_A - P_{\bar{v}})e^{-(D_L/(V_c \cdot \beta))t} \tag{13.29}$$

The change in partial pressure of the gas in the lung capillary can therefore be described by a single exponential. For oxygen, Eq. (13.29) can be written as:

$$P_cO_2(t) = P_AO_2 - (P_AO_2 - P_{\bar{v}}O_2)e^{-(D_LO_2/(V_c \cdot \beta))t} \tag{13.30}$$

and can be used to simulate the change in PO_2 in the lung capillary blood, accounting for the fact that Henry's law is not obeyed, meaning that the relationship between concentration and pressure is the nonlinear ODC, such that β varies with PO_2. Figure 13.3A illustrates the change in PO_2 in the lung capillary blood (P_cO_2) as it equilibrates with P_AO_2, for both: varying values of β (solid curve), according to the relationship between oxygen content and pressure illustrated in Figure 13.3B; and for a constant value of $\beta = 0.8$ mmol/L kPa (dotted curve), a value of β that occurs at the lower end of the pressure content curve seen in Figure 13.3B. It can be clearly seen that β decreases with increasing PO_2, causing more rapid equilibration between P_cO_2 and P_AO_2. Under normal circumstances, equilibrium is reached within 0.25 s, one third of the normal capillary transit time (0.75 s) [5].

A modification of the original Bohr method was made by Riley et al. [8] and Riley and Cournand [9]. Their approach eliminated Bohr's assumption that arterial blood and blood leaving the lung capillaries were at the same PO_2, making it possible to investigate the situation in which both a diffusion abnormality and a pulmonary shunt are present. Riley et al. [8] and Riley and Cournand [9] performed an experiment in which patients were studied at two inspired oxygen fractions, to achieve arterial oxygen saturations of 82% and 95%. At each F_IO_2, estimates of P_AO_2 were obtained from the alveolar air equation (Eq. (13.8)). The resultant $P_AO_2 - P_aO_2$ differences were then partitioned into $P_AO_2 - P_cO_2$ and $P_cO_2 - P_aO_2$ differences using Bohr integration. In effect, the overall $P_AO_2 - P_aO_2$ disorder was partitioned into that due to a diffusion abnormality and that due to pulmonary shunt.

As illustrated in Figure 13.2, the oxygen diffusion capacity D_LO_2 is a composite measure of the ability of oxygen to travel from the alveoli to the lung capillary plasma and then to bind with hemoglobin in the erythrocyte. Estimation of a constant D_LO_2 is therefore a mean across the whole capillary, which assumes that the speed of binding of oxygen to hemoglobin is constant. This assumption is invalidated by two physiological mechanisms that work in opposite directions. First, as the hemoglobin is saturated with oxygen, its ability to bind further oxygen is reduced. Second, each molecule of hemoglobin binds four molecules of oxygen with the rate constants of these reactions varying such that binding of the fourth molecule is significantly quicker than the other three [10]. Staub et al. [11]

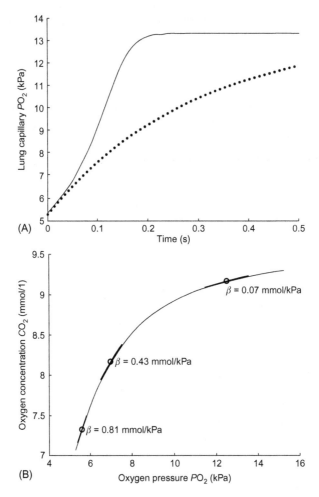

Figure 13.3 (A) Model predicted simulations (iterative solution of Eq. (13.30)) illustrating the change in PO_2 in lung capillary blood (P_cO_2) when equilibrated with alveolar P_AO_2, using values of $D_LO_2 = 12.5$ mmol/(min kPa) (40 mL/(min mmHg)), $V_c = 0.075$ L, $P_AO_2 = 13.3$ kPa (100 mmHg), and $P_{\bar{v}}O_2 = 5.3$ kPa (40 mmHg). *Solid line*, solution of Eq. (13.30) varying values of β according to (B); *dotted line*, solution of Eq. (13.30) with constant $\beta = 0.8$ mmol/L kPa. (B) Oxygen pressure (PO_2), concentration (CO_2) curve for blood. Plotted using Eq. (13.26) with Hb = 9.3 mmol/L, $\alpha_{O_2} = 0.01$ mmol/(L kPa). In Eq. (13.26) the ODC is the oxygen dissociation curve for blood implemented as described by Siggaard-Andersen et al. [6] (i.e., that included in the oxygen status algorithm (Version 3, [7])). β, the slope of the CO_2, PO_2 curve decreases with increasing values of PO_2.

proposed a model, previously derived for carbon monoxide [12], that accounted for these two effects. In doing so, the total oxygen diffusion capacity D_LO_2 was divided into terms describing the diffusion capacity across the blood gas barrier (D_MO_2) and the diffusion capacity associated with oxygen binding to hemoglobin

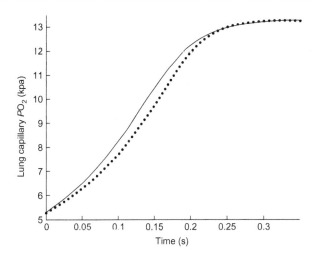

Figure 13.4 Model predicted simulations illustrating the change in P_cO_2 when equilibrated with $P_AO_2 = 13.3$ kPa (100 mmHg). *Dashed line*, solution of Eq. (13.30) using a constant value of $D_LO_2 = 9.0$ mmol/(min kPa) (29 mL/[min mmHg]), varying β according to Figure 13.3B; *solid line*, solution of Eq. (13.32) varying values of θ according to Staub et al. [11] while assuming a constant value of $D_MO_2 = 12.5$ mmol/(min kPa) (40 mL/[min mmHg]), and varying β according to Figure 13.3B. For both lines, $V_c = 0.075$ L and $P_{\bar{v}}O_2 = 5.3$ kPa (40 mmHg).

$(V_{c\theta})$. θ is the rate of oxygen binding in the blood per unit time, for a given pressure of oxygen in the blood and for the number of liters of blood. The equation proposed by Staub et al. [11] was:

$$\frac{1}{D_LO_2} = \frac{1}{D_MO_2} + \frac{1}{V_c \cdot \theta} \tag{13.31}$$

Staub et al. [11] described the relationship between θ and the oxygen saturation (SO_2). θ was found to decrease with increasing oxygenation when the blood was more than 75% saturated with oxygen, decreasing from 810 (mmol O_2/[min kPa L]), (2.6 [mL O_2/(min mmHg mL)]) at 75% oxygen saturation to 160 (mmol O_2/[min kPa L]), (0.5 [mL O_2/(min mmHg mL)]) at 98% oxygen saturation.

An overall measure of D_LO_2 is therefore more appropriately explained as a higher value of D_LO_2 (increased oxygen diffusion capacity) during the early stages of gas exchange, and a smaller value of D_LO_2 during the latter stages. This effect is illustrated in Figure 13.4, which shows the change in P_cO_2 for a constant D_LO_2 (dashed line) and for varying values of D_LO_2 (solid line) given by the relationship between θ and SO_2 described by Staub et al. [11]. By substituting for D_LO_2 in Eq. (13.30) using Eq. (13.31), an expression can be obtained by describing oxygenation of the lung capillary blood:

$$P_cO_2(t) = P_AO_2 - (P_AO_2 - P_{\bar{v}}O_2)e^{-(1/D_MO_2)+(1/(V_c \cdot \theta))(1/(V_c \cdot \beta \cdot x))t} \tag{13.32}$$

where oxygen diffusion abnormalities exist, high values of θ imply that these abnormalities are present in the alveolar capillary membrane, while low values imply that the abnormality occurs in the blood, most likely due to oxygen binding to hemoglobin. While the original study of Staub et al. [11] reported rather low initial values of θ of 2.6 mL O_2/(min mmHg mL blood), but more recent studies [13] have reported values of θ as high as 3.9 mL O_2/(min mmHg mL blood), implying a greater diffusion abnormality in the lung capillary membrane than was previously believed.

13.2.1 Gas Exchange Abnormalities Described by Heterogeneity of Diffusion/Perfusion $D/\dot{Q}$

Figure 13.1 illustrates a lung with a single compartment involved in gas exchange. This model is that applied by Riley et al. [8] and Riley and Cournand [9], and assumes that all nonshunted blood passes through this single lung compartment. Piiper et al. [14], when using a similar approach to Riley in estimating the shunt and diffusion abnormalities in anesthetized dogs, found that the model with a single gas exchange compartment was inadequate to describe alveolar—arterial oxygen differences. To account for these differences, Piiper et al. [14] proposed the concept of a diffusion/perfusion ($D/\dot{Q}$) mismatch in a heterogeneous lung, which is a lung with a number of compartments involved in gas exchange, each with varying diffusion properties ($D_{L1}O_2$, $D_{L2}O_2$) and with varying fractions of the total nonshunted blood ($\dot{Q}_1, \dot{Q}_2$) flowing through these compartments. This picture is illustrated for two compartments in Figure 13.5.

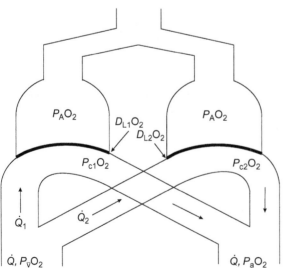

Figure 13.5 A conceptual model of the lungs including two compartments involved in gas exchange. These compartments have different diffusion capacities ($D_{L1}O_2$, $D_{L2}O_2$) and perfusion properties ($\dot{Q}_1, \dot{Q}_2$) resulting in varying partial pressures of oxygen flowing from the lung capillaries ($P_{c1}O_2$, $P_{c2}O_2$) for each compartment.

For oxygen transport, the diffusion/perfusion mismatch ($D_L O_2/\dot{Q}$) can be represented mathematically by modifying Eq. (13.30), substituting for transit time (t). For blood leaving the lung capillaries, t can be expressed as:

$$t = \frac{\text{Volume of pulmonary capillary blood}}{\text{Rate of blood flow through the capillary}} = \frac{V_c}{\dot{Q}}$$

Substituting for t in Eq. (13.30) gives:

$$P_c O_2 = P_A O_2 - (P_A O_2 - P_{\bar{v}} O_2)e^{(-1/\beta)\cdot(D_L O_2/\dot{Q})} \tag{13.33}$$

which describes $P_c O_2$ as an exponential function of the solubility (β) and the $D/\dot{Q}$ ratio. For a model with two compartments involved in gas exchange (Figure 13.5), Eq. (13.33) can be written to describe the partial pressure of oxygen in blood from each of the compartments ($P_{c1} O_2$, $P_{c2} O_2$).

Piiper [15] and Piiper and Scheid [16] have used the term equilibration index to describe the ratio $D/(\dot{Q} \cdot \beta)$, using this index to quantify the extent to which gas exchange abnormalities can be partitioned into a diffusion or perfusion problem. This index is particularly applicable where β is constant (i.e., for inert gases or for oxygen during hypoxia where the slope of the oxygen dissociation curve is almost constant).

This section has given a mathematical description of oxygen diffusion. In the pulmonary physiology laboratory, the diffusion properties of the lungs are more frequently estimated using carbon monoxide ($D_L CO$). Unlike oxygen, the partial pressure of CO in lung capillary plasma can be assumed to be negligible so that diffusion into the blood depends solely on the driving pressure in the alveoli.

Numerous methods exist for estimating $D_L CO$. These include methods where a steady-state PCO is obtained in the alveoli, a single breath is taken, or experiments involving rebreathing. While the details of these techniques are beyond the scope of this chapter, they are all based on applying Fick's first law of diffusion, which for carbon monoxide can be written as:

$$D_L CO = \frac{\dot{V}CO}{P_A CO - P_c CO}$$

where the lung capillary partial pressure of carbon monoxide ($P_c CO$) is assumed to be zero. Estimates of $D_L O_2$ are then made by multiplying $D_L CO$ by 1.23 to account for the different density and solubility of the two gases.

13.3 Models of Ventilation—Perfusion Mismatch

Much of our present understanding of how ventilation—perfusion relationships can be measured through the use of physiological models stems from the seminal work by the groups of Rahn and coworkers [17—19] and Riley et al. [8] and Riley and

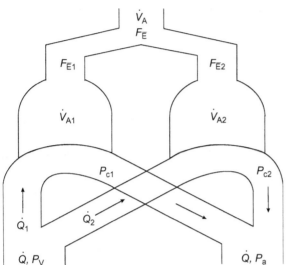

Figure 13.6 A conceptual model of the lungs, including two compartments involved in gas exchange. These compartments have different ventilation ($\dot{V}_{A1}$, $\dot{V}_{A2}$) and perfusion ($\dot{Q}_1$, $\dot{Q}_2$) properties, resulting in varying end capillary partial pressures (P_{c1}, P_{c2}).

Cournand [9]. Figure 13.1 illustrates the model by Riley and coworkers, representing a lung with a single compartment involved in gas exchange. This model is a simplification of the true situation where different regions of the lungs have varying ventilation ($\dot{V}$) and perfusion ($\dot{Q}$). Figure 13.6 illustrates a conceptual model of the lungs with two compartments involved in gas exchange, each with a different ventilation/perfusion ($\dot{V}/\dot{Q}$) ratio.

The two-compartment model illustrated in Figure 13.6 has four parameters ($\dot{V}_{A1}$, $\dot{V}_{A2}$, $\dot{Q}_1$, $\dot{Q}_2$), the unique identification of which is not possible from routine clinical measurements of O_2 and CO_2 in expired air and blood. Rahn and Riley and their coworkers showed that measurement of gas contents of alveolar air and arterial and mixed venous blood at varying levels of inspired O_2 provides information as to the $\dot{V}/\dot{Q}$ matching in the lungs, allowing identification of parameters of simple models as that depicted in Figure 13.1 [8,9,17−19]. Lenfant and Okubo [20,21] proposed a method for identifying continuous distributions of $\dot{V}$ and $\dot{Q}$ (which means an infinite number of $\dot{V}_A$ and $\dot{Q}$ parameters in a model, as illustrated in Figure 13.6) from measured changes in arterial oxygen saturation with increasing inspired oxygen fraction during nitrogen washout. However, this method was based on the assumption that changes in inspired oxygen fraction do not affect ventilation and perfusion distributions, which Lenfant [22] showed may not be correct. Furthermore, the method did not allow for separation of causes of impairment of gas exchange into that due to diffusion limitation and that due to ventilation−perfusion mismatch. Use of inert gases instead of oxygen, as done by the Multiple Inert Gas Elimination Technique (MIGET), allows circumvention of these limitations [23], and the MIGET has enabled estimation of parameters describing multicompartment $\dot{V}/\dot{Q}$ heterogeneity. The MIGET has been widely applied in clinical and experimental research, and includes both a mathematical model of $\dot{V}/\dot{Q}$ heterogeneity and an experimental technique using multiple inert gases as tracers.

A detailed description of the experimental technique is beyond the scope of this chapter. The mathematical model included in the MIGET illustrates how the effects of $\dot{V}/\dot{Q}$ heterogeneity can be mathematically formulated, and is described below.

13.3.1 The Multiple Inert Gas Elimination Technique

In the MIGET, multiple inert gases are simultaneously infused into a vein. Inert gases are used as tracers and, as such, eliminate the need for perturbing the underlying oxygen system, which may change the nature of the gas exchange abnormality. The MIGET assumes no diffusion limitation exists for inert gases, such that end lung capillary blood is at equilibrium with the alveoli. The key to the MIGET is that gases with a higher solubility are retained more readily in blood during its transport through lung capillaries. By choosing tracer gases with a wide range of solubilities and by measuring their concentration both in expired air and venous–arterial blood, a more complete description of the $\dot{V}/\dot{Q}$ characteristics of the lungs can be obtained.

The mathematical formulation used in the MIGET analysis can be derived from a mass balance equation similar to the Fick principle of blood flow (Eq. (13.13)) and the alveolar air equation, which is for a single tracer gas in a single alveolar gas exchange compartment:

$$\dot{V} = \dot{Q}(C_c - C_v) \tag{13.34}$$

$$P_A = F_I \cdot P_B - \frac{\dot{V} \cdot P_B}{\dot{V}_A} \tag{13.35}$$

For inert gases, inspired fraction (F_I) is zero. Eliminating $\dot{V}$ in Eqs. (13.34) and (13.35) and replacing concentrations with pressures using Henry's law (Eq. (13.27)) gives:

$$\frac{\dot{V}_A \cdot P_A}{P_B} = \beta \cdot \dot{Q}(P_{\bar{v}} - P_c) \tag{13.36}$$

Assuming no diffusion limitation $P_A = P_c$ so that:

$$\frac{\dot{V}_A \cdot P_c}{P_B} = \beta \cdot \dot{Q}(P_{\bar{v}} - P_c) \tag{13.37}$$

This equation can be rearranged to describe the retention (R) of the gas as blood passes through the lung capillary, which means the $P_c/P_{\bar{v}}$ ratio:

$$R = \frac{P_c}{P_{\bar{v}}} = \frac{\beta \cdot P_B}{\dot{V}_A/\dot{Q} + (\beta \cdot P_B)} \tag{13.38}$$

which is the equation originally described by Wagner et al. [23], modified to include barometric pressure (P_B).

High solubility gases are retained more readily in the blood such that as $\beta \to \infty$ the retention $(R) \to \beta/\beta = 1$. For gases with low solubility, the retention depends more upon the $\dot{V}/\dot{Q}$ ratio of the lung unit. Excretion (E) is defined in a similar way to retention as the ratio of alveolar to venous gas pressures, $P_A/P_{\bar{v}}$.

The retention of inert gas in a multicompartmental lung can be described by considering the mass of gas in blood leaving each of the perfusion compartments. For the two-compartment model illustrated in Figure 13.6, this can be described as:

$$\dot{Q} \cdot \beta \cdot P_a = \dot{Q}_1 \cdot \beta \cdot P_{c1} + \dot{Q}_2 \cdot \beta \cdot P_{c2} \tag{13.38}$$

Dividing through by $P_{\bar{v}}$ and canceling β in each of the terms gives:

$$\dot{Q}\frac{P_a}{P_{\bar{v}}} = \dot{Q}_1 \frac{P_{c1}}{P_{\bar{v}}} + \dot{Q}_2 \frac{P_{c2}}{P_{\bar{v}}} \tag{13.39}$$

giving:

$$\dot{Q} \cdot R = \dot{Q}_1 \cdot R_1 + \dot{Q}_2 \cdot R_2 \tag{13.40}$$

For multiple compartments $(i = 1{:}n)$ the overall retention (R), which is the ratio of pressure in arterial to venous blood, can be described as:

$$R = \frac{1}{\dot{Q}}\sum_{i=1}^{n} \dot{Q}_i \cdot R_i = \frac{1}{\dot{Q}}\sum_{i=1}^{n} \dot{Q}_i \cdot \frac{\beta}{\overline{V}_{Ai}/\dot{Q}_i + \beta} \tag{13.41}$$

Excretion is described in a similar way:

$$E = \frac{1}{\overline{V}}\sum_{i=1}^{n} \overline{V}_i \cdot R_i = \frac{1}{\overline{V}}\sum_{i=1}^{n} \overline{V}_{Ai} \cdot \frac{\beta}{\overline{V}_{Ai}/\dot{Q}_i + \beta} \tag{13.42}$$

Measurements of retention $(P_a/P_{\bar{v}})$ and excretion $(P_A/P_{\bar{v}})$ are obtained for each inert gas (usually 6) by sampling blood and expired gas. By dividing the lungs into a number of compartments, usually 50, each with a fixed $\dot{V}/\dot{Q}$ ratio, Eqs. (13.41) and (13.42) are fitted to measured values of R and E by varying the perfusion $(\dot{Q}_i)$ and ventilation $(\dot{V}_i)$ of each of the compartments. Results of these analyses are then reported as distributions of blood flow and ventilation across the range of possible $\dot{V}/\dot{Q}$ regions in the lungs.

13.4 Application of Mathematical Models of Ventilation, Perfusion, and Diffusion

13.4.1 Application of Models in Physiological and Clinical Experimentation

Despite the research into diffusion abnormalities, the standard method of describing pulmonary gas exchange in experimental studies is the MIGET, which is considered to be a reliable technique for quantifying shunt, physiological dead space, and the distribution of $\dot{V}/\dot{Q}$ ratios in the lungs. Indeed, in studies using the MIGET, the concept of diffusion impairment is seldom required to describe abnormalities in gas exchange, except in cases of pulmonary fibrosis [24], exercise, or mild exercise during hypoxia [25−27]. A brief summary of the application of the MIGET in human subjects in anesthesiology, intensive care medicine, and pulmonary medicine now follows, with emphasis given to major improvements in understanding enabled by the application of this technique.

In anesthesiological research, the MIGET has been used to describe gas exchange abnormalities following different types of anesthesia: inhalation [28−32], intravenous [33,34], and epidural [35,36]; and after numerous interventions during anesthesia: surgery [37−39], variation in positive end-expiratory pressure (PEEP) [29,40], different modes of mechanical ventilation [41], CO_2 pneumoperitoneum [42], or infusion of inotropic agents [43]. The findings of these studies consistently show an increased shunt and $\dot{V}/\dot{Q}$ mismatch following anesthesia, with increases in PEEP reducing the shunt fraction.

Investigations using the MIGET in intensive care have provided understanding of the pathophysiology of disorders such as acute respiratory distress syndrome (ARDS) and pneumonia and the effects of therapeutic interventions in patients with these severe disorders. The application of the MIGET in studying ARDS and pneumonia has been reviewed by Melot [44], the main finding being an increased shunt and increased perfusion of low $\dot{V}/\dot{Q}$ areas to explain severe hypoxemia seen in these patients. The use of MIGET to study therapeutic intervention in these patients has shown that PEEP improves oxygenation via a reduction in shunt fraction [45−47]; turning patients from supine to prone position reduces shunt [48,49]; inhalation of prostacyclin or nitric oxide causes a redistribution of blood flow to regions with normal $\dot{V}/\dot{Q}$ [50−52]; and using small tidal volumes in permissive hypercapnia causes an increase in shunt fraction [53].

The MIGET has been used extensively in pulmonary laboratories to describe gas exchange in patients with chronic pulmonary disease as reviewed by Agustí and Barbera and Rodriguez-Roisin and MacNee [54,55]. The major finding of studies in clinically stable chronic obstructive pulmonary disease (COPD) has been that these patients have severe $\dot{V}/\dot{Q}$ mismatching with diffusion limitation and shunt having limited influence on arterial hypoxemia [24,55−58]. A study during exacerbations of COPD showed augmentation of $\dot{V}/\dot{Q}$ mismatching and negligible intrapulmonary shunt [59]. In studies of patients with pulmonary fibrosis, $\dot{V}/\dot{Q}$

mismatch accounts for 80% of the hypoxemia at rest, and only 60% during exercise [24], indicating severe diffusion abnormalities in this group of patients.

While the MIGET has found widespread application as an experimental tool, its use as a routine clinical tool has been somewhat limited [60,61]. This is largely due to the cost and complexity of the technique. The MIGET requires a very systematic and detailed technical procedure that involves preparation of an infusate containing six inert gases, sampling blood and gas after 30 min of infusion, and analyzing these using gas chromatography. The complexity and invasive nature of the original technique has led to development of less invasive experimental techniques intended to introduce the MIGET into clinical practice [62]. The gas chromatography analysis of inert gas partial pressures necessary for MIGET requires that inert gases first are extracted into a gas phase from a blood sample. This time-consuming process may be avoided by using micropore membrane inlet mass spectrometry instead of gas chromatography [63].

13.4.2   Application of Models in Routine Clinical Practice

Estimation of the parameters of multicompartmental models, including $D'/\dot{Q}$ or $\dot{V}/\dot{Q}$ heterogeneity, is seldom routinely performed outside the pulmonary laboratory. In the surgical theater, intensive care unit, recovery room, or more general wards, the clinician usually relies upon other measurements to assess the patient's gas exchange problems. For the measurement of oxygenation problems and the prevention of hypoxemia, which can lead to ischemic organ dysfunction, these measurements include single measurements of arterial oxygen tensions and the P_aO_2/F_IO_2, calculations of alveolar–arterial oxygen difference (A–aO$_2$), or—as the clinical state of the art—estimation of the shunt fraction using information obtained from sampling of mixed venous blood through a pulmonary artery catheter [64]. For measurement of CO_2 gas exchange, these measurements include measurement of arterial carbon dioxide tension, expired levels of CO_2, and calculation of physiological or alveolar dead space. These measurements are an attempt to describe the "clinical" gas exchange problem and have insufficient detail to interpret problems in physiological terms. However, it has been shown that the P_aO_2/F_IO_2 ratio, and even the estimation of shunt, cannot adequately describe the "clinical" picture seen in patients when the inspired oxygen fraction is varied, which has been shown to vary significantly with clinical changes in inspired oxygen fraction [65–67]. This situation is illustrated for shunt in Figure 13.7, where the variation in oxygen saturation of arterial blood is plotted against expired oxygen fraction for one postoperative patient, and the shunt equation (Eq. (13.11)) is fitted to each data point. No single value of shunt adequately describes this patient, the estimated shunt increasing at lower values of end tidal oxygen fraction ($F_{E'}O_2$).

The realization that a "shunt-only" model cannot fit data when F_IO_2 is varied was recognized by Riley et al. [8] and Riley and Cournand [9] and later by King et al. [68]. To solve this problem, these authors divided the oxygenation problem into that due to an alveolar–lung capillary drop in the partial pressure of oxygen, and that due to pulmonary shunt. To estimate two parameters describing the

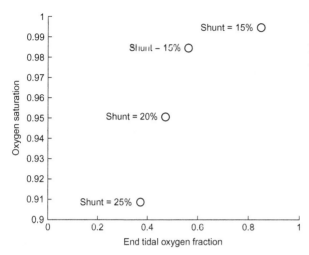

Figure 13.7 The shunt fraction estimated for one patient at varying values of $F_{E'}O_2$.

oxygenation problem then required no more than obtaining routine measurements of blood gases and ventilation at varying inspired oxygen fractions. However, use of oxygen as tracer is based on the assumption that variation in inspired oxygen fraction does not affect lung physiology, which, as described in Section 13.3, was the very reason for the development of MIGET. Changes in inspired oxygen fraction have been shown to affect lung physiology through absorption atelectasis at high levels of oxygen [69] and hypoxic pulmonary vasoconstriction [70,71]. However, such studies have been performed with maximal values of oxygen and maximal responses to changes in inspired oxygen fraction. Consequently, if identification of model parameters can be performed for inspired oxygen fractions below 0.80 and for small steps in inspired oxygen fraction, little evidence exists that variation in oxygen affects lung physiology significantly [72]. The two-parameter models proposed by Riley et al. and King et al., although a relatively poor description of the physiology, represent a substantial improvement over a shunt-only model, as they describe the effects of varying F_IO_2, a routine therapeutic intervention in mechanically ventilated patients.

Two-parameter mathematical models of oxygen transport have been formulated with the oxygenation problem being described as shunt combined with either a resistance to oxygen diffusion (labeled R_{diff}) [73−75] or with a $P_IO_2 - P_cO_2$ difference due to $\dot{V}/\dot{Q}$ mismatch [75−79]. These model representations have been shown to provide identical fits to routine blood gas and ventilatory data obtained by varying F_IO_2[80], but focus has in recent years been on models of shunt and $\dot{V}/\dot{Q}$ mismatch, as it has been shown that these factors are predominant in the majority of circumstances (as outlined in Section 13.4.1). The clinical relevance of two parameters describing shunt and $\dot{V}/\dot{Q}$ mismatch is illustrated in Figure 13.8 for the shunt and $\dot{V}/\dot{Q}$ mismatch model of Kjærgaard et al. [75].

As illustrated in Figure 13.8, increases in the pulmonary shunt parameter results in a vertical depression of the plateau of the $F_{E'}O_2/S_aO_2$ curve, while

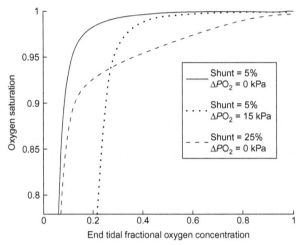

Figure 13.8 Model predicted arterial oxygen saturations for three hypothetical cases. *Solid line*, a normal subject with shunt = 5% and no $\dot{V}/\dot{Q}$ mismatch $\Delta PO_2 = 0$ kPa; *dotted line,* a hypothetical patient with $\dot{V}/\dot{Q}$ mismatch; *dashed line*, a hypothetical patient with a shunt disorder.

abnormalities in the second parameter (ventilation/perfusion ($\dot{V}/\dot{Q}$) mismatch, or oxygen diffusion resistance (R_{diff})) cause a lateral displacement of the $F_{E'}O_2/S_aO_2$ curve. The lateral displacement of the $F_{E'}O_2/S_aO_2$ curve can be seen as a more clinically significant problem as it describes a situation in which large changes in oxygen saturation can occur for only small changes in F_1O_2. In Figure 13.8, $\dot{V}/\dot{Q}$ mismatch is reported as ΔPO_2, which is the drop in oxygen pressure from ventilated alveoli to pulmonary capillary blood prior to mixing with shunted venous blood ($P_AO_2 - P_cO_2$). ΔPO_2 can therefore be understood as the extra oxygen pressure required at the mouth to counter oxygenation problems due to $\dot{V}/\dot{Q}$ mismatch; a ΔPO_2 of 15 kPa means air plus approximately 15% inspired O_2 ($F_1O_2 = 36\%$) is required.

The two-parameter model of Sapsford et al. [77] has been shown to fit data from normal subjects; patients before and after thoracotomy [77,78], patients during [77,79] and after [79] abdominal surgery, in neonates with pulmonary failure [81], and in infants with bronchopulmonary dysplasia [82,83]. The two-parameter model of Kjærgaard et al. has been shown to fit data from normal subjects [84], patients before [75,84−86] and after [75,84−87] surgery, patients presenting in intensive care [66,84], and patients with incompensated heart failure studied before and after diuretic therapy [84]. Furthermore, the model of Kjærgaard et al. has been shown to fit MIGET inert gas retention and excretion measurements [72,88] and to simulate oxygenation comparable to the mathematical model of MIGET [88] in lung injury animal models, and the model has been applied in a decision support system for management of mechanical ventilation [89−91].

Oxygen as a tracer provides most information about lung regions with low $\dot{V}/\dot{Q}$ ratios, and shunt as the shape of the $F_{E'}O_2/S_aO_2$ curve is affected primarily by these factors. In contrast, carbon dioxide is affected primarily by lung regions with high $\dot{V}/\dot{Q}$. CO_2 has been used as tracer in combination with oxygen by inclusion of end tidal and arterial CO_2 measurements allowing a third gas exchange parameter to be identified in models with structure as that illustrated in Figure 13.6

[92−94]. Vidal Melo et al. and Loepkky et al. used O_2 and CO_2 as tracers combined with shunt measurement by Eq. (13.11) to identify $\dot{V}/\dot{Q}$ of two compartments and a common diffusion impairment parameter in healthy subjects, COPD patients, and mechanically ventilated dogs with varying interventions [92,93]. Karbing et al. identified shunt and $\dot{V}/\dot{Q}$ of two compartments in intensive care patients, and compared these fits with the two-parameter model of Kjærgaard et al. [94]. The three-parameter model was necessary for accurate descriptions of O_2 and CO_2 gas exchange in some of these intensive care patients, the three-parameter model allowing description of larger ranges of ventilation-to-perfusion ratios [94]. Accurate description of CO_2 gas exchange requires description of the transport of CO_2 in blood, which means the blood acid−base chemistry, including the competitive binding of oxygen and carbon dioxide to hemoglobin [95].

Figure 13.9 illustrates a fit of the model by Karbing et al. to three different patients. Parameters are presented as shunt, ΔPO_2, and ΔPCO_2. ΔPCO_2 describes the increase in carbon dioxide pressure from ventilated alveoli to capillary blood, hence describing the carbon dioxide gas transport problem arising due to high $\dot{V}/\dot{Q}$

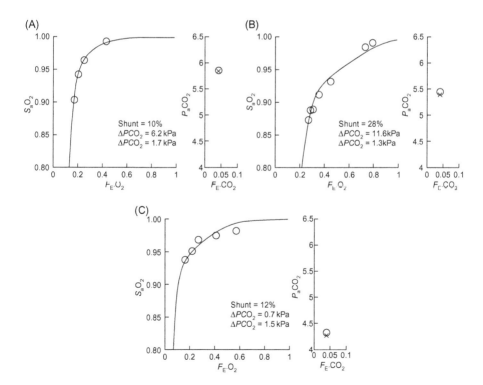

Figure 13.9 F_EO_2/S_aO_2 and $F_{E'}CO_2/P_aCO_2$ data (circles) and model fits (solid line and crosses) to data from (A) an intensive care patient, (B) an intensive care patient with severe gas exchange problems, and (C) a postoperative cardiac surgery patient.

regions. $\Delta PCO_2 > 0$ kPa signifies insufficient removal of CO_2, and potential need for increasing minute ventilation.

These two- and three-parameter models have yet to be routinely used in clinical practice, but with pulse oximetry technology enabling noninvasive estimation of the $F_{E'}O_2/S_aO_2$ curve for the case of two-parameter models and the addition of capnography and a single arterial blood gas measurement for three-parameter models, the estimation of the parameters of such models is a relatively simple task which might be performed as a part of clinical practice.

References

[1] Hughes J. Pulmonary gas exchange. Eur Respir Mon 2005;31:106−26.
[2] Lumb A. Nunn's applied respiratory physiology. Edinburgh, London, New York, Oxford, Philadelphia, St Louis, Sydney, Toronto: Churchill Livingstone; 2010.
[3] Bohr C. Über die spezifische tätigkeit der lungen bei der respiratorischen gasaufnahme und ihr verhalten zu der durch die alveolarwand stattfindenden gasdiffusion. Skandinavisches Archiv für Physiologie 1909;22:221−80.
[4] Siggaard-Andersen O. The acid-base status of the blood. Copenhagen: Munksgaard; 1974.
[5] Wagner P. Diffusion and chemical reaction in pulmonary gas exchange. Physiol Rev 1977;57:257−312.
[6] Siggaard-Andersen O, Wimberley P, Gøthgen I, Siggaard-Andersen M. A mathematical model of the hemoglobin-oxygen dissociation curve of human blood and of the oxygen partial pressure as a function of temperature. Clin Chem 1984;30:1646−51.
[7] Siggaard-Andersen M, Siggaard-Andersen O. Oxygen status algorithm, version 3, with some applications. Acta Anaesthesiol Scand 1995;39:13−20.
[8] Riley R, Cournand A, Donald K. Analysis of factors affecting partial pressures of oxygen and carbon dioxide in gas and blood of lungs: methods. J Appl Physiol 1951;4:102−20.
[9] Riley R, Cournand A. Analysis of factors affecting partial pressures of oxygen and carbon dioxide in gas and blood of lungs: theory. J Appl Physiol 1951;4:77−101.
[10] Staub NC, Bishop J, Forster R. Velocity of O_2 uptake by human red blood cells. J Appl Physiol 1961;16:511−6.
[11] Staub N, Bishop J, Forster R. Importance of diffusion and chemical reaction rates in O_2 uptake in the lung. J Appl Physiol 1962;17:21−7.
[12] Roughton F, Forster R. Relative importance of diffusion and chemical reaction rates in determining rate of exchange of gases in the human lung, with special reference to true diffusing capacity of pulmonary membrane and volume of blood in the lung capillaries. J Appl Physiol 1957;11:290−302.
[13] Yamaguchi K, Nguyen-Phu D, Scheid P, Piiper J. Kinetics of O_2 uptake and release by human erythrocytes studied by a stopped-flow technique. J Appl Physiol 1985;58:1215−24.
[14] Piiper J, Haab P, Rahn H. Unequal distribution of pulmonary diffusing capacity in the anesthetized dog. J Appl Physiol 1961;16:499−506.
[15] Piiper J. Diffusion-perfusion inhomogeneity and alveolar−arterial O_2 diffusion limitation: theory. Respir Physiol 1992;87:349−56.

[16] Piiper J, Scheid P. Model for capillary–alveolar equilibration with special reference to O_2 uptake in hypoxia. Respir Physiol 1981;46:193–208.

[17] Fenn WO, Rahn II, Otis AB. A theoretical study of the composition of the alveolar air at altitude. Am J Physiol 1946;146:637–53.

[18] Haab P, Piiper J, Rahn H. Attempt to demonstrate the distribution component of the alveolar–arterial oxygen pressure difference. J Appl Physiol 1960;15:235–40.

[19] Farhi L, Rahn H. A theoretical analysis of the alveolar–arterial O_2 difference with special reference to the distribution effect. J Appl Physiol 1955;7:699–703.

[20] Okubo T, Lenfant C. Distribution function of lung volume and ventilation determined by lung N_2 washout. J Appl Physiol 1968;24:658–67.

[21] Lenfant C, Okubo T. Distribution function of pulmonary blood flow and ventilation–perfusion ratio in man. J Appl Physiol 1968;24:668–77.

[22] Lenfant C. Effect of high FiO_2 on measurement of ventilation/perfusion distribution in man at sea level. Ann NY Acad Sci 1965;121:797–808.

[23] Wagner P, Saltzman H, West J. Measurement of continuous distributions of ventilation–perfusion ratios—theory. J Appl Physiol 1974;36:588–99.

[24] Agustí AG, Roca J, Gea J, Wagner PD, Xaubet A, Rodriguez-Roisin R. Mechanisms of gas-exchange impairment in idiopathic pulmonary fibrosis. Am J Respir Crit Care Med 1991;143:219–25.

[25] Wagner PD, Gale GE, Moon RE, Torre-Bueno JR, Stolp BW, Saltzman HA. Pulmonary gas exchange in humans exercising at sea level and simulated altitude. J Appl Physiol 1986;61:260–70.

[26] Torre-Bueno J, Wagner P, Saltzman H, Gale G, Moon R. Diffusion limitation in normal humans during exercise at sea level and simulated altitude. J Appl Physiol 1985;58:989–95.

[27] Rice AJ, Thornton AT, Gore CJ, Scroop GC, Greville HW, Wagner H, et al. Pulmonary gas exchange during exercise in highly trained cyclists with arterial hypoxemia. J Appl Physiol 1999;87:1802–12.

[28] Tokics L, Hedenstierna G, Svensson L, Brismar B, Cederlund T, Lundquist H, et al. V/Q distribution and correlation to atelectasis in anesthetized paralyzed humans. J Appl Physiol 1996;81:1822–33.

[29] Bindslev L, Hedenstierna G, Santesson J, Gottlieb I, Carvallhas A. Ventilation–perfusion distribution during inhalation anaesthesia. Acta Anaesthesiol Scand 1981;25:360–71.

[30] Lundh R, Hedenstierna G. Ventilation–perfusion relationships during halothane anaesthesia and mechanical ventilation. Effects of varying inspired oxygen concentration. Acta Anaesthesiol Scand 1984;28:191–8.

[31] Hedenstierna G, Tokics L, Strandberg Å, Lundquist H, Brismar B. Correlation of gas exchange impairment to development of atelectasis during anaesthesia and muscle paralysis. Acta Anaesthesiol Scand 1986;30:183–91.

[32] Gunnarsson L, Strandberg Å, Brismar B, Tokics L, Lundquist H, Hedenstierna G. Atelectasis and gas exchange impairment during enflurane/nitrous oxide anaesthesia. Acta Anaesthesiol Scand 1989;33:629–37.

[33] Anjou-Lindskog E, Broman L, Broman M, Holmgren A, Settergren G, Ohqvist G. Effects of intravenous anesthesia on VA/Q distribution: a study performed during ventilation with air and with 50% oxygen, supine and in the lateral position. Anesthesiology 1985;62:485–92.

[34] Rothen H, Sporre B, Engberg G, Wegenius G, Hedenstierna G. Airway closure, atelectasis and gas exchange during general anaesthesia. Br J Anaesth 1998;81:681–6.

[35] Lundh R, Hedenstierna G, Johansson H. Ventilation—perfusion relationships during epidural analgesia. Acta Anaesthesiol Scand 1983;27:410—6.

[36] Hachenberg T, Holst D, Ebel C, Pfeiffer B, Thomas H, Wendt M, et al. Effect of thoracic epidural anaesthesia on ventilation—perfusion distribution and intrathoracic blood volume before and after induction of general anaesthesia. Acta Anaesthesiol Scand 1997;41:1142—8.

[37] Hedenstierna G, Mebius C, Bygdeman S. Ventilation—perfusion relationship during hip arthroplasty. Acta Anaesthesiol Scand 1983;27:56—61.

[38] Lundh R, Hedenstierna G. Ventilation—perfusion relationships during anaesthesia and abdominal surgery. Acta Anaesthesiol Scand 1983;27:167—73.

[39] Hachenberg T, Tenling A, Nyström S, Tyden H, Hedenstierna G. Ventilation—perfusion inequality in patients undergoing cardiac surgery. Anesthesiology 1994;80:509—19.

[40] Tokics L, Hedenstierna G, Strandberg A, Brismar B, Lundquist H. Lung collapse and gas exchange during general anesthesia: effects of spontaneous breathing, muscle paralysis, and positive end-expiratory pressure. Anesthesiology 1987;66:157—67.

[41] Yu G, Yang K, Baker A, Young I. The effect of bi-level positive airway pressure mechanical ventilation on gas exchange during general anaesthesia. Br J Anaesth 2006;96:522—32.

[42] Andersson L, Lagerstrand L, Thörne A, Sollevi A, Brodin L, Odeberg-Wernerman S. Effect of CO_2 pneumoperitoneum on ventilation—perfusion relationships during laparoscopic cholecystectomy. Acta Anaesthesiol Scand 2002;46:552—60.

[43] Hachenberg T, Karmann S, Pfeiffer B, Thomas H, Gründling M, Wendt M. The effect of dopexamine on ventilation—perfusion distribution and pulmonary gas exchange in anesthetized, paralyzed patients. Anesth Analg 1998;86:314—9.

[44] Melot C. Contribution of multiple inert gas elimination technique to pulmonary medicine. 5. Ventilation—perfusion relationships in acute respiratory failure. Thorax 1994;49:1251—8.

[45] Dantzker D, Brook C, Dehart P, Lynch J, Weg J. Ventilation—perfusion distributions in the adult respiratory distress syndrome. Am Rev Respir Dis 1979;120:1039—52.

[46] Matamis D, Lemaire F, Harf A, Teisseire B, Brun-Buisson C. Redistribution of pulmonary blood flow induced by positive end-expiratory pressure and dopamine infusion in acute respiratory failure. Am Rev Respir Dis 1984;129:39—44.

[47] Ralph DD, Robertson HT, Weaver LJ, Hlastala MP, Carrico CJ, Hudson LD. Distribution of ventilation and perfusion during positive end-expiratory pressure in the adult respiratory distress syndrome. Am Rev Respir Dis 1985;131:54—60.

[48] Pappert D, Rossaint R, Slama K, Gruning T, Falke KJ. Influence of positioning on ventilation—perfusion relationships in severe adult respiratory distress syndrome. Chest 1994;106:1511—6.

[49] Bein T, Reber A, Metz C, Jauch K, Hedenstierna G. Acute effects of continuous rotational therapy on ventilation—perfusion inequality in lung injury. Intensive Care Med 1998;24:132—7.

[50] Rossaint R, Falke KJ, Lopez F, Slama K, Pison U, Zapol WM. Inhaled nitric oxide for the adult respiratory distress syndrome. N Engl J Med 1993;328:399—405.

[51] Walmrath D, Schneider T, Pilch J, Grimminger F, Seeger W. Aerosolised prostacyclin in adult respiratory distress syndrome. Lancet 1993;342:961—2.

[52] Bender K, Alexander J, Enos J, Skimming J. Effects of inhaled nitric oxide in patients with hypoxemia and pulmonary hypertension after cardiac surgery. Am J Crit Care 1997;6:127—31.

[53] Feihl F, Eckert P, Brimioulle S, Jacobs O, Schaller M, Melot C, et al. Permissive hypercapnia impairs pulmonary gas exchange in the acute respiratory distress syndrome. Am J Respir Crit Care Med 2000;162:209–15

[54] Agustí A, Barbera JA. Contribution of multiple inert gas elimination technique to pulmonary medicine. 2. Chronic pulmonary diseases: chronic obstructive pulmonary disease and idiopathic pulmonary fibrosis. Thorax 1994;49:924–32.

[55] Rodriguez-Roisin R, MacNee W. Pathophysiology of chronic obstructive pulmonary disease. Eur Respir Mon 2006;38:177–200.

[56] Wagner PD, Dantzker DR, Dueck R, Clausen JL, West JB. Ventilation–perfusion inequality in chronic obstructive pulmonary disease. J Clin Invest 1977;59:203–16.

[57] Sandek K, Andersson T, Bratel T, Lagerstrand L. Ventilation–perfusion inequality in nocturnal hypoxaemia due to chronic obstructive lung disease (COLD). Clin Physiol 1995;15:499–513.

[58] Ross A, Santos C, Roca J, Torres A, Felez MA, Rodriguez-Roisin R. Effects of PEEP on VA/Q mismatching in ventilated patients with chronic airflow obstruction. Am J Respir Crit Care Med 1994;149:1077–84.

[59] Barbera J, Roca J, Ferrer A, Felez M, Diaz O, Roger N, et al. Mechanisms of worsening gas exchange during acute exacerbations of chronic obstructive pulmonary disease. Eur Respir J 1997;10:1285–91.

[60] Wagner PD, Hedenstierna G, Bylin G. Ventilation–perfusion inequality in chronic asthma. Am J Respir Crit Care Med 1987;136:605–12.

[61] Wagner PD. The multiple inert gas elimination technique (MIGET). Intensive Care Med 2008;34:994–1001.

[62] Roca J, Wagner PD. Contribution of multiple inert gas elimination technique to pulmonary medicine. 1. Principles and information content of the multiple inert gas elimination technique. Thorax 1994;49:815–24.

[63] Baumgardner JE, Choi IC, Vonk-Noordegraaf A, Frasch HF, Neufeld GR, Marshall BE. Sequential V(A)/Q distributions in the normal rabbit by micropore membrane inlet mass spectrometry. J Appl Physiol 2000;89:1699–708.

[64] Wandrup J. Oxygen uptake in the lungs: shortcuts in clinical assessment of pulmonary oxygenation. Blood Gas News 1992;1:3–5.

[65] Gowda MS, Klocke RA. Variability of indices of hypoxemia in adult respiratory distress syndrome. Crit Care Med 1997;25:41–5.

[66] Karbing DS, Kjærgaard S, Smith BW, Espersen K, Allerød C, Andreassen S, et al. Variation in the PaO_2/FiO_2 ratio with FiO_2: mathematical and experimental description, and clinical relevance. Crit Care 2007;11:R118.

[67] Allardet-Servent J, Forel J, Roch A, Guervilly C, Chiche L, Castanier M, et al. FIO_2 and acute respiratory distress syndrome definition during lung protective ventilation. Crit Care Med 2009;37(202–7):e4–6.

[68] King T, Weber B, Okinaka A, Friedman S, Smith J, Briscoe W. Oxygen transfer in catastrophic respiratory failure. Chest 1974;65:40S–4S.

[69] Edmark L, Kostova-Aherdan K, Enlund M, Hedenstierna G. Optimal oxygen concentration during induction of general anesthesia. Anesthesiology 2003;98:28–33.

[70] Mélot C, Naeije R, Hallemans R, Lejeune P, Mols P. Hypoxic pulmonary vasoconstriction and pulmonary gas exchange in normal man. Respir Physiol 1987;68:11–27.

[71] Brimioulle S, Julien V, Gust R, Kozlowski JK, Naeije R, Schuster DP. Importance of hypoxic vasoconstriction in maintaining oxygenation during acute lung injury. Crit Care Med 2002;30:874–80.

[72] Rees SE, Kjærgaard S, Andreassen S, Hedenstierna G. Reproduction of MIGET retention and excretion data using a simple mathematical model of gas exchange in lung damage caused by oleic acid infusion. J Appl Physiol 2006;101:826−32.

[73] Andreassen S, Egeberg J, Schröter M, Andersen P. Estimation of pulmonary diffusion resistance and shunt in an oxygen status model. Comput Methods Programs Biomed 1996;51:95−105.

[74] Andreassen S, Rees SE, Kjærgaard S, Thorgaard P, Winter SM, Morgan CJ, et al. Hypoxemia after coronary bypass surgery modeled by resistance to oxygen diffusion. Crit Care Med 1999;27:2445−53.

[75] Kjærgaard S, Rees SE, Nielsen JA, Freundlich M, Thorgaard P, Andreassen S. Modelling of hypoxaemia after gynaecological laparotomy. Acta Anaesthesiol Scand 2001;45:349−56.

[76] Rees SE, Kjærgaard S, Thorgaard P, Malczynski J, Toft E, Andreassen S. The automatic lung parameter estimator (ALPE) system: non-invasive estimation of pulmonary gas exchange parameters in 10−15 minutes. J Clin Monit Comput 2002;17:43−52.

[77] Sapsford D, Jones J. The PIO$_2$ vs. SpO$_2$ diagram: a non-invasive measure of pulmonary oxygen exchange. Eur J Anaesthesiol 1995;12:369−74.

[78] De Gray L, Rush E, Jones J. A noninvasive method for evaluating the effect of thoracotomy on shunt and ventilation−perfusion inequality. Anaesthesia 1997;52:630−5.

[79] Roe P, Gadelrab R, Sapsford D, Jones J. Intra-operative gas exchange and postoperative hypoxaemia. Eur J Anaesthesiol 1997;14:203−10.

[80] Rees SE, Rutledge G, Andersen P, Andreassen S. Are alveolar block and ventilation−perfusion mismatch distinguishable in routine clinical data? In: Proceedings of the European society of computers in anaesthesia and intensive care conference. Erlangen Germany; September 18−19, 1997.

[81] Smith H, Jones J. Non-invasive assessment of shunt and ventilation/perfusion ratio in neonates with pulmonary failure. Arch Dis Child Fetal Neonatal Ed 2001;85: F127−32.

[82] Quine D, Wong CM, Boyle EM, Jones JG, Stenson BJ. Non-invasive measurement of reduced ventilation−perfusion ratio and shunt in infants with bronchopulmonary dysplasia: a physiological definition of the disease. Arch Dis Child Fetal Neonatal Ed 2006;91:F409−14.

[83] Rowe L, Jones JG, Quine D, Bhushan SS, Stenson BJ. A simplified method for deriving shunt and reduced VA/Q in infants. Arch Dis Child Fetal Neonatal Ed 2010;95: F47−52.

[84] Kjærgaard S, Rees S, Malczynski J, Nielsen JA, Thorgaard P, Toft E, et al. Non-invasive estimation of shunt and ventilation−perfusion mismatch. Intensive Care Med 2003;29:727−34.

[85] Rasmussen BS, Laugesen H, Sollid J, Gronlund J, Rees SE, Toft E, et al. Oxygenation and release of inflammatory mediators after off-pump compared with after on-pump coronary artery bypass surgery. Acta Anaesthesiol Scand 2007;51:1202−10.

[86] Rasmussen B, Sollid J, Rees SE, Kjærgaard S, Murley D, Toft E. Oxygenation within the first 120 h following coronary artery bypass grafting. influence of systemic hypothermia (32°C) or normothermia (36°C) during the cardiopulmonary bypass: a randomized clinical trial. Acta Anaesthesiol Scand 2006;50:64−71.

[87] Kjærgaard S, Rees SE, Gronlund J, Nielsen EM, Lambert P, Thorgaard P, et al. Hypoxaemia after cardiac surgery: clinical application of a model of pulmonary gas exchange. Eur J Anaesthesiol 2004;21:296−301.

[88] Rees SE, Kjærgaard S, Andreassen S, Hedenstierna G. Reproduction of inert gas and oxygenation data: a comparison of the MIGET and a simple model of pulmonary gas exchange. Intensive Care Med 2010;36:2117−24.

[89] Rees SE. The intelligent ventilator (INVENT) project: the role of mathematical models in translating physiological knowledge into clinical practice. Comput Methods Programs Biomed 2011;104:S1−29.

[90] Karbing DS, Allerød C, Thorgaard P, Carius AM, Frilev L, Andreassen S, et al. Prospective evaluation of a decision support system for setting inspired oxygen in intensive care patients. J Crit Care 2010;25:367−74.

[91] Allerød C, Rees SE, Rasmussen BS, Karbing DS, Kjærgaard S, Thorgaard P, et al. A decision support system for suggesting ventilator settings: retrospective evaluation in cardiac surgery patients ventilated in the ICU. Comput Methods Programs Biomed 2008;92:205−12.

[92] Vidal Melo M, Loeppky J, Caprihan A, Luft U. Alveolar ventilation to perfusion heterogeneity and diffusion impairment in a mathematical model of gas exchange. Comput Biomed Res 1993;26:103−20.

[93] Loeppky JA, Caprihan A, Altobelli SA, Icenogle MV, Scotto P, Vidal Melo MF. Validation of a two-compartment model of ventilation/perfusion distribution. Respir Physiol Neurobiol 2006;151:74−92.

[94] Karbing DS, Kjærgaard S, Andreassen S, Espersen K, Rees SE. Minimal model quantification of pulmonary gas exchange in intensive care patients. Med Eng Phys 2011;33:240−8.

[95] Rees SE, Andreassen S. Mathematical models of oxygen and carbon dioxide storage and transport: the acid−base chemistry of blood. Crit. Rev. Biomed. Eng. 2005;33: 209−64.

Appendix A—Glossary

Each symbol is constructed from three parts: the quantity, the location, and the substance. These three parts can take the following values:

Quantity: pressure (P), concentration (C), saturation (S), volume (V), fraction (F), and diffusion capacity (D). Flows are denoted by a point over the quantity term.
Location: arterial blood (a), venous blood (v), lung capillary blood (c), alveoli (A), inspired gas (I), expired gas (E), alveolar−lung capillary membrane (M). An overbar is used to represent a mixed pool, a "′" is used to represent an end tidal (E') expired gas.
Substance: oxygen (O_2), carbon dioxide (CO_2), water (H_2O), hemoglobin (Hb).

Examples of this nomenclature are arterial oxygen pressure P_aO_2, inspired carbon dioxide fraction F_ICO_2, flow of oxygen (oxygen consumption) $\dot{V}O_2$, flow of blood (cardiac output) $\dot{Q}$, mixed venous oxygen concentration $C_{\bar{v}}O_2$, end tidal oxygen fraction $F_{E'}O_2$.

Gases are expressed at three temperatures and pressures: standard (STP—0°C, 101 kPa), ambient (ATP), and body (BTP). Gases are also expressed as either dry (D) or saturated with water (S). Wet and dry gases at ambient or body temperature and pressure are represented using the subscripts ATPS, ATPD, BTPS, BTPD, respectively.

Exceptions to these rules are listed here:

Tidal volume	V_T
Dead space	V_D
Anatomical dead space	V_{Dana}
Alveolar dead space	V_{DA}
Physiological dead space	V_{Dp}
Barometric pressure	P_B
Respiratory frequency	f
Alveolar oxygen diffusion resistance	R_{diff}
Arteriovenous shunt	shunt
Respiratory quotient	R
Diffusion capacity alveoli to blood	D_L
Rate of oxygen binding to blood	θ
Solubility coefficient of gas in blood	β
Oxygen dissociation curve	ODC

Appendix B—Calculations Necessary to Convert Inspired Gas at ATPD to BTPS

Addition of Water

On addition of water to dry inspired gas, volume is increased such that:

$$V_{T,ATPS} = V_{T,ATPD} + V_{H_2O,added} \tag{B.1}$$

where $V_{H_2O,added}$ is the volume of water added during inspiration. The fraction of H_2O in fully saturated inspired gases ($V_{H_2O,ATPS}$) is approximately 6%, so that $V_{H_2O,added}$ can be calculated as follows:

$$V_{H_2O,added} = F_{H_2O,ATPS} \cdot V_{T,ATPS} \tag{B.2}$$

substituting for $V_{H_2O,added}$ in Eq. (B.1) gives:

$$V_{T,ATPS} = V_{T,ATPD} + F_{H_2O,ATPS} \cdot V_{T,ATPS} \tag{B.3}$$

which can be rearranged to give an expression describing the tidal volume of wet gases in terms of the dry inspired gases:

$$V_{T,ATPS} = \frac{V_{T,ATPD}}{1 - F_{H_2O,ATPS}} \tag{B.4}$$

Addition of water to the inspired gas lowers the fraction of all other gases in the inspiration. An equation describing the relationship between the fraction of oxygen

(O_2) in inspired gases before and after addition of water can be derived by considering the volume of O_2 in the dry ($F_IO_{2,ATPD} \cdot V_{T,ATPD}$) and humidified ($F_IO_{2,ATPS} \cdot V_{T,ATPS}$) gas. This volume remains constant such that:

$$F_IO_{2,ATPD} \cdot V_{T,ATPD} = F_IO_{2,ATPS} \cdot V_{T,ATPS} \tag{B.5}$$

Substituting for $V_{T,ATPS}$ in Eq. (B.4) gives:

$$F_IO_{2,ATPD} \cdot V_{T,ATPD} = F_IO_{2,ATPS} \frac{V_{T,ATPD}}{1 - F_{H_2O,ATPS}}$$

which when rearranged gives an expression describing the fraction of oxygen in wet gas in terms of the dry inspired gases:

$$F_IO_{2,ATPS} = F_IO_{2,ATPD}(1 - F_{H_2O,ATPS}) \tag{B.6}$$

Warming of Inspired Gases

Warming of the inspiration causes the gases to expand, but does not change the fraction of each gas in the inspiration, which means $F_{I,BTPS} = F_{I,ATPS}$ for all gases. The increase in volume can be described by considering the gas laws, such as Charles' law and Boyle's law, which together state that for a fixed mass of gas, the pressure multiplied by the volume is proportional to the temperature:

$$P \cdot V = R \cdot T \tag{B.7}$$

where R is the universal gas constant. Equation (B.7) can be written for humidified inspired gas at both ambient (AT) and body (BT) temperatures:

$$P_B \cdot V_{T,ATPS} = R \cdot AT \tag{B.8}$$

$$P_B \cdot V_{T,BTPS} = R \cdot BT \tag{B.9}$$

Dividing Eq. (B.8) by Eq. (B.9) gives an expression for the tidal volume of the inspired gases at body temperature in terms of the tidal volume at ambient temperature, that is:

$$V_{T,BTPS} = V_{T,ATPS} \frac{BT}{AT} \tag{B.10}$$

14 Mathematical Models for Computational Neuroscience

Mauro Ursino, Filippo Cona and Elisa Magosso

Department of Electrical, Electronic, and Information Engineering, University of Bologna, Bologna, Italy

14.1 Introduction

Understanding brain functioning represents one of the most challenging aspects of contemporary scientific research. In this rapidly developing field, it is essential to have quantitative instruments at one's disposal, able to summarize experimental data and qualitative hypotheses into a rigorous setting, to provide a theoretical description of the complexity of the mechanisms involved, and to generate accurate hypotheses that can be checked in actual or future biological tests. Computational neuroscience is playing this fundamental role today: its emphasis is on how biological neurons work and, above all, on how neurons reciprocally interact to provide those kinds of behavior that have a cognitive role in animals and humans.

Neurons differ from all other cells in the body, since they can propagate and receive information very quickly and at large distances (via action potentials or spikes); in this way, they can monitor and control all other cells in the body, including themselves, in a very short temporal scale. Networks of such cells are the fundamental elements for any information processing in the body (from simple reflexes in primitive animals to complex tasks in superior animals and human beings). Two aspects are essential here: the working of the individual neuron (the tie of the central neural system) and the topological organization of neuron connections (the architecture of the system).

Accordingly, the first step in constructing a neural network model for computational neuroscience consists in the choice of the constituent neural units. The function of each unit is to collect information from other parts (i.e. from other units or from the internal and external environment) and to fire an output signal (which can be represented as individual spikes or as a firing rate). The precision at which any single unit is described depends on the objectives of the model and on its level of detail. Generally, models that comprise just a few neurons and are devoted to the analysis of microcircuits include subtle details at the cell level (such as ionic currents) to capture the characteristics and diversity of individual neurons and their specific role; conversely, models devoted to the analysis of cognitive phenomena at

Modelling Methodology for Physiology and Medicine. DOI: http://dx.doi.org/10.1016/B978-0-12-411557-6.00014-8

a high level of abstraction neglect individual neuron characteristics, focusing on a more conceptual response. Of course, many differences and intermediate steps can be found between these two extremes.

In the following, we will first describe some of the main formalisms adopted to simulate individual neurocomputational units. Then we will move to networks of neurons and their cognitive behavior.

14.2   Models of Individual Neural Units

14.2.1   The Hodgkin–Huxley Model

Models of this type incorporate an accurate description of single ionic channels involved in the generation of action potentials. Figure 14.1 shows the electric analog of a cell. As it is shown, the input is the synaptic current, which can depolarize the membrane until the neuron eventually fires, or can hyperpolarize it, causing inhibition. The equation is

$$C\frac{dV}{dt} + g_L(V - V_0) + \sum_k I_k = 0 \tag{14.1}$$

where C is the membrane capacity, g_L is the leakage conductance, I_k represents a generic current flowing through a membrane conductance, and the sum is extended to all ionic currents.

In order to simulate the genesis of an action potential, the dynamics of voltage-dependent sodium and potassium channels are explicitly incorporated in these models. As it is well known from voltage clamp experiments [1], sodium channels immediately open (thus causing sodium entrance into the membrane) when the cell depolarizes, but with a transient dynamics; potassium channels open, too, during depolarization, but with a slower dynamics.

Each current is described through a reversal or equilibrium potential (say E), a maximal conductance parameter g_{max}, two gating variables, m and h, representing activation and inactivation (sometimes h is omitted, as in the case of the potassium channels), and two integer exponents, p and q. A generic current carried by the conductance at membrane potential V can be written

$$I = g_{max}m^p h^q(V - E) \tag{14.2}$$

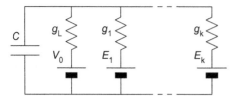

Figure 14.1 Electric analog of the neuron, where C is the membrane capacity, g_L is the leakage conductance, V_0 is the equilibrium voltage at rest, and g_k and E_k are ionic conductances and ionic reverse potentials, respectively.

where a different equation similar to Eq. (14.2) should be written for each current I_k in Eq. (14.1). The gating variables m and h vary between 0 and 1, and, for what concerns voltage-dependent currents, are described through differential equations of similar forms

$$\frac{dm}{dt} = \alpha_m(V)(1 - m) - \beta_m(V)m \quad \text{and} \quad \frac{dh}{dt} = \alpha_h(V)(1 - h) - \beta_h(V)h \qquad (14.3)$$

The functions $\alpha(V)$ and $\beta(V)$ are the voltage-dependent opening and closing rates, respectively.

The same basic formalism is used to represent the synaptic currents (i.e. the currents that the neuron receives from other neurons in the network through synapses). In this case, however, the opening and closing rates are described as a function of the neurotransmitter released at the presynaptic terminal (hence, as a function of the action potentials of the presynaptic neuron).

The result of Eqs. (14.1)–(14.3) is the genesis of an action potential if the overall current is positive enough to induce depolarization over a given threshold. Models of this type now also include some significant variants, with more detailed ionic currents (for instance depending on Ca2+), propagation along axons, or a distinction between the dendrites and the cell body (see Ref. [2] for more information).

14.2.2 The Integrate and Fire Model

Models of the Hodgkin-Huxley (HH) type are extremely useful to study the role of individual neurons within microcircuits, but are also onerous from a computational point of view. Moreover, the large number of parameters (which may differ substantially from one neuron type to another) often makes it difficult to arrive at a simple synthesis of the obtained results. At a superior abstraction level, one can find the so-called integrate and fire (IF) neuron. In this model, voltage-dependent currents are neglected, and the action potential is described as a simple stereotyped waveform. Hence, the model concentrates merely on the dynamical events occurring below the action potential threshold and can be mimicked as a membrane capacity with a parallel leakage conductance, charged (or uncharged) by the synaptic currents. The final schema is that of a leakage integrator: the voltage potential integrates the current until, once it reaches the threshold, it fires a spike, and is reset. Variants of this model include a varying threshold, to account for the presence of the absolute and refractory periods. It is worth noting that this kind of model has a long historical tradition, being originally formulated in the early 1990s by Lapicque [3], well before the development of the HH formalism. While outputs in the HH models are the overall voltage potentials waves, with a realistic shape, outputs of the IF model are the instants at which the neuron fires, since the individual spikes are stereotyped.

Consequently, a further simplification of an IF neuron consists in describing the output as a series of Dirac impulses (let us name this output $\rho_i(t)$, where the subscript i refers to the ith neuron in a hypothetical network). We have

$$\rho_i(t) = \sum_k \delta(t - t_{ki}) \tag{14.4}$$

where t_{ki} represents the instant at which the neuron i in the network fires its kth spike.

Of course, neurons communicate via synapses. In HH models, this communication is represented through the synaptic currents I_k in Eq. (14.1). However, in simpler IF models, an impulse response is often used to summarize the effect of a presynaptic spike on the postsynaptic voltage. Assuming that synaptic inputs do not interact with each other, and they behave linearly, the overall membrane potential of neuron i (say $v_i(t)$) can be expressed as follows:

$$v_i(t) = \sum_{j=1}^{N} \int_{-\infty}^{t} w_{ij} h_j(\tau) \rho_j(t - \tau) d\tau = \sum_{j=1}^{N} w_{ij} h_j(t) * \rho_j(t) \tag{14.5}$$

where w_{ij} represents the strength of the synapse from neuron j to neuron i, $*$ denotes convolution, $h_j(t)$ represents the synaptic impulse response of neuron j, and the sum is extended to all neurons that make synapses to neuron i. It is worth noting that $v_i(t)$ should be compared with the threshold to determine the instant, t_{ki}, of the next spike for neuron i. The choice of the exact form for $h_j(t)$ depends on model objectives and computational costs, and will be described below.

14.2.3 Firing Rate Models

An important simplification, often used in network models, consists in replacing the train of spikes in Eq. (14.4) with its instantaneous rate (say $r_i(t) = E\{\rho_i(t)\}$, where $E\{\}$ denotes the expectation operator, i.e. $\rho_i(t)$ is treated as a random process). Similarly, the membrane potential in Eq. (14.5) is replaced with an ensemble mean of the instantaneous potential (say $u_i(t) = E\{v_i(t)\}$). As a consequence, Eq. (14.5) can be rewritten as follows:

$$u_i(t) = E\left\{ \sum_{j=1}^{N} w_{ij} h_j(t) * \rho_j(t) \right\} = \sum_{j=1}^{N} w_{ij} h_j(t) * E\{\rho_j(t)\} = \sum_{j=1}^{N} w_{ij} h_j(t) * r_j(t) \tag{14.6}$$

Here we exploited linearity of the synapse description to interchange expectation summation and convolution.

In order to complete this simplified model without using individual spikes, but just using the instantaneous rates, we need an equation that allows the computation of $r_i(t)$ as a function of the mean potential $u_i(t)$. A frequent assumption is that the instantaneous firing rate is a monotonically increasing function of potential, with lower threshold and upper saturation. The first reflects the existence of a threshold to generate action potentials; the second the existence of a refractory period. An equation frequently used is the sigmoidal function:

$$r_i(t) = \frac{r_{i\text{max}}}{1 + e^{-b(u_i(t) - u_0)}} \tag{14.7}$$

where $r_{i\text{max}}$ represents the maximum firing rate, and b and u_0 are parameters that set the slope and the position of the sigmoid.

As specified above, the form of the synapse impulse response depends on the objectives and the degree of details in the model. Generally, synaptic kinetics differs depending on the specific neuron type. In particular, excitatory and inhibitory neurons have different synaptic dynamics. A quite precise form for the impulse response is the so-called α function ($h(t) = (t/t_{\text{peak}})e^{-t/t_{\text{peak}}}$), representing a second-order dynamics. A simpler form is the first-order response ($h(t) = (1/\tau)e^{-t/\tau}$).

A strong simplification, often adopted in the most abstract neural network models, assumes that the impulse responses of all synapses have the same first-order expression (hence, the same τ). In this case, Eq. (14.6) can be easily expressed as follows, by computing the time derivatives of both members:

$$\tau \frac{du_i(t)}{dt} = -u_i(t) + \sum_{j=1}^{N} w_{ij} r_j(t) \tag{14.8}$$

Equations (14.7) and (14.8) (written for all neurons in a network) represent a common way to build neural network models at a high level of abstraction.

14.3 Networks of Neurons

Once the model of the individual computational unit has been established, the following step consists in the construction of a network of units. The fundamental assumption here is that the structure of the network strongly determines its function. In particular, one must establish the following major points, which strongly affect the overall model behavior:

- a pattern of connectivity among units;
- a learning rule whereby patterns of connectivity are modified by experience;
- an environment within which each system must operate.

The first point specifies which kind of architecture the network implements. Connections are, of course, realized by means of synapses among neurons, (i.e. parameters w_{ij} in Eq. (14.6) or (14.8)). The most important distinction is between feed-forward networks, where information propagates just in one direction, and the feedback networks, which incorporate reciprocal connections (or loops) among neurons, and in which network dynamics play an essential role.

The second point represents, perhaps, the most peculiar and fascinating aspects of neural networks (i.e. the capacity to learn and adapt on the basis of past experience). In all networks, adaptation is realized by modifying the weights of synapses linking neurons.

Finally, the environment feeds the network with input patterns or episodes that the network has to learn or to respond to, and may provide a reward or a punishment to the network response.

In the following, we will consider three distinct classes of networks, which can be used to simulate important cognitive processes in the brain and which differ for what concerns both the architecture and adopted learning rules: these are the associative networks, the competitive self-organized networks, and the error-correction networks (that we consider just in the case of reinforcement learning). For each type, possible neurophysiological counterparts will be considered.

14.3.1 Associative Memories and Hebbian Learning

The basic idea behind these networks is that associations among past events can be learned in a single shot, and subsequently recovered when necessary, starting from a partial cue. The first idea of association was formulated by the great philosopher and psychologist William James who, in his "Psychology: briefer course" [4] says: "When two elementary brain processes have been active together or in immediate succession, one of them on recurring, tends to propagate its excitement on the other." This idea received a very popular formulation, in terms of neural activity and synaptic changes, only after many decades, when Donald Hebb, in his Organization of Behavior (1949) [5], proposed that synaptic learning (at least for what concerns synapse reinforcement) depends on the coincidence of presynaptic and postsynaptic activity: "When an axon of cell A is near enough to excite a cell B and repeatedly or persistently takes part in firing it, some growth process or metabolic change takes place in one or both cells such that A's efficiency, as one of the cells firing B, is increased." Indeed, this rule is too crude to obtain a realistic network and would lead to instability since it assumes just synaptic long-term potentiation. In order to have a correct understanding, we need to know that, when some synapses reinforce, others weaken. Synapse weakening (often referred to as long-term depression) occurs when the presynaptic and postsynaptic activities are anticorrelated.

In order to account for both potentiation and depression, the Hebb rule in rate models often assumes the following form:

$$dw_{ij} = \gamma(r_i - \theta_i)(r_j - \theta_j) \tag{14.9}$$

where γ is a learning factor, and θ_i and θ_j are plasticity thresholds which affect the direction of synaptic change. A similar form can be given, of course, also for models with spiking neurons, taking into account the relative timing of the presynaptic and postsynaptic spikes (see Refs. [2,6]).

In many cases, the presynaptic and postsynaptic thresholds in Eq. (14.9) are set equal to the average firing rates of the neuron over the training period. We have

$$dw_{ij} = \gamma(r_i - E\{r_i\})(r_j - E\{r_j\}) \tag{14.10}$$

According to Eq. (14.10), the average change in synaptic weight is proportional to the covariance of the presynaptic and postsynaptic firing: hence Eq. (14.10) is also known as *the covariance rule*. Conversely, if one assumes $\theta_i - \theta_j - 0$, one obtains a *correlation-based rule*:

$$dw_{ij} = \gamma r_i r_j \tag{14.11}$$

The latter, however, is unstable since it implies just synaptic potentiation, without any synaptic weakening.

Let us consider now a first example of an associative network employing Hebbian learning, which is named *the hetero-associator*. It consists of a feed-forward connection between two layers of linear units—an input layer consisting of N_x neurons and an output layer of N_y neurons. The input is often referred to as *the conditioned stimulus* (Figure 14.2A). The goal of a hetero-associative memory is to learn mappings between input–output pairs, such that the memory produces the appropriate output in response to a given input pattern. Let us denote with X the vector of input activities (i.e. x_i, $i = 1, 2, \ldots, N_x$ are the firing rates of the input neurons) and with Y the vector of output activities (i.e. y_i, $i = 1, 2, \ldots, N_y$ are the firing rates of the output neurons). Assuming for the sake of simplicity a linear relationship instead of Eq. (14.7), and computing the output patterns in steady-state conditions by using equations analogous to Eq. (14.8), we can write $y_i(t) = \sum_{j=1}^{N_x} w_{ij} x_j(t)$. In vector form, we have

$$Y = WX \tag{14.12}$$

where W is a $N_y \times N_x$ array of synapses.

In what follows we will consider a training phase, in which the network learns from experience, and a test or exploitation phase, in which the network exploits the previous experience. Although in nature they are often superimposed, we will treat them as distinct. Let us assume that, during training, Eq. (14.12) does not hold, but

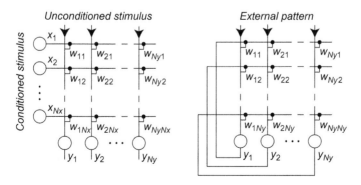

Figure 14.2 Schematic diagram showing the structure of a hetero-associative network (A) and of an auto-associative network (B).

the output is forced to assume a given value (say Y^k) via a second "unconditioned input." At the same time, the conditioned stimulus is given a value X^k. Application of the crude Hebb rule (Eq. (14.11)) leads to the following change in the synapse matrix:

$$\Delta W^k = \gamma Y^k X^{k^T} \tag{14.13}$$

If, before training, the synapse matrix is equal to zero, the combination of M consecutive "input—output" pairs leads to the following values for W after training:

$$W = \gamma \sum_{k=1}^{M} Y^k X^{k^T} \tag{14.14}$$

Let us now consider the exploitation phase, in which the network does not receive any unconditioned stimulus, but is free to respond to the conditioned stimulus, according to Eq. (14.12). What is the network response if one of the previous stimuli (for instance X^p) is given as input? Of course, before training, the response would be identically zero, since we assumed a null synapse matrix. After training, Eqs. (14.12) and (14.14) give

$$Y = \gamma \sum_{k=1}^{M} Y^k X^{k^T} X^p = \gamma \sum_{\substack{k=1 \\ k \neq p}}^{M} Y^k \langle X^k X^p \rangle + \gamma ||X^p||^2 Y^p \tag{14.15}$$

where $\langle \cdot \rangle$ denotes the inner product. It is worth noting that the first term in the right-hand member of Eq. (14.15) is a noise term, while the second represents the desired output, which returns the correct pattern Y^p apart for a multiplicative constant (in linear networks, the patterns are always given independently of a constant term).

In order to obtain the desired behavior (i.e. the network produces the output pattern Y^p in response to the conditioned input X^p), the conditioned stimuli must represent an orthogonal set. In fact, if this condition holds, Eq. (14.15) furnishes

$$Y = (\text{constant}) Y^p \tag{14.16}$$

This model makes the prediction that the outer product associators work better with input representations that are as orthogonal as possible. Indeed, several networks in the neuroscience literature analyze the problem of orthogonalization of input patterns in cortical processing (see Ref. [6]).

An important variant of the associative memory is autoassociation. In this particular network, the aim is to recover a pattern, previously stored, starting from its partial or corrupted information (i.e. it represents a *content addressable memory*). The main difference, compared with the hetero-associator, is that during training the input and output information are coincident (i.e. the conditioned stimulus is equal to the output stimulus that should be recovered). This implies the presence of a feedback as illustrated in Figure 14.2B.

Let us assume that, in the linear associator presented above, $Y^k = X^k$. The synapse matrix after this training becomes $W = \gamma \sum_{k=1}^{M} Y^k Y^{k^T}$. Let us assume that a pattern Y' is given to the network, in which a part of its elements is the same as in Y^p, while the other elements are identically zero. Let us construct a second vector, Y'', with the same length but with only the elements of Y^p that were set at zero in Y'. Of course, $Y^p = Y' + Y''$ and moreover, $Y''^T Y' = 0$. If Y' is the input to the autoassociator, the output becomes

$$Y = \gamma \sum_{\substack{k=1 \\ k \neq p}}^{M} Y^k Y^{k^T} Y' = \gamma \sum_{\substack{k=1 \\ k \neq p}}^{M} Y^k \langle Y^k Y' \rangle + \gamma Y^p (Y' + Y'')^T Y'$$

$$= \gamma \sum_{\substack{k=1 \\ k \neq p}}^{M} Y^k \langle Y^k Y' \rangle + \gamma ||Y'||^2 Y^p \tag{14.17}$$

Once again, the first term is noise, while the second provides the requested output. The fundamental assumption is still orthogonality. Let us assume that Y' is orthogonal not only to Y'', but also to all other patterns Y^k used during training. We then obtain

$$Y = (\text{constant}) Y^p$$

This behavior summarizes the reconstructive property of the autoassociator (i.e. it can recover the overall information starting from a part of it).

The two previous exempla considered the case of linear associators, assuming that neurons work in the linearity region. Of course, much more subtle and interesting behaviors can be obtained using nonlinearities (for instance, the sigmoidal relationship as in Eq. (14.7)).

A fundamental nonlinear autoassociator network was developed by Hopfield in the early 1980s [7]. It consists of N neurons, which may be completely interconnected (although, in real cases, just a small fraction of synapses are actually different from zero). Let us consider the continuous version of this model, assuming equations similar to Eqs. (14.7) and (14.8) for individual neurons. The main assumptions of the model are that: (i) the matrix of synapses is symmetrical (i.e. $w_{ij} = w_{ji}$ with eventual self-connections all positive or null; $w_{ii} \geq 0$); (ii) each neuron has a nonlinear activation function; (iii) the inverse of the activation function exists. It is worth noting that Eq. (14.7) satisfies the last two conditions, while the first can be satisfied by the Hebb rule, provided the pre- and postsynaptic thresholds, θ_i and θ_j, are equal. Under these conditions, Hopfield demonstrated that the overall network possesses a Lyapunov function (that is generally named "Energy function") that decreases monotonically during the temporal evolution of the network. As a consequence, all trajectories in the phase space must converge to minima of the energy function and stop at such points. This is the reason why these kinds of networks are often referred to as *attractor networks*.

The fundamental point in the Hopfield model, however, is that if synapses are trained with Hebbian rules, the patterns used during training are stored as minima of the energy function; hence, they represent equilibrium points in the network, surrounded by basins of attraction. A learning rule often adopted in Hopfield nets is the following one:

$$dw_{ij} = \gamma(r_i - a)(r_j - a) \tag{14.18}$$

where a represents the sparseness of the network (i.e. the average percentage of neurons that are active in a given moment; common values for physiological networks are in the range $0.05-0.1$).

It may be demonstrated that patterns are stored as equilibrium points if they are orthogonal (that is the same condition we found on linear associators above); if patterns are not orthogonal, some interferences may occur between them, which may lead to incorrect storage or incorrect recovery. A thorough analysis of the storage capacity of these networks and their tolerance to noise has been performed by several authors (see Ref. [8]); these works demonstrate that, in case of random patterns, the capacity of Hopfield networks depends on the ratio between the number of stored patterns (say M) and the number of neurons (say N). A robust storage with a good attraction basin can be realized if $M/N \simeq 0.01-0.02$. A catastrophe, which leads to the overall loss of memory, occurs if $M/N > 0.14$ (which is named the 14% catastrophe by some authors).

The possibility to store patterns as equilibrium points with an attractor dynamics allows these networks to work as a *content addressable memory*; this means that, in order to recover a stored pattern, the network must be supplied with similar content (i.e. a distorted pattern, or one lacking some parts). It is worth noting that all of the information is stored in the totality of synapses, and that the dynamics of trajectories converging toward an equilibrium point is exploited to reconstruct the overall pattern starting from partial or corrupted information, provided the initial state of the network lies within the attraction basin.

There are several regions in the brain that work in a way somewhat similar to Hopfield nets; these are especially important for episodic and working memory. The most popular is certainly the hippocampus: it consists of subcortical structures (one in the left and the other in the right hemisphere) placed within the medial temporal lobe. Several results, both from clinical patients and animal experiments, demonstrated that the hippocampus plays a pivotal role in the memorization of recent episodic events (see Ref. [9]). For instance, patients with a diffuse hippocampal lesion lose their capacity to store recent episodes of life. The hippocampus is involved in the memorization of spatial information, for instance, in rats. Moreover, the anatomical structure of the hippocampus exhibits some subregions (named the CA3) in which neurons exhibit significant feedback connections, as hypothesized in the Hopfield model. The validity of the Hebb rule was demonstrated in rats' hippocampal preparations *in vitro*, where both long-term potentiation and long-term depression were noticed [10].

However, the situation in the hippocampus is somewhat more intricate than that assumed by a simple attractor network. Recent results [11] have shown that the activity of neural cells in the hippocampus is not constant (as predicted by the Hopfield network at equilibrium), but exhibits the superimposition of two oscillatory patterns: a gamma rhythm (>30 Hz) placed over a slower theta rhythm (about 4 Hz). These rhythms can be simulated by rate models similar to those described previously, but including the presence of feedback loops between excitatory and inhibitory populations with different synaptic kinetics (more details can be found in Refs. [12,13]). A popular hypothesis is that the gamma rhythm is used to recover several episodes simultaneously (using time division among the different gamma oscillators) while the chain of episodes is reset at the beginning of any new theta period. Recent models have shown that, using the combination of autoassociation and hetero-association, up to seven episodes can be recovered within each theta period, in an assigned order [14] (see Figure 14.3 for an example).

Besides episodic memory, attractor networks are also possibly involved in working memory (a kind of short-term memory with limited capacity, where we store items from seconds to minutes, to solve immediate problems). Working memory is mostly located in the prefrontal cortex [9].

What are the main limitations of the Hopfield model? First, this memory has a very limited storage capacity (M/N of the order of a few percent in the case of random patterns). In order to have a good memory capacity, the input patterns must be as orthogonal as possible. Second, these memories are not invariant to simple transformations, like translations or scaling (a change of image dimension, for instance) since these transformations strongly modify the input characteristics in the state space. In order to solve these limitations, the brain adopts two main strategies: (i) orthogonalization of patterns before memorization, which is often realized using a sparse code [6]. Indeed, there are regions in the brain (for instance, granule cells in the dentate gyrus, a region from where information enters the hippocampus) where sparsification occurs. (ii) Objects are stored as a collection of features in order to achieve a kind of description that is invariant with respect to spatial transformations. Feature extraction represents the basis for semantic memory; this is performed in the "what" processing system, located in the temporal lobe [9], which makes use of competitive networks of the kind described below.

Finally, it is worthwhile to consider which kind of pathological alterations may occur in case of damage of an attractor network. These memories are quite robust against synapse damage, but completely lose their memory at a catastrophic point. These aspects resemble those occurring in late stages of Alzheimer disease. Furthermore, some authors [15] recently hypothesized that alterations in the attraction basin of these networks, induced by a reduction of neurotransmitter efficiency, can explain some "positive," "negative," and "cognitive" symptoms of

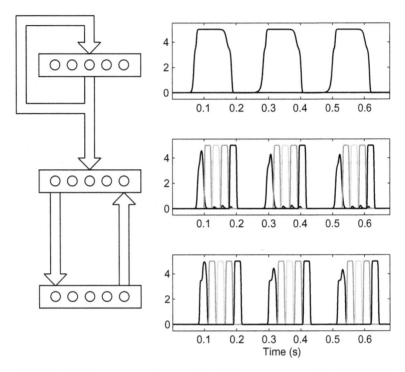

Figure 14.3 Recall of a sequence of events (five different events are plotted in this figure) obtained using a network consisting of an auto-associative net in series with a hetero-associative net [14]. Neurons in the first layer oscillate in the theta range, reconstruct incomplete patterns, and transmit information to the second and third layers. Neurons in these layers oscillate in the gamma range. An overall sequence of events is recalled in the last layer, starting from the first event, using temporal division in the gamma frequency band. At each theta cycle, the overall sequence is reset. Oscillations are induced via a feedback between excitatory and inhibitory neural groups.

schizophrenia. An example is shown in Figure 14.4: this illustrates the alterations in an attraction basin caused by a reduction in NMDA and/or GABA efficacy.

14.3.2 Self-Organized Networks

A different kind of network, which is used with distinct objectives compared with associators, is represented by networks including a kind of competition among neurons; for this reason, these are named *competitive networks*. While associative networks are trained via single shots of information, in which patterns or episodes are immediately stored, in these networks information is stored through a prolonged training process, which is aimed at extracting the main statistical properties of the environment. For this reason, these networks are also referred to as *self-organized networks* and are frequently used in perceptual problems. The objective here is to improve and summarize information arriving from the external world.

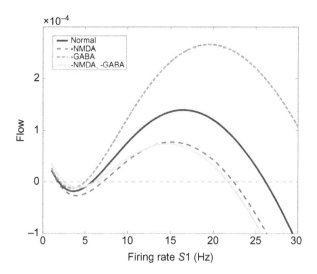

Figure 14.4 This figure represents the force that drives a Hopfield network toward one of the stable attractors. The model was an attractor IF network described in Loh et al. [15]. The stable/unstable attractor states are at crossings with the flow = 0 value on the flow axis with a negative/positive derivative, respectively. A modulation of the synapses labeled as (−NMDA) and (−GABA) corresponds to a reduction of 4.5% and 9% of the efficacies, respectively. *Source*: Reproduced from Ref. [15].

In order to understand the role of competition within a network, let us consider the structure depicted in Figure 14.5. This is frequently encountered in the cerebral cortex. Neurons are connected with other neurons in the same net with *lateral synapses*: synapses with proximal neurons are excitatory, whereas those with more distal neurons are inhibitory. Very distal neurons are not connected in appreciable ways. The pattern of synapses thus realizes a sort of *Mexican hat*, which is illustrated in the bottom of Figure 14.5.

In the case of a monodimensional lattice of neurons (considered here for simplicity), the network dynamics can be described through the following differential equations:

$$\tau \frac{du_i(t)}{dt} = -u_i(t) + \sum_{j=1}^{N} L_{ij} r_j(t) + I_i$$

$$I_i = \sum_{k=1}^{M} w_{ik} x_k(t)$$

(14.19)

where X represents the input vector, with dimension M, L_{ij} represents the lateral synapses among neurons in the network, and the synapses w_{ik} link the ith neuron with the different components of the input vector, thus establishing its receptive field. Of course, Eq. (14.7) completes the model.

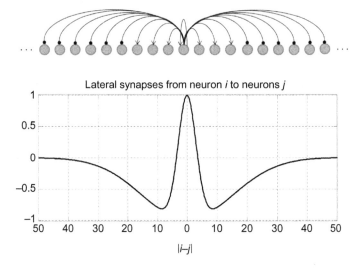

Figure 14.5 Schematic diagram of a competitive network (A) : proximal neurons interact via excitatory synapses and send inhibitory synapses to more distal neurons. This implements a Mexican hat disposition of synapses (B).

It is worth noting that, due to the presence of lateral synapses among neurons, these networks present a feedback dynamics, which—as in the case of the Hopfield network—is essential for reaching a final steady-state condition. The dynamics of these networks exhibit some essential properties: (i) if excitation is stronger than inhibition (due to cooperation among neurons), activity in the cooperating neurons increases up to the maximum saturation level; (ii) if inhibition prevails over excitation (due to a competition among neurons), neuron activity falls to a negligible value; (iii) in an intermediate condition, some neurons can reach a stable intermediate activity.

The last mode can be used in different kinds of information processing, in particular to improve some properties of the input signal. Let us consider, for instance, a single chain of neurons with moderate competition among them (the same example, of course, may be done using a bi-dimensional array; we use a single chain here just for simplicity). Let us assume that neurons have a spatial arrangement: each neuron is characterized by a specific receptive field (i.e. it receives input just from a specific portion of space); moreover, neurons are in the spatial register (i.e. proximal neurons receive their input from proximal portions of space; the composite eye of an insect is a good example). As illustrated in Figure 14.6A, a moderate cooperation and competition among neurons results in an improvement in contour detection: the edges of the input signal are emphasized, whereas constant levels (or background levels) are attenuated. In the case of vision, this results in an improved perception of borders, which is essential in form recognition; in the case of tactile perception, it results in an improved resolution to proximal body stimuli.

If competition among neurons is stronger (a dynamics which is often referred to as *the winner takes all*), competitive networks lead to the formation of "bubbles"

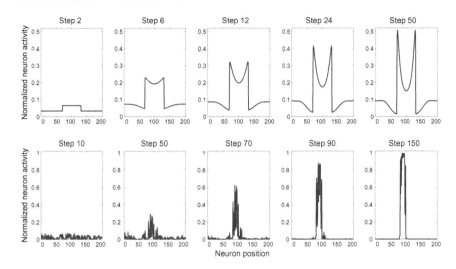

Figure 14.6 Two exempla of simulation results obtained with a competitive network (200 neurons) in the case of moderate competition (A, upper panels) and strong competition (B, bottom panels). Panels in each row show some snapshots of the overall network activity at different instants of the simulations. Neuron activities are normalized to 1. (A) simulates the response to a square input (between positions 70 and 130); the moderate competition results in contrast enhancement. (B) simulates the response to a random input: just one bubble of neurons wins the strong competition, signaling the position of the stronger input, while all other neurons are inhibited.

of active neurons (the winners) surrounded by large inhibited regions with no activity (the losers). This is a classical way through which the cortex solves the categorization problem, i.e. it assigns the input to a given category (represented by the position of the bubble). An example of strong competition is provided in Figure 14.6B, in which the network responds to a random input. Only a group of adjacent neurons wins the competition, causing the activation of a bubble, while all other neurons are inhibited.

But how can these networks reveal categories in a self-organized way, starting from the statistics of the external environment? In the exemplum with weak competition above (as in the insect eye), we assumed fixed receptive fields (i.e. fixed entering synapses w_{ik}). Conversely, in order to solve the categorization problem, synapses that connect neurons with input patterns (i.e. with the environment) must be plastic. A variant of the Hebb rule is often adopted here, named *the Oja rule*. It consists of a Hebb rule (with potentiation only) joined with a forgetting factor: an active neuron reinforces its entering synapses according to correlation with the input, but loses a portion of the synapses previously learned. A typical form of the Oja rule is as follows:

$$dw_{ij} = \gamma(r_i \cdot r_j - r_i^2 \cdot w_{ij}) \tag{14.20}$$

The first term represents a classic Hebbian potentiation, based on the correlation between the presynaptic and the postsynaptic activity. The second is a "forgetting factor": the postsynaptic neuron forgets a portion of the previous synapses (let us remember that the learning factor γ must be significantly smaller than 1), and this loss is stronger the higher the postsynaptic activity. Kohonen [16] demonstrated that these networks, with strong competition, develop topological maps: the prototype of each category is stored in the vector of synapses entering a neuron, and proximal neurons in the network are associated with similar prototypes. It is worth noting that these networks can learn continuously, and that the balance between plasticity and stability is regulated by the learning rate γ in Eq. (14.20).

Topological maps can be found in various regions of the cortex, with special emphasis on perceptual and/or motor problems: maps are present at different levels of the visual processing stream (for instance, maps of line orientation in the primary visual cortex, the color map in V4, etc.), in the auditory cortex (where one can observe a tonotopic map based on sound frequency), in the somatosensory cortex, and in the primary motor cortex (the famous homunculi, which maps the body, see Ref. [9]). Finally, it is worth noting that these feature maps are essential for the development of semantic memory; indeed, most present theories assume that the semantics of objects are represented in the mind through a collection of features, which can naturally be extracted through a self-organization process and represented in topological maps.

14.3.3 Error-Correction Networks

In previous sections, we considered the problems of episodic (or working) memory and the organization of external perception in feature maps. A further aspect that should be represented in the network of neurons is the capacity to learn a behavior or a skill via trial and error. While previous networks learned an episode in a single shot (associative nets) or were able to learn the statistical regularities of feature occurrences through prolonged training via a self-organization process (self-organized networks), those considered here learn via an error-correction procedure. The most popular networks of this kind use supervised training, in which an omniscient teacher provides the correct response to any trial. Since these networks are not physiologically well founded (although very popular in information engineering), in the following we will concentrate on a different kind of training, named *reinforcement learning*. Here a network (let us assume for the moment a feed-forward net with a single output neuron, for the sake of simplicity) responds to a given input via a given response, and then may receive or not receive a reward from the environment. It is worth noting that the environment does not provide the correct solution to the problem (as in the case of supervised learning), but simply provides a reward or a punishment. How the network learns a correct behavior (or, more generally, a behavior that maximizes the reward and minimizes the punishment) is a classical problem in behavioral psychology.

In order to fix our ideas, let us consider a network that receives an input vector stimulus, $X(t)$ ($X = x_1, x_2, \ldots, x_N$) and produces a single output, $y(t)$. Since the dynamics can be neglected in a feed-forward net, we can consider just the steady-state condition in Eqs. (14.7) and (14.8), and so

$$y(t) = S\left(\sum_{j=1}^{N} w_j x_j(t) + n(t)\right) = S(u(t) + n(t)) \tag{14.21}$$

where $S(\ )$ represents the sigmoidal function in Eq. (14.7), $u(t) = \sum_{j=1}^{N} w_j x_j(t)$ is an "action value" and should be adjusted according to a learning algorithm, and $n(t)$ is a noise term (whose function will be justified below). We wish that, after training, $y(t)$ provides a score on the appropriateness of a given action in response to the stimulus X (i.e. the higher $y(t)$, the higher the probability that this action is actually accomplished). To this end, the action value $u(t)$ should provide an estimate of the reward connected to the given action. Of course, as it will be generalized below, we need a neuron (or a group of neurons) for each action.

First, let us consider the case named *static action choice*, in which a reward or punishment immediately follows the action. Each time the action is performed, following a given input stimulus $X(t)$, the subject receives a reward (say $r(t)$). Positive values signify a reward, negative or null values of $r(t)$ signify a punishment (or no reward). The synapse w_j connecting the neuron to the stimulus x_j is then modified according to the following rule, which is a version of the well-known *delta rule* used in error-correction neural networks:

$$\Delta w_j(t) = \gamma \delta(t) e_j(t) \tag{14.22}$$

where $\delta(t)$ provides an indication of the error that the network is accomplishing at the given moment, while $e_j(t)$ is known as the *eligibility signal*. The latter indicates that, when certain conditions hold for the input pathway connecting the stimulus x_j to the neurons, that pathway becomes eligible to have its weight modified. Different choices can be given for the eligibility signal. If we assume that Eq. (14.22) holds whenever the given action has been accomplished, good choices are $e_j(t) = x_j(t)$ and $\delta(t) = r(t) - u(t)$.

As a consequence, Eq. (14.22) becomes

$$\Delta w_j(t) = \gamma(r(t) - u(t))x_j(t) \tag{14.23}$$

which is also known as the *Rescorla–Wagner rule*. The meaning of this rule is quite simple. Let us assume that a given stimulus x_j is positive at the moment of the action. If the action value, $u(t)$, is smaller than the reward, the synapses w_j is increased, otherwise it is decreased. At the end of prolonged training, the variable $u(t)$ provides an estimate of the average reward attainable when the action is accomplished in response to that given stimulus. In particular, if we assume that

$x_j = 1$ or $x_j = 0$ in each trial (binary stimuli), Eq. (14.23) provides a synaptic change only in the presence of the input stimulus. Hence, the average value turns out to be

$$E\{\Delta w_j(t)\} = \gamma E\{r(t) - u(t)\} \tag{14.24}$$

and so, at equilibrium, $E\{u(t)\} = E\{r(t)\}$, in other words $u(t)$ fluctuates about the mean value of the reward. It is worth noting that, in a stochastic environment, $r(t)$ is a random process (i.e. it exhibits random fluctuations); the environment does not always reward or punish the same stimulus-action couple!

The previous analysis suggests that, the higher the average reward of a given action following a given stimulus, the higher $u(t)$ and thus the neuron output.

Let us now consider a more complex variant of the previous problem, when a reward is not instantaneous, but can occur later in time (this is also known as *the temporal credit assignment problem*). In this case, $u(t)$ must not simply estimate the average present reward, but the average total future reward (say $R(t)$), which is defined as follows $R(t) = \sum_{k=0}^{\infty} \alpha^k r(t + k\Delta t))$, where α is a temporal discount factor (with $0 < \alpha < 1$) which describes how the value of a delayed reward is attenuated. To this end, $\delta(t)$ in Eq. (14.22) should be the difference between the action value and the predicted total future reward (i.e. $\delta(t) = R(t) - u(t)$). The big problem here is that we do not know the future reward values, but just the present. However, future rewards can be progressively estimated during the trials, considering the following idea, first proposed by Barto et al. [17]. Let us assume that $u(t)$ is already a good estimator of $R(t)$. Then, the following identity holds

$$u(t) \cong \sum_{k=0}^{\infty} \alpha^k r(t + k\Delta t) = r(t) + \sum_{k=1}^{\infty} \alpha^k r(t + k\Delta t) = r(t) + \alpha \cdot u(t + \Delta t)$$

$$\tag{14.25}$$

The idea is that the difference between the left- and right-hand members of Eq. (14.25) should be progressively minimized by the learning procedure. To this end, Eq. (14.23) becomes

$$\Delta w_j(t) = \gamma(r(t) + \alpha \cdot u(t + \Delta t) - u(t))x_j(t) \tag{14.26}$$

In Eq. (14.26), $\alpha \cdot u(t + \Delta t) - u(t)$ considers the difference between two successive estimates. For this reason, Eq. (14.26) is also named *the temporal difference rule*.

According to the previous equations, the output neuron provides a high response in the case of a high reward to the given action and vice versa. Hence, it tries to decide whether that particular action should be performed or avoided, in response to a given input. Let us now consider the case in which different possible actions are in competition. In that case, one may consider a network with multiple output neurons, each trained with an algorithm similar to the previous one. Two main differences, however, are of value: (i) neurons compete reciprocally via lateral inhibitory synapses (i.e. each neuron tries to inhibit all others and excites itself according to *winner takes all* dynamics). After a brief transient period, just the neuron with the greatest input (i.e. the neuron with the highest reward) remains maximally excited, while outputs of all other neurons fall to zero. The winner neuron signals the action choice; (ii) only the synapses entering the winner neuron must be modified, on the basis of the reward (or punishment) received by that particular action. To this end, eligibility traces of a neuron with output $y(t)$ can be modified as $e_j(t) = y(t) \cdot x_j(t)$. In this way, just the neuron with high output (the winner) modifies its synapses, while synapses entering all other neurons are not modified. A variant of the eligibility trace, which considers a temporal dynamics, can be found in Ref. [17].

According to the previous schema, on any occasion the action with maximal average reward is selected in response to a given input; this signifies that, on all occasions, the subject is *exploiting* its previous knowledge to maximize reward. This strategy, however, exhibits a major drawback when the environment is nonstationary (i.e. the statistics of punishments and rewards is changing with time). In a nonstationary environment, a subject must be able to modify its strategy quite rapidly, to meet the environmental changes. In the network presented above, the subject would continue to accomplish the same choice in response to a given input, until the average reward falls below the reward of another possible choice; this would require too much time. In order to improve the search of an optimal choice in the nonstationary case, it is useful to introduce some noise in the neurons (this is the meaning of the $n(t)$ term in Eq. (14.21)). Thanks to the presence of noise, the neuron output to a given stimulus becomes a random variable (i.e. the output is exploring new possibilities that can contradict previous knowledge). Psychologists often talk about a trade-off between *exploration* and *exploitation*; if the noise term is very high, there is an excess of exploration of new choices; if the noise term is too low, the network continuously exploits old knowledge with a poor or slow capacity to adapt to a modified condition.

A psychological test frequently used to investigate the effect of error in choice tasks is the Flanker test, where the subject must provide a response on the basis of a central card, in the presence of distractors (the flankers). The presence of flankers causes a strong competition among possible alternative choices, requiring focused attention by the subject. Results of a connectionist model [18], similar to that described above, are shown in Figure 14.7. The model predicts the existence of correct responses, uncorrected responses, and corrected errors, whose percentage depends on the attention level and on the presence of noise.

There are various aspects of cognitive neuroscience that can be studied with the reinforced paradigm explained above. Some portions of the prefrontal cortex (and, in particular, a region located proximally to it, the anterior cingulate gyrus) are involved in the problem of conflict resolution and action decision. Indeed, patients with a prefrontal lesion may exhibit a poor capacity to explore new choices in a nonstationary environment, as demonstrated by psychological tests [19]. The basal ganglia (a subcortical region directly implicated in movement initiation and termination) are good candidates for this kind of network, too [6]. In any case, we need

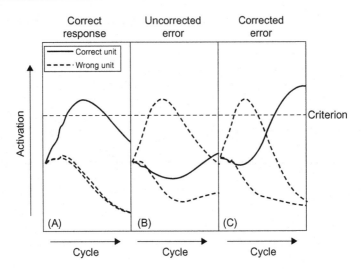

Figure 14.7 Time course of neuron responses in a network simulating the Flanker test [18] for trials with correct initial responses (A), trials with uncorrected errors (B), and trials with corrected errors (C). Solid lines indicate the activation of the neuron representing a correct response; dashed lines indicate the activation of neurons representing a wrong response. A response is selected whenever the activation of a unit exceeds the response criterion (dashed horizontal line).
Source: Reproduced from Ref. [18] with permission from the American Psychological Society.

a signal that communicates the reward and allows synapse changes. Dopaminergic neurons from the midbrain are thought to provide a learning signal to the prefrontal cortex and basal ganglia, relevant for planning and motor control. Depression and Parkinson's disease are both pathological problems strictly related to a dysfunction of the dopaminergic system.

14.4 Conclusions

The aim of the present chapter was to provide some essential elements on computational neuroscience. Of course, the field is so wide, and so much new knowledge has been gathered in recent years, that many important aspects have been necessarily neglected, and others have been just touched on lightly. The interested reader can find many more details from excellent texts on the subject [2,6,9].

A few aspects deserve some final comments. First, a cognitive behavior occurs in these networks as an "emergent property," which is the product of the topological structure of these networks (i.e. of the reciprocal interconnections among neurons) and, above all, of the learning procedure adopted for synaptic changes. There is a strict interdependency between the network topology, the training rule, and the final emergent cognitive behavior. The three major exempla presented in this work (episodic memory, category formation via self-organization, and action choice), although not exhaustive, represent three fundamental problems in cognitive neuroscience. Second, we observed that a single neural unit can be described with a different level of detail, depending on model objectives. The three network types presented here have been illustrated with reference to firing rate neurons (for the sake of simplicity), but similar properties would emerge, using a more detailed formalism, too, (i.e. using IF neurons or HH neurons). This creates the possibility of building multiscale models, ranging from the cellular or subcellular level (particularly ionic channels) up to the activity of entire brain regions or behavioral responses in human psychology. Of course, these different scales must communicate reciprocally: the solution at one level may be functional for another level, and all levels must be integrated into a coherent, noncontradictory, and exhaustive framework. This is one of the main functions of modern physiological models.

Finally, we wish to stress that, thanks to the enormous expansion of neuroscience in the present day, some essential problems of human thought, such as memory, the formation of concepts and language, the motivations for our choices, up to the problem of consciousness, are also becoming the subject of modern scientific investigation. Within this fascinating field, computational models are going to play an increasing role. Indeed, the solution of cognitive neuroscience problems requires a thoroughly multidisciplinary approach, with the engagement, not only of psychologists, neuroscientists, and philosophers, but also mathematicians and biomedical engineers.

References

[1] Hodgkin AL, Huxley AF. A quantitative description of membrane current and its application to conduction and excitation in nerve. J Physiol 1952;117(4):500−44.
[2] Dayan P, Abbott LF. Theoretical neuroscience. Cambridge, MA: The MIT Press; 2001.
[3] Lapicque L. Recherches quantitatives sur l'excitation électrique des nerfs traitée comme une polarisation. J Physiol Pathol Général 1907;9:620−35.
[4] James W. Psychology: briefer course. Cambridge, MA: Harvard University Press; 1985.
[5] Hebb DO. The organization of behavior. New York, NY: Wiley; 1949.
[6] Trappenberg T. Fundamentals of computational neuroscience. Oxford: Oxford University Press; 2010.
[7] Hopfield JJ. Neurons with graded response have collective computational properties like those of two-state neurons. Proc Natl Acad Sci USA 1984;81(10):3088−92.
[8] Amit DJ. Modeling brain function: the world of attractor neural networks. Cambridge: Cambridge University Press; 1989.
[9] Rolls E, Treves A. Neural networks and brain function. Oxford: Oxford University Press; 1998.
[10] Kelso SR, Ganong AH, Brown TH. Hebbian synapses in hippocampus. Proc Natl Acad Sci USA 1986;83(14):5326−30.
[11] Dragoi G, Buzsáki G. Temporal encoding of place sequences by hippocampal cell assemblies. Neuron 2006;50(1):145−57.

[12] Jansen BH, Rit VG. Electroencephalogram and visual evoked potential generation in a mathematical model of coupled cortical columns. Biol Cybern 1995;73(4):357—66.

[13] Ursino M, Cona F, Zavaglia M. The generation of rhythms within a cortical region: analysis of a neural mass model. Neuroimage 2010;52(3):1080—94.

[14] Cona F, Ursino M. A multi-layer neural-mass model for learning sequences using theta/gamma oscillations. Int J Neural Syst 2013;23(3):1—18.

[15] Loh M, Rolls ET, Deco G. A dynamical systems hypothesis of schizophrenia. PLoS Comput Biol 2007;3(11):e228.

[16] Kohonen T. Self-organized formation of topologically correct feature maps. Biol Cybern 1982;43(1):59—69.

[17] Barto AG, Sutton RS, Anderson CW. Neuronlike adaptive elements that can solve difficult learning control problems. IEEE Trans Syst Man Cybern 1983;SMC-13 (5):834—46.

[18] Steinhauser M, Maier M, Hübner R. Modeling behavioral measures of error detection in choice tasks: response monitoring versus conflict monitoring. J Exp Psychol Hum Percept Perform 2008;34(1):158—76.

[19] Clark L, Cools R, Robbins TW. The neuropsychology of ventral prefrontal cortex: decision-making and reversal learning. Brain Cogn 2004;55(1):41—53.

15 Insulin Modelling

Morten Gram Pedersen and Claudio Cobelli

Department of Information Engineering, University of Padova, Padova, Italy

Insulin, the only glucose-lowering hormone, stimulates glucose uptake by peripheral tissues and inhibits hepatic glucose output. Insulin is secreted from the beta-cells, which are located in the endocrine part of the pancreas in microorgans called islets of Langerhans, in response to various nutrients with glucose being the physiologically most important stimulus. It is now generally acknowledged that deficient insulin secretion plays a major role in the development of diabetes mellitus, and a large research effort has been devoted to studies of beta-cell function in health and disease.

In this chapter, we will review a number of mathematical models that have been used to obtain insight from data ranging from single beta-cells to clinical measurements. We will show how it is possible to relate these models operating on disparate physiological scales with so-called multiscale modelling.

15.1 Dynamics of Insulin Secretion

The major pathway of glucose stimulated insulin secretion involves metabolism of the sugar, which raises the intracellular ATP concentration at the expense of the ADP levels. The higher ATP-to-ADP ratio closes ATP-sensitive potassium (K(ATP)-) channels in the plasma membrane, which leads to depolarization of the cell and electrical activity. Voltage-sensitive calcium channels contribute to the electrical pattern, and when they are open, calcium flows into the cell, where it triggers fusion of insulin-containing granules with the plasma membrane. These exocytotic events allow insulin to escape to the extracellular space and eventually to enter the blood stream. Besides this "triggering pathway," another and less understood "amplifying pathway" modifies the amount of insulin that is secreted in response to a given calcium concentration during a glucose challenge [1] (Figure 15.1).

In response to a step in glucose concentration, insulin is secreted in a characteristic biphasic pattern consisting of a first phase lasting $\sim$5 min followed by a sustained second phase where insulin is released in distinct pulses with a period of $\sim$5 min. First phase secretion is blunted and the second phase is reduced in early phases of diabetes [2]. Moreover, the oscillatory response is disturbed in

Modelling Methodology for Physiology and Medicine. DOI: http://dx.doi.org/10.1016/B978-0-12-411557-6.00015-X

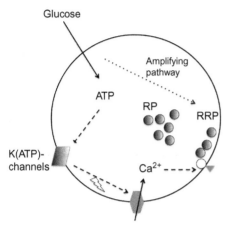

Figure 15.1 A schematic overview of the major signaling pathways underlying glucose-stimulated insulin secretion. See main text for details. The figure indicates the reserve pool (RP) and the readily releasable pool (RRP) of insulin granules.

prediabetes [3]. Hence, these dynamical responses are considered signatures of healthy beta-cell function, and their disturbance has been suggested to be a predictor of diabetes [2,3].

Both biphasic and oscillatory secretion patterns are inherent to the beta-cells, and not a result of systemic regulation, since they are consistently found in isolated pancreases [4−6], pancreatic islets [7−10], and even isolated beta-cells [11,12]. It seems to be generally recognized that the oscillations in insulin secretion are a result of oscillations in beta-cell metabolism, membrane potential, and calcium levels [13,14]. Although the underlying mechanisms are still debated, increasing evidence indicates that oscillations in metabolism play a major pacemaking role [13,15]. The cellular mechanisms underlying biphasic insulin secretion are also still a subject of research. It has been suggested that the two phases are attributable to distinct pools of insulin granules with different release capacities [16,17]. Alternatively, the dynamics of signals underlying the release of insulin have been suggested to cause the biphasic secretion pattern [18,19]. Indeed, the intracellular Ca^{2+} concentration shows a characteristic phasic behavior in response to a step in the extracellular glucose concentration [20,21].

15.2 Cellular Modelling of Beta-Cell Function

An entire branch of mathematical modelling of beta-cells has focused on oscillatory behavior in metabolism, membrane potential, and Ca^{2+} levels underlying pulsatile insulin secretion. We will refer to excellent reviews [13,22−24] discussing these aspects in greater detail, and here focus on models of insulin secretion.

15.2.1 Phenomenological Modelling

Considering that the task of the beta-cell is to secrete insulin, surprisingly little effort has been devoted to mathematical modelling of the cell biology of insulin secretion, compared to the focus on the other aspects of beta-cell physiology mentioned above. However, in the 1970s, Grodsky and coworkers [5,25] and Cerasi et al. [18] developed models of the pancreatic insulin response to various patterns of glucose stimuli. Because of the limited knowledge of beta-cell biology at that time, these early models were phenomenological. Only recently has our knowledge of the events leading to exocytosis of insulin granules reached a level that allows us to formulate mechanistically based models.

Cerasi et al. [18] suggested that the dynamics of insulin secretion was due to time-dependent inhibitory and potentiating signals, still unidentified [26], which act on insulin secretion at different times. For biphasic insulin secretion, inhibition is responsible for creating the nadir after the first phase peak, while potentiation acts later to produce the second phase. An alternative view on the signal hypothesis is that the dynamics of intracellular Ca^{2+}, the triggering signal for insulin release, dictates the dynamics of insulin secretion [19]. In particular, Ca^{2+} shows a phasic pattern in response to a step in glucose concentration [20,21], which could drive biphasic insulin secretion. However, it should be noted that when Ca^{2+} is kept at a constant and high level, insulin secretion is still biphasic [27]. Thus, other signals or mechanisms must be operating in addition to Ca^{2+}. Future research will likely provide insight into these hypothetical signals as well as in the control and role of Ca^{2+} dynamics in biphasic insulin release.

As an alternative to Cerasi's signal hypothesis, Grodsky and colleagues [5,16] proposed that insulin was located in "packets," plausibly insulin-containing granules, but possibly entire beta-cells. In this model, part of the insulin is stored in a reserve pool, while other insulin packets belong to a labile and releasable pool. The rapid release of the labile pool results in the first phase of insulin secretion [5,16], while the reserve pool is responsible for the sustained second phase. The distinction between reserve and "readily releasable" insulin has been at least partly confirmed when the packets are identified with granules [28,29].

However, a modification of this conceptually simple model is needed to explain the so-called staircase experiment in which the glucose concentration is increased in consecutive steps, each step giving rise to a peak of insulin. Grodsky [5] assumed that the labile pool is heterogeneous in the sense that the packets in the pool have different thresholds with respect to glucose beyond which they release their content. The resulting mathematical model can reproduce the staircase and many other experiments. Although there has been no support of granules having different thresholds [26], Grodsky [5] mentioned that cells apparently have different thresholds based on electrophysiological measurements [30]. In the following, we provide an update of Grodsky's model based on the idea and recent data of cell-to-cell heterogeneity with respect to their activation threshold.

15.2.2 Mechanistic Pancreatic Modelling

We [31] have built a model of beta-cell function starting from the idea that the insulin-containing granules belong to different pools [5,17,32], as suggested from biological experiments [27–29,33]. Insulin is synthesized in the endoplasmic reticulum and packed in secretory granules. These newly prepared insulin granules enter a large internal "reserve pool" from which the granules are mobilized to the cell membrane. This is likely to happen via messengers such as Ca^{2+} [34] and ATP, which drives the motor protein myosin V [35]. Mobilized granules are either reinternalized or proceed to dock to the membrane. Once docked, the granules can then undergo priming, a process that involves acidification of the granules [36]. The primed granules belong to the readily releasable pool (RRP) and can fuse with the cell membrane in response to raised intracellular Ca^{2+} levels when calcium channels open at stimulatory glucose levels. In the model, glucose controls fusion (which is triggered by calcium influx not included in the model) and regulates mobilization from the reserve pool to the cell membrane.

The model was simplified by considering the reserve pool constant (infinite). We thus neglected synthesis of new granules and changes in the size of the reserve pool due to mobilization, reinternalization, and crinophagy. This hypothesis is supported by the observation that more than 80% of the granules are located in the center of the cell [17], and the fact that modifying the biosynthesis rate appears to have little effect on secretion over the first 2 h after an increased glucose stimulus [37]. In addition, the mobilized and docked granules were merged into a single "docked pool." The model is summarized in Figure 15.2.

We were interested in confronting our simulation model with experiments performed on the intact pancreas. Therefore, we considered all the beta-cells in a pancreas. In agreement with earlier electrophysiological findings [30], recent Ca^{2+} imaging experiments [38,39] showed that the beta-cells have different glucose thresholds for triggering Ca^{2+} influx. Above this threshold, the intracellular Ca^{2+} concentration changes little with glucose [38,39]. Of interest, very recent imaging studies of exocytotic events in intact islets show that beta-cells are recruited to release insulin at different glucose concentrations [40].

Based on these observations, we distinguish between silent and active beta-cells. We denote by $\Phi(g)$ the fraction of cells that are active at a given extracellular glucose concentration (g), which is a sigmoidal function of g [30,38,39]. The density function describing the fraction of cells with threshold between g and $g + dg$ is the mathematical derivative $\phi(g) = d\Phi/dg$. All rate constants modelling the granule pool dynamics are assumed identical in all cells. Thus, the only difference between

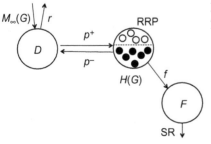

Figure 15.2 Overview of the mechanistic pancreatic model [31].

cells is whether they are active and release insulin as a result of exocytosis, or are silent and, hence, their readily releasable granules do not fuse.

This distinction therefore leads to a heterogeneous total pancreatic RRP, in which a part of the readily releasable granules are located in silent cells and another part are located in active cells and undergo exocytosis. We denote the part of the RRP in active cells by H.

As in Ref. [41], we assume that mobilization occurs with no delay with rate

$$M_\infty(G) = M_0 + cG^n/(G^n + (K_M)^n) \tag{15.1}$$

The delay in mobilization is not needed to reproduce the characteristic biphasic profile in response to a step in glucose concentration. Indeed, we [31] had to use a very short delay in M in order to reproduce the data from O'Connor et al. [25], further justifying the assumption of removing the delay in M.

To derive the equations describing the various pools, we need to account for the fluxes in and out of each pool. The docked pool D is refilled by mobilization with rate $M_\infty(G)$, and by "unpriming," i.e., a process in which readily releasable granules lose their release capacity, with rate constant p^-. Granules leave the docked pool due to undocking and reinternalization, with rate constant r, and because of priming modelled with the rate constant p^+. This leads to

$$dD/dt = M_\infty(G) + p^-\text{RRP} - (r + p^+)D \tag{15.2}$$

The dynamics of the granules in the entire RRP are described by

$$d\text{RRP}/dt = p^+D - p^-\text{RRP} - fH \tag{15.3}$$

where the last term describes fusion of granules located in the RRP in active cells, i.e., in H, and f is the fusion constant.

The modelling of the part of the RRP in active cells (H) is slightly more tricky. Unpriming occurs with rate p^-H. The priming flux is assumed identical in all cells. This implies that the flux of granules entering $H(G)$ due to priming is proportional to $\Phi(G)$, and equal to $p^+D \, \Phi(G)$. H also changes when G changes due to activation or deactivation of cells. Intuitively, this term should be proportional to dG/dt, because the faster G increases, the more rapidly do new cells activate, and their RRP enters H at the same rate. The total RRP in cells with activation threshold G is given by $h(G) = dH/dG$. One thus arrives at the following equation (see also Ref. [42] for a more mathematical treatment),

$$dH(G)/dt = p^+D\Phi(G) - (f + p^-)H(G) + h(G)dG/dt \tag{15.4}$$

The pool of fused granules F is described as follows:

$$dF/dt = fH - mF \tag{15.5}$$

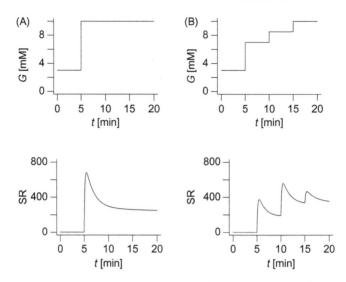

Figure 15.3 Simulations of biphasic secretion (A) and the staircase protocol (B) using the mechanistic model [31] shown in Figure 15.2. Parameters as in Ref. [41].

where m is the rate constant of insulin release. Finally, the secretion rate is given by

$$SR = mF + SR_b \tag{15.6}$$

where SR_b is the basal secretion rate. As shown in Figure 15.3, the model reproduces satisfactorily both biphasic secretion after a step in the glucose concentration and the three peaks in response to the staircase protocol. The latter is due to the recruitment of more cells at each step of glucose, which then release their RRP.

15.2.3 Mechanistic Single-Cell Modelling

15.2.3.1 Models to Simulate

Other models [37,43−45] describe details of the dynamics of granules as they move between different pools in single beta-cells. Some of these [43,44] include the control of exocytosis by Ca^{2+}, a necessity for coupling the granule model to models of metabolism, electrical activity, and calcium handling, in this way obtaining a model that includes all steps from glycolysis via calcium dynamics to exocytosis and insulin release. Using such a model it was suggested [44] that a highly Ca^{2+}-sensitive pool, described by capacitance measurements [46,47], could correspond to newcomer granules seen in total internal reflection fluorescence (TIRF) microscopy experiments [48], illustrating how mathematical models can be used to integrate data obtained with different techniques and on different timescales. Subsequently, this link between Ca^{2+} sensitivity and fusion mode was partly confirmed experimentally [49]. As increasing insight in the molecular mechanisms regulating the dynamics of granules on their way to exocytosis becomes available, such detailed models can be used to simulate e.g., defects of regulations or expression of exocytotic proteins, or to computationally study interventions targeting these molecular mechanisms by varying rate constants correspondingly.

15.2.3.2 Models to Test Hypotheses

In contrast to these large-scale simulation models that include most of the mechanisms relevant for granule maturation and exocytosis, simpler models including only parts of these processes have been used to investigate the plausibility of hypotheses based directly on a limited set of experimental data.

Pedersen [50] used such a relatively simple model to study the effect of palmitate exposure on Ca^{2+} and exocytosis [51]. The question was whether palmitate disrupted Ca^{2+}-channel clustering or, possibly in addition, moved granules away from Ca^{2+}-channels. Modelling simulations showed that no clear conclusion could be drawn from capacitance measurements using the patch clamp technique, but it was argued that imaging data [51] favored the interpretation that palmitate exposure dissociated granules from Ca^{2+}-channels. More generally and based on simple assumptions, the conclusions that can be extracted from a certain protocol of capacitance measurements have been investigated in detail with mathematical modelling [52].

The control of exocytosis by Ca^{2+} in human beta-cells was recently studied with a mathematical model of Ca^{2+} in a submembrane compartment and a single pool of granules [53]. The model was used to investigate the exocytotic responses to short depolarizations in single human beta-cells [54,55], and showed that the experimental data do not favor the proposed interpretation of a certain type of Ca^{2+} channel preferably controlling exocytosis in human beta-cells. The simulations suggested, rather, that L- and P/Q-type Ca^{2+} channels contribute to a similar extent to insulin secretion in humans. Since the data investigated was on a subsecond timescale, the slower processes of granule dynamics, such as docking and priming, were not included in the model. Few data on these processes are available in human beta-cells.

15.3 Whole-Body Modelling of Beta-Cell Function

In contrast to the mechanistic subcellular models described in the previous sections, mathematical modelling is also of great importance for the assessment of beta-cell function in humans. In this section, we will describe the minimal model approach to extract information from three widely used clinical tests: the intravenous glucose tolerance test (IVGTT), the mixed-meal tolerance test (MTT), and the oral glucose tolerance test (OGTT). However, we first address the question of hepatic insulin extraction, which must be overcome in order to estimate pancreatic insulin secretion.

15.3.1 C-Peptide Modelling

Insulin is secreted into the portal vein and enters the liver where a substantial part of insulin is extracted. For this reason, insulin secretion cannot be isolated from hepatic insulin extraction when inferred from plasma insulin concentrations, since plasma data reflect the fraction of pancreatic secretion that appears in plasma, denoted as posthepatic insulin secretion and approximately equal to 50% of pancreatic secretion. This problem can be bypassed if C-peptide concentration is measured during the perturbation and used to estimate insulin secretion, since C-peptide is secreted equimolarly with insulin [56], but it is extracted by the liver to a negligible extent [57]. Plasma C-peptide concentration thus reflects C-peptide plasma rate of appearance, which, apart from the rapid liver dynamics, is a good measure of C-peptide pancreatic secretion, which in turn coincides with insulin pancreatic secretion.

To be identified on plasma C-peptide measurements, the secretion models must be integrated into a model of whole-body C-peptide kinetics. The widely used model [58] is shown in Figure 15.4. Model equations are conveniently expressed in terms of C-peptide concentration above basal in the two compartments, denoted as CP_1 and CP_2 (pmol/L):

$$dCP_1(t)/dt = -[k_{01} + k_{21}]CP_1(t) + k_{12}CP_2(t) + SR(t), \quad CP_1(0) = 0 \qquad (15.7)$$

$$dCP_2(t)/dt = k_{21}CP_1(t) - k_{12}CP_2(t), \quad CP_2(0) = 0 \qquad (15.8)$$

where k_{01}, k_{21}, k_{12} (min^{-1}) are transfer rate parameters and SR (pmol/L min) is secretion above basal, normalized by the volume of distribution of compartment 1, to be described according to the models presented in the following section.

Parameters of C-peptide kinetics are usually determined, without loss of accuracy, with the population approach [59].

15.3.2 Clinical Tests

Models of insulin secretion enable evaluation of beta-cell function following intravenous injection of a bolus of glucose (e.g., during an IVGTT [60]) or after oral

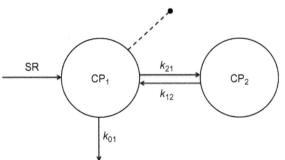

Figure 15.4 The two-compartment C-peptide model; compartment 1, accessible to measure, represents plasma and rapidly equilibrating tissues, compartment 2 represents tissues in slow exchange with plasma.

ingestion of nutrients (MTT/OGTT) [61]. The oral perturbations are without doubt more physiological than the intravenous tests, with the MTT being superior to the OGTT because of the presence of proteins and fat. Oral ingestion of nutrients leads to the release of gut hormones, in particular glucagon-like peptide 1 (GLP-1) and gastric inhibitory peptide (also known as glucose-dependent insulinotropic polypeptide or GIP), which enhance insulin secretion. The fact that insulin secretion is greater after oral delivery of glucose compared to intravenous injection even when glucose profiles are matched is denoted as the incretin effect, and is largely due to GLP-1 and GIP, which are therefore known as incretins [62]. Giving a mechanistic description of pancreatic insulin secretion as a function of plasma glucose concentration has the advantage to provide quantitative indices of beta-cell function. In addition to providing estimates of beta-cell function, insulin action and hepatic insulin extraction also can be assessed by models during both IVGTT and MTT/OGTT [32].

15.3.3 IVGTT Minimal Model

In the minimal modelling methodology, the glucose—insulin system is decomposed into two subsystems, where we here look at the insulin subsystem with plasma glucose, G, considered the "input" (assumed known) and C-peptide, C, the "output" (assumed noisy). Since the secretion model is identified on C-peptide measurements taken in plasma, it must be integrated into a model of whole-body C-peptide kinetics, which has two compartments [58], as described above.

During an IVGTT, insulin secretion [60] is modelled by (Figure 15.5A)

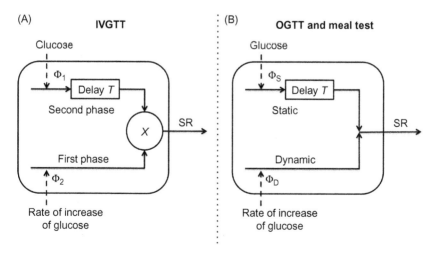

Figure 15.5 (A) IVGTT secretion minimal model [60]. (B) MTT/OGTT secretion minimal model [61].

$$SR = mX + SR_b \tag{15.9}$$

where SR_b is basal secretion, and X is the readily releasable insulin described by

$$dX/dt = Y - mX, \quad X(0) = X_0 \tag{15.10}$$

where X_0 is the amount of insulin released immediately after the glucose stimulus and Y is the provision of new insulin, which depends on the glucose level:

$$dY/dt = (Y_\infty(G) - Y)/T, \quad Y(0) = 0 \tag{15.11}$$

with

$$Y_\infty(G) = \max\{0, \beta \cdot (G - h)\} \tag{15.12}$$

In other words, insulin secretion consists of two components: first and second phase secretion. First phase secretion is portrayed by a rapidly turning-over compartment (2 min) and supposedly represents exocytosis of previously primed, readily releasable insulin secretory granules. It exerts derivative control, since it is proportional to the rate of increase of glucose from basal up to the maximum through a parameter, Φ_1, which defines the first responsivity index:

$$\Phi_1 = X_0/\Delta G \tag{15.13}$$

with ΔG the difference between peak and basal glucose concentration.

Second phase insulin secretion is believed to be derived from the provision and/ or docking of new insulin secretory granules, reaching the releasable pool with a delay time constant, T. The provision processes are assumed to occur in response to a given (i.e., proportional to) glucose concentration, through a parameter $\Phi_2 = \beta$, which defines the second phase responsivity index.

The mathematical meaning of Φ_2 and T can be explained by analyzing model parameters under above-basal step increase of glucose: provision tends with time constant T toward a steady state, which is linearly related to the glucose step size through the parameter Φ_2. In addition to Φ_1 and Φ_2, a basal responsivity index can also be calculated, Φ_b. Finally, a single total index of stimulated beta-cell responsivity, Φ^{IVGTT}, can be derived by combining Φ_1 and Φ_2. All these three ingredients (i.e., derivative term, delay, and the presence of a RRP) have been shown to be necessary for the model to accurately describe C-peptide data [60].

15.3.4 MTT/OGTT Minimal Model

Beta-cell function can also be assessed from an oral test, such as an MTT or an OGTT. An oral test differs from the IVGTT in several important aspects, including the route of delivery with the associated incretin hormone secretion, the more physiological and smoother changes in glucose, insulin, and C-peptide concentrations, and, during a mixed meal, the presence of nonglucose nutrients stimulation (i.e., amino acids and fat). Various models have been proposed to assess beta-cell function during MTT/OGTT [61,63−65]. All of them share the model of C-peptide kinetics described in Section 15.3.2, but differ on the assumption of how glucose controls the secretion. The oral C-peptide minimal model proposed by Breda et al. ([61]; Figure 15.5B) maintains basically all of the previous model ingredients employed in the IVGTT model, with the exception of the fast releasable pool of insulin (X, which is not evident under these conditions) to describe the data. In particular, both a rate of change of glucose component of insulin secretion and a delay between glucose stimulus and beta-cell response have been shown to be necessary to fit the data [66].

Model equations are

$$SR = Y + SR_d + SR_h \tag{15.14}$$

with Y described by Eqs. (15.11) and (15.12) denoting the static component of insulin secretion, and SR_d, the dynamic component of insulin secretion:

$$SR_d = K \cdot \max\{0, \, dG/dt\} \tag{15.15}$$

Thus, this model features a dynamic component that senses how rapidly the glucose concentration increases, and a static component that represents the release of insulin that, after a delay, occurs in proportion to the prevailing glucose concentration. Similarly to the IVGTT, where first, Φ_1, and second phase, Φ_2, beta-cell responsivity indices were defined, from the oral model dynamic, Φ_D ($=K$), and static Φ_S ($=\beta$) responsivity indices can be derived. A basal responsivity index, Φ_b, can be obtained as the ratio between basal secretion and basal glucose, and a total responsivity index, Φ^{ORAL}, which combines Φ_D and Φ_S can also be derived.

The model has been successfully used during "up&down" intravenous glucose infusion [67] and for describing hyperglyccmic clamp C-peptide data as well as meals [68], thus providing further independent evidence for its validity. In contrast to the IVGTT model where the derivative component of first phase secretion was operative only during the first few minutes as the plasma glucose concentration increased from a "basal" to a "maximal" concentration, the relatively gradual pattern of glucose appearance observed during oral tests necessitated the presence of a secretion component proportional to glucose rate of change that contributed to the model for the first 60−90 min. In addition, and similar to the IVGTT, a component of insulin secretion proportional to glucose, characterized by a delay time T (presumably reflecting at least in part the time it takes for new granules to reach the releasable pool), that contributed throughout the experimental period was also necessary. Of note, Φ_D is markedly different from its IVGTT counterpart Φ_1. In fact, during an IVGTT, the first phase component only contributed during the first 4−6 min, while the proportional component is operating for the rest of test. Thus, it is probable that Φ_D and Φ_1 are assessing different aspects of the insulin secretory pathway, a question that will be addressed below.

The model proposed by Hovorka et al. [63] assumes an instantaneous linear control of glucose on insulin secretion; i.e., there is no delay between glucose stimulus and beta-cell response. The model proposed by Cretti et al. [64] describes insulin secretion with the static component of glucose control of the C-peptide minimal model; thus, it is characterized by a delay but does not include any dynamic (i.e., rate of change, glucose control). Interestingly, the same authors have recently included a dynamic control to describe first phase secretion in a subsequent report [69]. The model proposed by Mari et al. [65], similar to the oral model shown in Figure 15.5B, has both a proportional component and a component responsive to the rate of change of glucose, but there is no delay between glucose signaling and the supplying of new insulin to the circulation. The authors choose to account for the expected inability of a proportional plus derivative glucose control to account for C-peptide measurements with a time-varying term correcting only the static component of insulin secretion, which has been called the potentiation factor. In simple words, the potentiation factor is a time-varying correction term that mathematically compensates for the proportional plus derivative description deficiency.

15.4 Multiscale Modelling of Insulin Secretion

Minimal models are useful tools for the study of clinical data by estimating parameters from model fitting to the observed data, in particular when based on C-peptide measurements as discussed above [32]. These models must be fairly simple to allow parameter identification, but should at the same time reflect truthfully the underlying biology down to the cellular events underlying insulin secretion. That is, the models should be minimal in the sense that further reduction would make the model nonphysiological and/or prevent an accurate description of the data.

One way to make the coupling between events at different physiological levels consists of the use of multiscale modelling. In Pedersen et al. [42], such an approach was used to gain insight into the mechanistic interpretation of parameters in models of insulin secretion during oral tolerance tests (MTT/OGTT) [61,63−65]. Multiscale modelling has also been applied to the IVGTT [41], and to a minimal oral model [70] of GLP-1 action on secretion [71].

For this purpose, appropriate cellular models built from a mechanistic description of well-defined subcellular events must be analyzed and simplified. As described in Section 15.2.2, we developed a mathematical model of insulin granule dynamics that incorporated cell-to-cell heterogeneity into the glucose threshold for cell activation and secretion [31]. In the following section, we analyze this model with a multiscale approach to get insight into the biological meaning of minimal model parameters based on our previous work [41,42]. In addition, our mechanistic model was recently suggested as the core of a bio-inspired artificial pancreas [72], further underlining the need for a thorough multiscale understanding of the cellular model.

15.4.1 Multiscale Analysis of the IVGTT

We are interested in comparing the two expressions for the secretion rate, Eq. (15.9) for the minimal model and Eq. (15.6) for the mechanistic model during an IVGTT. Inspection readily reveals that we need to relate F to X. In the following, this coupling is done by approximating F by a variable X^* with kinetics similar to that of X.

First we note that the last term in Eq. (15.4) describes derivative control (i.e., secretion does not only depend on the glucose level, but also on its rate of change). As discussed in Section 15.3.4, this fact is important during oral and meal tolerance tests and included in several oral minimal secretion models. However, during an IVGTT, the glucose concentration decreases monotonically (Figure 15.6A) in contrast to the rising glucose level during the first phase of an oral test. We have previously argued that in our mechanistic model derivative control is negligible when dG/dt is negative [42]. For the IVGTT, this claim is further supported by a more careful analysis as in Ref. [41]. Hence, derivative control can be neglected during an IVGTT because of the decreasing glucose profile, and the fraction of the RRP located in active cells can be described by

$$dH(G)/dt = p^+ D\Phi(G) - (f + p^-)H(G) \tag{15.16}$$

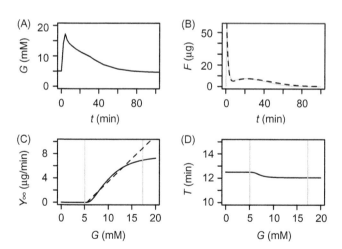

Figure 15.6 (A) Mean average plasma glucose profile in a cohort of 204 healthy subjects during an IVGTT. (B) The fused pool F as a function of time (full gray curve) and the approximation X^* (dashed black curve) in response to the glucose profile in panel A. Parameters as in Ref. [41]. (C) The steady-state mobilization rate $Y^*_\infty(G)$ (full curve) is compared with the linear function $\beta(G - h)$ with $h = G_b = 5$ mM. The vertical gray lines indicate the basal and maximal glucose levels from panel (A). (D) The delay parameter T^* as a function of G. The vertical gray lines indicate the basal and maximal glucose levels from panel (A).

In order to derive the large-scale minimal model, we invoke quasi-steady-state approximations based on an evaluation of the timescales of the different processes. The timescale for H is $1/(f + p^-)$, which is of the order of seconds due to rapid fusion; this is much faster than other timescales of the model and of the glucose dynamics during the IVGTT. Hence, we can assume that H is in a quasi-steady state:

$$H(G) = p^+ D\Phi(G)/(f + p^-) \tag{15.17}$$

Similarly, we can assume that the entire RRP is in a steady state, which after some algebra leads to [41]:

$$dD/dt = \rho(G)[M_\infty(G)/\rho(G) - D] \tag{15.18}$$

$$dF/dt = fp^+ D\Phi(G)/(f + p^-) - mF \tag{15.19}$$

where

$$\rho(G) = r + fp^+ \Phi(G)/(f + p^-) \tag{15.20}$$

Define $X^* = F$ and $Y^* = [fp^+ \Phi(G)/(f + p^-)] \cdot D$. Then,

$$dX^*/dt = Y^* - mX^* \tag{15.21}$$

$$dY^*/dt = (Y^*_\infty(G) - Y^*)/T^* \tag{15.22}$$

with $T^* = 1/\rho(G)$ and

$$Y^*_\infty(G) = [fp^+ \Phi(G)/(f + p^-)] \cdot M_\infty(G)/\rho(G) \tag{15.23}$$

Note the analogy between the models in Eqs. (15.10)–(15.11) and (15.21)–(15.22). The initial condition for Y^* is $Y^* = 0$, because $\Phi(G_b) = 0$. When G rises to G_{max} rapidly after the glucose bolus, a part of the RRP equal to $H(G_{max})$ fuses rapidly and enters F within seconds. Thus, the initial condition for $X^* = F$ is

$$X^*(0) = X^*_0 = H(G_{max}) = p^+ D(0)\Phi(G_{max})/(f + p^-) \tag{15.24}$$

Figure 15.6B shows a typical pattern of the pool F and its approximation X^*, which shows excellent correspondence. The asymptotic function for mobilization $Y^*_\infty(G)$ is plotted in Figure 15.6C together with the corresponding linear function $\max\{0, \beta(G - h)\}$ from Eq. (15.12). Note the good correspondence over most of the glucose range attained during an IVGTT (indicated by vertical gray lines).

The minimal model Eq. (15.11) has a constant delay T, whereas the delay T^* in Eq. (15.22) depends on G. However, T^* is nearly constant (Figure 15.6D) because

r is an order of magnitude greater than p^+, justifying the minimal model assumption of constant delay. With the parameters used here, we find that the delay T^* is of the order of 12 min, in reasonable agreement with Toffolo et al. [60], who found a delay of approximately 15 min.

15.4.2   Multiscale Analysis of the MTT/OGTT

Similar to the analysis presented above for the IVGTT, we have investigated how the three main components of oral minimal models—derivative control, proportional control, and delay—are related to the subcellular events [42]. As for the IVGTT, the simplification of the mechanistic model [31] consists in quasi-steady-state assumptions based on timescale analysis. During the MTT/OGTT, release is rapid (timescale $1/m$ ~1.5 min) compared to glucose dynamics (timescale of tens of minutes), in contrast to the beginning of the IVGTT in which the glucose levels change on a timescale comparable to release. Hence, during the MTT/OGTT, the pool of fused granules F is also in a quasi-steady state and its dynamics cannot be inferred. The overall result of the analysis is that during an MTT/OGTT, the secretion rate is approximately equal to [42]:

$$\text{SR}(t) \approx [f^+/(f^+ + p^-)] \cdot [p^+ D(t; \tau)\Phi(G(t)) + h(G, t)\text{d}G/\text{d}t] + \text{SR}_b \qquad (15.25)$$

Hence secretion is, besides basal secretion SR_b, composed of a static and a dynamic term, in close correspondence with the minimal oral model (15.14). Static secretion includes a delay because of the time needed for mobilization and docking. In addition, it depends on priming and the number of active cells.

Dynamic secretion is related to $h(G,t)$, which in the minimal model is described by the parameter K in Eq. (15.15). As for the IVGTT, it can be argued that h is negligible when G decreases [42], which supports the minimal model assumption of a contribution from $\text{d}G/\text{d}t$ only when G increases (Eq. (15.15)). Breda et al. [61] and Toffolo et al. [67] considered a variant of the model in which K decreased with glucose, and in some cases this extended model was needed to fit the data satisfactorily [67]. Interestingly, the glucose-dependence of K corresponds well to the mechanistic model since h is bell-shaped and peaks near the basal glucose level and then decreases [42].

15.5   Conclusion

We have here reviewed mathematical models of insulin secretion useful for a span of experimental situations: from the analysis of exocytosis, to secretion from groups of islets or the isolated pancreas, and all the way to minimal models needed for the investigation of beta-cell function in a clinical setting. Importantly, these minimal models can be given mechanistic underpinning by the use of multiscale modelling.

The biology underlying insulin secretion during an IVGTT or an MTT/OGTT is obviously the same, while the minimal secretion models, although similar, are not identical. We have shown that our model description of the cellular events underlying glucose-stimulated insulin secretion [31] simplifies to our IVGTT minimal model [60] when the beta-cells respond to a IVGTT glucose profile (Section 15.4.1; [41]). Similarly, in our previous work [42] we showed that the beta-cell model [31] reduces to our MTT/OGTT minimal secretion model [61] when subjected to a typical glucose stimulus seen during an MTT/OGTT, as discussed in Section 15.4.2. Thus, depending on the clinical setting, a single mechanistic beta-cell model simplifies to either the IVGTT or the MTT/OGTT minimal secretion models needed for parameter identifiability in tests of beta-cell function. This fact justifies, on the one hand, the differences between the two minimal models that represent the same underlying biology but under different conditions, and, on the other hand, highlights why the two minimal models have a structural similarity.

In the IVGTT multiscale analysis, we found that the delay parameter T^* is nearly constant and, surprisingly, approximately equal to the inverse of the reinternalization rate (i.e., to $1/r$). The parameter $T^* = 1/\rho(G)$ reflects the time needed for the docked pool D to respond to changes in G (see Eq. (15.18)), and when granule movement to and from the membrane is substantial, this time constant is mainly controlled by the reinternalization rate. A recent study using total internal reflection fluorescence imaging experiments has indeed suggested such frequent movement [73].

The IVGTT minimal model parameter X_0 corresponds to the amount of the RRP that is released when glucose increases to G_{max} (see Eq. (15.24)), and the first phase index $\Phi_1 = X_0/(G_{max}-G_b)$ is hence related to the function $h = \partial H/\partial G$. For the MTT/OGTT minimal model, we found a similar relation between the dynamic index Φ_D and the function h, again reflecting that the two minimal models share similarities because they are reflecting the same underlying biology. However, note that during the rising phase, which lasts ~60 min during an MTT/OGTT, mobilization, docking, and priming increase the RRP. This means that the dynamic term described with $h(G,t)$ in Eq. (15.25) increases over time. Hence, the dynamic index Φ_D reflects not only the size of the RRP at the basal state, but also how it changes because of granule maturation. These observations support the speculations in Section 15.3.4 on whether Φ_1 and Φ_D assess different aspects of beta-cell function.

The second phase index $\Phi_2 = \beta$ in Eq. (15.12) is approximately equal to $dY^*_\infty(G)/dG$ (Figure 15.6C). Using that $f \gg p^-$ and $\rho(G) \approx r$, we find that

$$\beta \approx dY^*_\infty(G)/dG \approx (p^+/r) \cdot d/dG[\Phi(G)M_\infty(G)] \qquad (15.26)$$

Hence, the second phase index $\Phi_2 = \beta$ reflects the combined effect of mobilization, cell recruitment, and the strength of priming versus reinternalization (i.e., the net effect of the processes that lead to an increased amount of readily releasable insulin). The same processes influence the static index Φ_S as discussed in Section 15.4.2.

Future work will address how dynamic signals such as Ca^{2+} interact with granule pool dynamics to shape the patterns of insulin secretion. The insight that will be obtained from these studies should be linked to the minimal models with multiscale modelling as outlined here. Multiscale modelling of glucagon secretion and incretins will also be important future subjects to address. Finally, the cellular models should, as far as possible, be built on data from human islets and beta-cells.

References

[1] Henquin JC. Regulation of insulin secretion: a matter of phase control and amplitude modulation. Diabetologia 2009;52(5):739−51.
[2] Gerich JE. Is reduced first-phase insulin release the earliest detectable abnormality in individuals destined to develop type 2 diabetes? Diabetes 2002;51(Suppl. 1): S117−21.
[3] Pørksen N. The *in vivo* regulation of pulsatile insulin secretion. Diabetologia 2002;45:3−20.
[4] Curry DL, Bennett LL, Grodsky GM. Dynamics of insulin secretion by the perfused rat pancreas. Endocrinology 1968;83(3):572−84.
[5] Grodsky GM. A threshold distribution hypothesis for packet storage of insulin and its mathematical modeling. J Clin Invest 1972;51:2047−59.
[6] Stagner JI, Samols E, Weir GC. Sustained oscillations of insulin, glucagon, and somatostatin from the isolated canine pancreas during exposure to a constant glucose concentration. J Clin Invest 1980;65:939−42.
[7] Gilon P, Shepherd RM, Henquin JC. Oscillations of secretion driven by oscillations of cytoplasmic $Ca2+$ as evidences in single pancreatic islets. J Biol Chem 1993;268 (30):22265−8.
[8] Hellman B, Gylfe E, Bergsten P, Grapengiesser E, Lund PE, Berts A, et al. Glucose induces oscillatory $Ca2+$ signalling and insulin release in human pancreatic beta cells. Diabetologia 1994;37(Suppl. 2):S11−20.
[9] Roper MG, Shackman JG, Dahlgren GM, Kennedy RT. Microfluidic chip for continuous monitoring of hormone secretion from live cells using an electrophoresis-based immunoassay. Anal Chem 2003;75(18):4711−7.
[10] Dishinger JF, Reid KR, Kennedy RT. Quantitative monitoring of insulin secretion from single islets of Langerhans in parallel on a microfluidic chip. Anal Chem 2009;81 (8):3119−27.
[11] Michael DJ, Xiong W, Geng X, Drain P, Chow RH. Human insulin vesicle dynamics during pulsatile secretion. Diabetes 2007;56(5):1277−88.
[12] Dyachok O, Idevall-Hagren O, Sågetorp J, Tian G, Wuttke A, Arrieumerlou C, et al. Glucose-induced cyclic AMP oscillations regulate pulsatile insulin secretion. Cell Metab 2008;8(1):26−37.
[13] Bertram R, Sherman A, Satin LS. Metabolic and electrical oscillations: partners in controlling pulsatile insulin secretion. Am J Physiol Endocrinol Metab 2007;293(4): E890−900.
[14] Tengholm A, Gylfe E. Oscillatory control of insulin secretion. Mol Cell Endocrinol 2009;297(1−2):58−72.
[15] Merrins MJ, Fendler B, Zhang M, Sherman A, Bertram R, Satin LS. Metabolic oscillations in pancreatic islets depend on the intracellular $Ca2+$ level but not $Ca2+$ oscillations. Biophys J 2010;99(1):76−84.

[16] Grodsky GM, Curry D, Landahl H, Bennett L. Further studies on the dynamic aspects of insulin release *in vitro* with evidence for a two-compartmental storage system. Acta Diabetol Lat 1969;6(Suppl. 1):554−78.

[17] Rorsman P, Renström E. Insulin granule dynamics in pancreatic beta cells. Diabetologia 2003;46(8):1029−45.

[18] Cerasi E, Fick G, Rudemo M. A mathematical model for the glucose induced insulin release in man. Eur J Clin Invest 1974;4(4):267−78.

[19] Eliasson L, Abdulkader F, Braun M, Galvanovskis J, Hoppa MB, Rorsman P. Novel aspects of the molecular mechanisms controlling insulin secretion. J Physiol 2008;586 (14):3313−24.

[20] Valdeolmillos M, Santos RM, Contreras D, Soria B, Rosario LM. Glucose-induced oscillations of intracellular Ca2+ concentration resembling bursting electrical activity in single mouse islets of Langerhans. FEBS Lett 1989;259:19−23.

[21] Henquin JC, Nenquin M, Stiernet P, Ahren B. *In vivo* and *in vitro* glucose-induced biphasic insulin secretion in the mouse: pattern and role of cytoplasmic Ca2+ and amplification signals in beta-cells. Diabetes 2006;55:441−51.

[22] Sherman A. Contributions of modeling to understanding stimulus-secretion coupling in pancreatic beta-cells. Am J Physiol 1996;271(2 Pt 1):E362−72.

[23] Pedersen MG. Contributions of mathematical modeling of beta cells to the understanding of beta-cell oscillations and insulin secretion. J Diabetes Sci Technol 2009;3(1):12−20.

[24] Bertram R, Sherman A. Negative calcium feedback: the road from Chay-Keizer. In: Coombes S, Bressloff PC, editors. Bursting: the genesis of rhythm in the nervous system. Singapore: World Scientific; 2005. p. 19−48.

[25] O'Connor MDL, Landahl H, Grodsky GM. Comparison of storage- and signal-limited models of pancreatic insulin secretion. Am J Physiol 1980;238:R378−89.

[26] Nesher R, Cerasi E. Modeling phasic insulin release: immediate and time-dependent effects of glucose. Diabetes 2002;51(Suppl. 1):S53−9.

[27] Henquin JC, Ishiyama N, Nenquin M, Ravier MA, Jonas JC. Signals and pools underlying biphasic insulin secretion. Diabetes 2002;51(Suppl. 1):S60−7.

[28] Daniel S, Noda M, Straub SG, Sharp GW. Identification of the docked granule pool responsible for the first phase of glucose-stimulated insulin secretion. Diabetes 1999;48(9):1686−90.

[29] Olofsson CS, Göpel SO, Barg S, Galvanovskis J, Ma X, Salehi A, et al. Fast insulin secretion reflects exocytosis of docked granules in mouse pancreatic B-cells. Pflugers Arch 2002;444(1−2):43−51.

[30] Dean PM, Matthews EK. Glucose-induced electrical activity in pancreatic islet cells. J Physiol 1970;210(2):255−64.

[31] Pedersen MG, Corradin A, Toffolo GM, Cobelli C. A subcellular model of glucose-stimulated pancreatic insulin secretion. Philos Transact A Math Phys Eng Sci 2008;366 (1880):3525−43.

[32] Cobelli C, Toffolo GM, Dalla Man C, Campioni M, Denti P, Caumo A, et al. Assessment of beta-cell function in humans, simultaneously with insulin sensitivity and hepatic extraction, from intravenous and oral glucose tests. Am J Physiol Endocrinol Metab 2007;293(1):E1−15.

[33] Eliasson L, Renström E, Ding WG, Proks P, Rorsman P. Rapid ATP-dependent priming of secretory granules precedes Ca(2+)-induced exocytosis in mouse pancreatic B-cells. J Physiol 1997;503(Pt 2):399−412.

[34] Jing X, Li DQ, Olofsson CS, Salehi A, Surve VV, Caballero J, et al. CaV2.3 calcium channels control second-phase insulin release. J Clin Invest 2005;115(1):146−54.

[35] Ivarsson R, Jing X, Waselle L, Regazzi R, Renström E. Myosin 5a controls insulin granule recruitment during late-phase secretion. Traffic 2005;6(11):1027–35.

[36] Barg S, Huang P, Eliasson I, Nelson DJ, Obermuller S, Rorsman P, et al. Priming of insulin granules for exocytosis by granular Cl(-) uptake and acidification. J Cell Sci 2001;114(Pt 11):2145–54.

[37] Bertuzzi A, Salinari S, Mingrone G. Insulin granule trafficking in beta-cells: mathematical model of glucose-induced insulin secretion. Am J Physiol Endocrinol Metab 2007;293(1):E396–409.

[38] Jonkers FC, Henquin JC. Measurements of cytoplasmic Ca2+ in islet cell clusters show that glucose rapidly recruits beta-cells and gradually increases the individual cell response. Diabetes 2001;50:540–50.

[39] Heart E, Corkey RF, Wikstrom JD, Shirihai OS, Corkey BE. Glucose-dependent increase in mitochondrial membrane potential, but not cytoplasmic calcium, correlates with insulin secretion in single islet cells. Am J Physiol Endocrinol Metab 2006;290:E143–8.

[40] Low JT, Mitchell JM, Do OH, Bax J, Rawlings A, Zavortink M, et al. Glucose principally regulates insulin secretion in mouse islets by controlling the numbers of granule fusion events per cell. Diabetologia 2013; [in press]

[41] Pedersen MG, Cobelli C. Multiscale modelling of insulin secretion during an intravenous glucose tolerance test. Interface Focus 2013;3:20120085

[42] Pedersen MG, Toffolo GM, Cobelli C. Cellular modeling: insight into oral minimal models of insulin secretion. Am J Physiol Endocrinol Metab 2010;298(3):E597–601.

[43] Chen Y, Wang S, Sherman A. Identifying the targets of the amplifying pathway for insulin secretion in pancreatic beta-cells by kinetic modeling of granule exocytosis. Biophys J 2008;95(5):2226–41.

[44] Pedersen MG, Sherman A. Newcomer insulin secretory granules as a highly calcium-sensitive pool. Proc Natl Acad Sci USA 2009;106(18):7432–6.

[45] Stamper IJ, Wang X. Mathematical modeling of insulin secretion and the role of glucose-dependent mobilization, docking, priming and fusion of insulin granules. J Theoret Biol 2013;318:210–25.

[46] Wan QF, Dong Y, Yang H, Lou X, Ding J, Xu T. Protein kinase activation increases insulin secretion by sensitizing the secretory machinery to Ca2+. J Gen Physiol 2004;124(6):653–62.

[47] Yang Y, Gillis KD. A highly Ca2+-sensitive pool of granules is regulated by glucose and protein kinases in insulin-secreting INS-1 cells. J Gen Physiol 2004;124(6):641–51.

[48] Ohara-Imaizumi M, Nishiwaki C, Kikuta T, Nagai S, Nakamichi Y, Nagamatsu S. TIRF imaging of docking and fusion of single insulin granule motion in primary rat pancreatic beta-cells: different behaviour of granule motion between normal and Goto-Kakizaki diabetic rat beta-cells. Biochem J 2004;381(1):13–8.

[49] Ohara-Imaizumi M, Aoyagi K, Nakamichi Y, Nishiwaki C, Sakurai T, Nagamatsu S. Pattern of rise in subplasma membrane Ca2+ concentration determines type of fusing insulin granules in pancreatic beta cells. Biochem Biophys Res Commun 2009;385(3):291–5.

[50] Pedersen MG. Insulin secretory granules enter a highly calcium-sensitive state following palmitate-induced dissociation from calcium channels: a theoretical study. J Neuroendocrinol 2010;22(12):1315–24.

[51] Hoppa MB, Collins S, Ramracheya R, Hodson L, Amisten S, Zhang Q, et al. Chronic palmitate exposure inhibits insulin secretion by dissociation of Ca(2+) channels from secretory granules. Cell Metab 2009;10(6):455–65.

[52] Pedersen MG. On depolarization-evoked exocytosis as a function of calcium entry: possibilities and pitfalls. Biophys J 2011;101(4):793−802.

[53] Pedersen MG, Cortese G, Eliasson L. Mathematical modeling and statistical analysis of calcium-regulated insulin granule exocytosis in β-cells from mice and humans. Prog Biophys Mol Biol 2011;107(2):257−64.

[54] Braun M, Ramracheya R, Bengtsson M, Zhang Q, Karanauskaite J, Partridge C, et al. Voltage-gated ion channels in human pancreatic beta-cells: electrophysiological characterization and role in insulin secretion. Diabetes 2008;57(6):1618−28.

[55] Braun M, Ramracheya R, Johnson PR, Rorsman P. Exocytotic properties of human pancreatic beta-cells. Ann NY Acad Sci 2009;1152:187−93.

[56] Zawalich WS, Zawalich KC. Effects of glucose, exogenous insulin, and carbachol on C-peptide and insulin secretion from isolated perifused rat islets. J Biol Chem 2002;277(29):26233−7.

[57] Polonsky KS, Rubenstein AH. C-peptide as a measure of the secretion and hepatic extraction of insulin. Pitfalls and limitations. Diabetes 1984;33(5):486−94.

[58] Eaton RP, Allen RC, Schade DS, Erickson KM, Standefer J. Prehepatic insulin production in man: kinetic analysis using peripheral connecting peptide behavior. J Clin Endocrinol Metab 1980;51(3):520−8.

[59] Van Cauter E, Mestrez F, Sturis J, Polonsky KS. Estimation of insulin secretion rates from C-peptide levels. Comparison of individual and standard kinetic parameters for C-peptide clearance. Diabetes 1992;41(3):368−77.

[60] Toffolo G, Grandi FD, Cobelli C. Estimation of beta-cell sensitivity from intravenous glucose tolerance test C-peptide data. Knowledge of the kinetics avoids errors in modeling the secretion. Diabetes 1995;44(7):845−54.

[61] Breda E, Cavaghan MK, Toffolo G, Polonsky KS, Cobelli C. Oral glucose tolerance test minimal model indexes of beta-cell function and insulin sensitivity. Diabetes 2001;50(1):150−8.

[62] Yabe D, Seino Y. Two incretin hormones GLP-1 and GIP: comparison of their actions in insulin secretion and cell preservation. Prog Biophys Mol Biol 2011;107(2):248−56.

[63] Hovorka R, Chassin L, Luzio SD, Playle R, Owens DR. Pancreatic beta-cell responsiveness during meal tolerance test: model assessment in normal subjects and subjects with newly diagnosed noninsulin-dependent diabetes mellitus. J Clin Endocrinol Metab 1998;83(3):744−50.

[64] Cretti A, Lehtovirta M, Bonora E, Brunato B, Zenti MG, Tosi F, et al. Assessment of beta-cell function during the oral glucose tolerance test by a minimal model of insulin secretion. Eur J Clin Invest 2001;31(5):405−16.

[65] Mari A, Schmitz O, Gastaldelli A, Oestergaard T, Nyholm B, Ferrannini E. Meal and oral glucose tests for assessment of beta-cell function: modeling analysis in normal subjects. Am J Physiol Endocrinol Metab 2002;283(6):E1159−66.

[66] Breda E, Toffolo G, Polonsky KS, Cobelli C. Insulin release in impaired glucose tolerance: oral minimal model predicts normal sensitivity to glucose but defective response times. Diabetes 2002;51(Suppl. 1):S227−33.

[67] Toffolo G, Breda E, Cavaghan MK, Ehrmann DA, Polonsky KS, Cobelli C. Quantitative indexes of beta-cell function during graded up&down glucose infusion from C-peptide minimal models. Am J Physiol Endocrinol Metab 2001;280(1):E2−10.

[68] Steil GM, Hwu CM, Janowski R, Hariri F, Jinagouda S, Darwin C, et al. Evaluation of insulin sensitivity and beta-cell function indexes obtained from minimal model analysis of a meal tolerance test. Diabetes 2004;53(5):1201−7.

[69] Weiss R, Caprio S, Trombetta M, Taksali SE, Tamborlane WV, Bonadonna R. Beta-cell function across the spectrum of glucose tolerance in obese youth. Diabetes 2005;54(6):1735−43.

[70] Dalla Man C, Micheletto F, Sathananthan A, Rizza RA, Vella A, Cobelli C. A model of GLP-1 action on insulin secretion in nondiabetic subjects. Am J Physiol Endocrinol Metab 2010;298(6):E1115−21.

[71] Pedersen MG, Dalla Man C, Cobelli C. Multiscale modeling of insulin secretion. IEEE Trans Biomed Eng 2011;58(10):3020−3.

[72] Herrero P, Georgiou P, Oliver N, Johnston DG, Toumazou C. A bio-inspired glucose controller based on pancreatic β-cell physiology. J Diabetes Sci Technol 2012;6 (3):606−16.

[73] Hatlapatka K, Matz M, Schumacher K, Baumann K, Rustenbeck I. Bidirectional insulin granule turnover in the submembrane space during K(+) depolarization-induced secretion. Traffic 2011;12(9):1166−78.

16 Glucose Modelling

Chiara Dalla Man and Claudio Cobelli

Department of Information Engineering, University of Padova, Padova, Italy

16.1 Introduction

Since the 1960s [1], several mathematical models have been proposed to describe the glucose system in different experimental conditions (e.g., in steady- and non-steady states), with and without glucose tracers [2–4]. These can be found in the original publications as well as in a number of books [2]. Among them, the minimal model of glucose kinetics [5] was very successful, having being used so far in hundreds of publications. It describes the interaction between glucose and insulin after an intravenous glucose tolerance test (IVGTT) (i.e., a glucose bolus injected intravenously (IV)) and provides an estimate of insulin sensitivity (i.e., an index quantifying the control exerted by insulin on glucose production and utilization). The choice of an IV test facilitates system modelling. In fact, during IVGTT, the system is perturbed by a known input. In contrast, oral glucose tolerance tests (OGTTs), which would be more appealing than the IV ones since they are more physiological and easier to perform, are more difficult to model since the input of the system (i.e., the rate of appearance of glucose in plasma (Ra)) is unknown.

The aim of this chapter is to present the development of glucose modelling in the last decade, mainly focusing on describing system behavior during an oral glucose test (e.g., the OGTT and the mixed meal tolerance test (MTT)). The models belong to two classes: models to measure (Section 16.2) and models to simulate (Section 16.3), which have different objectives and characteristics.

Models to measure are usually "minimal" (i.e., they have a simple structure and a reasonably small number of parameters to describe the system functions [3]). They are generally useful for the quantification of specific metabolic relationships in the system. In contrast, models to simulate are usually "maximal" (i.e., they are comprehensive descriptions attempting to fully reproduce the behavior of the system), and thus usually are large scale, are nonlinear, are of a high order, and have a large number of parameters. These models are generally useful for system simulation [3].

Modelling Methodology for Physiology and Medicine. DOI: http://dx.doi.org/10.1016/B978-0-12-411557-6.00016-1

16.2   Oral Glucose Minimal Models

A first attempt to use oral tests to estimate insulin sensitivity has been the integral method proposed in Ref. [6]. There, the authors overcome the lack of knowledge of Ra by integrating the minimal model equations under the assumption that Ra is an anticipated version of plasma glucose concentration and that glucose has come back to its basal state at the end of the experiment. This assumption renders the use of the methods difficult in impaired glucose tolerant and type 2 diabetic subjects, for which the above conditions are not met. This limitation can be overcome by switching from an integral to a differential approach by explicitly modelling Ra [7]. This approach is used in the oral glucose minimal model, which allows the estimation of the unknown Ra together with other model parameters, including net insulin sensitivity (S_I). Moreover, if the MTT/OGTT is labeled with a tracer, the so-called tracer oral minimal model can be used to assess disposal insulin sensitivity (S_I^*) [8].

The oral glucose minimal models [7,8] couple the classic minimal model of glucose kinetics [5] with a parametric description of Ra by solving the system input estimation problem as a parameter estimation problem.

16.2.1   Model Equations

The general formulation of the oral minimal model after an oral glucose perturbation (Figure 16.1) is

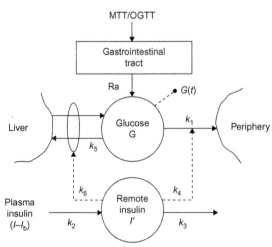

Figure 16.1 The oral glucose minimal model. Parameters k_i are linked to the parameters of Eq. (16.1) as follows: $S_G = k_1 + k_5$, $p_2 = k_3$, $p_3 = k_2 \cdot (k_4 + k_6)$. I' is remote insulin related to X as $X = k_2 \cdot (k_4 + k_6)I'$. Ra($t$) is the rate of appearance of the glucose in plasma after an oral ingestion. *Source*: Adapted from Ref. [7].

$$\begin{cases} \dot{G}(t) = -[S_G + X(t)] \cdot G(t) + S_G \cdot G_b + \dfrac{Ra(t, \alpha)}{V} & G(0) = G_b \\ \dot{X}(t) = -p_2 \cdot X(t) + p_3 \cdot [I(t) - I_b] & X(0) = 0 \end{cases} \tag{16.1}$$

where G is plasma glucose concentration, I is plasma insulin concentration, suffix "b" denotes basal (pre-test) values, X is insulin action on glucose production and disposal, V is distribution volume, and S_G, p_2, and p_3 are model parameters. Specifically, S_G is the fractional (i.e., per unit distribution volume) glucose effectiveness, measuring glucose ability *per se* to promote glucose disposal and inhibit glucose production; p_2 is the rate constant describing the dynamics of insulin action; and p_3 is the parameter governing the magnitude of insulin action like in Ref. [5]. The insulin sensitivity index is given by Ref. [7]:

$$S_I = \frac{p_3}{p_2} \cdot V \; (dL/kg/min \text{ per } \mu U/mL) \tag{16.2}$$

The model describes Ra as a piecewise-linear function with a given number of break points. Since the rate of glucose absorption varies more rapidly in the first portion of the test (Figure 16.3), intervals are shorter at the beginning and longer toward the end. A preliminary analysis indicated that eight intervals—and thus eight break points—are a good compromise between model flexibility and the number of parameters to be estimated from the data. The break points are allocated at 0, 13, 25, 35, 60, 90, 120, 180, and 420. The expression for Ra is thus given by

$$Ra(t, \alpha) = \begin{cases} \alpha_{i-1} + \dfrac{\alpha_i - \alpha_{i-1}}{t_i - t_{i-1}} \cdot (t - t_{i-1}) & \text{for } t_{i-1} \leq t < t_i, \; i = 1, \ldots, 8 \\ 0 & \text{otherwise} \end{cases} \tag{16.3}$$

In the original work [7], three different parametric descriptions of Ra were proposed: piecewise linear, spline, and convolution. No differences were found either in the ability of fitting the data, in the reconstruction of Ra, or in the estimated insulin sensitivity. Given its simplicity, the piecewise-linear model was used thereafter.

If the oral glucose load is labeled with a glucose tracer, the exogenous glucose (G_e) (i.e., the glucose concentration due to MTT/OGTT only) can be calculated (Eq. (16.7)) and a labeled version of the oral minimal model, OMM* (Figure 16.2), can be identified on G_e data, to derive S_I^* (i.e., the effect of insulin on glucose disposal only) and Ra ([8]). The model equations are

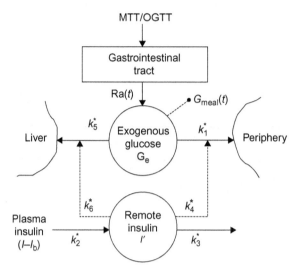

Figure 16.2 The labeled oral glucose minimal model. Parameters k_i^* are linked to the parameters of Eq. (16.4) as follows: $S_G^* = k_1^* + k_5^*$, $p_2^* = k_3^*$, $p_3^* = k_2^* (k_4^* + k_6^*)$. I' is remote insulin related to X^* as $X^* = k_2^* (k_4^* + k_6^*)I'$. Ra(t) is the rate of appearance of the glucose in plasma after an oral ingestion. *Source*: Adapted from Ref. [8].

$$
\begin{cases}
\dot{G}_e(t) = -\left[S_G^* + X^*(t)\right] \cdot G_e(t) + \dfrac{\mathrm{Ra}(t, \alpha)}{V} & G_e(0) = 0 \\[2mm]
\dot{X}^*(t) = -p_2^* \cdot X^*(t) + p_3^* \cdot [I(t) - I_b] & X(0) = 0
\end{cases}
\tag{16.4}
$$

where S_G^* is fractional glucose effectiveness measuring glucose ability *per se* to promote glucose disposal, X^* insulin action on glucose disposal, with p_2^* and p_3^* rate constants describing respectively its dynamics and magnitude. OMM* allows for the estimation of Ra and insulin sensitivity on glucose disposal, S_I^* defined as

$$
S_I^* = \frac{p_3^*}{p_2^*} \cdot V \; (\mathrm{dL/kg/min \; per \; \mu U/mL})
\tag{16.5}
$$

Ra is that already described for OMM (Eq. (16.3)).

Table 16.1 Reference Values of OMM and OMM* Nonidentifiable Parameters for MTT and OGTT

Parameters	MTT		OGTT	
	Value	Unit	Value	Unit
V	1.45	dL/kg	1.34	dL/kg
S_G	0.025	min^{-1}	0.028	min^{-1}
V^*	1.60	dL/kg	1.48	dL/kg
S_G^*	0.0118	min^{-1}	0.0067	min^{-1}
f	0.90	Dimensionless	0.87	Dimensionless

16.2.2 Model Identification

16.2.2.1 Identifiability

OMM and OMM* are *a priori* nonidentifiable [7,8]. The *a priori* knowledge necessary for their identification can be found in Table 16.1. If tracer glucose data are not available, and thus OMM is identified alone, a Bayesian prior on p_2 is needed to improve numerical identifiability of the model. On the other hand, if the MTT/OGTT is labeled, OMM and OMM* can be simultaneously identified. Since they share the same Ra, this allows for relaxing the necessity of using Bayesian priors for p_2 and p_2^*. A constraint is also imposed to guarantee that the area under the estimated Ra equals the total amount of ingested glucose, D, multiplied by the fraction that is actually absorbed, f (Table 16.1), and divided by the body weight (BW):

$$\int_0^\infty \mathrm{Ra}(t)\mathrm{d}t = \frac{Df}{\mathrm{BW}} \tag{16.6}$$

By solving Eq. (16.6) for one of the α_i describing Ra, the number of unknowns is reduced by 1.

Finally, oral tracer measurements can provide information as to when Ra begins to rise in each subject: if tracer concentration is 0 up to time t_i and is different from 0 at time t_{i+1}, then one can safely assume that Ra is 0 up to t_i.

16.2.2.2 Parameter Estimation

OMM and OMM* are usually numerically identified by nonlinear least squares (see Chapter 5) (e.g., implemented in SAAM II (Simulation Analysis and Modelling software [9] or MATLAB™)). Measurement error on glucose data is assumed to be independent, Gaussian, with zero mean and known standard deviation (CV = 2%); error on exogenous glucose data must be calculated by propagating the error on oral tracer measurements. Insulin concentration is the model forcing function and is assumed to be known without error.

16.2.3 Validation Against the Oral Reference Model Method

The main limitation of OMM and OMM* is that there is the need to fix some parameters to population values. It is thus important to assess the validity of S_I, S_I^*, and Ra profile with those obtained with an independent method. Here we discuss the oral reference model method. This consists in the use of the classic minimal model [5] driven, in each subject, by a model-independent estimate of Ra (Ra^{ref}). Ra^{ref} can be obtained by the triple tracer protocol [10], as briefly described below.

16.2.3.1 The Triple Tracer Protocol

Each subject undergoing to a triple tracer protocol receives a mixed meal labeled with [1-^{13}C]-glucose (G^*), thus allowing one to derive the exogenous (i.e., coming from the meal) glucose (G_e) as

$$G_e = G^* \cdot \left(1 + \frac{1}{z_{meal}}\right) \tag{16.7}$$

where z_{meal} is the tracer-to-tracee ratio in the meal.

In addition, two tracers ($[6,6\text{-}^2H_2]$-glucose and $[6\text{-}^3H]$-glucose) are infused IV at variable rate mimicking, respectively, the endogenous glucose production (EGP) and Ra. For OMM validation, only two of the three tracer plasma concentrations are needed, that is, the $[1\text{-}^{13}C]$-glucose labeling the meal and the $[6\text{-}^3H]$-glucose mimicking Ra.

The rate of appearance of the meal tracer (Ra^{*ref}) is calculated using Steele's (one-compartment) or Radziuk's (two-compartment) equation [10,11]. For instance, the one-compartment equation is

$$Ra^{*ref} = \frac{F_{3H}(t)}{SA(t)} - p \cdot V \cdot \frac{G^*(t)}{SA(t)} \cdot \frac{dSA(t)}{dt} \tag{16.8}$$

where F_{3H} is the infusion rate of $[6\text{-}^3H]$-glucose, $SA = [6\text{-}^3H]\text{-glucose}/G^*$ is the specific activity, p is the pool fraction (equal to 0.65), and V is the distribution volume of glucose.

Given that SA is maintained almost constant (clamped), Eq. (16.8) provides an essentially model-independent estimate of the Ra^{*ref}[10,11]. In fact, if SA is well-clamped, a two-compartment model would give virtually the same results. However, calculation of the derivative of SA is not trivial, since SA is the ratio of two variables affected by noise. In Ref. [10], the SA profile was smoothed using stochastic regularized deconvolution [12] (see Chapter 3 for details), which also provides the first derivative of the smoothed signal. Ra^{ref} can finally be calculated as

$$Ra^{ref} = Ra^{*ref} \cdot \left(1 + \frac{1}{z_{meal}}\right) \tag{16.9}$$

16.2.3.2 The Reference Models

Ra^{ref} is used both to validate the estimated Ra and as known input of the reference model (RM) and labeled reference (RM*) models [8,13]:

$$\begin{cases} \dot{G}(t) = -\left[S_G^{ref} + X(t)\right] \cdot G(t) + S_G^{ref} \cdot G_b + \dfrac{Ra^{ref}(t)}{V} & G(0) = G_b \\ \dot{X}(t) = -p_2^{ref} \cdot X(t) + p_3^{ref} \cdot [I(t) - I_b] & X(0) = 0 \end{cases} \tag{16.10}$$

$$\begin{cases} \dot{G}_e(t) = -\left[S_G^{*\text{ref}} + X^*(t)\right] \cdot G_e(t) + \dfrac{\text{Ra}^{\text{ref}}(t)}{V} & G_e(0) = 0 \\ \dot{X}^*(t) = -p_2^{*\text{ref}} \cdot X^*(t) + p_3^{*\text{ref}} \cdot [I(t) - I_b] & X^*(0) = 0 \end{cases} \tag{16.11}$$

Identification of RM and RM* provided reference values for OMM and OMM* parameters (Section 16.2.2), denoted, as in Eqs. (16.10) and (16.11), with a superscript "ref."

16.2.3.3 Results

Validation was performed using data of 88 healthy subjects who underwent a triple tracer mixed meal protocol [8,13]. The OMM and OMM* described plasma glucose and exogenous glucose satisfactory. S_I and S_I^* are estimated with good precision (CV~5−10%). They also agree with their reference counterpart ($S_I = 11.68 \pm 0.73$ vs. $S_I^{\text{ref}} = 11.55 \pm 0.68 \times 10^{-4}$ dL/kg/min per μU/mL, $P =$ ns, $R = 0.86$, $P < 0.0001$; $S_I^* = 9.64 \pm 0.80$ vs. $S_I^{*\text{ref}} = 9.24 \pm 0.63 \times 10^{-4}$ dL/kg/min per μU/mL, $P =$ ns, $R = 0.80$, $P < 0.0001$). The mean profile of Ra provided by the triple tracer protocol is comparable with the reference one (Figure 16.3).

16.2.4 Validation Against the Euglycemic−Hyperinsulinemic Clamp Method

The next model validation step consisted in the comparison of model derived S_I and S_I^* with those obtained with the euglycemic−hyperinsulinemic clamp [14].

16.2.4.1 The Labelled Euglycemic−Hyperinsulinemic Clamp

The labeled euglycemic−hyperinsulinemic clamp consists of a tracer equilibration period ($t = 0−120$ min) and a glucose clamp period ($t = 120−300$ min). During the

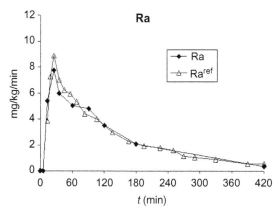

Figure 16.3 Average ($N = 88$) Ra estimated with OMM and OMM* (black diamonds) versus Ra^{ref} estimated with the triple tracer protocol (white triangles). *Source:* Adapted from Ref. [8].

tracer equilibration period a primed continuous infusion of [^{2}H$_2$]-glucose is administered and plasma samples are frequently collected to measure plasma glucose, tracer glucose, and insulin concentrations. Insulin is infused at 25 mU/m^2 min from time 120 to 300 min. A variable intravenous infusion of 20% glucose, containing 1.4% of [^{2}H$_2$]-glucose, is also administered to maintain glucose concentration at its basal level in each individual; the infusion is adjusted according to the arterialized glucose concentration determined every 5 min.

Insulin sensitivity (S_I^{clamp}) is calculated from plasma glucose and insulin concentrations and glucose rate of infusion as

$$S_I^{clamp} = \frac{GIR_{ss}}{G_{ss} \cdot \Delta I} \tag{16.12}$$

where GIR_{ss} is the steady-state (average in the last 40 min) glucose infusion rate, G_{ss} the steady-state glucose concentration, and ΔI the difference between end-test and basal insulin concentration.

Similarly, disposal insulin sensitivity is calculated from [^{2}H$_2$]-glucose and insulin plasma concentrations, and from tracer glucose rate of infusion as

$$S_I^{*clamp} = \frac{GIR_{ss}^*}{z_{ss} \cdot G_{ss} \cdot \Delta I} \tag{16.13}$$

where GIR_{ss}^* (mg/kg/min) is the difference between steady-state (average in the last 40 min) tracer infusion rate at the end of the clamp period and during the tracer equilibration period, z_{ss} is the end-test tracer-to-tracee ratio (i.e., [^{2}H$_2$]-glucose/glucose), G_{ss} is the steady-state glucose concentration, and ΔI is the difference between end-test and basal insulin concentration.

16.2.4.2   Results

S_I and S_I^* were compared to clamp indices in 21 subjects with different degrees of glucose tolerance [15]. Oral models and clamp insulin sensitivity were well correlated: $R = 0.81$, $P < 0.001$ for S_I versus S_I^{clamp} and $R = 0.70$, $P < 0.001$ for S_I^* versus S_I^{*clamp}. S_I was lower than S_I^{clamp} by 34%: 8.08 versus 13.66×10^{-4} dL/kg/min per μU/mL ($P = 0.0002$), while S_I^* was similar to S_I^{*clamp}: 8.17 versus 8.84×10^{-4} dL/kg/min per μU/mL ($P = 0.52$).

The correlation shown between OMM and clamp S_I is similar to that observed between IVGTT minimal model S_I and clamp [16]. The lower S_I values versus the clamp have also been reported for IVGTT. The reason for this discrepancy can be found either in the inadequacy of the minimal model description of glucose kinetics (single pool), and/or in the different experimental conditions of the two tests and/or in the possible nonlinearity of insulin action. However, insulin action on glucose disposal estimated by OMM*, S_I^*, was virtually identical to that obtained during the clamp, S_I^{*clamp}. This may indicate that the glucose disposal component of OMM is more correctly described than the glucose production one. Thus, a possible explanation for the 34% underestimation of OMM S_I (in comparison to S_I^{clamp}) might be an inadequate description, in the OMM, of the control of glucose and insulin on EGP. This potential inaccuracy may also explain the physiologically implausible finding that S_I^* was greater than S_I in a significant percentage of subjects, during MTT, OGTT, and IVGTT studies [17−19].

16.2.5 Comparison with IVGTT

The 88 subjects also received a labeled IVGTT [20], thus making it possible to compare oral and IV estimates of net and disposal insulin sensitivity. Estimates of both net and disposal insulin sensitivity provided by OMM and OMM*, respectively, were compared with those provided by IVGTT models. Albeit the tests are different and performed on different days, the measures correlate well: $r = 0.65$, $P < 0.0001$ for S_I and $r = 0.67$, $P < 0.0001$ for S_I^*; however, average values were found significantly different in both cases ($S_I^{IVGTT} < S_I$; $S_I^{*IVGTT} < S_I^*$).

16.2.6 Reduced Oral Protocols

One of the advantages of OMM is the possibility of applying the "differential equation" approach to a reduced oral test protocol in which the observation period, originally of 300 and 420 min for OGTT and MTT, respectively, was shortened to 180 or 120 min with a consequent reduction of the number of samples from 11 in OGTT and 21 in MTT, to 8 or 7. The major obstacle to overcome when applying OMM to a short protocol is that there is the need to know the fraction of the ingested dose which is absorbed in the first 180 or 120 min (f_{180}, f_{120}) to correctly apply the constraint of Eq. (16.6). Unfortunately, in the literature, the knowledge on these parameters is scarce also due to their high intra- and intersubject variabilities. The need for fixing these values in each subject was solved by using a strategy, often used in chemotherapy pharmacokinetics, that is, to assume that from 120 min onward—beyond the shortened observation period—Ra decays exponentially at a known rate [21]:

$$
Ra(t) = \begin{cases} \alpha_{i-1} + \dfrac{\alpha_i - \alpha_{i-1}}{t_i - t_{i-1}} \cdot (t - t_{i-1}) & \text{for } t_{i-1} \le t \le t_i,\ i = 1,\ldots,5,\ \text{and } t \le 120 \text{ min} \\ \alpha_5 \cdot e^{-(t-t_5)/T} & \text{for } t > 120 \text{ min} \end{cases}
$$

$$(16.14)$$

where the break points were chosen at 0, 10, 30, 60, 90, and 120 min and T is the time constant of the exponential function.

As a matter of fact, the area under Ra for $t > 120$ min only depends on α_5 and T with α_5 estimated from 0 to 120 min data and T fixed to a population value, depending on oral dose composition ($T = 120$ min for MTT, $T = 60$ min for OGTT):

$$
\int_{120}^{\infty} Ra_{meal}(t)dt = \alpha_5 T
$$

$$(16.15)$$

16.2.6.1 Comparison with Full Protocol

S_I estimated from a reduced protocol (120 min and 7 samples) was compared with that obtained with the full protocol in healthy subjects studied with MTT, healthy and impaired glucose tolerant subjects studied with OGTT, and type 2 diabetic subjects studied with MTT.

16.2.6.1.1 MTT in Healthy Subjects

Insulin sensitivity identified from full (420 min and 21 samples) and reduced (120 min and 7 samples) protocols compared very well: $S_I^{420\text{-}21} = 11.86 \pm 0.72$; $S_I^{120\text{-}7} = 11.38 \pm 0.83 \times 10^{-4}$ dL/kg/min per μU/mL (not significantly different by Wilcoxon signed rank test). They also were strongly correlated: $R = 0.86$, $P < 0.0001$ for 420-21 versus 120-7.

16.2.6.1.2 OGTT in Healthy and Impaired Glucose Tolerant Subjects

Insulin sensitivity identified from full (300 min and 11 samples) and reduced (120 min and 7 samples) protocols compared very well: $S_I^{300\text{-}11} = 13.66 \pm 0.87$; $S_I^{120\text{-}7} = 14.04 \pm 1.02 \times 10^{-4}$ dL/kg/min per μU/mL (not significantly different by Wilcoxon signed rank test). They also correlated very strongly: $R = 0.89$, $P < 0.0001$.

16.2.6.1.3 Meal in Diabetic Subjects

Estimation of insulin sensitivity in diabetic patients is more difficult even with a full protocol due to the scarce *a priori* knowledge on the nonidentifiable parameters of the model. When reduced protocols were tested in type 2 diabetics, results were less satisfactory than those obtained in normal and impaired glucose tolerant subjects. Insulin sensitivity identified from full (360-20) and reduced protocol of 240 min and 9 samples (240-9) compared well with the full protocol S_I: $S_I^{360\text{-}20} = 5.49 \pm 0.70$; $S_I^{240\text{-}9} = 4.87 \pm 0.78 \times 10^{-4}$ dL/kg/min per μU/mL (not significantly different by Wilcoxon signed rank test); they also correlated well: $R = 0.90$, $P < 0.0001$. Comparison between full and 180−8 was not completely satisfactory: albeit S_I was similar on average, $S_I^{360\text{-}20} = 5.49 \pm 0.70$; $S_I^{180\text{-}8} = 5.38 \pm 0.79 \times 10^{-4}$ dL/kg/min per μU/mL (not significantly different by Wilcoxon signed rank test), the correlation between the two was not excellent: $R = 0.74$, $P < 0.0001$. Finally, it was not possible to estimate accurately S_I with the 120-7 protocol, since in many subjects the estimate collapsed to zero and in the remaining subjects the estimates were not precise (mean CV $= 344 \pm 144\%$); this is likely due to the much slower dynamics of glucose and insulin concentrations, which, in type 2 diabetics, don't reach their peak before $t = 120$.

16.2.7 EGP: Lesson from the Oral Minimal Models

Theoretically, using the unlabeled and labeled models described above to measure, respectively, net (i.e., periphery+liver) and peripheral indices, one should be able to calculate liver indices from the difference between the two.

However, as already anticipated in Section 16.2.4, liver indices derived in this way are often unreliable (negative). This was found also during an IVGTT using the classic minimal model [17–19]. Caumo et al. [17] suggested that the above inconsistencies were symptoms of model error related, for example, to the description of glucose and insulin effect on EGP. In fact, both the IVGTT and the oral minimal models assume that insulin action on the liver has the same time course of insulin action on glucose disposal. Moreover, EGP suppression includes a term linearly dependent on glucose and a term equal to the product of glucose concentration and insulin action (i.e., it assumes that the effect of insulin on the liver is glucose mediated). The EGP model incorporated in the minimal model and alternative descriptions have been assessed against virtually model-independent estimate of EGP provided by a triple tracer meal protocol [10] to test their ability to accurately describe the known EGP profile and, thereby, to simultaneously assess liver insulin sensitivity (S_I^L) and glucose effectiveness (GE^L).

The model better describing the EGP time course according to Ref. [22] follows.

16.2.7.1 The Model

According to Ref. [22], EGP can be described by the following model:

$$EGP(t) = EGP_b - k_G \cdot [G(t) - G_b] - X^L(t) - X^{Der}(t) \quad EGP(0) = EGP_b$$

(16.16)

where EGP_b is basal endogenous glucose production and k_G is liver glucose effectiveness.

X^L is liver insulin action, defined as

$$\begin{cases} \dot{X}^L(t) = -k_1 \cdot X^L(t) + k_1 \cdot X'(t) & X^L(0) = 0 \\ \dot{X}'(t) = -k_1 \cdot [X'(t) - k_2 \cdot (I(t) - I_b)] & X'(0) = 0 \end{cases}$$

(16.17)

with k_1 accounting for the delay of liver insulin action versus plasma insulin and k_2 a parameter governing its efficacy. X^{Der} is a surrogate of portal insulin, which anticipates insulin and glucose patterns, and was demonstrated to significantly improve model ability to fit the rapid suppression of EGP occurring immediately after a meal:

$$X^{Der}(t) = \begin{cases} k_{GR} \cdot \dfrac{dG(t)}{dt} & \text{if } \dfrac{dG(t)}{dt} \geq 0 \\ 0 & \text{if } \dfrac{dG(t)}{dt} < 0 \end{cases}$$

(16.18)

where k_{GR} is a parameter governing the magnitude of glucose derivative control.

An index of liver insulin sensitivity (S_I^L) can be derived from model parameters as follows:

$$S_I^L = -\left.\frac{\partial EGP}{\partial I}\right|_{ss} \cdot \frac{1}{G_b} = \frac{k_2}{G_b} \tag{16.19}$$

where the symbol $|_{ss}$ indicates that the derivative of EGP is calculated in steady state.

The model has been assessed against data of 20 healthy subjects [22].

16.2.7.2 Validation Against Euglycemic—Hyperinsulinemic Clamp

The model was validated by comparing S_I^L with liver insulin sensitivity measured with a labeled euglycemic—hyperinsulinemic clamp (S_I^{Lclamp}) in 24 subjects with different degrees of glucose tolerance [23]. S_I^{Lclamp} is derived from total and peripheral indices of Eqs. (16.12) and (16.13) as

$$S_I^{Lclamp} = S_I^{clamp} - S_I^{*\ clamp} \tag{16.20}$$

Correlation between S_I^{Lclamp} and S_I^L was good ($r = 0.72$, $P < 0.0001$) with S_I^{Lmeal} being lower than S_I^{Lclamp} (4.60 ± 0.64 vs. $8.73 \pm 1.07 \times 10^{-4}$ dL/kg/min per μU/mL, $P < 0.01$). It is noteworthy that the correlation improved to 0.80, $P < 0.001$ if the normal fasting glucose subjects were considered ($n = 15$), while correlation was lower in the impaired fasting glucose subjects ($r = 0.56$, $P = 0.11$) with a nonsignificant P value likely due to the limited sample size.

16.3 Oral Glucose Maximal Models

This class of large-scale models is an important research tool to test theories, to assess the empirical validity of models, to serve as teaching aid, or to perform *in silico* experiments. In particular, the use of these models to simulate the system can be of enormous value. In fact, it is not always possible, convenient, or ethically appropriate to perform an experiment on the glucose system in humans. In such cases, simulation offers an alternative way of *in silico* experimenting on the system.

In this section, we describe in detail a whole-body meal model of glucose—insulin—glucagon interaction, whose peculiarity is its ability to predict the interperson variability observed in the desired population (e.g., healthy or type 1 diabetic subjects). We start with the healthy state simulator, since it was the one first proposed [24] to reproduce the data of a mixed meal in a population of 204 healthy subjects and, more recently, updated to account for hypoglycemia and

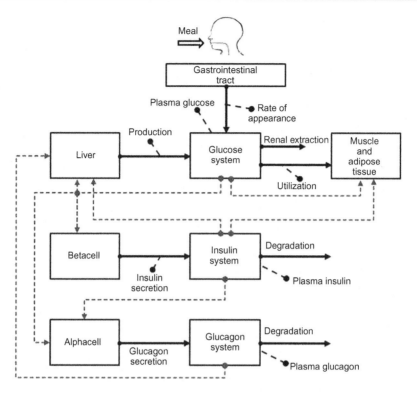

Figure 16.4 A scheme of the model of the glucose−insulin−glucagon system included in the healthy state simulator. Continuous lines denote material fluxes, while dashed lines are control signals. Input and measurements are indicated by an arrow and bulleted dashed lines, respectively.

counterregulation [25]. Then, we describe the type 1 diabetes simulator, which was accepted by the FDA as a substitute for preclinical (animal) trials for certain insulin treatments, including the artificial pancreas [26,27].

16.3.1 Meal Healthy State Simulator

A scheme of the glucose regulatory control system during a meal is shown in Figure 16.4, which puts in relation the measured plasma concentrations (i.e., glucose, insulin, and glucagon), the glucose fluxes (i.e., glucose rate of appearance, production, utilization, and renal extraction), and the insulin and glucagon fluxes (i.e., secretion and degradation).

Glucose kinetics is described with a two-compartment model [28]:

$$\begin{cases} \dot{G}_p(t) = \text{EGP}(t) + \text{Ra}(t) - U_{ii}(t) - E(t) - k_1 \cdot G_p(t) + k_2 \cdot G_t(t) & G_p(0) = G_{pb} \\ \dot{G}_t(t) = -U_{id}(t) + k_1 \cdot G_p(t) - k_2 \cdot G_t(t) & G_t(0) = G_{tb} \\ G(t) = \dfrac{G_p}{V_G} & G(0) = G_b \end{cases}$$

$$(16.21)$$

where G_p is glucose mass in plasma and rapidly equilibrating tissues, and G_t in slowly equilibrating tissues, respectively, G is plasma glucose concentration, suffix b denotes basal state, EGP is endogenous glucose production, Ra is glucose rate of appearance in plasma, E is renal excretion, U_{ii} and U_{id} are insulin-independent and insulin-dependent glucose utilizations, respectively, V_G is distribution volume of glucose, and k_1 and k_2 are rate parameters.

At basal steady-state, endogenous production, EGP_b, equals glucose disappearance (i.e., the sum of glucose utilization and renal excretion, which is virtually zero in normal subjects), $U_b + E_b$.

Parameter values of V_G, k_1, k_2 are reported in Table 16.2 (glucose kinetics).

Insulin kinetics is described with a two-compartment model [29]:

$$\begin{cases} \dot{I}_l(t) = -(m_1 + m_3(t)) \cdot I_l(t) + m_2 I_p(t) + S(t) & I_l(0) = I_{lb} \\ \dot{I}_p(t) = -(m_2 + m_4) \cdot I_p(t) + m_1 \cdot I_l(t) & I_p(0) = I_{pb} \\ I(t) = \dfrac{I_p}{V_I} & I(0) = I_b \end{cases}$$

$$(16.22)$$

where I_p and I_l are insulin masses in plasma and the liver, respectively, I is plasma insulin concentration, S is insulin secretion, V_I is distribution volume of insulin, and m_1, m_2, and m_4 are rate parameters. Degradation, D, occurs both in the liver and in the periphery. Peripheral degradation has been assumed to be linear (m_4). Hepatic extraction of insulin, HE (i.e., the insulin flux that leaves the liver irreversibly divided by the total insulin flux leaving the liver), is time varying and, according to Ref. [30], is linked to insulin secretion, S, as

$$\text{HE}(t) = -m_5 \cdot S(t) + m_6 \quad \text{HE}(0) = \text{HE}_b \qquad (16.23)$$

Thus one has

$$m_3(t) = \frac{\text{HE}(t) \cdot m_1}{1 - \text{HE}(t)} \qquad (16.24)$$

In addition, given that the liver is responsible for 60% of insulin clearance in steady state, one has

Table 16.2 Average Parameter Values of the Healthy State Simulator

Processes	Parameter	Value	Unit
Glucose kinetics	V_G	1.88	dL/kg
	k_1	0.065	min^{-1}
	k_2	0.079	min^{-1}
Insulin kinetics	V_I	0.05	L/kg
	m_1	0.190	min^{-1}
	m_2	0.484	min^{-1}
	m_4	0.194	min^{-1}
	m_5	0.0304	min per kg/pmol
	m_6	0.6471	Dimensionless
	HE_b	0.6	Dimensionless
Rate of appearance	k_{max}	0.0558	min^{-1}
	k_{min}	0.0080	min^{-1}
	k_{abs}	0.057	min^{-1}
	k_{gri}	0.0558	min^{-1}
	f	0.90	Dimensionless
	a	0.00013	mg^{-1}
	b	0.82	Dimensionless
	c	0.00236	mg^{-1}
	d	0.010	Dimensionless
Endogenous	k_{p1}	2.70	mg/kg/min
Production	k_{p2}	0.0021	min^{-1}
	k_{p3}	0.009	mg/kg/min per pmol/L
	k_{p4}	0.0618	mg/kg/min per pmol/kg
	k_i	0.0079	min^{-1}
	ς	0.05	mg/kg/min per (ng/L)
	k_H	0.093	min^{-1}
Utilization	F_{cns}	1	mg/kg/min
	V_{m0}	2.50	mg/kg/min
	V_{mx}	0.047	mg/kg/min per pmol/L
	K_{m0}	225.59	mg/kg
	p_{2U}	0.0331	min^{-1}
	r_1	2.10	Dimensionless
	r_2	1.084	Dimensionless
Secretion	K	2.30	pmol/kg per (mg/dL)
	α	0.050	min^{-1}
	β	0.11	pmol/kg/mirn per (mg/dL)
	γ	0.5	min^{-1}
Renal excretion	k_{e1}	0.0005	min^{-1}
	k_{e2}	339	mg/kg
Glucagon kinetics	n	0.22	min^{-1}
Glucagon	σ	0.41	ng/L/min
Secretion	δ	3.01	ng/L per (mg/dL)
	ρ	0.86	min^{-1}

$$m_2 = \left(\frac{S_b}{I_{pb}} - \frac{m_4}{1 - HE_b}\right) \cdot \frac{1 - HE_b}{HE_b} \tag{16.25}$$

$$m_4 = \frac{2}{5} \cdot \frac{S_b}{I_{pb}} \tag{16.26}$$

with S_b and D_b standing for basal secretion and degradation, respectively.

HE_b was fixed to 0.6 and is reported together with V_I, m_1, m_2, m_4, m_5, and m_6 in Table 16.2 (insulin kinetics).

The functional description of EGP in terms of plasma glucose, insulin, and glucagon is reported in Ref. [25]; it comprises a plasma glucose signal, a delayed and an anticipated versus plasma insulin signal, and a delayed (vs. plasma) glucagon signal:

$$
\begin{aligned}
EGP(t) &= k_{p1} - k_{p2} \cdot G_p(t) - k_{p3} \cdot I_d(t) \\
&\quad + k_{p4} \cdot I_{po}(t) + \xi \cdot X^H(t) \\
EGP(0) &= EGP_b
\end{aligned}
\tag{16.27}
$$

where I_{po} is the amount of insulin in the portal vein and I_d is a delayed insulin signal realized with a chain of two compartments:

$$
\begin{cases}
\dot{I}_1(t) = -k_i \cdot [I_1(t) - I(t)] & I_1(0) = I_b \\
\dot{I}_d(t) = -k_i \cdot [I_d(t) - I_1(t)] & I_d(0) = I_b
\end{cases}
\tag{16.28}
$$

and X^H is the delayed glucagon signal on EGP:

$$
\begin{cases}
\dot{X}^H(t) = -k_H \cdot X^H(t) + k_H \cdot \max[(H(t) - H_b), 0] \\
X^H(0) = 0
\end{cases}
\tag{16.29}
$$

where H is plasma glucagon concentration, k_{p1} is the extrapolated EGP at zero glucose and insulin, k_{p2} liver glucose effectiveness, k_{p3} parameter governing amplitude of insulin action on the liver, k_{p4} parameter governing amplitude of portal insulin action on the liver and k_i rate parameter accounting for delayed plasma insulin action, ξ parameter governing amplitude of glucagon action, and k_H rate parameter accounting for delay between plasma glucagon and action. EGP is also constrained to be nonnegative.

A physiological model of glucose intestinal absorption is also incorporated in the model [31]. Briefly, it describes the glucose transit through the stomach and intestine by assuming the stomach to be represented by two compartments, while a single compartment is used to describe the gut:

$$\begin{cases} Q_{sto}(t) = Q_{sto1}(t) + Q_{sto2}(t) & Q_{sto}(0) = 0 \\ \dot{Q}_{sto1}(t) = -k_{gri} \cdot Q_{sto1}(t) + D \cdot \delta(t) & Q_{sto1}(0) = 0 \\ \dot{Q}_{sto2}(t) = -k_{empt}(Q_{sto}) \cdot Q_{sto2}(t) + k_{gri} \cdot Q_{sto1}(t) & Q_{sto2}(0) = 0 \\ \dot{Q}_{gut} = -k_{abs} \cdot Q_{gut}(t) + k_{empt}(Q_{sto}) \cdot Q_{sto2}(t) & Q_{gut}(0) = 0 \\ Ra(t) = \dfrac{f \cdot k_{abs} \cdot Q_{gut}(t)}{BW} & Ra(0) = 0 \end{cases} \qquad (16.30)$$

where Q_{sto} is amount of glucose in the stomach (solid phase, Q_{sto1}, and liquid phase, Q_{sto2}), Q_{gut} glucose mass in the intestine, k_{gri} rate constant of grinding, $k_{empt}(Q_{sto})$ rate constant of gastric emptying assessed to be a nonlinear function of Q_{sto}, k_{abs} rate constant of intestinal absorption, f fraction of intestinal absorption that actually appears in plasma, D amount of ingested glucose, BW body weight, and Ra rate of glucose appearance in plasma.

The model describing glucose utilization by body tissues during a meal (both insulin-independent and insulin-dependent) is based on the literature reports [32−35] and assumes that glucose utilization is made up of two components. Insulin-independent utilization takes place in the first compartment, is constant, and represents glucose uptake by the brain and erythrocytes (F_{cns}):

$$U_{ii}(t) = F_{cns} \qquad (16.31)$$

Insulin-dependent utilization takes place in the remote compartment and depends nonlinearly (Michaelis Menten) on glucose in the tissues [34,35]:

$$\text{Uid}(t) = \frac{[V_{m0} + V_{mx} \cdot X(t) \cdot (1 + r_1 \cdot \text{risk})] \cdot G_t(t)}{K_{m0} + G_t(t)} \qquad (16.32)$$

with

$$\dot{X}(t) = -p_{2U} \cdot X(t) + p_{2U}[I(t) - I_b] \quad X(0) = 0 \qquad (16.33)$$

where I is plasma insulin and p_{2U} is rate constant of insulin action on the peripheral glucose utilization.

Insulin action is then modulated in hypoglycemia by a nonlinear function (called *risk*) [25]:

$$\text{risk} = \begin{cases} 0 & \text{if } G \geq G_b \\ 10 \cdot [f(G)]^2 & \text{if } G_{th} \leq G < G_b \\ 10 \cdot [f(G_{th})]^2 & \text{if } G < G_{th} \end{cases} \qquad (16.34)$$

with G_b patient basal glucose, G_{th} the hypoglycemic threshold (set at 60 mg/dL),

$$f(G) = \log\left(\frac{G}{G_b}\right)^{r_2} \tag{16.35}$$

and r_1, r_2 are model parameters.

The model used to describe beta-cell insulin secretion is that proposed in Ref. [36]. The model equations are

$$S(t) = \gamma \cdot I_{po}(t) \tag{16.36}$$

$$\dot{I}_{po}(t) = -\gamma \cdot I_{po}(t) + S_{po}(t) \quad I_{po}(0) = I_{pob} \tag{16.37}$$

$$S_{po}(t) = \begin{cases} Y(t) + K \cdot \dot{G}(t) + S_b & \text{for } \dot{G} > 0 \\ Y(t) + S_b & \text{for } \dot{G} \leq 0 \end{cases} \tag{16.38}$$

and

$$\dot{Y}(t) = \begin{cases} -\alpha \cdot [Y(t) - \beta \cdot (G(t) - h)] & \text{if } \beta \cdot (G(t) - h) \geq -S_b \\ -\alpha \cdot Y(t) - \alpha \cdot S_b & \text{if } \beta \cdot (G(t) - h) < -S_b \end{cases}; \quad Y(0) = 0 \tag{16.39}$$

where γ is the transfer rate constant between portal vein and liver, K is pancreatic responsivity to glucose rate of change, α is delay between plasma glucose and insulin secretion, β is pancreatic responsivity to glucose, and h is threshold level of glucose above beta cells initiate to produce new insulin (h is set to the basal glucose concentration G_b to guarantee system steady state in basal condition).

Glucose excretion by the kidney occurs if plasma glucose exceeds a certain threshold and can be modeled by a linear relationship with plasma glucose:

$$E(t) = \begin{cases} k_{e1} \cdot [G_p(t) - k_{e2}] & \text{if } G_p(t) > k_{e2} \\ 0 & \text{if } G_p(t) \leq k_{e2} \end{cases} \tag{16.40}$$

where k_{e1} is glomerular filtration rate and k_{e2} is renal threshold of glucose.

Glucagon kinetics are described with a one-compartment linear model [25]:

$$\dot{H}(t) = -n \cdot H(t) + SR_H(t) \quad H(0) = H_b \tag{16.41}$$

where H is plasma hormone concentration, SR_H is glucagon secretion rate, and n is glucagon clearance rate.

Glucagon secretion is made up of a static (SR_H^s) and a dynamic component (SR_H^d) [25]:

$$SR_H(t) = SR_H^s(t) + SR_H^d(t) \tag{16.42}$$

The static component is described by the differential equation:

$$\dot{SR}_H^s(t) = -\rho \cdot \left[SR_H^s(t) - \max\left(\frac{\sigma \cdot [G_b - G(t)]}{I(t) - I_b + 1} + SR_H^b, 0 \right) \right] \tag{16.43}$$

where G is plasma glucose, G_b its basal value, I plasma insulin concentration, I_b its basal value, σ the alpha-cell responsivity to glucose level, $1/\rho$ the delay between static glucagon secretion and plasma glucose.

The dynamic component is given by

$$SR_H^d(t) = \delta \cdot \max\left(-\frac{dG(t)}{dt}, 0\right) \tag{16.44}$$

where $dG(t)/dt$ is the glucose rate of change and δ the alpha-cell responsivity to glucose rate of change.

16.3.1.1 Model Identification

The identification of large-scale models is difficult, since they need a rich database usually requiring complex experiments. In this case, identification was made possible by the availability of triple tracer meal data in 204 healthy individuals [37], which provided virtually model-independent estimates of glucose and insulin fluxes in the system. Moreover, in order to describe system behavior in hypoglycemia (which very rarely occurs in healthy individuals after a meal) the database of a pioneering study investigating the role of glucagon and other hormones in human counterregulation [38] has been used. The flux information is key to developing and testing maximal models with confidence. In fact, with only plasma

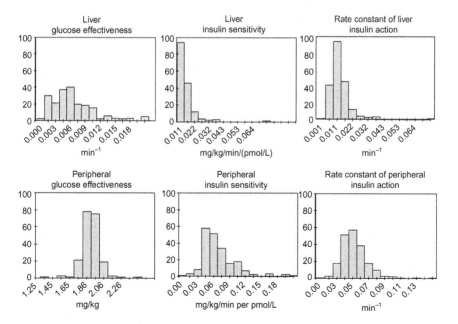

Figure 16.5 Distributions of the most important parameters of the healthy state simulator.

concentrations available, one can certainly describe their dynamics well, but compensations among different model errors are likely to occur. Thanks to this rich flux plus concentration data, the model was identified by resorting to a subsystem partitioning and forcing function strategy, which minimizes structural uncertainties in modelling the various subsystems. For instance, the glucose production model (Eqs. (16.27) and (16.28)) was numerically identified using the measured glucose production (EGP) as output and plasma glucose, insulin, and glucagon as known input. The other subsystems of Figure 16.4 have been identified following a similar strategy (more details can be found in Refs. [24,25]).

From the 204 subject model parameters, their joint probability distribution in the healthy population was reconstructed (Figure 16.5). Since most parameters were approximately log-normally distributed, this probability distribution is uniquely defined by the average vector and the covariance matrix of the log-transformed parameters. Given the joint distribution, any desired number of virtual subjects can be generated by randomly sampling the appropriate number of realizations of the log-transformed parameter vector from the multivariate normal distribution.

16.3.2 Meal Type 1 Diabetes Simulator

A type 1 diabetes simulator is particularly useful for preclinical testing of control strategies in artificial pancreas studies. In fact, large-scale simulation models may account better for intersubject variability than small-size animal trials and may allow for more extensive testing of the limits and robustness of control algorithms [26,27]. A type 1 diabetes simulator requires the model of insulin secretion to be substituted with a model of exogenous insulin delivery. Several models have been proposed to describe subcutaneous insulin transport [39]; in Refs. [26,27] a two-compartment model is used which approximates nonmonomeric and monomeric insulin fractions in the subcutaneous space [40]:

$$\begin{cases} \dot{I}_{sc1}(t) = -(k_d + k_{a1}) \cdot I_{sc1}(t) + IIR(t) & I_{sc1}(0) = I_{sc1ss} \\ \dot{I}_{sc2}(t) = k_d \cdot I_{sc1}(t) - k_{a2} \cdot I_{sc2}(t) & I_{sc2}(0) = I_{sc2ss} \end{cases} \tag{16.45}$$

with I_{sc1}, I_{sc2} amounts of the nonmonomeric and monomeric insulin in the subcutaneous space, respectively, IIR exogenous insulin infusion rate, k_d rate constant of insulin dissociation, k_{a1} and k_{a2} rate constants of nonmonomeric and monomeric insulin absorption, respectively. The rate of appearance of insulin in plasma (R_i) is thus

$$R_i(t) = k_{a1} \cdot I_{sc1}(t) + k_{a2} \cdot I_{sc2}(t) \tag{16.46}$$

Intersubject variability was assumed to be the same as the healthy state (same covariance matrix), but certain clinically relevant modifications were introduced in the average vector (e.g., higher basal glucose concentration, lower insulin clearance, and higher glucose production). In addition to virtual subjects, the simulator is equipped with a model describing subcutaneous glucose sensing.

In January 2008, the first version of the simulator (S2008 [26]) was accepted by the Food and Drug Administration (FDA) as a substitute for animal trials for preclinical testing of control strategies in artificial pancreas studies, and has been largely employed by the JDRF Artificial Pancreas Consortium to test the robustness of the closed-loop control algorithms. Recently, the FDA has also accepted the new version of the simulator that includes new features to better describe hypoglycemia and counterregulation (S2013 [27]).

In silico testing of a control algorithm helps to test extreme situations and to rule out inefficient scenarios, but it is obviously only a prerequisite, not a substitute, for human clinical trials.

16.3.2.1 Validation Against Data

To validate the type 1 diabetes simulator, a database of 24 T1DM subjects was used who received dinner and breakfast on two occasions (i.e., in open- and closed-loop modes) for a total of 96 meal glucose profiles [41]. Measured plasma glucose profiles were compared with those simulated in 100 *in silico* adults and continuous glucose error grid analysis (CG-EGA [42]) was used to assess clinical validity of the simulated traces. In addition, the most common outcome measures obtained in real and simulated experiments have been compared. In particular, each measured plasma glucose profile has been compared with those simulated in the 100 *in silico* subjects undergoing the same experimental scenario (meals, basal insulin, and

Table 16.3 CG-EGA Results on Data Versus S2013 Simulator

Hypoglycemia			Euglycemia			Hyperglycemia		
Accurate	Benign	Bad	Accurate	Benign	Bad	Accurate	Benign	Bad
87.5%	0.7%	11.8%	98.9%	0.6%	0.5%	98.0%	0.7%	1.3%

Table 16.4 Outcome Measures in Real Versus Simulated Experiments

	Data	S2013
Mean(G)	156.9 ± 41.3	157.0 ± 46.5 (0.652)
IQR(G)	71.2 [50.9−96.1]	62.0 [45.6−88.4] (0.169)
LBGI	0.59 [0.02−2.22]	0.40 [0.00−2.54] (0.935)
HBGI	4.85 [1.79−8.34]	4.19 [1.55−8.14] (0.892)

Values are mean ± SD for normally distributed variables (*P* value from paired *t*-test) or median [IQR] for not normally distributed variables (*P* value from Wilcoxon signed rank test).

boluses). Among the 100 simulated profiles, the one that best fitted the data is selected and compared with the glucose profile using CG-EGA. In addition, the distribution of mean glucose (MEAN(G)), intrasubject interquartile range (IQR(G)), low- and high-blood glucose indices (LBGI, HBGI) [43], obtained in real and simulated experiments, have also been compared.

Results of CG-EGA on measured data versus simulations obtained with the S2013 (Table 16.3) show a very good performance in euglycemia, hyperglycemia, and hypoglycemia (percentage in accurate + benign zones: 99.5% in euglycemia; 98.7% in hyperglycemia; 88.2% in hypoglycemia).

These results are confirmed by those reported in Table 16.4. The comparison between data and simulations shows that all of the outcome measures, MEAN(G), IQR(G), LBGI, and HBGI, are not significantly different in real and simulated trials (P value > 0.05).

16.3.3 Future developments

Healthy state and a type 1 diabetes simulators have been presented that are based on a large-scale model of the glucose—insulin—glucagon systems. A prediabetes and type 2 diabetes simulator, usable for assessing the efficacy of various drug therapies before performing experiments in humans, are under development. The model proposed to describe glucose—insulin—glucagon interaction in these diseases is likely to have the same structure as that proposed for describing that in healthy individuals, but the numerical values of model parameters will be different to account for the pathological state. Such values can be derived using an identification strategy similar to that described in the previous section. This is made possible by the availability of triple tracer mixed meal data in these populations [44—46].

16.4 Conclusion

In this chapter, the developments in glucose modelling of the last decade have been presented. The novelty in this research area was mainly the development of models able to describe system behavior during oral tests, such as MTT and OGTT. We focused first on the class of the so-called minimal (parsimonious) models usable for estimating important parameters describing the system behavior and its efficiency (e.g., insulin sensitivity). Model assumption and equations have been provided and details on model identification have been reported to guide the reader in their use. The validity of the models was then presented by reporting the most significant results obtained in relevant clinical studies. We then discussed the class of maximal (large-scale) models. In particular, the healthy state and type 1 diabetes simulators have been described in detail, focusing on the methodology employed for model development, identification, and validation.

References

[1] Bolie VW. Coefficients of normal blood glucose regulation. J Appl Physiol 1961;16: 783−8.

[2] Caumo A, Simeoni M, Cobelli C. Glucose modeling. In: Carson E, Cobelli C, editors. Modelling methodology for physiology and medicine. London, UK: Academic Press; 2001.

[3] Cobelli C, Dalla Man C, Sparacino G, Magni L, De Nicolao G, Kovatchev BP. Diabetes: models, signals, and control. IEEE Rev Biomed Eng 2009;2:54−96.

[4] Cobelli C, Dalla Man C, Pedersen MG, Bertoldo A, Toffolo G. Multi-scale modeling of the glucose system. IEEE Trans Biomed Eng 2013; [in press].

[5] Bergmann RN, Ider YZ, Bowden CR, Cobelli C. Quantitative estimation of insulin sensitivity. Am J Physiol: Endocrinol Metab 1979;236(6):667−77.

[6] Caumo A, Bergman RN, Cobelli C. Insulin sensitivity from meal tolerance tests in normal subjects: a minimal model index. J Clin Endocrinol Metab 2000;85 (11):4396−402.

[7] Dalla Man C, Caumo A, Cobelli C. The oral glucose minimal model: estimation of insulin sensitivity from a meal test. IEEE Trans Biomed Eng 2002;49(5):419−29.

[8] Dalla Man C, Caumo A, Basu R, Rizza R, Toffolo G, Cobelli C. Measurement of selective effect of insulin on glucose disposal from labelled glucose oral test minimal model. Am J Physiol: Endocrinol Metab 2005;289:909−9014.

[9] Barrett PH, Bell BM, Cobelli C, Golde H, Schumitzky A, Vicini P, et al. SAAM II: simulation, analysis, and modelling software for tracer and pharmacokinetic studies. Metabolism 1998;47(4):484−92.

[10] Basu R, Di Camillo B, Toffolo G, Basu A, Shah P, Vella A, et al. Use of a novel triple tracer approach to asses postprandial glucose metabolism. Am J Physiol: Endocrinol Metab 2003;284(1):55−69.

[11] Radziuk J, Norwich KH, Vranic M. Experimental validation of measurements of glucose turnover in nonsteady state. Am J Physiol: Endocrinol Metab 1978;234:84−93.

[12] De Nicolao G, Sparacino G, Cobelli C. Nonparametric input estimation in physiological systems: problems, methods, case studies. Automatica 1997;33:851−70.

[13] Dalla Man C, Caumo A, Basu R, Rizza R, Toffolo G, Cobelli C. Minimal model estimation of glucose absorption and insulin sensitivity from oral test: validation with a tracer method. Am J Physiol: Endocrinol Metab 2004;287:637−43.

[14] De Fronzo RA, Tobin JD, Andres R. Glucose clamp technique: a method for quantifying insulin secretion and resistance. Am J Physiol: Endocrinol Metab 1979;237: 214−23.

[15] Dalla Man C, Yarasheski KE, Caumo A, Robertson H, Toffolo G, Polonsky KS, et al. Insulin sensitivity by oral glucose minimal models: validation against clamp. Am J Physiol: Endocrinol Metab 2005;289:954−9.

[16] Saad MF, Anderson R, Laws A, Watanabe R, Kades W, Sands YCR, et al. A comparison between the minimal model and the glucose clamp in the assessment of insulin sensitivity across the spectrum of glucose tolerance. Diabetes 1994;43:1114−21.

[17] Caumo A, Vicini P, Zachwieja J, Avogaro A, Yarasheski K, Bier D, et al. Undermodelling affects minimal model indexes: insights from a two-compartment model. Am J Physiol: Endocrinol Metab Endocrinol Metab 1999;276:1171−93.

[18] Cobelli C, Pacini G, Toffolo G, Sacca L. Estimation of insulin sensitivity and glucose clearance from minimal model: new insights from labelled IVGTT. Am J Physiol: Endocrinol Metab 1986;250:591−8.

[19] Vicini P, Caumo A, Cobelli C. Glucose effectiveness and insulin sensitivity from the minimal models: consequences of undermodelling assessed by Monte Carlo simulation. IEEE Trans Biomed Eng 1999;46:130−7.

[20] Basu R, Breda E, Oberg A, Powell C, Dalla Man C, Arora P, et al. Mechanisms of age-associated deterioration in glucose tolerance: contribution of alterations in insulin secretion, action and clearance. Diabetes 2003;52:1738−48.

[21] Dalla Man C, Campioni M, Polonsky KS, Basu R, Rizza RA, Toffolo G, et al. Two-hour seven-sample oral glucose tolerance test and meal protocol: minimal model assessment of beta-cell responsivity and insulin sensitivity in nondiabetic individuals. Diabetes 2005;54:3265−73.

[22] Dalla Man C, Toffolo G, Basu R, Rizza RA, Cobelli C. Use of labelled oral minimal model to measure hepatic insulin sensitivity. Am J Physiol: Endocrinol Metab 2008;295(5):1152−9.

[23] Dalla Man C, Piccinini F, Basu R, Basu A, Rizza RA, Cobelli C. Modelling hepatic insulin sensitivity during a meal: validation against the euglycemic hyperinsulinemic clamp. Am J Physiol: Endocrinol Metab 2013;304(8):819−25.

[24] Dalla Man C, Rizza RA, Cobelli C. Meal simulation model of the glucose-insulin system. IEEE Trans Biomed Eng 2007;54(10):1740−9.

[25] Micheletto F. Dalla Man C., Vella A., Cobelli C. A Model of Glucagon Secretion and Action in Healthy Subjects. Proceedings of 10th Diabetes Technology Meeting (DTM), pp. A105, Bethesda, MD, USA, November 11−13, 2010.

[26] Kovatchev BP, Breton M, Dalla Man C, Cobelli C. In silico preclinical trials: a proof of concept in closed-loop control of type 1 diabetes. J Diabetes Sci Technol 2009;3(1):44−55.

[27] Dalla Man C, Micheletto F, Lv D, Breton M, Kovatchev B, Cobelli C. The UVa/Padova type 1 diabetes simulator: new features. J Diabetes Sci Technol, in press.

[28] Vicini P, Caumo A, Cobelli C. The hot IVGTT two-compartment minimal model: indexes of glucose effectiveness and insulin sensitivity. Am J Physiol Endocrinol Metab 1997;273:1024−32.

[29] Ferrannini E, Cobelli C. The kinetics of insulin in man. I. General aspects. Diabetes Metab Rev 1987;3:335−63.

[30] Meier JJ, Veldhuis JD, Butler PC. Pulsatile insulin secretion dictates systemic insulin delivery by regulating hepatic insulin extraction in humans. Diabetes 2005;54: 1649−56.

[31] Dalla Man C, Camilleri M, Cobelli C. A system model of oral glucose absorption: validation on gold standard data. IEEE Trans Biomed Eng 2006;53(12 Pt 1):2472−8.

[32] Rizza RA, Mandarino LJ, Gerich JE. Dose−response characteristics for effects of insulin on production and utilization of glucose in man. Am J Physiol Endocrinol Metab 1982;240:630−9.

[33] Prager R, Wallace P, Olefsky JM. *In vivo* kinetics of insulin action on peripheral glucose disposal and hepatic glucose output in normal and obese subjects. J Clin Invest 1986;78:472−81.

[34] Yki-Jarvinen H, Young AA, Lamkin C, Foley JE. Kinetics of glucose disposal in whole body and across the forearm in man. J Clin Invest 1987;79:1713−9.

[35] Nielsen MF, Basu R, Wise S, Caumo A, Cobelli C, Rizza RA. Normal glucose-induced suppression of glucose production but impaired stimulation of glucose disposal

in type 2 diabetes: evidence for a concentration-dependent defect in uptake. Diabetes 1998;47:1735−47.

[36] Toffolo G, Breda E, Cavaghan MK, Ehrmann DA, Polonsky KS, Cobelli C. Quantitative indexes of beta-cell function during graded up & down glucose infusion from C-peptide minimal models. Am J Physiol Endocrinol Metab 2001;280(1):2−10.

[37] Basu R, Dalla Man C, Campioni M, Basu A, Klee G, Jenkins G, et al. Effect of age and sex on postprandial glucose metabolism: difference in glucose turnover, insulin secretion, insulin action, and hepatic insulin extraction. Diabetes 2006;55: 2001−20014.

[38] Rizza RA, Cryer PE, Gerich JE. Role of glucagon, catecholamines, and growth hormone in human glucose counterregulation. J Clin Invest 1979;64:62−71.

[39] Nucci G, Cobelli C. Models of subcutaneous insulin kinetics. A critical review. Comput Methods Programs Biomed 2000;62:249−57.

[40] Dalla Man C, Raimondo DM, Rizza RA, Cobelli C. GIM, simulation software of meal glucose-insulin model. J Diabetes Sci Technol 2007;1:323−30.

[41] Visentin R, Dalla Man C, Kovatchev B, Cobelli C. Clinical validity of the UVA/ PADOVA type 1 diabetes simulator. Diabetes Technol Ther 2013;.

[42] Kovatchev BP, Cox DJ, Gonder-Frederick LA, Clarke WL. Evaluating the accuracy of continuous glucose-monitoring sensors: continuous glucose-error grid analysis illustrated by TheraSense Freestyle Navigator data. Diabetes Care 2004;27:1922−11928.

[43] Kovatchev BP, Cox DJ, Gonder-Frederick LA, Young-Hyman D, Schlundt D, Clarke WL. Assessment of risk for severe hypoglycemia among adults with IDDM: validation of the low blood glucose index. Diabetes Care 1998;21:1870−11875.

[44] Bock G, Dalla Man C, Campioni M, Chittilapilly E, Basu R, Toffolo G, et al. Pathogenesis of prediabetes: mechanisms of fasting and postprandial hyperglycemia in people with impaired fasting glucose and/or impaired glucose tolerance. Diabetes 2006;55:3536−49.

[45] Basu A, Dalla Man C, Basu R, Toffolo G, Cobelli C, Rizza RA. Effects of type 2 diabetes on insulin secretion, insulin action, glucose effectiveness and postprandial glucose metabolism. Diabetes Care 2009;32(5):866−72.

[46] Dalla Man C, Bock G, Giesler PD, Serra DB, Ligueros Saylan M, Foley JE, et al. Dipeptidyl peptidase-4 inhibition by vildagliptin and the effect on insulin secretion and action in response to meal ingestion in type 2 diabetes. Diabetes Care 2009;32(1). 14−8.

17 Blood–Tissue Exchange Modelling

Paolo Vicini[a] and James B. Bassingthwaighte[b]

[a]Pharmacokinetics, Dynamics and Metabolism, Pfizer worldwide Research and Development, San Diego, CA, [b]Department of Bioengineering, University of Washington, Seattle, WA, USA

17.1 Introduction

Recent years have seen a tremendous surge of interest in multiscale analysis of complex biological systems in a variety of fields [1,2]. Current issues and problems in multiscale modelling are not new, motivated as they are by the analysis of many complicated biological systems. It can be argued that mammalian blood–tissue exchange is the archetypal multiscale system. Blood–tissue exchange takes place, thanks to elementary units, the capillaries, arranged in a functional network of remarkable complexity. Vascular trees irrigating organs *in vivo* branch out to ever smaller functional units, with the capillaries at the end of the scale. Within a human organ, about 25% of the blood is in arteries, 25% in capillaries, and 50% in veins; in mice the conduit vessel (arteries, veins) fraction is smaller, in elephants larger.

The capillaries have the smallest diameters, just a little smaller than red blood cells (RBC), and are remarkably dense, enough that no cell is more than a few microns away from the oxygen-carrying hemoglobin in the RBC. Skeletal muscle, for example, a highly perfused organ, has a density of 300 million capillary per kg tissue, and their length ranges from 100 micrometers to a few millimeters [3]. The capillary level is where exchange of drugs or hormones occurs between the circulation and the tissues. Moving from the microscale to the macroscale, things become quickly more complicated. It has been known for a long time that flows to different tissues are non-uniform, and that the regional flows (per gram of tissue) within any selected organ are also heterogeneous. The *heterogeneity of flow* (local perfusion) and of nutrients reaching the capillaries [4] is, in normal tissues, governed by the tissue's needs, but with arterial obstruction perfusion is compromised. The capillaries drain together into venules and into the organ's outlet veins (Figure 17.1). The dependence of nutrient availability on local flow suggests that local transport and metabolism phenomena at the capillary level are also heterogeneous.

The functional metabolic units of tissue are composed of the arterioles branching pairwise from arteries and feeding into the dense capillary network, the parenchymal cells of the organ, and the venous drainage. The capillary networks are peculiar to each tissue, serving all the cells to minimize diffusion distances relative to cell metabolic rates.

Modelling Methodology for Physiology and Medicine. DOI: http://dx.doi.org/10.1016/B978-0-12-411557-6.00017-3

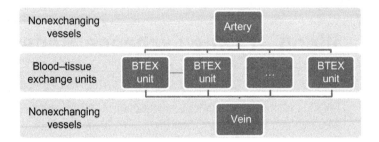

Figure 17.1 The figure shows a simplified scheme of how heterogeneity of the vascular tree at the organ level is partitioned between inlet arteries and outflow veins. At the level of the single blood–tissue exchange unit, substances are transported and metabolized.

Capillaries are essentially tubes, with the tube surface (capillary membrane) formed by endothelial cells less than 0.5 μm thick. Substances can travel across the membrane in two ways. Small hydrophilic molecules can cross the endothelial barrier through pores or gaps where the endothelial cells are not joined tightly. Lipid soluble substances dissolve in the membrane to get across. Larger molecules and charged solutes are usually ferried across by solute selective transporters: this is called facilitated diffusion, or, if it requires energy, active transport. Once substances escape the capillary blood, gaining access to the interstitial fluid (ISF), they can be taken up by the parenchymal cells (this term is generic for the principal cell type within the organ), where they may be metabolized (as in the breakdown of a drug or the oxidation of glucose to carbon dioxide) or may act on receptors or enzymes to influence the dynamics of the cell function (Figure 17.2).

Noninvasive quantification of local convective blood flow and the other elementary processes is difficult *in vivo*. Some direct information has been obtained on transcapillary exchange rates in the brain and other organs by optical imaging and estimating concentrations. Indirect approaches use mathematical modelling to infer the kinetics and mechanisms of the processes, most often using tracers. A particularly powerful approach is the use of multiple tracers simultaneously, each reflecting one particular functional step in blood–tissue exchange, *the multiple indicator dilution (MID) technique*, applicable *in vivo* and *in vitro*.

17.2   Theory and Experimental Approaches

Indicator dilution and MID experiments are based on a few basic principles, chiefly mass balance: at its simplest, what goes in comes out, unless it is retained; so, the better statement is: What goes in is equal to the sum of what is retained, plus what comes out. The indicator dilution method has been the standard for measuring flow, through a pipe or through a segment of the circulation, since the mid-1800s. An amount q_0, say measured in mg, is injected at an upstream point; the

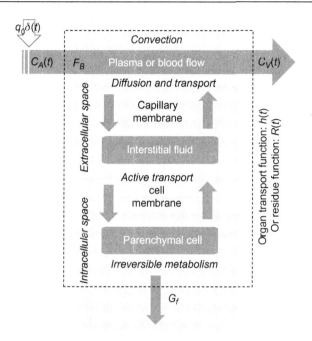

Figure 17.2 Elementary steps of blood–tissue exchange for a generic substance. Plasma and ISF together comprise the extracellular space. Molecules traverse capillaries by convective flow and during passage can diffuse across the capillary wall through gaps to the ISF space. They must then permeate the cell membrane to enter the parenchymal cell. These processes can be passive or active, reversible or irreversible. At the single organ level, a pulse input of amplitude (dose) q_0 is given at the organ arterial inlet. Molecules are subject to convective flow F_B. The substance may be metabolized irreversibly, where G_f is the fraction removed in a single pass through the organ: Consumption $= G_f \times F_B \times C_A$. Venous outflow concentration is labeled $C_V(t)$. The temporal shape of the organ transport function, $h(t)$, or the residue function, $R(t)$, summarizes the information available, but the details within the dashed box are usually not discernible from a single organ transport or residue function.

concentration is measured at a downstream point continuously or as a series of samples. The product of the flow, F (in units of mL/s), times the outflow concentration, $C(t)$, in mg/mL, integrated over time until all of the indicator comes out, must equal q_0. When F is constant and all of q_0 comes out, we have:

$$F = \frac{q_0}{\int_0^\infty C(t)\mathrm{d}t}$$

The schematic in Figure 17.2, with the associated parameter values, generalizes upon this and illustrates how to account for retention within the organ:

$$q_0 = F \int_0^\infty C(t)\mathrm{d}t + q_0 R(t) \tag{17.1}$$

where the *residue function* $R(t)$ is the fraction of the injected tracer still remaining within the organ at time t. This mass-balance statement holds true for tracer or non-tracer and is true no matter what the mechanism of the retention is.

The transport function, $h(t)$, is the probability density function of transit times through the organ and is the fraction of dose injected at the inflow emerging from the outflow at time t. When the input is a very brief pulse at time zero, $\delta(t)$ (Dirac delta function or impulse input), then $h(t)$ is given directly by the outflow concentration–time curve $C_v(t)$ normalized to unit area:

$$h(t) = \frac{C_v(t)}{q_0} \qquad (17.2)$$

with the units being fractions per second.

If the molecule of interest is subject to various transport and metabolism processes as it travels through the vascular tree, we may want to consider using a *tracer*. The second principle to state is a definition of tracer. A tracer is a molecule that is sufficiently similar to (ideally, chemically indistinguishable from) the original substance under study, but where one or more atoms in the structure have been replaced with isotopes (*labels*) that are in low natural abundance and thus suitable to tracking [5,6]. Depending on the properties of the label, tracers can be stable or radioactive. The "tracer" concentration is negligible compared to that for the "mother substance" (the "tracee"). This requirement is essential for examining reactions or binding or transmembrane transport for substances whose reaction rate depends on concentration. When the rate constant $k(C)$ changes with C, then the rate of change of concentration of a tracer $C^*(t)$, dC^*/dt changes with C and thereby tracks the effects of C. When C is constant, and $C^*(t)$ is much smaller (e.g., 1/10,000th of C), then changes in tracer concentration have no affect on the rate constant $k(C)$, i.e.:

$$k(C) = k(C + C^*)$$

for all values of C^*. Then, the injection of tracer C^* can be used to determine $k(C)$, since C^* has no measurable effect, and the rate of reaction of C^* is completely governed by C.

Conversely, a single tracer experiment cannot determine how $k(C)$ changes with C; this requires repeating the tracer experiment at different constant levels of C, to capture the full nonlinear behavior of the tracee system. When tracers are in minuscule doses, with the tracee remaining in steady state, the system is well described by linear time-invariant mathematical models, thus greatly simplifying the required mathematical formalism [7,8].

Various tracer-based techniques have been proposed for the measurement of elementary transport and metabolism fluxes, each with its own advantages, but also limitations and challenges. The main challenge is separately quantifying the individual blood–tissue exchange processes. Irreversible uptake, either metabolism or permanent binding, is, on the surface, the easiest to measure. Arterial–venous differences of tracer and tracee can be measured and interpreted according to the Fick principle [9], which provides an estimate of either organ flow or metabolism from the concentration differences at the inlet (artery) and the outlet (vein) of the organ. The Fick equation is used, for example, for estimating flow from the arteriovenous differences of a substance whose uptake can be measured. Having, for example, measured the oxygen uptake from an oxygen filled bag (in mL O_2 per minute), one can estimate the flow, mL/s, through the lung:

$$F = \frac{\text{Oxygen consumption}}{C_{\text{art}} - C_{\text{vein}}}$$

where C_{art} and C_{vein} are the concentrations in the pulmonary artery and the left atrium, respectively. However informative, application of the Fick principle gives information on net processes and not unidirectional rates and cannot separate the elementary processes we would like to understand (Figure 17.2). Reversible transport within the organ, for example, cannot be quantified exclusively through arterial–venous differences. Multifactorial pathologies, such as diabetes and insulin resistance, are probably due to a combination of cellular transport and metabolic defects. These need to be studied separately if they are to be understood and manipulated through therapeutic intervention.

Noninvasive methods based on imaging techniques have been used extensively to try to further dissect the roles of the elementary processes of blood–tissue exchange at the organ level. As opposed to measurements at inlet and outlet concentrations, these employ various tracers and appropriate detectors to measure the time-varying amount of tracer residing in the organ at each point in time. The amount remaining (the residue function) is the dose injected minus the integral of the cumulative outflow (Eq. (17.1)). The shape of the residue function reflects the composite view of all of the processes (metabolic, transport, etc.) and generally requires mathematical models for quantitative interpretation. Most often, the interpretation requires other data from other experiments to define the elementary processes of transport and metabolism very precisely. The major source of the difficulty is that usually one can image only a single tracer at a time, so the study does not have the power of the MID technique, to be demonstrated below.

The MID method was introduced by Chinard et al. [10]; it provides great flexibility. The idea is to employ multiple tracers simultaneously to characterize different aspects of the exchange and metabolic processes for the substance of interest, each tracer with different molecular characteristics, but with approximately the same molecular weight and similar molecular structure. Here is an example using four tracers simultaneously as controls or references substances, while characterizing D-glucose transport: (1) one confined to the vascular plasma (e.g. albumin), (2) L-glucose, which penetrates the interendothelial cell clefts with exactly the same permeability as D-glucose, but remains confined to ISF and plasma and cannot enter cells, (3) 2-deoxy-D-glucose, which enters cells using the same transporter as D-glucose, but is metabolized via hexokinase only to the -6-phosphate, and (4) D-glucose, which can be fully metabolized. If these four substances are simultaneously injected as a bolus upstream of the organ (in an artery that transports blood flow entering the organ) and measured downstream (from a vein that collects the flow leaving the organ) at several time points, the different shapes of the dilution curves provide detailed information through modelling analysis. Albumin and L-glucose extraction defines the heterogeneity of flows within the organ and permeability-surface area product (a passive process) for transport through the clefts. The deoxy- and D-glucose enter the cells on the same transporter, then the deoxy-glucose is trapped as deoxy-glucose-6-PO_4 and remains trapped, not refluxing (in the absence of glucose phosphatase) or metabolizing further. Thus using chemically similar solutes, each highlighting different steps in the blood–tissue exchange cascade, allows an experimentalist to quantitate each step of transport and metabolism from the dilution curves comparisons [11].

Once the outflow tracer curves have been measured, their interpretation can be conducted in a variety of ways, from least to most sophisticated. At its simplest, analysis of appearance times, mean transit times, and the times of peak concentrations suggest how the solutes travel and exchange. While mean transit times are difficult to calculate because of the challenges in modelling the long observed tails for solutes entering cells, they are exact: each solute's mean transit time is its volume of distribution divided by the flow, and is independent of the mechanisms of transport so long as the solute is *not metabolized*. At the other extreme lie detailed, plausible, and physiologically reasonable mathematical models of the organ. If the experiment has been well designed, the models' parameters can be determined by data fitting procedures and they indirectly measure the elementary processes of interest. Clearly, since individual blood–tissue exchange steps are determined from comparing two tracers, it is important that the organ space that is common between the two is accessed similarly. For example, if glucose is the substance of interest, methyl-glucose would be an appropriate intracellular volume reference as it is not metabolized, and L-glucose is a reference for the extracellular space (plasma + ISF); the difference between L-glucose and methyl-glucose, deoxy-glucose, or D-glucose informs on active cellular transport, on the reasonable assumption that cleft permeation, interstitial diffusion, and volume of the ISF is similarly accessed by these tracers [12].

As a methodological note, since the shape of the tracer outflow curves is of utmost importance to interpreting the results and the time constants of interest can be very short, considerable attention has been devoted to accounting for delay and dispersion introduced by catheters or other sampling methods that alter the results by filtering, delaying, or smoothing the outflow curves [13,14].

17.3 Models of Blood–Tissue Exchange

17.3.1 Input–Output Models

The complexity of multiple indicator dilution datasets can initially be tackled just by inspecting their gross features. An input–output, stochastic, or "black box" model, as it has been defined [8,15], can be used for interpretation. While the Fick equation is suitable only for steady state, we now want to observe transients in order to get at the kinetics or even nonlinear dynamics. Assuming that the tracer(s) are infused as a bolus pulse, we can define a "pseudo-transport function" $h'(t)$, which differs from $h(t)$ (Eq. (17.2)) and has area scaled up from the outflow curve, $C_v(t)$:

$$h'(t) = \frac{F_B}{(1 - G_f)q_0} C_v(t) \tag{17.3}$$

where $C_v(t)$ (a function of time) is the venous outflow concentration–time curve and G_f is the fraction of substance irreversibly metabolized (again going back to the scheme of the system shown in Figure 17.2). In the context of the input–output models, $h'(t)$ in Eq. (17.3) can be interpreted as a normalized version of the observed concentration $C_v(t)$. The shape of this $C_v(t)$ is not the same as the curve shape in the absence of consumption since fractional extraction scales the curves nonuniformly over time. By fiat, the area under the curve (the integral from 0 to infinity) of $h'(t)$ is still unity [8], and if F_B and q_0 are known, the value of the fractional retention G_f (the substance's fractional uptake due to irreversible sequestering metabolism) can be estimated, but it is not a reaction rate. The product of flow times G_f gives an estimate of the average consumption for the organ, and is therefore equivalent to the Fick method as used to estimate consumption when F is known.

An alternative way to measure tracer kinetics, as was discussed, is based on using residue functions from imaging experiments (Eq. (17.1)). These measure the amount of tracer within a given region of interest (ROI), a volume element of tissue, or the whole organ. The "residue function," $R(t)$, is related to $h(t)$ (Figure 17.2) as shown here:

$$R(t) = G_f + (1 - G_f)\left[1 - \int_0^t h(\tau)d\tau\right] \tag{17.4}$$

Given the injected amount from the initial post-injection value, an estimate of G_f can also be recovered from the residue function. Just as with the Fick equation, the information is limited. Also, in this case, the detailed information that the shape of $h(t)$ brings (including convection, transport, and metabolism) is blurred in the integration of the curve. While the information that can be gained from a single tracer is limited, adding multiple tracers and a system model that aids in data interpretation potentially brings great rewards. Mathematical models that have been used in this context include both compartmental and distributed formalisms. We will discuss distributed models since they historically predated the compartmental approach, and remain the more biologically realistic class of models, and then briefly compare the two classes of models.

17.3.2 Distributed Parameter Models of the Single Capillary

In spatially distributed models, tracer concentrations are functions of both spatial and temporal coordinates. While Bohr was the first to describe the effects of one

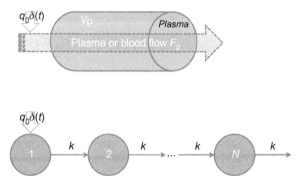

Figure 17.3 The one-region single-capillary blood—tissue exchange model (top panel) and a stirred tanks-in-series compartmental model of intravascular tracer distribution (bottom panel). The figure represents the idea that convection—diffusion problems can be solved numerically by alternative methods, partial differential equations (above), or ordinary differential equations (below). See text for details.

barrier on the gaseous exchange at the pulmonary level [16], the Nobel Prize winner Krogh [17,18] (working with Erlang) introduced the "Krogh cylinder" capillary-tissue model as a radial symmetric tube with uniform flow surrounded by stagnant immobile tissue into which solute diffused radially and was metabolized. He did not describe heterogeneity of intraorgan or intraregional flows, or the axial diffusion and dispersion due to velocity profiles and eddies, or the flow heterogeneity.

17.3.2.1 The Single-Capillary Model

Our mathematical modelling analysis begins with a limited case: a single capillary and a single tracer that cannot leave the vasculature. The temporal—spatial concentration profile in capillary plasma $c_p(x,t)$ (expressed in mass per volume, where x is space and t is time) along a three-dimensional capillary is influenced by: (1) convection, (2) axial diffusion, and (3) radial diffusion. Given that the radius of the capillary is small, around $r = 2.5\ \mu m$, it is as a first approximation reasonable to neglect radial diffusion since radial gradients will be dissipated by diffusion in several milliseconds. Thus, the fundamental equation of the distributed parameter model for the single capillary (Figure 17.3) is given by the classic equation describing convection and axial diffusion:

$$\frac{\partial c_p(x,t)}{\partial t} = -v(x)\frac{\partial c_p(x,t)}{\partial x} + D\frac{\partial^2 c_p(x,t)}{\partial x^2} \tag{17.5}$$

where v is the convection (measured in cm/s) and D the diffusion coefficient (cm^2/s or area per unit time). The inlet boundary condition has to balance the effects of diffusion versus convection so that there is material balance: at $x = 0$, $v \times (C_{in} - C$ $(x = 0)) = D \, dC/dx$. The outlet condition is $dC/dx = 0$ and $C_v = C(x = L)$. Early numerical algorithms to solve similar equations were given by Langmuir [19] and Brenner [20]. If convective velocity is not a function of position, such as when the capillary has a constant diameter, then $v(x) = v$. The diffusion term is critical to the fitting of data, as it accounts for intravascular dispersion due to the velocity profile, axial and radial mixing due to RBC rotation, eddies, and branching (that occurs about every $80–100 \, \mu m$ along capillaries in the heart), and, lastly, the pulsatility of the flow. If $D = 0$, diffusion is zero, then Eq. (17.5) is the case of *plug flow* or *piston flow*, i.e. *uniform velocity across the cross section*. This is not really correct, but the presence of RBC, whose diameters are comparable to that of capillaries, blunts the velocity profile and helps to justify the assumption.

Convective velocity can be expressed as a function of the volume of the capillary V_p (volume per mass unit), the flow of plasma F_p (volume per mass unit per time unit), and the length of the capillary L (length unit). When cross-sectional area is not a function of position, x, then:

$$v = \frac{F_p L}{V_p}$$

and therefore:

$$\frac{\partial c_p(x, t)}{\partial t} = -\frac{F_p L}{V_p} \frac{\partial c_p(x, t)}{\partial x} \tag{17.6}$$

This is a first-order partial differential equation. The system boundary conditions are as follows: at the inlet, $x = 0$, $v \times [C_{in}(t) - C(x = 0)] = D \, dC/dx$, and at the outlet, $x = L$, $dC/dx = 0$, and $C_v = C(L)$, where $C_{in}(t)$ is the input waveform at $L = 0$ (inlet) and $C_v(t)$ is the output at $x = L$ (outlet). Without axial diffusion, the convection then gives piston flow, a flat front space–time profile, whose transport function is $h(t) = \delta(t - V/F)$ or $\delta(t - L/v)$ a delayed Dirac delta function in accord with Eq. (17.6).

If the balancing act at the inlet is not considered, then mass balance is not obtained and material is lost from the capillary into the inflow and does not return, behaving as if it were a consumption process. Only if axial dispersion and diffusion are neglected can one use, for example, Laplace transforms, to get analytic solutions; these are derivable, and could in theory serve as verification tests to determine whether or not the model solutions are correct. This is not actually practical: as it has been shown, the analytical solution for the capillary–ISF–parenchymal cell model requires taking the roots and products of Bessel functions and a convolution integration, a computation which is more than 100-fold slower than obtaining a numerical solution with a good solver such as LSFEA [21]. If the endothelial cell is also considered, the "analytical equation" is even worse [22]: 10 sets of Bessel functions, 3 single convolution integrations, and 1 double convolution integration, over a million times slower to compute than solving the PDE numerically.

For formality's sake, consider Eq. (17.6) and apply the Laplace transform in the time domain only in the case of a pulse input, $u(t) = \delta(t)$. We will use s for the complex Laplace variable, uppercase letters for variables, and functions in the Laplace domain with and lowercase letters for the time domain. We obtain

$$sC_p(x, s) = -v\frac{\partial C_p(x, s)}{\partial x}$$

Solving this differential equation and evaluating the solution at $x = L$, we find

$$C_p(x, s)|_{x=L} = C_p(0, s)e^{-\frac{Ls}{v}} = e^{-\tau s} = Y(s)$$

Therefore, in the time domain, the output $y(t) = u(t - \tau)$. The output of the single-capillary model is then a delayed version of the input with delay $\tau = L/v$. For a generic input waveform $u(t)$, one has:

$$c_p(L, t) = \delta\left(t - \frac{V_p}{F_p}\right) \otimes u(t) = u\left(t - \frac{V_p}{F_p}\right)$$

where $\otimes$ is the convolution operator. Capillary length L does not alter the profile of outflow concentration when axial diffusion is negligible.

Others have considered a *flow-compartmental approximation* to the intravascular kinetics. These relate to the spatially distributed approach. If we assume (a) complete and instantaneous mixing and (b) kinetic homogeneity in the capillary bed, assumptions which are patently incorrect, the appropriate model would have a single, instantaneously mixed compartment from end to end (pool). As it has been shown before, such a model is not particularly interesting nor particularly useful in modelling the kinetics of a substance flowing along the capillary bed in a discrete fashion [23]. However, if we instead assume (a) incomplete mixing and (b) kinetic heterogeneity, we can create a model with an infinite number of compartments that produces a response equal (in the time domain) to the response of the distributed model of intravascular kinetics.

Evolution of this model is based on the hypothesis that a serial arrangement of regions which are individually well mixed (stirred tanks-in-series) composes the capillary: heterogeneity results from the juxtaposition of these well-mixed elements. In the compartmental setting, a delay line is a chain of compartments arranged as in Figure 17.3. A lag element is modeled as a chain of well-mixed spaces with a constant fractional flow rate, k, entering and leaving (called the tanks-in-series model). Successive integration yields the so-called Erlang density function, the response for the rate of outflow of such a chain of n compartments with rate constant k:

$$h(t) = k^n e^{-kt}\frac{t^{n-1}}{(n-1)!} \tag{17.7}$$

It has been shown (e.g. by Jacquez [24]) that the first moment (transit time) of such a function is

$$\int_0^{+\infty} \sigma h(\sigma)d\sigma = \frac{n}{k} = \tau \tag{17.8}$$

In general, we can intuitively see that the larger the n, the tighter the distribution of transit (lag) times within the system (Eq. (17.7)). In the limit of Eq. (17.7), as $n \to \infty$, a velocity front will be characterized by a single transit time τ. Using Laplace transforms again,

$$H(s) = k^n \frac{1}{(s+k)^n} = \left(\frac{k}{s+k}\right)^n$$

After some straightforward algebra and defining $y = n/(s\tau)$, we obtain

$$H(s) = \left(1 + \frac{1}{y}\right)^{-s\tau y}$$

and lastly, when $y \to \infty$ (equivalent to $n \to \infty$), we obtain

$$H(s) = e^{-s\tau}$$

which is, in the Laplace domain, the response of a pure delay. Therefore, *at the limit*, the response of an infinite chain of compartments is equal to the response (in time) of the distributed model.

It has been observed that a capillary-tissue model composed of as few as 5–30 similar segment volumes in series has an impulse response that can describe indicator dilution curves for albumin [25]. These numerical approaches to spatially distributed modelling are similar to serial compartmental approximations, but instead of continuous flow, use sliding segments: Bassingthwaighte et al. [26] computed their two-region model by using 20–60 capillary segments, each exchanging with an extravascular compartment, and with each time step replacing the volume of the fluid segment with the one just upstream, the Lagrangian sliding fluid element approach. It takes advantage of splitting the time and space independent variables [21]. Figure 17.4 displays a numerical simulation showing the values of the impulse response of a tanks-in-series model (Eq. (17.10)) for $\tau = 40$ and for different values of n. We see that the model response approaches a pure delay as the number of compartments increases; however, the number of compartments needs to be relatively large for the approximation to be good.

17.3.2.2   The Two-Region Capillary–ISF Model

In vivo, the capillaries are surrounded by ISF, a homogenous and stagnant medium lying between the capillary wall and the parenchymal cell membrane.

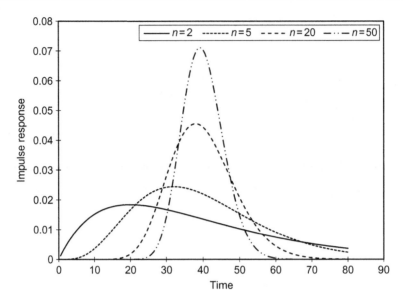

Figure 17.4 Solutions of the compartmental model of Figure 17.3 (Erlang distribution) for different values of the number of compartments in the delay chain, n. These right-skewed curves become nearly Gaussian at high n.

We can now extend the basic model to describe the kinetics of exchange between plasma and ISF. As a generality, the permeability of a membrane [27] is defined by the flux across the membrane divided by the product of the membrane surface area and the concentration difference across the membrane, i.e. flux per unit concentration difference per unit surface area. The capillary membrane is porous, with gaps, or "fissures" between the adjacent endothelial cells. These gaps can be distributed differently and have different width: these characteristics will influence transcapillary molecule passage. Assuming that the dimensions of capillary and tissue are uniform along the capillary length, and radial diffusion in ISF is rapid (with short distances through the ISF), we can consider a simplified two-region model of the concentration gradient through the capillary membrane (Figure 17.5). The explicit solution of this model was derived by Sheppard [5], Sangren and Sheppard [28], and by Rose and Goresky [29], while Bassingthwaighte et al. [21] preferred using efficient numerical algorithms that could include the axial dispersion as well.

We will indicate concentration in the capillary with $c_p(x, t)$, that in the ISF (referred to the interstitial volume, V_{isf}) with $c_{isf}(x, t)$. Concentration in the plasma is the basis for the calculation of extravascular concentrations: we will clarify this with the concept of *virtual volumes*. If $c_p(x, t)$ is the plasma concentration and V_p the plasma space where the substance can diffuse freely, the concentration $c'_{isf}(x, t)$ in the interstitial virtual volume V'_{isf} (mL/g), when equilibrium is reached, is equal

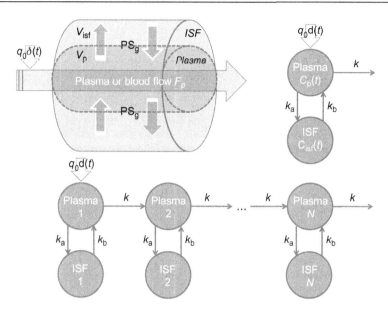

Figure 17.5 The two-region single-capillary blood–tissue exchange model (top left panel), a two-compartment, lumped parameter model of extracellular kinetics (top right panel), and a flow-compartmental model of transcapillary transport (bottom panel). See text for details.

to $c_p(x,t)$. More generally, however, the virtual volume V'_{isf} can be different from the "true" volume (in water equivalents), V_{isf}. For example, in the presence of accumulation at the membrane level, which can be accounted for by asymmetry in the transport or differences in the substance solubility in plasma and ISF, the ratio between virtual volume and true volume can vary from 1 and is generally given by $\gamma = V'_{isf}/V_{isf}$, where γ is the fraction of V_{isf} accessible to the substance. The virtual volume, therefore, represents a composite measure of various phenomena such as the presence of excluded volume, different solubilities, presence of binding sites on the capillary wall, and, eventually, asymmetric transport. The definition of virtual volume can be easily extended to all the extravascular regions.

The plasma region equation is the following:

$$\frac{\partial c_p(x,t)}{\partial t} = -\frac{F_p L}{V_p}\frac{\partial c_p(x,t)}{\partial x} - \frac{PS_g}{V_p}\left[c_p(x,t) - \frac{c_{isf}(x,t)}{\gamma}\right]$$

where PS_g is the permeability-surface area product (volume per time per unit tissue) of the capillary membrane. We can write it in terms of virtual concentrations:

$$\frac{\partial c_p(x,t)}{\partial t} = -\frac{F_p L}{V_p}\frac{\partial c_p(x,t)}{\partial x} - \frac{PS_g}{V_p}\left[c_p(x,t) - c'_{isf}(x,t)\right] \tag{17.9}$$

and analogously for the ISF equation:

$$\frac{\partial c_{isf}'(x,t)}{\partial t} = \frac{PS_g}{V_{isf}'}\left[c_p(x,t) - c_{isf}'(x,t)\right] \tag{17.10}$$

The system of two equations (Eqs. (17.9)−(17.10)) such defined describes transport through the endothelial gaps of the capillary membrane for a single capillary.

Sheppard [5] contains reviews of the explicit solutions for this two-region model available in the literature. We will solve first in the Laplace and then in the time domain, as we did previously for the single capillary. If we define the rate constants (min^{-1}) for solute leaving plasma and entering ISF (outward) versus entering from ISF (inward) as

$$k_a = \frac{PS_g}{V_p}$$

$$k_b = \frac{PS_g}{V_{isf}'}$$

respectively, we can rewrite Eqs. (17.9) and (17.10):

$$\frac{\partial c_p(x,t)}{\partial t} = -v\frac{\partial c_p(x,t)}{\partial x} - k_a\left[c_p(x,t) - c_{isf}'(x,t)\right]$$

$$\frac{\partial c_{isf}'(x,t)}{\partial t} = -k_b[c_{isf}'(x,t) - c_p(x,t)] \tag{17.11}$$

Taking the Laplace transform in time, we can write

$$sC_p(x,s) = -v\frac{\partial C_p(x,s)}{\partial x} - k_a\left[C_p(x,s) - C_{isf}'(x,s)\right]$$

$$sC_{isf}'(x,s) = -k_b[C_{isf}'(x,s) - C_p(x,s)]$$

After a few substitutions, we can solve the resulting differential equation for plasma concentration, $C_p(x,s)$ in the variable x and then evaluate the solution at $x = L$, thus obtaining the outflow of the model at the end of the capillary. In the frequency domain, we have

$$Y(s) = C_p(x,s)|_{x=L} = e^{-\frac{L}{v}\frac{s^2+(k_a+k_b)s}{s+k_b}}U(s) \tag{17.12}$$

Using partial fraction, expansion on the exponent and decomposing the inverse Laplace transform, it can be shown that the response in the Laplace domain is

$$Y(s) = e^{-\tau k_a}\sum_{n=0}^{+\infty}\frac{1}{n!}\left(\frac{\tau k_a k_b}{s+k_b}\right)^n e^{-\tau s}U(s)$$

The inverse of $Y(s)$ (Eq. 17.12) gives the response in time of the distributed model at capillary length $x = L$ to a generic input $u(t)$:

$$y(t) = e^{-\tau k_a}\left[\delta(t - \tau) + \sum_{n=1}^{+\infty} \frac{(\tau k_a k_b)^n (t - \tau)^{n-1} e^{-k_b(t-\tau)}}{n!(n - 1)!}\right] \otimes u(t) \qquad (17.13)$$

where $\otimes$ is the convolution operator. Until $t = \tau$, $y(t) = 0$. If we expand the variables in their original parameterization, again for the boundary condition $c_p(0,t) = u(t)$, the solution becomes:

$$c_p(L, t) = e^{-\frac{PS_g}{F_p}}\delta\left(t - \frac{V_p}{F_p}\right) + \sum_{n=1}^{+\infty} \frac{\left(\frac{V_p}{F_p}\frac{PS_g}{V_p}\frac{PS_g}{V'_{isf}}\right)^n \left(t - \frac{V_p}{F_p}\right)^{n-1} e^{-\frac{PS_g}{V'_{isf}}\left(t-\frac{V_p}{F_p}\right)-\frac{PS_g}{F_p}} 1(t - \tau)}{n!(n - 1)!}$$

$$(17.14)$$

where $1(t - \tau)$ is the step function (zero before τ and one at and after τ).

In Eq. (17.14), the first term on the right-hand side represents the *throughput fraction*, the solute molecules that flow directly through the capillary without ever leaving it, and is therefore equal to the intravascular response of the single capillary, scaled by the factor $e^{-\tau k_a}$ (basically a delayed version of the input), while the second summation term is the *tail function* describing the return (*backdiffusion*, or *backflux*) of the substance from the ISF to the capillary [23].

Given two tracers, one confined to plasma (R, intravascular) and the second escaping into the ISF (D, extravascular), we can define the *extraction* $E(t)$ as a function of their respective plasma concentrations at the outflow:

$$E(t) = \frac{c_R(t) - c_D(t)}{c_R(t)} = 1 - \frac{c_D(t)}{c_R(t)}$$

Assuming no return flux from ISF to plasma, then $c_R(t)$ and $c_D(t)$ are described by a simplification of the previous equations:

$$c_R(t) = \delta\left(t - \frac{V_p}{F_p}\right)$$

$$c_D(t) = e^{-\frac{PS_g}{F_p}}\delta\left(t - \frac{V_p}{F_p}\right)$$

and therefore

$$E(t) = 1 - e^{-\frac{PS_g}{F_p}} \qquad (17.15)$$

Equation (17.15) is the *Crone–Renkin equation* for estimating the permeability-surface product of the capillary membrane [27,30,31]. Thus, the distributed convection–permeation model without axial diffusion model of Figure 17.5 upper panel is the underlying model for the widely used Crone–Renkin equation [23], reinvented by the two investigators, Crone [30] and Renkin [31], independently and at the same time and without reference to the basic idea of Bohr [16] or the analytical solution of Sangren and Sheppard [28]. Both appreciated that capillaries, with lengths greater than 100 diameters and substantial gradients in concentration along the length, could not be treated as stirred tanks.

The most critical aspect of this equation is the hypothesis of absence of backdiffusion. Although it is widely used, its domain of validity is therefore restricted [32]. Another problem is that the single-capillary model (Equation (17.15)) needs to be put into multicapillary form to account for intratissue flow heterogeneity [4,23] with nonuniform flows [23,25,33].

When we modeled the capillary as a chain of plasma compartments, we showed that the model output, at the limit, converges to be the same as the spatially distributed formalism, but the dispersion is fixed at $1/\sqrt{n}$ for the serial tank model, and is completely flexible, determined by the dispersion coefficient D for the axially distributed model [34]. We can extend this to a situation where each plasma compartment is exchanging bidirectionally with another compartment representing ISF. As the number of compartments in the chain approaches infinity, would we obtain the same mathematical expression provided by the distributed model?

Let us first describe the case of the single plasma, single ISF compartment (Figure 17.5). We will maintain the same symbols as before for compartmental volumes of distribution (even if the meanings, clarified later, differ), for the outward and inward rate constants, k_a and k_b, and for the flow rate constant, k (irreversible exit from the plasma pool). Equations for the linear two-compartment model, with input $u(t)$ (having the dimensions of mass/time), are:

$$V_p \frac{dc_p(t)}{dt} = -(k + k_a)V_p c_p(t) + k_b V'_{isf} c'_{isf}(t) + u(t)$$

$$V'_{isf} \frac{dc'_{isf}(t)}{dt} = +k_a V_p c_p(t) + k_b V'_{isf} c'_{isf}(t)$$

(17.16)

The symbols k_a and k_b have the same meaning in these equations as in Eq. (17.11): the fractional permeability-surface area product. The flow term, $kV_p c_p(t)$, is not distributed in space. Remembering the definition of virtual concentration, Eq. (17.16), rearranged, reads:

$$\frac{dc_p(t)}{dt} = -kc_p(t) - \frac{PS_g}{V_p}\left[c_p(t) - c'_{isf}(t)\right] + \frac{u(t)}{V_p}$$

$$\frac{dc'_{isf}(t)}{dt} = -\frac{PS_g}{V'_{isf}}\left[c'_{isf}(t) - c_p(t)\right]$$

(17.17)

This equation is loosely similar to the corresponding two-region distributed-in-space model in Eqs. (17.9)−(17.10). Note, however, that the traditional comparison made in Ref. [23] between such a model and a two-region distributed model can be ambiguous, as we previously discussed for intravascular kinetics.

The general solution, in time, for the plasma concentration of this system, given an impulse input, is straightforward:

$$c_p(t) = \frac{(\lambda_1 + k_b)e^{\lambda_1 t} - (\lambda_2 + k_b)e^{\lambda_2 t}}{V_p(\lambda_1 - \lambda_2)}$$

where λ_1 and λ_2 are the eigenvalues of the two-compartment model, and the Laplace transform for $c_p(t)$ (i.e. the transfer function of the system, from concentration in plasma to concentration in plasma), expressed as a function of the rate constants, is

$$H(s) = \frac{s + k_b}{s^2 + (k + k_a + k_b)s + kk_b} \tag{17.18}$$

Let us now consider a chain of n plasma compartments, each exchanging with an extravascular compartment (rate constants defined as in Figure 17.5). For simplicity, we can redefine the input function $u(t)$ by dropping the term V_p; thus, the dimensions of $u(t)$ are mass/volume/time, comparable to that of the input of the distributed model (concentration per unit time). Therefore, from now on, we will consider concentrations also in the compartmental equations. The general solution for the transfer function of the system is

$$H(s) = \left[\frac{s + k_b}{s^2 + (k + k_a + k_b)s + kk_b} \right]^n \tag{17.19}$$

Again, using partial fraction expansion and calculating the limit as $n \to \infty$, it can be shown that the system transfer function is

$$H(s) = e^{-\tau \frac{s^2 + (k_a + k_b)s}{s + k_b}} \tag{17.20}$$

which is the same response as that of the equations of the two-region distributed model in Eq. (17.12).

Despite this theoretical comparison, as we have alluded to, fitting the multicompartmental two-region model to the distributed two-region model, even for simple circumstances, requires many compartments (e.g. the 109 compartments shown in Figure 17.6 are taken from Ref. [34]).

Attempts to fit high resolution indicator dilution data with the stirred tank model yield systematic errors in the estimation of membrane permeabilities (usually underestimated) and do not usually give good fits to the data because of the fixed relationship between the number of tanks and the dispersion. An example of

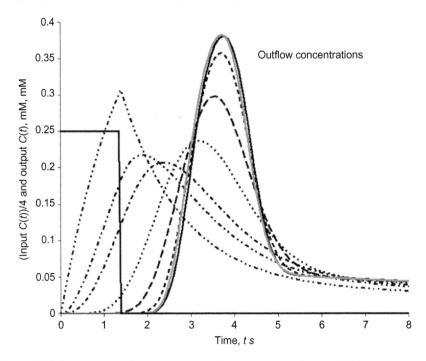

Figure 17.6 Fitting the serial-stirred tank model to the solution for the PDE-based model. The input function is given by the black 1.4 s duration square pulse (drawn at ¼ scale). The distributed model solution with $F = 1$ mL/(g min), $V_p = 0.05$ mL/g, $V_{isf} = 0.15$ ml/g, $PS = PS_{max}* C(t)/(KmPS + C)$ with $PS_{max} = 50{,}000$ mM $\times$ mL/(g $\times$ min) and $KmPS = 50{,}000$ mM so that the $PS = PSmax/KmPS = 1$ mL/(g $\times$ min). With 109 tanks (solid black curve), the serial tank model matches the PDE. Decreasing tanks numbers to 50 (short dashes), 20 (long dashes), 8 (dotted), 3 (dash-dot), 2 (dash-dot-dot), and 1 (dash-dot-dot-dot) gives exactly the same mean transit time, but displaces the peak to earlier times. (This is model #0046 at http://www.physiome.org, titled Anderson_JC_2007_fig12.)

high-resolution data is shown in Figure 17.7 for D- and L-glucose using a two-compartment model (N-tanks = 1 for vascular space). Fitting with the PDE-based model and accounting for flow heterogeneity gives a good fit (Figure 17.8).

Briefly, we have shown that while serial stirred tank models can be used as a type of numerical solver for a partial differential equation, they cannot be fitted to high-resolution data (not demonstrated here except for the extreme example of 1-tank in Figure 17.7). As the simulations in Figures 17.6–17.8 show, it is difficult to fit the serial tank algorithm to the distributed model and parameter estimates for permeability become biased to a variable degree. It is probable that augmenting the stirred tank ODE model by adding some backflux to upstream tanks would provide enough parametric flexibility to fit the data; this is not really worth doing since it would slow the computation further, and the ODE model with more than a few tanks is already slower than the PDE spatially distributed model.

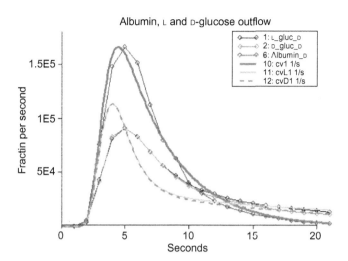

Figure 17.7 Misfitting of the 2-compartment (1 stirred tank for plasma), but properly minimizing the root mean square error results in the thin solid and thin dashed lines for L-glucose and D-glucose. The data are the black and light grey diamonds marking the L- and D- glucose curves; these are fairly precisely overlapped through the first 8 seconds, indicating that L- and D- glucose have the same permeability through the interendothelial clefts and that uptake of D-glucose by endothelial cells is really below detection levels even though they are glycolytic.
Source: Data from (11), available in models 126, 163, and 173 or search on Kuikka at http:// www.physiome.org.

17.3.2.3   The Three-Region, Capillary–ISF Cell Model

The extension of this situation to a third region, intracellular space, is conceptually straightforward. A substance accessing intravascular space, ISF, and intracellular volume will have to be described with a three-region model. The data on volumes in the different spaces have different variability: *in vitro* studies on anesthetized rabbits [35] measured the coefficients of variation around 26% for vascular volumes, 15% for extracellular volumes, and a remarkably low 3% for total water volume in the heart. This version, shown in Figure 17.9, is described by three partial differential equations:

$$\frac{\partial c_p(x,t)}{\partial t} = -\frac{F_p L}{V_p}\frac{\partial c_p(x,t)}{\partial x} - \frac{PS_g}{V_p}\left[c_p(x,t) - c_{isf}(x,t)\right]$$

$$\frac{\partial c_{isf}(x,t)}{\partial t} = -\frac{PS_g}{V'_{isf}}\left[c_{isf}(x,t) - c_p(x,t)\right] - \frac{PS_{pc}}{V'_{isf}}\left[c_{isf}(x,t) - c_{pc}(x,t)\right] \qquad (17.21)$$

$$\frac{\partial c_{pc}(x,t)}{\partial t} = -\frac{PS_{pc}}{V'_{pc}}\left[c_{pc}(x,t) - c_{isf}(x,t)\right] - \frac{G_{pc}}{V'_{pc}}c_{pc}(x,t)$$

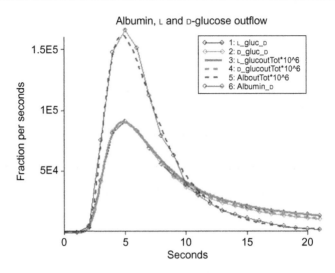

Figure 17.8 Fitting of the same L- and D-glucose and albumin multiple indicator dilution data using the axially distributed model with the same free parameters as the compartmental model used in Figure 17.7. The difference between the L-glucose curve (thick continuous line) and the D-glucose curve (thick dashed line) is due to uptake from the interstitium into the cardiac cells, and gives an estimate of their glucose consumption.

where PS_{pc} (volume per unit time per unit tissue) is the permeability-surface product of the parenchymal cell membrane; V'_{pc} (volume per unit tissue) is the cell's virtual volume of distribution, a sum of the cell water space plus any additional transient or steady-state adsorption; and G_{pc} is the consumption (G for gulosity, volume per unit time per unit tissue) is irreversible metabolism, when present. The primed V, V', is the virtual volume and can be larger (e.g. by binding of high solubility) or smaller (by molecular exclusion) than the actual aqueous volume. The values for the concentrations, c, are the actual concentrations and are not primed since they are the true driving forces for the fluxes. Given the efficiency of numerical solvers today, [21], numerical solutions for Eq. (17.21) are preferable to closed-form solutions [29], especially given the cost of adding axial diffusion in all regions [21], endothelial cell transport [22] and red blood cell kinetics, which require extending the plug flow model to two moving fronts [36,37], for RBC and plasma.

17.3.3 Whole-Organ Models

17.3.3.1 Importance of Flow Heterogeneity

The single-capillary, multiple-region models we have considered so far are the basic building blocks of more detailed models encompassing blood–tissue exchange at the organ level. This requires extending the single-capillary description

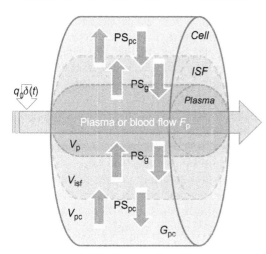

Figure 17.9 The three-region single-capillary blood–tissue exchange model. ISF is interstitial fluid, PC is parenchymal cell of the organ. The PSs are considered as linear conductances for tracer exchanges even when they may be saturable by mother solute. See text for further details.

to a network of interconnected capillaries [4]. The complication is that organ-level partitioning of inflow among capillaries, or groups of capillaries, is not uniform (i.e. it is *heterogeneous* [23]). It is extensive, in perfused tissue and *in vivo* in every organ studied, in heart, skeletal muscle, lung [38], and brain, and in every species, including for the heart, human dog, rabbit, cat, and sheep [39]. However, comparatively little progress has been made toward understanding the evolutionary advantage and the significance of such a heterogeneous flow distribution. This is often attributed to the fluidics of vascular branching, but that is likely only a secondary basis. The primary basis of vascular growth and the state of vasodilatation is most likely the metabolic requirements of the tissue locally. Heterogeneity may not be limited to flow, as others [40] have shown that local capillary permeability is also heterogeneous. Models of blood–tissue exchange at the whole-organ level need to account for heterogeneity of flow. Inaccurate model building can occur when heterogeneity is not accounted for, as has been shown before [4,11]. Since flow heterogeneity is essential for modelling outflow curves obtained with a multiple tracer dilution protocol [41], it is worth reviewing how it can be quantified and how it has been included in mathematical models of blood–tissue exchange. We will see that there are several ways to account for heterogeneity of flow, each with its own advantages and shortcomings.

17.3.3.2 Organ-Level Transit-Time Models

In general, heterogeneity is caused by the particular "physical distribution" of the capillary network, where capillaries of differing lengths, for example, will bring about different local transit times. A simple way is to consider the organ to be composed of multiple blood–tissue exchange units in parallel [4,25], all with the same input, using the experimental data on the heterogeneity of regional flows to estimate the local transit times [4,25]. A variant on this theme is to consider a part of the arterial and venous transit times to be coupled with the heterogeneity of capillary transit times, as proposed by Rose and Goresky [29], Goresky [42], and Goresky et al. [43]. The setting is a multiple tracer dilution experiment. They assumed that there was zero dispersion along each pathway, and that all of the dispersion of the transport function $h(t)$ occurred because of different pathway transit times. They used the data from an intravascular and an extracellular tracer together, assuming injection at time $t = 0$ of a pulse of tracer $\delta(t)$ and the further assumption that the tracer appearance times at the outflow can be partitioned in a fraction due exclusively to transit through nonexchanging large vessels and another due to transit through the blood−tissue-exchanging capillary network. The large vessels are nondispersive, but each large vessel transit time uniquely determined the transit time for the connecting capillary [29]. The simplest case is that of a reference tracer, which—being intravascular—purely reflects convection. Thus, a tracer molecule appearing at the outflow $x = L$ is characterized by a transit time t and is representative of a particular path through the organ, which can be written as

$$t = \tau_{LV}(t) + \tau_C(t)$$

where τ_{LV} is the transit time in the large vessels and τ_C is that for the linked capillaries.

If we now assume that $\tau_{LV}(t)$ and $\tau_C(t)$ are *linear or constant* functions of t and define the reference tracer appearance time t_{app}, we can write

$$\begin{aligned}
\tau_C(t) &= \tau_C(t_{app}) + b(t - t_{app}) \\
\tau_{LV}(t) &= \tau_{LV}(t_{app}) + d(t - t_{app})
\end{aligned} \tag{17.22}$$

where b and d are appropriate constants. Given that $\tau_C(t)$ and $\tau_{LV}(t)$ are increasing functions of t, and must be minimal (τ_C^{min} and τ_{LV}^{min}) at t_{app}, summing Eq. (17.22), we have

$$\begin{aligned}
\tau_C(t) &= \tau_C^{min} + b(t - t_{app}) \\
\tau_{LV}(t) &= \tau_{LV}^{min} + (1 - b)(t - t_{app})
\end{aligned}$$

The range of values of b will dictate the structure of the blood−tissue exchange model, and thus, after Rose and Goresky, we can have two extreme and one intermediate condition:

I. $b = 0 \rightarrow \tau_C(t) = \tau_C^{min}$ for all the capillaries and $\tau_{LV}(t) = \tau_{LV}^{min} + (t - t_{app})$ (i.e. the observed outflow dispersion is entirely due to the large vessels);

II. $b = 1 \rightarrow \tau_C(t) = \tau_C^{min} + (t - t_{app})$ and $\tau_{LV}(t) = \tau_{LV}^{min}$ (i.e. the observed outflow dispersion is entirely due to the capillaries);

III. $0 < b < 1$ (both capillaries and large vessels have variable transit times).

A mixture of variable transit times (Model III), especially in absence of independent information, is the most physiologically likely [29,44]. It is possible to formulate an organ-level model as an extension of the single-capillary perfusion model.

Using the same notation as in Eq. (17.6), the convection equation for the single capillary is

$$\frac{\partial c(x,t)}{\partial t} - - v \frac{\partial c(x,t)}{\partial x}$$

with $c(0,t) = \delta(t)$ (pulse input at capillary inlet), $c(L,t) = h_R(t)$ (measured reference profile at capillary outlet), and v is convection velocity. Plug flow provides a basis for neglecting axial diffusion in the capillaries, while the ~ 15–20% input waveform dispersion values seen in large vessels [45] are not accounted for and will contribute to the observed heterogeneity of transit times. We will see that independent information may be necessary to resolve this.

The organ-level response (transit-time distribution) can then be formulated as the aggregate (weighted) sum of all of the pathways. For a reference (intravascular) tracer:

$$h(t) = \int_0^{+\infty} u[L, t - \tau_{LV}(s)]w(s)ds \tag{17.23}$$

where $s = \tau_{LV}(s) + \tau_C(s)$. The shape term $u[L, t - \tau_{LV}(s)]$ accounts for the single-capillary response and the function $w(s)$ is an appropriate weighting function accounting for transit-time heterogeneity. Under these circumstances, the heterogeneity of transit times through the organ is described by the impulse response of a reference tracer:

$$h_R(t) = w(t)$$

This implies that the observed dispersion of the outflow of an intravascular tracer is entirely due to flow heterogeneity. Since this may not be true in practice, due to large vessels also contributing to the observed waveform, the effects of flow heterogeneity *per se* may be overestimated. This is also assuming that any distortion of the measured profile can be corrected by appropriate input reconstruction methods [13,14].

A description of the organ-level response of an extracellular tracer can be described by assuming, as we have done before, that capillaries are perfused by locally varying flow, but are identical as far as their exchange with the extravascular space. The response of a single capillary without diffusion to a pulse input of a diffusible tracer is [28]

$$h_D(L,t) = e^{-\tau k_a}\delta(t - \tau) + \sum_{n=1}^{+\infty} \frac{(\tau k_a k_b)^n (t-\tau)^{n-1} e^{-k_b(t-\tau)-\tau k_a}}{n!(n-1)!}$$

where $\tau = \tau_C(t)$ is the capillary transit time and other symbols are as defined before. The extension to the whole organ is given by the integral equation [29]:

$$h_D(L,t) = e^{-\tau k_a}w(t) + \int_{t_{app}}^t e^{-k_b(t-s)-\tau(s)k_a}w(s) \sum_{n=1}^{+\infty} \frac{[\tau(s)k_a k_b]^n (t-s)^{n-1}}{n!(n-1)!}ds$$

This approach to whole-organ modelling is parsimonious and attributes all of the dispersion at the outflow to transit-time heterogeneity in piston-flow vessels, large and small. But its great contribution was to incorporate the idea of using the intravascular and extravascular tracer transit times, together with the transmembrane extraction, to estimate the parameters for the heterogeneity modelling. In contrast, the Bassingthwaighte [25] approach, using the experimental data on microspheres to give the flow distributions, is in one sense a stronger technique, because it brings independent data to bear on the analysis even though it ignores the sensible idea of transit-time coupling between large and small vessels. We will see how other approaches have used independent information to relax strict proportionality between large vessel and capillary transit times, in order to account better for dichotomous branching and dispersion in the network.

17.3.3.3   Parallel Network Organ Models

Independent determination of local blood flow distribution can be accomplished in a few ways. For *in vitro* studies, invasive methods such as microsphere deposition [23] and autoradiography [46] have been used. Functional regional flow heterogeneity *in vivo* can be determined using imaging approaches such as [^{15}O] water positron emission tomography [47], or tracers with prolonged retention times, such as NH_3[48] or Rb [49] being deposited regionally to give measures of local flow. "Parallel network models" [4] require assuming that the organ flow distribution can be summarized with a discrete approximation of N parallel pathways (Figure 17.10), with every pathway including a diffusive and a blood–tissue exchange unit [4]. Every pathway provides a compact description of subregions (possibly not adjacent) within the organ, each one characterized by a fraction f_i, $i = 1, \ldots, N$ (or regional relative flow, $f_i = F_i/F_p$) of the total organ inflow F_p, and a fractional mass w_i (which can be interpreted as the percentage of the organ mass perfused by f_i). Total outflow, i.e. the weighted (with f_i and w_i) sum of the responses of the single capillaries, provides the comprehensive organ-level response.

The relative flow distribution histogram can be defined as having widths Δf_1, $\Delta f_2, \ldots, \Delta f_N$:

$$\Delta f_1 = f_2 - f_1$$
$$\Delta f_i = \frac{f_{i-1} + f_{i+1}}{2}, \quad i = 2, \ldots, N-1$$
$$\Delta f_N = f_N - f_{N-1}$$

Its area and mean are both normalized to 1: $\sum_{i=1}^{N} w_i \, \Delta f_i = 1$ and $\sum_{i=1}^{N} f_i w_i \, \Delta f_i = 1$.

In the same way, prior information is used to characterize flow heterogeneity [23], a priori knowledge about large vessel dispersion can be used to define an organ-level model that accounts for flow heterogeneity (Δf_i, $i = 1, \ldots, N$), capillary response (the distributed model capillary response, $h_{C,i}(t)$, $i = 1, \ldots, N$), and vessel

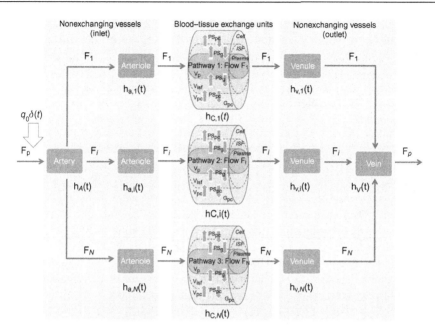

Figure 17.10 Structure of the discrete distributed parameter model by Bassingthwaighte et al. Individual blood–tissue exchange units are described as in Figure 17.9.

dispersion (characterized by a $h_{LV}(t)$ transport function equal for all pathways). The large vessel impulse response can be modeled (e.g. with a second-order differential operator [50]) if other data are not available. Convolving the individual responses and summing them, appropriately weighted by flow distribution, gives the composite waveform:

$$h(t) = h_{LV}(t) \otimes \sum_{i=1}^{N} w_i f_i \Delta f_i h_{C,i}(t)$$

The assumption of a single $h_{LV}(t)$ reflects Goresky's Model II (constant large vessel transit time, variable capillary transit time). The model can be further generalized by separating arterial $h_A(t)$ and venous $h_V(t)$ large vessel dispersions, in addition to variable small artery $h_{a,i}(t)$, $i = 1, \ldots, N$ and vein $h_{v,i}(t)$, $i = 1, \ldots, N$, responses in every pathway:

$$h(t) = h_A(t) \otimes \left[\sum_{i=1}^{N} w_i f_i \Delta f_i h_{a,i}(t) \otimes h_{c,i}(t) \otimes h_{v,i}(t) \right] \otimes h_v(t) \qquad (17.24)$$

This formulation is, of course, more general with respect to Model III, but prior information and assumptions are necessary at every step of the way. This approach is general, and any of the various model forms can be used. The data are either multiple indicator outflow dilution curves over time or the residue functions, not normalized, but quantifiable in calibratable units, for PET or CT or density for MRI. The fundamental problem with imaging is that only one tracer is measured at a time, except in very special circumstances. Compartmental models have also been proposed to characterize multiple tracer dilution studies *in vivo*, and we will review them next.

17.3.3.4 Compartmental Organ Models

Compartmental models are a very flexible class of mathematical models for complex biological systems, based on conservation of mass, even if incompatible with the anatomy. They have been well described elsewhere, together with methodological aspects of their formulation, identification, and validation [7,51,52]. The compartmental approach is based on partitioning the system under study in a finite number of states (compartments), which can represent spatially homogeneous entities of chemical species. A compartmental model for a tracer perturbation can be described by a system of linear ordinary differential equation:

$$q(t) = Kq(t) + u(t)$$
$$c(t) = V^{-1}q(t)$$

where $q(t)$ is the vector of the tracer masses in the compartments, K is the compartmental matrix, $u(t)$ is the input to the system, and $c(t)$ is the measured concentration, in the volume(s) of distribution characterized by the matrix V. System nonlinearity is a straightforward extension of this notation. Physiologically based pharmacokinetic (PBPK) models borrow from principles of compartmental models to represent in a parsimonious manner the interconnections of organs and flows in different species, as a basis for interspecies scaling and translation in risk assessment [53], toxicology [54], and drug discovery [55]. While these approaches have historically been based on lumped parameter ordinary differential equations [56], distributed model (partial differential) equations have been proposed for this purpose as well [57,58]. Several software tools have been proposed to implement and build compartmental and PBPK models [59,60] from data.

As far as blood−tissue exchange is concerned, it is well known that compartmental models, due to their lumped approach to describing spatial heterogeneity, are ill suited to describing organ-level and diffusive phenomena, where mixing may be slow or nonexistent [61]. We will describe the application of a compartmental model (Figure 17.11) to transmembrane transport [12] and phosphorylation [62] of glucose in humans. A multiple tracer dilution study in human skeletal muscle included dilution curves of three tracers, extracellular (L-[^{3}H]-glucose) permeant, nonmetabolizable ([^{14}C]-3-O-methyl-D-glucose), and permeant, metabolizable ([^{3}H]-glucose). The model provided *in vivo* physiological estimates of rate constants of glucose transport into and out of the cell, in addition to irreversible intracellular glucose metabolism. Analysis of reversible cellular transport parameters

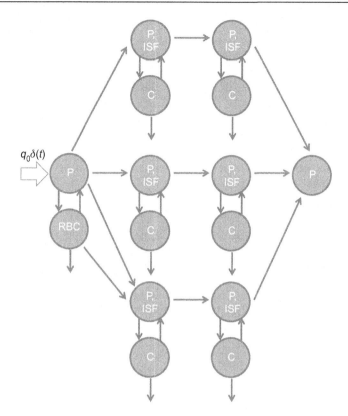

Figure 17.11 Compartmental model of transport and metabolism of glucose in human skeletal muscle. Parameters relative to the compartments (in gray) labeled with P (plasma) and P,ISF (plasma and interstitial fluid) are estimated from extracellular tracer data, while those (in white) relative to the cell (C) are estimated from permeant tracer data. The compartment labeled with RBC represents RBC kinetics.

demonstrated a defect in insulin control of both glucose cellular uptake and phosphorylation in type-2 diabetes [63,64]. Therefore, the model was used to demonstrate that cellular transport plays a very important role in the insulin resistance associated with this pathology. In addition, using steady-state measures of glucose concentration and blood flow, the glucose fluxes in and out of the cell and glucose phosphorylation flux can be estimated, apart from other variables of physiological interest, such as the intracellular concentration of glucose, in normal and diabetic subjects [65]. The cellular transport parameters estimated by the model have also been independently validated [62].

It is interesting to compare and contrast the "compartmental" and "distributed" approach to modelling flow heterogeneity. The compartmental model we described (Figure 17.11) used three parallel chains to describe the extracellular tracer (combined intravascular and interstitial space), each chain including two compartments

(P + IS). The cellular (C) and RBC compartments describe the spaces interested by the cellular, nonmetabolizable tracer, while the metabolizable tracer model includes irreversible intracellular phosphorylation losses emanating from the cellular compartments. The compartmental model used a parsimonious approach to heterogeneity of flow, by essentially lumping plasma and ISF in a single space (not well mixed, though, since the parallel chains of compartments account for heterogeneity of flow and transit times). This likely provides a good fit to the data only because temporal resolution of the measurements is not sufficiently high to completely resolve the convection profile. We should also point out that the compartmental model does not allow one to conclude much on extracellular kinetics (mean transit time and residence time). The extracellular tracer model does not have an exact physiological counterpart, since it is basically equivalent to a lumped description of the transit times through the system made of plasma and ISF. These transit times are the final result of capillary and large vessel convective kinetics, of transcapillary transfer and flow heterogeneity, and the model is a lumped description of all of these processes. All this makes the compartmental model particularly tailored to a specific question: estimation of glucose transport and metabolism parameters. In contrast, while the organ-level distributed models we considered are more general, they require more *a priori* information to be identified from data. The choice of the most appropriate framework needs to include experimental design and model purpose considerations.

17.4   Identification of Blood–Tissue Exchange Models

The numerical, or *a posteriori*, identification of distributed parameter models is most often performed via nonlinear weighted least squares algorithms [66]. The parameter estimation algorithm minimizes the weighted sum of squared residuals (differences between models and data) using Levenberg–Marquardt type techniques [67,68], refined using derivative-free approximations [66]. Parameter confidence ranges (expressed as standard errors or coefficient of variation) can be calculated directly from the sensitivity functions for the free parameters as previously described [69]. Parameters for which estimates have been derived from anatomic studies or rate constants taken from the literature pose problems. Failing to account for their indeterminacy results in underestimating the variance of parameters optimized. Even when parameters in the particular experiment have been measured directly (e.g. flow), the error inherent in the measurement is not acknowledged or accounted for in the covariance matrix from which the confidence limits are calculated, and those limits are inevitably too narrow [70]. The covariance matrix, at best, provides only a local linear approximation. A current approach is to supplement this analytic calculation with a broader analysis using a Monte Carlo approach [71,72]. The method (see http://www.physiome.org/jsim/models/webmodel/ NSR/CorrelationOfParameters/ for details) is to add noise that reflects the properties of the experimental measurement error (e.g. Gaussian with a few percent coefficient of variation) to the optimized solution (the best fit) at the time of each data point, then reoptimize the model to obtain a best fit to the noise-added pseudo-data curve, doing this 100 times (or other chosen practical number). This provides the basis for calculating the probability density function of the parameter values for the 100 optimization runs and from which confidence ranges can be obtained, and it also provides the correlations between all the pairs of free parameters [73]. A more general method incorporating any known means and covariances of parameters not being optimized, such as the one described in Ref. [72], would be an improvement.

17.5 Applications

The complexity of the distributed parameter model of blood–tissue exchange was the primary motivation for the design of a software program, SIMCON, to implement the simulation and identification of this class of models [74]. The initial SIMCON infrastructure evolved in different software systems, increasingly flexible and user-friendly: XSIM [75], GENTEX [76], and, most recently, JSIM [77]. Distributed models have branched out from their original application to blood–tissue exchange and are considered an important component of large-scale, multinational physiological modelling projects such as the cardiac physiome [78].

The structure of spatially distributed models is rather general and allows one to describe the kinetics of a large number of substances. The most frequent applications have been in the transcapillary transport, both *in vivo* and *in vitro* [79,80]. The model has been applied to cardiac glucose transport [11] and receptors [81,82]. Various results on the cardiac and pulmonary kinetics of sodium [83], serotonin [84], and adenosine [85] exist. The *in vivo* kinetics of fatty acids [86] and uric acid [87] have been modeled. The models have been successfully applied to magnetic resonance data [88] and, more recently, to joint magnetic resonance and positron emission tomography databases [1]. Other applications have been in pharmacokinetics [89–91]. Given the central role of the liver in drug disposition, particular attention has been paid to realistic models of liver drug dispersion and diffusion [92–95]. Stirred tank and distributed model approaches can be found and have been researched in pharmaceutical literature. As we have mentioned, in experimental physiology literature, the kinetics of various substances in the liver have been intensely studied using this approach [42,89–91,96].

17.6 Conclusions

We have reviewed here the theoretical and practical fundamentals of the constituent equations of blood–tissue exchange, both at the level of a single capillary and of a whole perfused organ. We have described how the kinetics of substances flowing through organ capillary beds is shaped by elementary kinetic steps, including transcapillary exchange, transport through the cell membrane, and intracellular metabolism. Mathematical models provide a natural approach to the minimally invasive quantification of these individual processes, especially when these are not amenable to direct measurement. Multiple indicator dilution datasets, when analyzed with appropriate models of the system, provide parameter estimates giving insight into the physiological mechanisms involved. We described how models of the single capillary can be extended to capillary networks and organ-level heterogeneous flows. Lastly, we have attempted to draw a parallel between distributed and lumped (compartmental) parameter models, and we highlighted differences and commonalities. A general recommendation for physiological and pharmacologic systems analysis is that, when there are substantial arteriovenous differences in concentrations, there must be gradients along the capillaries and in tissues, and explicitly accounting for spatial distribution is the only way to avoid biases in estimates of kinetic parameters.

References

[1] Bassingthwaighte JB, Raymond GM, Butterworth E, Alessio A, Caldwell JH. Multiscale modeling of metabolism, flows, and exchanges in heterogeneous organs. Ann NY Acad Sci 2010;1188(1):111−20.
[2] Bassingthwaighte JB, Chizeck HJ. The physiome projects and multiscale modeling [Life sciences]. Signal Process Mag IEEE 2008;25(2):121−44.
[3] Lilloja S, Young AA, Culter CL, Ivy JL, Abbott GH, Zawadzki JK, et al. Skeletal muscle capillary density and fiber type are possible determinants of in vivo insulin resistance in man. J Clin Invest 1987;80:415−24.
[4] King RB, Raymond GM, Bassingthwaighte JB. Modeling blood flow heterogeneity. Ann Biomed Eng 1996;24:352−72.
[5] Sheppard CW. Basic principles of the tracer method. New York, London: Wiley; 1962.
[6] Zierler KL. Theory of use of indicators to measure blood flow and extracellular volume and calculation of transcapillary movement of tracers. Circ Res 1963;12(5):464−71.
[7] Carson ER, Cobelli C, Finkelstein L. Mathematical modeling of metabolic and endocrine systems: model formulation, identification, and validation. New York, NY: John Wiley & Sons; 1983.
[8] Norwich KH. Molecular dynamics in biosystems: the kinetics of tracers in intact organisms. Oxford: Pergamon Press; 1977.
[9] Fick A. Über die Messung der Blutquantums in den Herzventrikeln. Verhandl Phys Med Ges Wurzburg 1870;2: XVI.
[10] Chinard FP, Vosburgh GJ, Enns T. Transcapillary exchange of water and of other substances in certain organs of the dog. Am J Physiol 1955;183:221−34.
[11] Kuikka J, Levin M, Bassingthwaighte JB. Multiple tracer dilution estimates of D- and 2-deoxy-D-glucose uptake by the heart. Am J Physiol 1986;250:H29−42.
[12] Cobelli C, Saccomani MP, Ferrannini E, DeFronzo RA, Gelfand R, Bonadonna RC. A compartmental model to quantitate in vivo glucose transport in the human forearm. Am J Physiol 1989;257:E943−58.
[13] Goresky CA, Silverman M. Effect of correction of catheter distortion on calculated liver sinusoidal volumes. Am J Physiol 1964;207:883−92.

[14] Sparacino G, Vicini P, Bonadonna R, Marraccini P, Lehtovirta M, Ferrannini E, et al. Removal of catheter distortion in multiple indicator dilution studies: a deconvolution-based method and case studies on glucose blood–tissue exchange. Med Biol Eng Comput 1997;35:337–42.

[15] Lassen NA, Perl W. Tracer kinetic methods in medical physiology. New York: Raven Press; 1979.

[16] Bohr C. Über die spezifische Tatigkeit der Lungen bei der respiratorischen Gasaufnahme und ihr Verhalten zu der durch die Alveolarwand stattfindenden Gasdiffusion. Skand Arch Physiol 1909;22:221–80.

[17] Krogh A. The number and distribution of capillaries in muscles with calculations of the oxygen pressure head necessary for supplying the tissue. J Physiol 1919;52:409–15.

[18] Krogh A. The anatomy and physiology of capillaries. New Haven, CT: Yale University Press; 1922.

[19] Langmuir I. The velocity of reactions in gases moving through heated vessels and the effect of convection and diffusion. J Am Chem Soc 1908;30(11):1742–54.

[20] Brenner H. The diffusion model of longitudinal mixing in beds of finite length. Numerical values. ChEnS 1962;17(4):229–43.

[21] Bassingthwaighte JB, Chan IS, Wang CY. Computationally efficient algorithms for convection–permeation–diffusion models for blood–tissue exchange. Ann Biomed Eng 1992;20:687–725.

[22] Bassingthwaighte JB, Wang CY, Chan IS. Blood–tissue exchange via transport and transformation by capillary endothelial cells. Circ Res 1989;65:1–24.

[23] Bassingthwaighte J, Goresky C. Modeling in the analysis of solute and water exchange in the microvasculature. In: Renkin EM, Michel CC, editors. Handbook of physiology sect 2, The cardiovascular system, vol. IV, The microcirculation. Bethesda, MD: American Physiological Society; 1984. p. 549–626.

[24] Jacquez JA. Compartmental analysis in biology and medicine. Ann Arbor, MI: University of Michigan Press; 1985.

[25] Bassingthwaighte JB. Blood flow and diffusion through mammalian organs. Science 1970;167(3923):1347–53.

[26] Bassingthwaighte J, Knopp T, Hazelrig J. A concurrent flow model for capillary–tissue exchanges. Capillary permeability: Alfred Benzon Symp II. Copenhagen: Munksgaard; 1970. p. 60–80

[27] Crone C, Lassen N. Capillary permeability: Alfred Benzon Symp II. Copenhagen: Munksgaard; 1970.

[28] Sangren WC, Sheppard CW. A mathematical derivation of the exchange of a labeled substance between a liquid flowing in a vessel and an external compartment. Bull Math Biophys 1953;15:387–94.

[29] Rose CP, Goresky CA. Vasomotor control of capillary transit time heterogeneity in the canine coronary circulation. Circ Res 1976;39:541–54.

[30] Crone C. The permeability of capillaries in various organs as determined by the use of "indicator diffusion" method. Acta Physiol Scand 1963;58:292–305.

[31] Renkin EM. Transport of potassium 42 from blood to tissue in isolated mammalian skeletal muscles. Am J Physiol 1959;197:1205–10.

[32] Bass L, Robinson PJ. Capillary permeability of heterogeneous organs: a parsimonious interpretation of indicator diffusion data. Clin Exper Pharmacol and Physiol 1982;9:363–88.

[33] Bass L, Aisbett J. Extended theory of the early diffusion of multiple indicators: bounds on permeability ratios, with applications to intestinal capillaries. Clin Exper Pharmacol and Physiol 1985;12:387–406.

[34] Bassingthwaighte JB, Anderson JC. Tracers in physiological systems modeling. In: Mark D, Hanigan JN, Marsteller Casey L, editors. Mathematical modeling in nutrition and agriculture: Proc 9th international conference on mathematical modeling in nutrition, Roanoke, VA, August 14−17, 2006. Blacksburg, VA: Virginia Polytechnic Institute and State University; 2006. p. 125−59.

[35] Gonzalez F, Bassingthwaighte JB. Heterogeneities in regional volumes of distribution and flows in rabbit heart. Am J Physiol 1990;258:H1012−24.

[36] Li Z, Yipintsoi T, Bassingthwaighte J. Nonlinear model for capillary−tissue oxygen transport and metabolism. Ann Biomed Eng 1997;25(4):604−19.

[37] Beyer Jr RP, Bassingthwaighte JB, Deussen AJ. A computational model of oxygen transport from red blood cells to mitochondria. Comput Methods Programs Biomed 2002;67(1):39−54.

[38] Glenny RW, Robertson HT. Fractal modeling of pulmonary blood flow heterogeneity. J Appl Physiol 1991;70:1024−30.

[39] Iversen PO, Nicolaysen G. Heterogeneous blood flow distribution within single skeletal muscles in the rabbit: role of vasomotion, sympathetic nerve activity and effect of vasodilation. Acta Physiol Scand 1989;137:125−33.

[40] Caldwell JH, Martin GV, Raymond GM, Bassingthwaighte JB. Regional myocardial flow and capillary permeability-surface area products are nearly proportional. Am J Physiol 1994;267:H654−66.

[41] Vicini P, Bonadonna RC, Lehtovirta M, Groop L, Cobelli C. Estimation of blood flow heterogeneity in human skeletal muscle using intravascular tracer data: importance for modeling transcapillary exchange. Ann Biomed Eng 1998;26:764−74.

[42] Goresky CA. A linear method for determining liver sinusoidal and extravascular volumes. Am J Physiol 1963;204:626−40.

[43] Goresky CA, Ziegler WH, Bach GG. Capillary exchange modeling: barrier limited and flow limited distribution. Circ Res 1970;27:739−64.

[44] Rose CP, Goresky CA, Bach GG. The capillary and sarcolemmal barriers in the heart. An exploration of labeled water permeability. Circ Res 1977;41:515−33.

[45] King RB, Deussen A, Raymond GM, Bassingthwaighte JB. A vascular transport operator. Am J Physiol 1993;265:H2196−208.

[46] Stapleton DD, Moffett TC, Baskin DG, Bassingthwaighte JB. Autoradiographic assessment of blood flow heterogeneity in the hamster heart. Microcirculation 1995;2:277−82.

[47] Vicini P, Bonadonna RC, Utriainen T, Nuutila P, Raitakari M, Yki-Järvinen H, et al. Estimation of blood flow heterogeneity distribution in human skeletal muscle from positron emission tomography. Ann Biomed Eng 1997;25:906−10.

[48] Alessio A, Bassingthwaighte J, Glenny R, Caldwell J. Validation of an axially distributed model for quantification of myocardial blood flow using 13N-ammonia PET. J Nucl Cardiol 2013;20(1):64−75.

[49] Marshall RC, Taylor SE, Powers-Risius P, Reutter BW, Kuruc A, Coxson PG, et al. Kinetic analysis of rubidium and thallium as deposited myocardial blood flow tracers in isolated rabbit heart. Am J Physiol—Heart Circ Physiol 1997;272(3):H1480−90.

[50] Paynter HM. Methods results from MIT studies in unsteady flow. Boston Soc Civil Eng J 1952;39:120−4.

[51] Cobelli C, DiStefano JJ. Parameter and structural identifiability concepts and ambiguities: a critical review and analysis. Am J Physiol Regul, Integr Comp Physiol 1980;239 (1):R7−24.

[52] Jacquez JA, Simon CP. Qualitative theory of compartmental systems. SIAM Rev 1993;35(1):43.

[53] Clewell HJ, Gentry PR, Gearhart JM, Allen BC, Andersen ME. Comparison of cancer risk estimates for vinyl chloride using animal and human data with a PBPK model. Sci Total Environ 2001;274(1—3):37—66.

[54] Clewell HJ. Coupling of computer modeling with *in vitro* methodologies to reduce animal usage in toxicity testing. Toxicol Lett 1993;68(1—2):101—17.

[55] Rowland M, Balant L, Peck C. Physiologically based pharmacokinetics in drug development and regulatory science: a workshop report (Georgetown University, Washington, DC, May 29—30, 2002). AAPS Pharm Sci 2004;6(1):56—67.

[56] Nestorov I. Whole body pharmacokinetic models. Clin Pharmacokinet 2003;42:883—908.

[57] Oliver RE, Heatherington AC, Jones AF, Rowland MA. Physiologically based pharmacokinetic model incorporating dispersion principles to describe solute distribution in the perfused rat hindlimb preparation. J Pharmacokinet Biopharm 1997;25 (4):389—412.

[58] Oliver RE, Jones AF, Rowland M. A whole-body physiologically based pharmacokinetic model incorporating dispersion concepts: short and long time characteristics. J Pharmacokinet Pharmacodyn 2001;28(1):27—55.

[59] Barrett PHR, Bell BM, Cobelli C, Golde H, Schumitzky A, Vicini P, et al. SAAM II: simulation, analysis, and modeling software for tracer and pharmacokinetic studies. Metabolism 1998;47(4):484—92.

[60] D'Argenio DZ, Schumitzky A. ADAPT II user's guide: pharmacokinetic/pharmacodynamic systems analysis software. 4th ed. Los Angeles: Biomedical Simulations Resource; 1997:367 p.

[61] Zierler K. A critique of compartmental analysis. Ann Rev Biophys Bioeng 1981;10:531—62.

[62] Saccomani MP, Bonadonna RC, Bier DM, DeFronzo RA, Cobelli C. A compartmental model to measure the effects of insulin on glucose transport and phosphorylation in human skeletal muscle: a triple tracer study. Am J Physiol 1996;270:E170—85.

[63] Bonadonna RC, Saccomani MP, Seely L, Starick Zych K, Ferrannini E, Cobelli C, et al. Glucose transport in human skeletal muscle. The *in vivo* response to insulin. Diabetes 1993;42:191 8.

[64] Bonadonna RC, Del Prato S, Saccomani MP, Bonora E, Gulli G, Ferrannini E, et al. Transmembrane glucose transport in skeletal muscle of patients with non-insulin dependent diabetes. J Clin Invest 1993;92:486—94.

[65] Bonadonna RC, Del Prato S, Bonora E, Saccomani MP, Gulli G, Natali A, et al. Roles of glucose transport and glucose phosphorylation in muscle insulin resistance of NIDDM. Diabetes 1996;45:915—25.

[66] Chan IS, Bassingthwaighte JB, Goldstein AA. SENSOP: a derivative free solver for nonlinear least squares with sensitivity scaling. Ann Biomed Eng 1993;21:621—31.

[67] Levenberg K. A method for the solution of certain problems in least squares. Quart Appl Math 1944;2:164—8.

[68] Marquardt DW. An algorithm for least squares estimation of nonlinear parameters. J Soc Ind Appl Math 1963;11:431.

[69] Landaw EM, DiStefano JJ. Multiexponential, multicompartmental, and noncompartmental modeling. II. Data analysis and statistical considerations. Am J Physiol 1984;246(5):R665—77.

[70] Grove TM, Bekey GA, Haywood LJ. Analysis of errors in parameter estimation with application to physiological systems. Am J Physiol 1980;239:R390—400.

[71] Mosteller F, Tukey JW. Data analysis and regression. A second course in statistics. Addison-wesley series in behavioral science: quantitative methods, vol. 1. Reading, MA: Addison-Wesley; 1977.

[72] Vicini P, Cobelli C. Parameter estimation in distributed models of blood—tissue exchange: a Monte Carlo strategy to assess precision of parameter estimates. Ann Biomed Eng 1997;25:815—21.

[73] Yue H, Brown M, Knowles J, Wang H, Broomhead DS, Kell DB. Insights into the behaviour of systems biology models from dynamic sensitivity and identifiability analysis: a case study of an NF-κB signalling pathway. Mol Biosyst 2006;2(12):640—9.

[74] Anderson DU, Knopp TJ, Bassingthwaighte JB. SIMCON—simulation control to optimize man—machine interaction. Simul 1970;14(2):81—6.

[75] King RB, Butterworth EA, Weissman LJ, Bassingthwaighte JB. A graphical user-interface for computer-simulation. FASEB J. (9650 Rockville Pike, Bethesda, MD 20814-3998: Federation Amer Soc Exp Biol); 1995. p. A14.

[76] Bassingthwaighte JB, Raymond GM, Ploger JD, Schwartz LM, Bukowski TR. GENTEX a general multiscale model for *in vivo* tissue exchanges and intraorgan metabolism. Philos Trans R Soc Lond A Math Phys Eng Sci 2006;364 (1843):1423—42.

[77] Raymond GM, Butterworth EA, Bassingthwaighte JB. JSIM: mathematical modeling for organ systems, tissues, and cells. FASEB J. (9650 Rockville Pike, Bethesda, MD 20814-3998 USA: Federation Amer Soc Exp Biol); 2007. p. A827.

[78] Bassingthwaighte J, Hunter P, Noble D. The Cardiac Physiome: perspectives for the future. Exp Physiol 2009;94(5):597—605.

[79] Cousineau DF, Goresky CA, Rouleau JR, Rose CP. Microsphere and dilution measurements of flow and interstitial space in dog heart. J Appl Physiol 1994;77:113—20.

[80] Cousineau DF, Goresky CA, Rose CP, Simard A, Schwab AJ. Effects of flow, perfusion pressure, and oxygen consumption on cardiac capillary exchange. J Appl Physiol 1995;78:1350—9.

[81] Cousineau DF, Goresky CA, Rose CP, Schwab AJ. Cardiac microcirculatory effects of beta adrenergic blockade during sympathetic stimulation. Circ Res 1991;68:997—1006.

[82] Rose CP, Cousineau D, Goresky CA, De Champlain J. Constitutive nonexocytotic norepinephrine release in sympathetic curves of *in situ* canine heart. Am J Physiol 1994;266:H1386—94.

[83] Guller B, Yipintsoi T, Orvis AL, Bassingthwaighte JB. Myocardial sodium extraction at varied coronary flows in the dog: estimation of capillary permeability by residue and outflow detection. Circ Res 1975;37:359—78.

[84] Dupuis J, Goresky CA, Rouleau JL, Bach GG, Simard A, SA J. Kinetics of pulmonary uptake of serotonin during exercise in dogs. J Appl Physiol 1996;80:30—46.

[85] Wangler RD, Gorman MW, Wang CY, DeWitt DF, Chan IS, Bassingthwaighte JB, et al. Transcapillary adenosine transport and interstitial adenosine concentration in guinea pig hearts. Am J Physiol 1989;257:H89—106.

[86] Goresky CA, Stremmel W, Rose CP, Guirguis S, Schwab AJ, Diede HE, et al. The capillary transport system for free fatty acids in the heart. Circ Res 1994;74:1015—26.

[87] Kroll K, Bukowski TR, Schwartz LM, Knoepfler D, Bassingthwaighte JB. Capillary endothelial transport of uric acid in guinea pig heart. Am J Physiol 1992;262: H420—31.

[88] Wilke N, Kroll K, Merkle H, Wang Y, Ishibashi Y, Xu Y, et al. Regional myocardial blood volume and flow: first pass MR imaging with polylysine Gd-DTPA. J Magn Reson Imaging 1995;5:227–37.

[89] Pang KS, Barker III F, Schwab AJ, Goresky CA. [^{14}C]urea and 58Co EDTA as reference indicators in hepatic multiple indicator dilution studies. Am J Physiol 1990;259: G32–40.

[90] Pang KS, Barker III F, Schwab AJ, Goresky CA. Demonstration of rapid entry and a cellular binding space for salicylamide in perfused rat liver: a multiple indicator dilution study. J Pharmacol Exp Ther 1994;270:285–95.

[91] Pang KS, Barker III F, Schwab AJ, Goresky CA. Sulfation of acetaminophen by the perfused rat liver: the effect of red blood cell carriage. Hepatology 1995;22:267–82.

[92] Roberts MS, Rowland M. A dispersion model of hepatic elimination: 1. Formulation of the model and bolus considerations. J Pharmacokinet Biopharm 1986;14(3):227–60.

[93] Roberts M, Rowland M. A dispersion model of hepatic elimination: 2. Steady-state considerations-influence of hepatic blood flow, binding within blood, and hepatocellular enzyme activity. J Pharmacokinet Biopharm 1986;14(3):261–88.

[94] Roberts M, Rowland M. A dispersion model of hepatic elimination: 3. Application to metabolite formation and elimination kinetics. J Pharmacokinet Biopharm 1986;14 (3):289–308.

[95] Roberts M, Fraser S, Wagner A, McLeod L. Residence time distributions of solutes in the perfused rat liver using a dispersion model of hepatic elimination: 2. Effect of pharmacological agents, retrograde perfusions, and enzyme inhibition on evans blue, sucrose, water, and taurocholate. J Pharmacokinet Biopharm 1990;18(3):235–58.

[96] Kassissia I, Rose CP, Goresky CA, Schwab AJ, Bach GG, Guirguis S. Flow limited tracer oxygen distribution in the isolated perfused rat liver: effects of temperature and hematocrit. Hepatology 1992;16:763–75.

18 Physiological Modelling of Positron Emission Tomography Images

Alessandra Bertoldo and Claudio Cobelli

Department of Information Engineering, University of Padova, Padova, Italy

18.1 Introduction

Positron emission tomography (PET) has made it possible to detect accurately and noninvasively at the regional level (i.e., organ and tissue) the *in vivo* concentration of radio pharmaceuticals tagged with positron emitters. Image analysis can be made both qualitatively and quantitatively. For some clinical PET studies, a qualitative analysis is appropriate to answer the biological question, for example, when the localization of metabolic defects is the principal purpose of the study. However, often quantitative information is necessary and this requires the interpretation of the PET tracer measurement with a mathematical model of the underlying physiological process. PET kinetic modelling allows, for example, for the estimation of glucose metabolic rate and blood flow in the brain [1,2], in skeletal muscle [3−5], in the myocardium [6,7], and of receptor affinity in specific cerebral structures [8]. In this chapter, after some fundamentals on modelling strategies and on measurement error, we discuss the quantitation of three physiological processes, namely, glucose metabolism, blood flow, and ligand−receptor interaction.

18.2 Modelling Strategies

Various models have been proposed in the last 20 years to convert the radioactive tracer concentrations detected by a PET tomograph in an organ or tissue (more precisely, in a region of interest (ROI) or in a unit of image (pixel or voxel)) into measures of physiological parameters. The most frequently used belong to three model classes: compartmental models, input−output models, and graphical methods. While input−output models and graphical methods are essentially linear modelling techniques, thus usable for quantitation of physiological systems in steady state, compartmental models can be linear or nonlinear and can thus also

Modelling Methodology for Physiology and Medicine. DOI: http://dx.doi.org/10.1016/B978-0-12-411557-6.00018-5

describe their nonsteady-state behavior. In addition, the quantitative physiological portrait provided by compartmental models is richer than the one of input—output models and graphical methods. Clearly, all of this is at the expense of an increased modelling effort.

Compartmental models are widely employed for describing regional tracer kinetics, since the landmark contribution of Sokoloff and colleagues [9]. One has to postulate a linear or nonlinear structure in terms of number of compartments and their interconnections and resolve it from the data [10−12]. The structure must have firm and sound grounds in biochemistry and physiology since one usually describes the intimate function of the system (e.g., in terms of diffusion, transport, metabolism, and receptor—ligand binding). In the following, we will see several examples of linear and nonlinear compartmental models in discussing glucose metabolism, blood flow, and receptor—ligand interaction. For more details on the theory of compartmental models and their identification from the data, we refer the reader to Cobelli and Carson [10], Jacquez [11], and Cobelli and colleagues [12].

Input—output models give, at variance with compartmental models, only a black box representation of the physiological system. The most widely employed input—output model is the so-called spectral analysis (SA) method introduced by Cunningham and Jones [13] and then generalized by Bertoldo and Cobelli [14]. If the system is linear, the impulse response can be written as:

$$h(t) = \sum_{j=1}^{M} \alpha_j \, e^{-\beta_j t} \tag{18.1}$$

with $\beta_j \geq 0$, for every j, and the tissue tracer concentration $C_i(t)$ is simply the convolution of $h(t)$ with the plasma tracer concentration $C_p(t)$:

$$C_i(t) = \sum_{j=1}^{M} \alpha_j \int_0^t C_p(\tau) e^{-\beta_j (t-\tau)} \, d\tau \tag{18.2}$$

The estimation of α_j and β_j from the data provides useful insight into the system behavior. For the sake of reasoning, let us make a distinction between low, intermediate, and high eigenvalues β_j (also referred to as frequency components, thus the term spectral analysis).

The amplitude α, corresponding to the highest eigenvalue ($\beta \to \infty$) gives a measure of the vasculature within the ROI since $\alpha \int_0^t C_p(\tau) e^{-\beta(t-\tau)} \, d\tau \to \frac{\alpha}{\beta} C_p(t)$. The number of amplitudes α_i corresponding to the intermediate β_i gives the number of reversible compartments that can be distinguished in the tissue. However, nothing can be said in terms of compartment connectivity, e.g., two amplitudes at the intermediate frequencies do not establish whether the corresponding reversible tissue compartments are in parallel (heterogeneous tissue) or in cascade (homogeneous tissue), since these two structures are kinetically indistinguishable.

Finally, the amplitude α corresponding to the lowest eigenvalue ($\beta \rightarrow 0$) reveals the presence of an irreversible process within the region since $\alpha \int_0^t C_p(\tau) e^{-\beta(t-\tau)}$ $d\tau \rightarrow \alpha \int_0^t C_p(\tau) d\tau$.

Thus, the intermediate and low frequency components of the spectrum reflect the extravascular behavior of the tracer (i.e., the activity of the tracer within the tissue).

These models, by definition, cannot provide a physiological interpretation of the system, but are of tremendous help in the model selection process. In fact, if used in conjunction with sound parameter estimation techniques and parsimony criteria [14], they provide statistically sound, model-independent guidance to characterize the reversible and irreversible components of the system and to estimate the minimum number of compartments of the system. Sometimes, spectral analysis is also used to obtain kinetic parameters of the system [15−18]; however, in this case, it is associated with a specific compartmental or noncompartmental structure of the system and, thus, gives the same answer in terms of the underlying model.

Graphical methods are appealingly easy to use, and for this reason they are very popular in the quantitation of PET images. Generally, they estimate the physiological parameters by performing some simple calculations on the plasma and tissue time activity curves. Even if these methods are easy to implement, one has to keep in mind the hypotheses on which they are based. Graphical methods emerge from the simplification of a more complex system model, often a compartmental one, so violation of these simplifying assumptions can lead to an unreliable or an under/overestimate of the physiological parameters. Finally, it should be remembered that these models only allow the estimation of some macroscopic parameters and do not portray the microscopic nature of the physiological process (e.g., in a study with a PET glucose analogue, a graphical method can only provide the metabolic rate of glucose, but not the rate constants of transport into and out of the tissue). In the following, we will discuss two widely used graphical methods for quantifying glucose metabolism and ligand−receptor interaction.

18.3 PET Measurement Error

A reliable description of PET measurement error is crucial for sound model identification and, thus, for the estimation of the physiological parameter of interest. An important practical limitation of radionuclide imaging in general, and of PET in particular, is the small number of counts and the resulting large statistical uncertainty (noise). In digital scintigraphic images, each picture element (pixel or voxel in the three-dimensional (3D) reconstruction method) is characterized by a number of counts. From radioactive tracer theory, one understands that the random error of the number of counts in a given pixel is characterized by Poisson statistics (i.e., variance equal to the counts, if the pixel count is independent of that in the other pixels of the image). This is not true in PET. Specifically, in tomographic image reconstruction by filtered back-projection, the counts in a given bin of the projection image data are distributed among the pixels along the sampling line and subsequently removed from the inappropriate pixels by mathematical filtering. Thus, the arithmetic operations composing back-projection propagate the random error in the projection image data among all the pixels along the sampling line.

Although it is difficult to characterize in general the statistical uncertainty in tomographic images, there has been significant work in describing the noise in two-dimensional (2D) PET images [19–24], while less is available for 3D images [25,26]. One of the first contributions was that of Budinger and colleagues [21], in which the percent standard deviation (%SD) in a single uniform object in a tomographic image reconstructed by using a ramp filter was derived as:

$$SD = 1.2 \frac{(\text{Number of resolution cells in the object})^{3/4}}{(\text{Total number of events})^{1/2}} \times 100 \qquad (18.3)$$

where the factor 1.2 is related to the particular form of the filter function (a ramp here) and a resolution cell is a square area within the object whose sides are equal to the linear sampling distance. The formula also shows the relationship between the total number of counts required to achieve a specified %SD and the number of resolution cells in an object in a reconstructed tomographic image. So, for example, to achieve a 20% uncertainty in an image of $3 \cdot 10^3$ resolution cells, the formula predicts that $6 \cdot 10^6$ total events are necessary; however, if one increases to 10^4 the resolution cell number, $36 \cdot 10^6$ total events are required.

This simple equation has been widely used to describe ROI noise, but it does not explicitly account for the many factors that contribute to noise in both 2D and 3D PET images (e.g., attenuation, correction for random coincidences, correction for scattered radiation, and change in statistical quality due to radioactive decay during measurement with short-lived radionuclides). Alpert and colleagues [19] observed that in 2D PET, the measured coincidence can be considered as the sum of three factors: the true coincidence, the random coincidence, and the scatter measured values. Thus, the recovery of the true coincidence rate can be obtained at the expense of an increase in the statistical noise due to the subtraction of two "not true" values. However, prompt scatter is not easily measurable, and is usually minimized by appropriate shielding of the tomograph, whereas random coincidence rates can be routinely estimated during each experiment. Neglecting the prompt scatter contribution, the true coincidence values can be estimated by subtracting the random coincidences from the measured ones. Moreover, photon attenuation has the general effect of decreasing the signal, thereby increasing the relative noise level. A variance formula has been derived that explicitly includes all of these effects. In particular, if the filtered back-projection algorithm is employed, the reconstructed concentration of radioactivity, $\hat{C}$, and its variance, $\hat{\sigma}^2$, are given by:

$$\hat{C}(x,y) = \left(\frac{\pi}{m}\right)^2 \sum_{j=1}^{m} \int [\hat{p}_p(\varphi_j, u) - \hat{p}_r(\varphi_j, u)] A_M(\varphi_j, u) S(u) h(x' - u) du \qquad (18.4)$$

$$\hat{\sigma}^2(x,y) = \left(\frac{\pi}{m}\right)^2 \sum_{j=1}^{m} \int [\hat{p}_p(\varphi_j, u) - \hat{p}_r(\varphi_j, u)] A_M^2(\varphi_j, u) S^2(u) h^2(x' - u) du \qquad (18.5)$$

where ϕ is the projection angle, p the projection measurement, h the reconstruction filter, m the number of projections, A the attenuation correction, S the correction factor for detector nonuniformity (assumed to be noiseless), with subscripts p and r denoting measured and random coincidences, respectively.

Huesman [23,24] derived a formula to calculate 2D ROI variance directly from the projection data without image reconstruction, but usually ROIs are drawn on images long after data acquisition. So, to permit a more practical approach, Carson and colleagues [22] developed an approximation formula that determines the variance of ROI values without having to use the raw projection data like in the formula of Huesman. This formula was derived by applying a series of approximations to the filtered back-projection reconstruction algorithm, and, thus gives an approximation of the variance of an arbitrary ROI and not an exact value like the one derived by Alpert and colleagues [19]. In particular, it requires the filtered back-projection algorithm and accounts for radioactivity distribution, attenuation, randoms, scatter, deadtime, detector normalization, scan length, decay, and reconstruction filter. If F is the mean value of the n pixels in the ROI:

$$F = \frac{1}{n}\sum_{i=1}^{n} f_i \tag{18.6}$$

where f_i is the individual pixel value, the variance of the ROI value is given by:

$$\sigma^2(F) = \frac{1}{n^2}\sum_{i=1}^{n}\sum_{j=1}^{n} \text{Cov}(f_i, f_j) \cong \frac{\hat{V}}{n^2}\sum_{i=1}^{n}\sum_{j=1}^{n} \rho(d_{i,j}) \tag{18.7}$$

where σ^2 is the variance of the ROI, $\hat{V}$ the average pixel variance within the ROI, n the number of pixels in the ROI, Cov the covariance matrix, and $\rho(d)$ the predicted correlation as a function of interpixel distance. It is possible to show that:

$$\sigma^2(F) = \frac{\hat{V}}{n^2}\sum_{i=1}^{n}\sum_{j=1}^{n}$$

$$\times \frac{\sum_{\theta=1}^{n_\theta}\sum_{r=1}^{n_r} h_{r,\theta}^{(i)} h_{r,\theta}^{(j)} A_{r,\theta}^2 N_{r,\theta}^2 W_r^2 p_{r,\theta}}{\sqrt{\left[\sum_{\theta=1}^{n_\theta}\sum_{r=1}^{n_r} (h_{r,\theta}^{(i)})^2 A_{r,\theta}^2 N_{r,\theta}^2 W_r^2 p_{r,\theta}\right]\left[\sum_{\theta=1}^{n_\theta}\sum_{r=1}^{n_r} (h_{r,\theta}^{(j)})^2 A_{r,\theta}^2 N_{r,\theta}^2 W_r^2 p_{r,\theta}\right]}} \tag{18.8}$$

where $h_{r,\theta}^{(i)}$ is the reconstruction matrix value for the pixel i, $A_{r,\theta}$ the attenuation correction factor, $N_{r,\theta}$ the normalization correction factor, $W_{r,\theta}$ the wobble correction factor, and $p_{r,\theta}$ the total counts for angle θ and ray r.

All the above formulas require some knowledge of the tomographic device and of the reconstruction method. To overcome potential difficulties in using Eq. (18.5) or (18.8) to estimate the variance, Mazoyer and colleagues [27] presented a simple formula for the variance of measurement error:

$$\sigma^2_{C(t_k)} = \frac{C(t_k)}{t_k - t_{k-1}} = \frac{C(t_k)}{\Delta t_k} \tag{18.9}$$

where $C(t_k)$ represents the mean value of the activity, $a(t)$, over the time interval $\Delta t_k = t_k - t_{k-1}$, i.e., $a(t_k) = C(t_k) \cdot \Delta t_k$. Note that if $a(t_k)$ has a Poisson distribution, i.e., $\sigma^2_{a(t_k)} = a(t_k)$, $C(t_k)$ has a variance:

$$\sigma^2_{C(t_k)} = \frac{\sigma^2_{a(t_k)}}{\Delta t_k^2} = \frac{a(t_k)}{\Delta t_k^2} = \frac{C(t_k)}{\Delta t_k} \tag{18.10}$$

This approach was used by Delforge and colleagues [28−30] in studies on receptor−ligand systems and recently by ourselves in brain and myocardium glucose metabolism human studies [31]. This formula is appealing because it is independent from the particular PET scanner and reconstruction algorithm, but one has to remember that the formula is correct only if one also assumes that the noise of $C(t_k)$ is independent of the activities in the neighboring ROI or pixel area.

3D PET offers higher sensitivity than 2D PET because of the increased number of lines of response (LOR) detected in 3D PET when the septa are removed. However, this increase in sensitivity is not uniform throughout the whole image volume. In the region near both ends of the scanner, the sensitivity is lower than in the central regions because fewer oblique LOR are detected. Therefore, the noise also becomes a function of the spatial location of the concerned region. Pajevic and colleagues [26] have analyzed the noise characteristic of 2D and 3D images obtained from the General Electric Advance PET scanner. The results of this study show that 3D noise decrease by 9%, 15%, and 18% with respect to 2D for ROIs that cover 2, 4, and 10 slices, respectively, in the central regions of the scanner, and that the ROI noise ratio between 3D and 2D is independent of the transaxial dimension of the ROI.

18.4 Models of Regional Glucose Metabolism

The ideal tracer to quantitate regional glucose metabolism is $[^{11}C]$glucose. However, the interpretative model of its tissue activity must describe the complexity of $[^{11}C]$glucose kinetics (i.e., it has to explicitly account for its metabolic products along the glycolysis and glycogenosynthesis pathways). A rich compartmental model is probably the model of choice, but one is faced with the necessarily limited information content of PET data. A parsimonious model has been proposed by Blomqvist and colleagues [32,33], subsequently used by Powers and colleagues [34], and by Fanelli and colleagues [35].

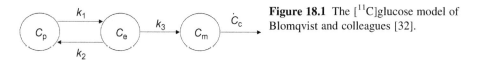

Figure 18.1 The $[^{11}C]$glucose model of Blomqvist and colleagues [32].

The difficulties in handling the complexity of ^{11}C-glucose kinetics has favored an alternative strategy inspired by the landmark model of 2-$[^{14}C]$deoxiglucose kinetics developed by Sokoloff and colleagues [9] in the brain. The elected tracer was $[^{18}F]$fluorodeoxyglucose ($[^{18}F]$FDG). $[^{18}F]$FDG is a glucose analogue, which competes with glucose for facilitated transport sites and with hexokinase for phosphorylation to $[^{18}F]$FDG-6-phosphate ($[^{18}F]$FDG-6-P). The advantage of this analogue is that $[^{18}F]$FDG-6-P is trapped in the tissue and released very slowly from the tissue. In other words, $[^{18}F]$FDG-6-P cannot be metabolized further, while glucose-6-P does so along the glycolysis and glycogenosynthesis pathways. The major disadvantage of $[^{18}F]$FDG is the necessity for correcting for the differences in transport and phosphorylation between the analogue $[^{18}F]$FDG and glucose. A correction factor called lumped constant (LC) can be employed to convert $[^{18}F]$ FDG fractional uptake (but not the microscopic $[^{18}F]$FDG transport rate parameters) to that of glucose. The value of LC is dependent on the type of tissue (brain, skeletal muscle, myocardium) and may also be dependent upon specific study conditions, such as insulin and competing substrate concentrations, or oxygen availability. Several studies are available on LC in the human brain [36,37] and myocardium [38,39] tissue, but only few in human skeletal muscle [40,41].

In Section 18.4.1, we first discuss the $[^{11}C]$glucose model and then the various $[^{18}F]$FDG models that have been proposed for regional studies of glucose metabolism. Glucose metabolism is assumed to be in steady state, and tracer theory predicts that $[^{11}C]$glucose and $[^{18}F]$FDG kinetics are described by linear time-invariant differential equations.

18.4.1 $[^{11}C]$Glucose Models

$[^{11}C]$Glucose is the ideal tracer to study regional glucose metabolism with PET since a rich parametric portrait can be obtained that includes the glucose transport and phosphorylation fluxes. However, $[^{11}C]$glucose modelling must account for the regional loss of all $[^{11}C]$metabolites, mainly $[^{11}C]CO_2$. In 1985, Blomqvist and colleagues [32] proposed a three-compartment model (Figure 18.1) to describe $[^{11}C]$ glucose kinetics in which the loss of tracer was explicitly considered. The model is described by the following equations:

$$\begin{aligned}
\dot{C}_e(t) &= k_1 C_p(t) - (k_2 + k_3)C_e & C_e(0) &= 0 \\
\dot{C}_m(t) &= k_3 C_e(t) - \dot{C}_c(t) & C_m(0) &= 0
\end{aligned} \tag{18.11}$$

where $C_p(t)$ is the arterial concentration of nonmetabolized $[^{11}C]$glucose in plasma, $C_e(t)$ the concentration of $[^{11}C]$glucose in tissue, $C_m(t)$ the concentration of the various [^{11}C]-labeled metabolic products, $\dot{C}_c(t)$ denotes the loss of [^{11}C]glucose metabolites, k_1 and k_2, respectively, the rate constants of [^{11}C]glucose forward and reverse transcapillary membrane transport, and k_3 the rate constant of [^{11}C]glucose metabolism. The dephosphorylation process was neglected in the model (this parameter is usually denoted k_4 in the literature, thus $k_4 = 0$), and it is also assumed that there are no recirculating labeled metabolites in arterial blood.

By considering that loss of labeled metabolites besides [^{11}C]CO$_2$ can, to a good approximation, be neglected, the average cumulated loss of [^{11}C]glucose at different times can be described as:

$$C_c(t) = f \int_0^t C_{av}(\tau) d\tau \tag{18.12}$$

where $C_{av}(t)$ is the arteriovenous difference of [^{11}C]CO$_2$ and f is the tissue perfusion (blood flow per unit mass of tissue). For model identification, $C_p(t)$ is used as the known noise-free input. The PET measurement is:

$$C_i(t) = (1 - V_b)(C_e(t) + C_m(t)) + V_b C_b(t) \tag{18.13}$$

where C_i is the total [^{11}C] activity in the tissue, C_b the [^{11}C] activity in blood, and V_b (unitless) accounts for the vascular volume present in the tissue. All the four model parameters k_1, k_2, k_3, V_b are a priori uniquely identifiable [10,12]. The model parameters can be estimated (e.g., as described by Bertoldo and colleagues [31]) by weighted nonlinear least squares minimizing the cost function

$$\text{WRSS}(p) = \sum_{j=1}^{N} w_j [C_i^{obs}(t_j) - C_i(p, t_j)]^2 \tag{18.14}$$

where C_i^{obs} is the measured PET datum, p the vector of unknown model parameters of dimension P, $C_i(p, t)$ the model prediction, and w_j the weight of the jth datum chosen optimally [10] as:

$$w_j = \frac{\Delta t_j}{C_i^{obs}(t_j)} \tag{18.15}$$

where Δt_j is the length of the scanning interval relative to $C_i^{obs}(t_j)$. Precision of the parameter estimates can be evaluated from the inverse of the Fisher information matrix $\mathbf{M}$ by:

$$\text{COV}(\hat{p}) = \gamma M^{-1} \tag{18.16}$$

where γ is an unknown proportionality constant estimated a posteriori [10,12] as:

$$\gamma = \frac{\text{WRSS}(\hat{p})}{N - P} \tag{18.17}$$

where WRSS($\hat{p}$) is the value of the cost function evaluated at the minimum.

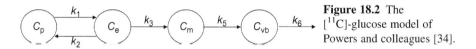

Figure 18.2 The [^{11}C]-glucose model of Powers and colleagues [34].

The model allows the calculation of the fractional uptake of [^{11}C]glucose:

$$K = \frac{k_1 k_3}{k_2 + k_3} \tag{18.18}$$

and the regional metabolic rate of glucose as:

$$\text{rGl} = \frac{k_1 k_3}{k_2 + k_3} C_{p\text{-}g} \tag{18.19}$$

where $C_{p\text{-}g}$ is the arterial plasma concentration of glucose.

Starting from the demonstration by Blomqvist and colleagues [33] that the rate of loss of [^{11}C]CO$_2$ is a constant fraction of the available amount of labeled metabolites, Powers and colleagues [34] introduced into the model of Figure 18.1 a fourth rate constant to describe the loss of all labeled metabolites. This model is shown in Figure 18.2.

The model equations are:

$$\begin{aligned}
\dot{C}_e(t) &= k_1 C_p(t) - (k_2 + k_3)C_e(t) & C_e(0) &= 0 \\
\dot{C}_m(t) &= k_3 C_e(t) - k_5 C_m(t) & C_m(0) &= 0 \\
\dot{C}_{vb}(t) &= k_5 C_m(t) - k_6 C_{vb}(t) & C_{vb}(0) &= 0
\end{aligned} \tag{18.20}$$

where C_p, C_e, C_m, k_1, k_2, and k_3 have the same meanings as before, $C_{vb}(t)$ represents the vascular concentration of the metabolites normalized to tissue volume, k_5 the rate constant describing [^{11}C]glucose metabolites leaving the tissue, and k_6 the rate constant of [^{11}C]glucose metabolites washout. In particular, k_6 is defined as f/V_b (i.e., the reciprocal of the vascular mean transit time).

The PET measurement is described by:

$$C_i(t) = (1 - V_b)(C_e(t) + C_m(t) + C_{vb}(t)) + V_b C_b(t) \tag{18.21}$$

where C_i is total [^{11}C] activity in the tissue and C_b is [^{11}C]glucose activity in blood. The model is a priori nonuniquely identifiable [10,12]. It becomes a priori uniquely identifiable if k_6 is assumed to be known. Since $k_6 = f/V_b$, one can use values for V_b and f previously obtained by quantitative analysis of PET images employing, respectively, the [^{15}O]CO and [^{15}O]H$_2$O tracers. The fractional uptake of [^{11}C]glucose, K, and the regional metabolic rate of glucose, rGl, can be estimated by using Eqs. (18.18) and (18.19). This model does not require venous sampling for [^{11}C]glucose metabolites and does not assume that the loss of labeled metabolites is proportional to regional blood flow (rBF). This model has been

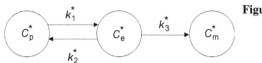

Figure 18.3 The 3K model.

recently utilized by Fanelli and colleagues [35] to measure brain glucose metabolic rate in poorly controlled insulin-dependent diabetes.

18.4.2 $[^{18}F]$ FDG Models

18.4.2.1 The 3K Model

The two-tissue three-compartment rate constants model, 3K, proposed by Sokoloff and colleagues [9] was originally developed for autoradiographic studies in the brain with 2-$[^{14}C]$deoxyglucose as tracer, and subsequently used for PET $[^{18}F]$FDG studies in the brain and other tissues and organs. The 3K model of $[^{18}F]$FDG kinetics of Sokoloff et al. is shown in Figure 18.3, where C_p is $[^{18}F]$FDG plasma arterial concentration, C_e the $[^{18}F]$FDG tissue concentration, C_m $[^{18}F]$FDG-6-P concentration in tissue, k_1 and k_2, respectively, the rate constants of $[^{18}F]$FDG forward and reverse transcapillary membrane transport, and k_3 the rate constant of $[^{18}F]$FDG phosphorylation. The asterisk recalls that we are dealing with a tracer analogue and not with an ideal tracer (e.g., the k_i of Figure 18.3 are different from the k_i of Figure 18.2).

The kinetics of $[^{18}F]$FDG in the tissue is described by:

$$
\begin{aligned}
\dot{C}_e(t) &= k_1 C_p(t) - (k_2 + k_3)C_e(t) \quad C_e(0) = 0 \\
\dot{C}_m(t) &= k_3 C_e(t) \quad C_m(0) = 0
\end{aligned}
\tag{18.22}
$$

At any time following the introduction of $[^{18}F]$FDG into the blood, the total concentration of radioactivity in the tissue, C_i, is equal to the sum of the concentrations of $[^{18}F]$FDG and $[^{18}F]$FDG-6-P:

$$
C_i(t) = C_e(t) + C_m(t)
\tag{18.23}
$$

With the limited spatial resolution of the PET scanner, however, measurement of radioactivity in an ROI includes the radioactivity in the blood volume present within the tissue, so the measured tracer activity is the sum of the tissue activities and a certain fraction, V_b, of the blood ^{18}F concentration, C_b:

$$
C_i(t) = (1 - V_b)(C_e(t) + C_m(t)) + V_b C_b(t)
\tag{18.24}
$$

In particular, V_b accounts in brain and skeletal muscle for the vascular volume present in the tissue ROI, while, for example, in the heart it mainly accounts for the spillover effects from blood to tissue (negligible in brain and skeletal muscle). All the four model parameters k_1, k_2, k_3, V_b are *a priori* uniquely identifiable [10,12].

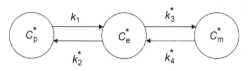

Figure 18.4 The 4K model.

The model allows the calculation of the fractional uptake of $[^{18}F]FDG$:

$$K = \frac{k_1 k_3}{k_2 + k_3} \tag{18.25}$$

Once K is known, one can calculate the regional metabolic rate of glucose as [1]:

$$rGl = \frac{k_1 k_3}{k_2 + k_3} \frac{C_{p\text{-}g}}{LC} \tag{18.26}$$

where $C_{p\text{-}g}$ is the arterial plasma glucose concentration and LC is the lumped constant (i.e., the factor that describes the relation between the glucose analogue $[^{18}F]$ FDG and glucose itself). LC is given by:

$$LC = \frac{E^{FDG}}{E^{GLU}} \tag{18.27}$$

where E^{FDG} and E^{GLU} are, respectively, the extraction of $[^{18}F]FDG$ and glucose. For the 3K model, LC becomes:

$$LC = \frac{k_1 k_3}{k_2 + k_3} \Big/ \frac{k_1 k_3}{k_2 + k_3} \tag{18.28}$$

18.4.2.2 The 4K Model

In 1979, Phelps and colleagues [1] proposed a modification of the 3K model by observing that after 120 min, following a pulse of $[^{18}F]FDG$, total tissue activity was declining, thus indicating a loss of product. The authors allowed dephosphory-lation of $[^{18}F]FDG$-6-P; thus, an additional rate constant was incorporated into the 3K model that becomes a four-rate constants model (4K). The 4K model is shown in Figure 18.4. The meaning of C_p, C_e, C_m, and of the rate constants is the same as for the 3K model, while k_4 is the dephosphorylation rate constant.

The model equations of tissue activity are:

$$\begin{aligned}
\dot{C}_e(t) &= k_1 C_p(t) - (k_2 + k_3)C_e(t) + k_4 C_m(t) \quad && C_e(0) = 0 \\
\dot{C}_m(t) &= k_3 C_e(t) - k_4 C_m(t) \quad && C_m(0) = 0
\end{aligned} \tag{18.29}$$

The PET measurement equation is the same as that of the 3K model (i.e., Eq. (18.24)).

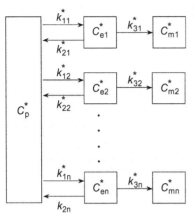

Figure 18.5 The TH model.

All model parameters $(k_1, k_2, k_3, k_4, V_b)$ are *a priori* uniquely identifiable [10,12]. Also, for the 4K model it is possible to obtain a measure of the fractional uptake of $[^{18}F]$FDG and of regional metabolic rate of glucose: they are given by Eqs. (18.25) and (18.26), respectively [1].

18.4.2.3   The Heterogeneous Model

Schmidt and colleagues [42,43] proposed a model, which takes into account the heterogeneous composure of a tissue. In fact, often the regions represented in the PET images are kinetically heterogeneous with respect to structure, blood flow, tracer transport, and metabolism, above all for the brain tissue where white matter and gray matter are difficult to separate in a single ROI. In addition, the limited spatial resolution of the PET scanners does not always permit an accurate delimitation of a homogeneous region. The model is an extension of the 3K model to a heterogeneous tissue and is shown in Figure 18.5. The heterogeneous tissue is assumed to consist of n smaller homogeneous subregions. $C_{e1}, C_{e2}, \ldots, C_{en}$ represent $[^{18}F]$FDG concentration in each homogeneous region, $C_{m1}, C_{m2}, \ldots, C_{mn}$ $[^{18}F]$FDG-6-P concentration in these same subregions, k_{1i}, k_{2i}, k_{3i} the corresponding rate constants for transport and phosphorylation of $[^{18}F]$FDG, and $C_p(t)$ arterial plasma concentration.

The model assumptions are the same as those of the 3K model with the only difference being that the tissue is now considered heterogeneous. The differential equations describing the rate of change of $[^{18}F]$FDG in each subregion are:

$$\frac{dC_{e1}(t)}{dt} = k_{11}C_p(t) - (k_{21} + k_{31})C_{e1}(t) \quad C_{e1}(0) = 0$$

$$\frac{dC_{e2}(t)}{dt} = k_{12}C_p(t) - (k_{22} + k_{32})C_{e2}(t) \quad C_{e2}(0) = 0$$

$$\vdots$$

$$\frac{dC_{en}(t)}{dt} = k_{1n}C_p(t) - (k_{2n} + k_{3n})C_{en}(t) \quad C_{en}(0) = 0$$

(18.30)

Consequently the rate of change of $[^{18}F]FDG$ in the heterogeneous tissue can be found by summing up the concentration in each subregion weighted by its relative tissue mass, $w_1, w_2, \ldots, w_n$ with $\sum w_i = 1$:

$$\frac{d\overline{C}_e(t)}{dt} = \sum_{i=1}^{n} w_i \frac{dC_{ei}(t)}{dt} = \left(\sum_{i=1}^{n} w_i k_{1i}\right) C_p(t) - \left[\sum_{i=1}^{n} w_i(k_{2i} + k_{3i}) C_{ei}(t)\right]$$

(18.31)

where $\overline{C}_e(t)$ gives the $[^{18}F]FDG$ concentration in the heterogeneous tissue. By defining:

$$\overline{k}_1 = \sum_{I=1}^{N} w_i k_{1i}$$

(18.32)

after a simple manipulation, one has:

$$\frac{d\overline{C}_e(t)}{dt} = \overline{k}_1 C_p(t) - \frac{\left[\sum_{i=1}^{n} w_i(k_{2i} + k_{3i}) C_{ei}(t)\right]}{\left[\sum_{i=1}^{n} w_i C_{ei}(t)\right]} \overline{C}_e(t)$$

(18.33)

Defining:

$$k_2(t) = \frac{\sum_{i=1}^{n} w_i k_{2i} C_{ei}(t)}{\sum_{i=1}^{n} w_i C_{ei}(t)}$$

(18.34)

as the parameter describing the efflux of $[^{18}F]FDG$ from the heterogeneous tissue to plasma, and

$$k_3(t) = \frac{\sum_{i=1}^{n} w_i k_{3i} C_{ei}(t)}{\sum_{i=1}^{n} w_i C_{ei}(t)}$$

(18.35)

as the parameter describing phosphorylation of $[^{18}F]FDG$ to $[^{18}F]FDG$-6-P, then $[^{18}F]FDG$ concentration in the tissue becomes:

$$\frac{d\overline{C}_e(t)}{dt} = \overline{k}_1 C_p(t) - (k_2(t) + k_3(t))\overline{C}_e(t)$$

(18.36)

The rate of change of $[^{18}F]FDG$-6-P in each subregion is given by:

$$\frac{dC_{m1}(t)}{dt} = k_{31} C_{e1}(t) \quad C_{m1}(0) = 0$$

$$\frac{dC_{m2}(t)}{dt} = k_{32} C_{e2}(t) \quad C_{m2}(0) = 0$$

$$\vdots$$

(18.37)

$$\frac{dC_{mn}(t)}{dt} = k_{3n} C_{en}(t) \quad C_{mn}(0) = 0$$

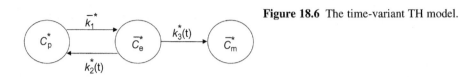

Figure 18.6 The time-variant TH model.

and thus the rate of change of [^{18}F]FDG-6-P in the heterogeneous tissue, $\overline{C}_m$, becomes:

$$\frac{\mathrm{d}\overline{C}_m(t)}{\mathrm{d}t} = k_3(t)\overline{C}_e(t) \tag{18.38}$$

Thus, the resulting model becomes time-variant (Figure 18.6):

$$\overline{C}_e = \overline{k}_1 C_p(t) - [k_2(t) + k_3(t)]\overline{C}_e(t) \quad \overline{C}_e(0) = 0 \tag{18.39}$$

The time-varying parameters for efflux and phosphorylation have been described by an exponential function as:

$$\begin{aligned} k_2(t) &= \overline{k}_2(1 + Ae^{-Bt}) \\ k_3(t) &= \overline{k}_3(1 + Ae^{-Bt}) \end{aligned} \tag{18.40}$$

The PET measurement is described by:

$$C_i(t) = (1 - V_b)(\overline{C}_e(t) + \overline{C}_m(t)) + V_b C_b(t) \tag{18.41}$$

All model parameters, $\overline{k}_1, \overline{k}_2, \overline{k}_3, A, B, V_b$, are *a priori* uniquely identifiable [10,12]. [^{18}F]FDG fractional uptake K can be calculated as [31]:

$$K = \sum_{j=1}^{N} w_j K_j = \sum_{j=1}^{N} w_j \frac{k_{1j}k_{3j}}{k_{2j} + k_{3j}} = \frac{\overline{k}_1 \overline{k}_3}{\overline{k}_2 + \overline{k}_3} \tag{18.42}$$

Finally, the regional metabolic rate of glucose is given by:

$$\mathrm{rGl} = \frac{\overline{k}_1 \overline{k}_3}{\overline{k}_2 + \overline{k}_3} \frac{C_{p\text{-}g}}{\mathrm{LC}} \tag{18.43}$$

18.4.2.4 The Patlak Graphical Method

Patlak and colleagues [44] have developed a graphical method to estimate the regional metabolic rate of glucose starting from a general compartmental model of blood–tissue exchange (i.e., no particular arrangement or number of compartments is assumed). We have to refer the reader to the original reference for the mathematical details and will focus here on the essential ingredients of the method. By using the tracer tissue and plasma measurements, one plots on a Cartesian plane the points:

$$x = \frac{\int C_p(\tau)d\tau}{C_p(t)}, \quad y = \frac{C_i(t)}{C_p(t)} \tag{18.44}$$

If there is a time t^* such that for all $t > t^*$ the amount in the exchangeable compartments is in equilibrium with the plasma tracer concentration, then, if the resultant curve for $t > t^*$ is a straight line, there is a completely irreversible region where a tracer can enter, but cannot move back into blood or into the exchangeable tissue compartments. If, on the contrary, the resultant curve for $t > t^*$ is a straight line parallel to the abscissa (with slope equal to 0), there is no irreversible region, but there are only reversible regions. Finally, if the resultant curve for $t > t^*$ is concave, this means that there is a noncompletely irreversible compartment so that there is a slow loss of the tracer from this compartment to blood.

This graphical method applies equally well to heterogeneous as well as to homogeneous tissues [45]. In particular, when the Patlak plot is used with [18F]FDG data, if one has a straight line with slope equal to m, then $k_4 = 0$; thus, 3K and TH models are correct and rGl can be calculated as:

$$rGl = m\frac{C_p}{LC} \tag{18.45}$$

If, on the contrary, the curve is concave, then $k_4 \neq 0$ and the quantitation of glucose metabolic rate is more complex.

However, the apparent simplicity of this method is limited by the difficulty of objectively determining the time t^* after which all of the tissue pools can be considered in equilibrium with the blood pool. An incorrect t^* value can, for example, lead to the conclusion of a product loss from the irreversible compartment (i.e., a concave curve for $t > t^*$), while this is only due to the lack of equilibration between tissue and plasma compartments. The t^* value changes with different tracers and different tissue regions. In some cases, t^* may exceed the time over which a particular experiment can be run and this makes the method unusable. Finally, it is worth noting that the Patlak method neither permits one to understand if the tissue is homogenous nor to estimate other physiological parameters of interest, such as the blood—tissue exchange parameters.

18.5 Models of [15O]H₂O Kinetics to Assess Blood Flow

Over the last years, various techniques and models for *in vivo* measurement of blood flow in humans have been developed using as PET tracers [15O]H₂O and

$[^{13}\text{N}]\text{NH}_3$. The basis is the Kety and Schmidt method [46], which utilizes low concentrations of a freely diffusible, chemically inert gas (N_2O) and is based on the Fick principle. If a chemically inert substance (i.e., one that is neither consumed nor produced by the organ or tissue) is introduced into the blood, the amount of the tracer in the organ depends on the difference in rates at which the tracer is brought to the organ by the arterial blood and removed from it by the venous blood:

$$dQ(t)/dt = F_A C_A(t) - F_V C_V(t) \tag{18.46}$$

where $Q(t)$ is the tracer amount in the organ or tissue, F_A and F_V are the steady-state rates of arterial inflow and venous outflow, respectively, $C_A(t)$ and $C_V(t)$ are the concentrations of the tracer in the arterial and venous blood.

In steady state, $F_A = F_V = F$ and the equation becomes:

$$dQ(t)/dt = F[C_A(t) - C_V(t)] \tag{18.47}$$

For tissues that are homogeneous with respect to the rate of perfusion and solubility of tracer, the following equation holds:

$$dQ_i(t)/dt = F_i[C_A(t) - C_{V_i}(t)] \tag{18.48}$$

where $Q_i(t)$ is the quantity of tracer in the homogeneous tissue i; F_i [mL/gr/min] is the steady-state rate of blood flow through tissue i; C_{V_i} is the venous tracer concentration in the homogenous tissue i. The difference between homogeneous and heterogeneous tissue is important if one considers, for example, brain studies. The brain is a heterogeneous organ with many component structures functioning more or less autonomously [47]. Consequently, blood flow differs in these different structures accordingly, with the variety of regulated local functions and the metabolic needs of the structural units.

For a chemically inert tracer, the difference between arterial and venous concentrations can be expressed as [48]:

$$[C_A(t) - C_{V_i}(t)] = m[C_A(t) - C_i(t)/\lambda_i] \tag{18.49}$$

where C_i is the tracer concentration in tissue i, λ_i [mL/mL] is the tissue/blood partition coefficient for the tracer in tissue i defined as:

$$\lambda_i = \lim_{t \to \infty} \frac{C_{\text{tissue}}(t)}{C_{\text{blood}}(t)} \tag{18.50}$$

and m is a constant between 0 and 1 that represents the effect of actors such as arteriovenous shunts' presence and capillary impermeability that limit the equilibration of the tissue with the blood. In absence of arteriovenous shunts and diffusion limitations of the tracer, $m = 1$.

Dividing Eq. (18.48) by the mass of tissue i, W_i [g], and using Eq. (18.49), one has:

$$\frac{dC_i(t)}{dt} = \frac{mF_i}{\lambda_i W_i}[\lambda_i C_A(t) - C_i(t)] = \frac{mF_i}{W_i}C_A(t) - \frac{mF_i}{\lambda_i W_i}C_i(t) \tag{18.51}$$

where F_i/W_i [mL/g/min] is the rate of blood flow per unit mass of tissue, and it is more properly denoted as perfusion, f_i:

$$\frac{dC_i(t)}{dt} = mf_i C_A(t) - \frac{m}{\lambda_i}f_i C_i(t) \tag{18.52}$$

Even if Kety and Schmidt have developed their method by using N_2O, the equations can also be applied when the tracer is not a gas; however, it is essential that the tracer be physiologically inert in the concentrations employed and capable of diffusing rapidly.

$[^{15}O]H_2O$ is the most frequently used positron emitting tracers for the measurement of blood perfusion by PET. By using $[^{15}O]H_2O$, it was possible to develop a simple, reliable, and noninvasive method to quantitate cerebral blood flow [2] and skeletal muscle blood flow [5] in humans. The method does not require the use of local anesthesia, and, consequently, it minimizes the possibility of functional alterations in the tissue/organ. However, $[^{15}O]H_2O$ tracer is not the only one used to measure blood flow with PET. Often for myocardial blood flow, it is preferable to use $[^{13}N]$-labeled ammonia, even if the compartmental model needed to describe the $[^{13}N]$ammonia ($[^{13}N]NH_3$) is more complex than the $[^{15}O]H_2O$ one. The advantage is that $[^{13}N]NH_3$ gives high-contrast cross-sectional images of the myocardium, since the tracer moves from the vascular space to tissue by both active transport (sodium−potassium pump) as well as by passive diffusion. Once inside the cells, this tracer is metabolized on the contrary of labeled water. These properties, together with the high tissue retention fraction, permit one to obtain high-contrast cross-sectional images of the myocardium.

$[^{15}O]H_2O$ tracer has the important property that once introduced into the tissue/organ, it is not involved in any biochemical reaction (inert tracer). Moreover, $[^{15}O]H_2O$ is close to being completely diffusible (even if Eichling and colleagues [49] showed that there is some limitation in the $[^{15}O]H_2O$ diffusion) so that the $m \approx 1$ approximation is reasonable.

The transport of $[^{15}O]H_2O$ across the capillary wall is quite fast. Furthermore, the cell membrane usually is not a large barrier for water transport. Therefore, the vascular, interstitial, and cellular spaces can be merged into a single compartment. At this point, it is simple to develop the compartmental model (Figure 18.7) to describe the $[^{15}O]H_2O$ tracer kinetics under the hypotheses that the tissue is homogeneous and that the tracer is physiologically inert at the concentrations employed and is capable of diffusing rapidly across the capillary wall.

By defining $k_1 = f_i$ [mL/g/min], $k_2 = (f_i/\lambda_i)$ [min^{-1}] one has:

$$\frac{dC_i(t)}{dt} = k_1 C_A(t) - k_2 C_i(t) \quad C_i(0) = 0 \tag{18.53}$$

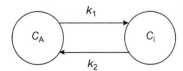

Figure 18.7 The compartmental model for the blood flow estimation with $[^{15}O]H_2O$.

where $C_i(t)$ is the tissue concentration of $[^{15}O]H_2O$ and $C_A(t)$ is its arterial plasma concentration. The knowledge of the two-tracer concentration curves allows the estimation of the two model parameters and, thus, of $k_1/k_2 = \lambda_i$.

The two parameters of physiological interest are f_i (reported as rBF) representing the blood flow value per unit of mass of tissue i (perfusion) and the value of the tissue/blood partition coefficient λ_i.

The use of ^{15}O-labeled water has the advantage that it allows it to carry out a series of repeated studies in quite a short experimental time and also to follow rapid changes in tissue activity after experimental maneuvers like, in the case of cerebral tissue, visual or sound stimuli. In dynamic studies with $[^{15}O]H_2O$, the PET procedure requires the injection at zero time of a tracer bolus, and the simultaneous start of the PET scans and arterial blood sampling (usually from a radial artery). Since ^{15}O has a half-life of 122.1 s, the total scan time is usually less than 3 min; as a result, PET scans and plasma samples have to be frequent enough to allow, respectively, for the correct detection of tracer kinetics and an accurate description of the input function. It is, however, difficult to obtain exactly the true input function in a human study, because the measured plasma arterial curve is affected by both external dispersion, due to the sampling system (blood withdrawal speed and catheter length), and by internal dispersion in the arm artery. Another practical problem that might degrade the quality of the measured input function is the time shift between the measured and the true plasma curve (i.e., the blood tracer concentration in the tissue ROI [2]). Summarizing, the peripherally sampled arterial curve, $C_c(t)$, is related to the true input function, $C_A(t)$, as:

$$C_c(t + \Delta t) = C_A(t) \otimes d(t) \tag{18.54}$$

where Δt is the time delay of the peripheral arterial curve relative to the true input function and $d(t)$ is the effective dispersion function usually described by using a single exponential function:

$$d(t) = \frac{1}{\tau}\exp\left(-\frac{t}{\tau}\right) \tag{18.55}$$

Consequently, the true input function that will be the forcing function of the model can be obtained from the peripheral arterial curve by using a deconvolution technique. Even if this technique is widely used to correct the input function for delay and dispersion, it should be noted that it is affected by some approximations. First of all, the external dispersion in the sampling system (catheter) depends not only on the withdrawal speed and catheter length, but also on other factors, such as the hematocrit value; in addition, both internal and external dispersions are described by a simple single exponential that is, probably, a rough approximation of reality.

Starting from these observations and trying to avoid the invasive arterial blood sampling, recently Watabe and colleagues [50] have proposed a new method for the calculation of the rBF that involves the elimination of the arterial input function term during cerebral studies. The model is based on the assumption of the existence of two distinct cerebral regions, ROI1 and ROI2, both described by using the compartmental model showed in Figure 18.7:

$$\frac{dC_{i1}(t)}{dt} = k_{11}C_A(t) - k_{21}C_{i1}(t) \quad C_{i1}(0) = 0 \tag{18.56}$$

$$\frac{dC_{i2}(t)}{dt} = k_{12}C_A(t) - k_{22}C_{i2}(t) \quad C_{i2}(0) = 0 \tag{18.57}$$

where it is assumed that the input function, $C_A(t)$, is the same for the two regions. By integrating Eqs. (18.56) and (18.57) twice from time 0 to T, one has:

$$\int_0^T C_{i1}(t)dt = k_{11}\int_0^T dt \int_0^t C_A(s)ds - k_{21}\int_0^T dt \int_0^t C_{i1}(s)ds \tag{18.58}$$

$$\int_0^T C_{i2}(t)dt = k_{12}\int_0^T dt \int_0^t C_A(s)ds - k_{22}\int_0^T dt \int_0^t C_{i2}(s)ds \tag{18.59}$$

From Eqs. (18.58) and (18.59), the arterial input function can be eliminated to give:

$$\int_0^T C_{i1}(t)dt = \frac{k_{11}}{k_{12}}\int_0^T C_{i2}(t)dt + k_{22}\int_0^T dt \int_0^t C_{i2}(s)ds - k_{21}\int_0^T dt \int_0^t C_{i1}(s)ds \tag{18.60}$$

The method provides regional cerebral blood flow values from the knowledge of the two tissue curves of ROI1 and ROI2, and does not require arterial blood sampling.

18.6   Models of the Ligand–Receptor System

PET allows the study of receptor density and radioligand affinity in the brain and myocardium. Quantification of the ligand–receptor system is of fundamental importance, not only to understand how the brain works (e.g., how it performs the various commands and reacts to stimuli), but also in the investigation of the pathogenesis of important diseases like Alzheimer's and Parkinson's. In recent years,

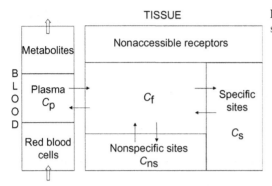

Figure 18.8 The ligand—receptor system.

PET has become an increasingly used tool to quantitate important parameters like the receptor density and the binding affinity of radioligands, and several models have been proposed of specific ligand—receptor interactions, including dopaminergic [51–53] and benzodiazepimergic [29,54–56] receptors in the brain, and muscarinic binding sites in the myocardium [57,58]. In Section 18.6.1, we discuss some of the most representative models.

18.6.1   The Three- and Two-Tissue Compartment Models

The ligand—receptor interactions can be schematized in Figure 18.8. A compartmental model reflecting the major kinetic events is shown in Figure 18.9, where C_p is the arterial plasma concentration corrected for metabolites, C_f the concentration of free ligand, C_{ns} the concentration of nonspecifically bound ligand, and C_s the concentration of specifically bound ligand.

The model equations are:

$$\begin{aligned}
\dot{C}_f(t) &= k_1 C_p(t) - (k_2 + k_3 + k_5)C_f(t) + k_4 C_s(t) + k_6 C_{ns}(t) & C_f(0) &= 0 \\
\dot{C}_s(t) &= k_3 C_f(t) - k_4 C_s(t) & C_s(0) &= 0 \\
\dot{C}_{ns}(t) &= k_5 C_f(t) - k_6 C_{ns}(t) & C_{ns}(0) &= 0
\end{aligned}$$

$$(18.61)$$

where k_1 [mL/mL/min] is the rate constant of transfer from plasma to free ligand tissue compartment, and k_2, k_3, k_4, k_5, k_6 [min^{-1}] are the rate constants of ligand transfer from tissue to plasma and inside the tissue.

To better understand the physiological meaning of parameters k_3 and k_4, let us assume that the binding of the ligand to the receptor site is describable as a bimolecular reaction:

$$L + R \underset{k_{off}}{\overset{k_{on}}{\rightleftharpoons}} LR$$

where L represents the ligand, R the receptor site, LR the binding product, k_{on} is the association rate of the ligand with the receptor sites, and k_{off} the dissociation

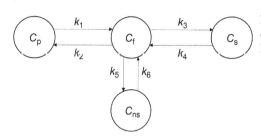

Figure 18.9 The three-tissue compartment model of the ligand−receptor system.

rate of the specifically bound reaction product. In the notation of Figure 18.9, C_f and C_s represent L and LR, respectively, thus:

$$\frac{dC_s(t)}{dt} = k_{on}C_f(t)C_r(t) - k_{off}C_s(t) \tag{18.62}$$

where C_r denotes the concentration of receptors. If B_{max} is the total number of available reactions sites, then:

$$B_{max} = C_s + C_r \tag{18.63}$$

and, if the ligand is present in tracer concentration, the concentration C_s is negligible and thus:

$$B_{max} \approx C_r \tag{18.64}$$

Equation (18.62) becomes:

$$\frac{dC_s(t)}{dt} = k_{on}B_{max}C_f(t) - k_{off}C_s(t) = k_3C_f(t) - k_4C_s(t) \tag{18.65}$$

with $k_3 = k_{on}B_{max}$ and $k_4 = k_{off}$.

An important parameter is also the equilibrium binding constant K_d, which is defined with the ligand−receptor reaction in steady state as:

$$K_d = \frac{C_s}{C_rC_f} = \frac{k_{on}}{k_{off}} \tag{18.66}$$

The PET measurement is the result of the tracer present in the tissue and of that present in the blood of the ROI. Consequently, the measurement equation is:

$$C_i(t) = (1 - V_b)(C_f(t) + C_{ns}(t) + C_s(t)) + V_bC_b(t) \tag{18.67}$$

where C_b is whole blood tracer concentration and V_b is the vascular volume.

The model is *a priori* only locally (nonuniquely) identifiable, in particular it admits two solutions for each parameter. To ensure unique identifiability, it is

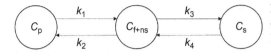

Figure 18.10 The two-tissue compartment model of the ligand−receptor system.

usually assumed that the exchange rates between the free tissue and nonspecific binding pools are sufficiently rapid (compared with the other rates of the model) so that the three-tissue compartment model of Figure 18.9 reduces to the two-tissue model of Figure 18.10, where $C_{f+ns}(t) = C_f(t) + C_{ns}(t)$ is the free and nonspecific binding tracer concentration.

The model equations are:

$$\dot{C}_{f+ns}(t) = k_1 C_p(t) - (k_2 + k_3)C_{f+ns}(t) + k_4 C_s(t) \quad C_{f+ns}(0) = 0$$
$$\dot{C}_s(t) = k_3 C_{f+ns}(t) - k_4 C_s(t) \quad\quad\quad\quad\quad\quad\quad C_s(0) = 0 \tag{18.68}$$

with:

$$k_3 = k_{on} B_{max} f \tag{18.69}$$

where f is given by:

$$f \equiv \frac{C_f}{C_{f+ns}} = \frac{C_f}{C_f + C_{ns}} = \frac{C_f}{C_f\left(1 + \frac{C_{ns}}{C_f}\right)} = \frac{1}{1 + \frac{k_5}{k_6}} \tag{18.70}$$

The measurement equation becomes:

$$C_i(t) = (1 - V_b)(C_{f+ns}(t) + C_s(t)) + V_b C_b(t) \tag{18.71}$$

where C_b is whole blood tracer concentration.

The model of Figure 18.10 is *a priori* uniquely identifiable and in addition to k_1, k_2, k_3, k_4, V_b, it is also possible to estimate the binding potential (BP):

$$BP = \frac{B_{max}}{K_d} = \frac{k_3}{k_4} \tag{18.72}$$

18.6.2 The Reference Tissue Models

The ligand−receptor models described above require the knowledge of the plasma labeled ligand concentration, which is the forcing function of the models for their identification. Recently, a method has been described by Lammertsma and colleagues [59,60] that allows the quantification of receptor kinetics without measuring the arterial input function. This method relies on the presence of a region without specific binding of the ligand, which can be considered as reference for all the other regions.

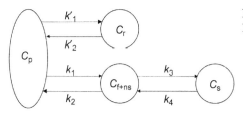

Figure 18.11 The reference ligand–receptor model.

The model is shown in Figure 18.11 where C_r is the concentration in the reference tissue and C_p is the plasma concentration corrected for metabolites and assumed to be the same for both of the regions.

The model equations are:

$$\dot{C_r}(t) = k_1' C_p(t) - k_2' C_r(t) \qquad\qquad C_r(0) = 0$$
$$\dot{C}_{f+ns}(t) = k_1 C_p(t) - (k_2 + k_3)C_{f+ns}(t) + k_4 C_s(t) \quad C_{f+ns}(0) = 0 \qquad (18.73)$$
$$\dot{C_s}(t) = k_3 C_{f+ns}(t) - k_4 C_s(t) \qquad\qquad C_s(0) = 0$$

with

$$k_3 = k_{on} B_{max} f \qquad\qquad (18.74)$$

$$k_4 = k_{off} \qquad\qquad (18.75)$$

where f has the expression of Eq. (18.70).

The model is *a priori* uniquely identifiable if one defines R_1 as:

$$R_1 = \frac{k_1}{k_1'} \qquad\qquad (18.76)$$

and assumes that volume of distribution of the not specifically bound tracer in both tissues is the same, that is:

$$V_d = \frac{C_{f+ns}}{C_p} = \frac{C_r}{C_p} \qquad\qquad (18.77)$$

and thus:

$$\frac{k_1}{k_2} = \frac{k_1'}{k_2'} \qquad\qquad (18.78)$$

with these assumptions the model output can be written as:

$$C_{f+ns}(t) + C_s(t) = R_1 C_r(t) + a \int_0^t C_r(\tau) e^{-c(t-\tau)} \, d\tau + b \int_0^t C_r(\tau) e^{-d(t-\tau)} \, d\tau \qquad (18.79)$$

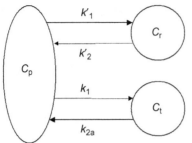

Figure 18.12 The simplified reference ligand−receptor model.

where C_r is the model input and a, b, c, and d are combinations of parameters R_1, k_2, k_3, k_4.

The measurement equation is:

$$C_i(t) = C_{f+ns}(t) + C_s(t) \tag{18.80}$$

The four model parameters R_1, k_2, k_3, k_4 can be estimated together with their precision by using nonlinear least squares as previously described. The model also provides the binding potential BP as:

$$\text{BP} = \frac{k_3}{k_4} \tag{18.81}$$

If the tracer kinetics in the target region are such that it is difficult to distinguish between free and specifically bound compartments, the reference tissue model can be simplified as in Figure 18.12.

The equations are:

$$\begin{aligned}\dot{C}_r(t) &= k_1' C_p(t) - k_2' C_r(t) \quad C_r(0) = 0 \\ \dot{C}_t(t) &= k_1 C_p(t) - k_{2a} C_t(t) \quad C_t(0) = 0\end{aligned} \tag{18.82}$$

where $C_t(t) = C_{f+ns}(t) + C_r(t)$ is the total tracer concentration in the tissue and k_{2a} the apparent rate constant of transfer from the specifically bound compartment to plasma related to the parameters k_2, k_3, k_4 of the model of Figure 18.11 as:

$$k_{2a} = \frac{k_2}{1 + \text{BP}} = \frac{k_2}{1 + k_3 k_4} \tag{18.83}$$

Under the assumption that Eqs. (18.76) and (18.78) still hold, the equation corresponding to Eq. (18.79) becomes:

$$C_t(t) = R_1 C_r(t) + \left[k_2 - \frac{R_1 k_2}{1 + \text{BP}}\right] \int_0^t C_r(\tau) e^{-\frac{k_2}{1+\text{BP}}(t-\tau)} \, d\tau \tag{18.84}$$

The measurement equation is:

$$C_i(t) = C_t(t) \tag{18.85}$$

By using nonlinear least squares, the three model parameters R_1, k_2, BP can be estimated together with precision.

18.6.3 A Nonlinear Model of the Ligand–Receptor System

All of the linear compartmental models discussed above do not permit an estimation of the individual values of B_{max}, k_{on}, and K_d. To do so, it is necessary to move from a single-labeled ligand experiment to a protocol also including an injection of unlabeled ligand. Under these circumstances, most receptor sites become occupied, C_s in Eq. (18.62) is not negligible, and thus k_3 is no longer a constant. Equation (18.62) becomes:

$$\frac{dC_s(t)}{dt} = k_{on}(B_{max} - C_s(t))C_f(t) - k_{off}C_s(t) = k_3(t)C_f(t) - k_{off}C_s(t) \tag{18.86}$$

with

$$\begin{aligned}
k_3(t) &= k_{on}(B_{max} - C_s(t)) \\
k_4 &= k_{off}
\end{aligned} \tag{18.87}$$

There is thus the need to also describe the unlabeled ligand kinetics. The model is shown in Figure 18.13.

It is nonlinear and its equations are:

$$\begin{aligned}
\dot{C}_{f+ns}(t) &= k_1 C_p(t) - (k_2 + k_3(t))C_{f+ns}(t) + k_4 C_s(t) & C_{f+ns}(0) &= 0 \\
\dot{C}_s(t) &= k_3(t)C_{f+ns}(t) - k_{off}C_s(t) & C_s(0) &= 0 \\
\dot{C}'_{f+ns}(t) &= k_1 C'_p(t) - (k_2 + k_3(t))C'_{f+ns}(t) + k_4 C'_s(t) & C'_{f+ns}(0) &= 0 \\
\dot{C}'_s(t) &= k_3(t)C'_{f+ns}(t) - k_{off}C'_s(t) & C'_s(0) &= 0
\end{aligned} \tag{18.88}$$

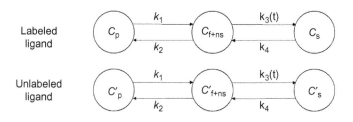

Figure 18.13 The labeled and unlabeled ligand two-tissue compartment model of the ligand–receptor system.

where C_p, C_{f+ns}, and C_s have the same meaning as before and C'_p, C'_{f+ns}, C'_s are, respectively, the unlabeled ligand plasma concentration (corrected for metabolites), free plus nonspecifically bound unlabeled ligand concentration, and specifically bound unlabeled ligand concentration.

The measurement equation is still given by Eq. (18.71).

The model is *a priori* uniquely identifiable from the measurements C_i, C_p, and C'_p of a labeled plus unlabeled ligand injection experiment (C_b can be calculated from C_p by using well-known formulas involving hematocrit). In particular, one can estimate k_1, k_2, k_{on}, B_{max}, k_{off}, V_b parameters and, thus, the ligand affinity K_d. Often, to improve *a posteriori* (numerical) identifiability of the model, the experimental protocol involves several injections of labeled and/or unlabeled ligand (e.g., Ref. 58 a tracer dose of [^{11}C]flumazenil, a specific ligand of benzodiazepine receptors, followed by two cold injections of flumazenil and by a co-injection of labeled and unlabeled flumazenil at a subsequent time).

18.6.4   *The Logan Graphical Method*

All of the ligand−receptor models presented above require numerical identification (e.g., by nonlinear least squares) from the data to quantitate the receptor system. Simpler approaches based on graphical methods have been developed and, because of their simplicity, they are frequently employed for system quantification. A popular graphical method for BP quantification has been proposed by Logan and colleagues [61]. Under the assumption that the two-tissue compartment model of Figure 18.10 accurately describes tracer kinetics in the tissue, the method uses the plasma and tissue tracer concentration curves and proposes (see Logan and colleagues [61] for details) plotting on the Cartesian plane the points of coordinates:

$$x = \frac{\int C_p(\tau)d\tau}{C_i(t)}, \quad y = \frac{\int_0^t C_i(\tau)d\tau}{C_i(t)} \tag{18.89}$$

If there exists a time t^* such that the tissue regions are in equilibrium with plasma, then the plot yields, after time t^*, a straight line with slope m equal to:

$$m = \frac{k_1}{k_2} + \frac{k_1 k_3}{k_2 k_4} + V_b = V_d\left(1 + \frac{B_{max}}{K_d}\right) + V_b = V_d(1 + BP) + V_b \tag{18.90}$$

Under the assumption of V_d known and V_b known or negligible (i.e., for the brain tissue $V_b \cong 3 - 4\%$), it is possible to estimate BP from the m value.

In 1996, Logan and colleagues [62] also proposed another graphical method for determining the distribution volume ratio (which is a linear function of receptor availability) that does not require blood sampling. As in the model of Lammertsma and colleagues [59,60], in this case, one uses plasma arterial activity instead of data from a region not containing specific receptors, but an additional assumption is needed related to the tissue to plasma efflux constant that is fixed to a population value.

18.7   The Way Forward

In the previous sections, the basis of PET modelling has been addressed. The identification of model parameters at ROI level approach is commonly used to quantify PET images by using nonlinear estimators, since ROI time activity curves are characterized by a good signal-to-noise ratio (SNR). However, the use of ROI time activity curves causes a loss of spatial resolution. Consequently, in the last few years, methods for voxel level analysis have been developed in order to produce parametric maps having the same spatial resolution as the original PET image, but addressing, at the same time, the low SNR presence at voxel level. However, this renders the use of nonlinear estimators difficult and unwieldy because of their computational cost. To improve robustness and/or computational efficiency of voxel-wise quantification level via compartmental modelling, several solutions have been proposed in literature. The most recent are represented by the basis function methods as used in [63,64] Global Two-Stage [65,66] and Bayesian estimation [67,68].

In addition, several attempts have been made to make the experimental protocol required for a gold standard quantitative analysis less invasive. In particular, quantitative PET often requires that the input function be measured, typically via arterial cannulation. Image-derived input function (IDIF) is an attractive noninvasive alternative to arterial sampling. Thus, several attempts have been made in the last few years in order to understand the reliability of the IDIF approach with alternative findings, depending on the PET tracer. In a recent article [69], different methods have been compared using different tracers. The conclusions were that IDIF can be successfully implemented only with a minority of PET tracers. However, PET modellers are still working to solve this important issue and, recently, a new contribution has been published that derived the IDIF based on calculation of the Pearson correlation coefficient between the time activity curves [70].

18.8   Conclusion

PET images interpreted with physiological models allow the noninvasive quantification at the organ level of important physiological processes such as glucose metabolism, blood flow, and receptor density and affinity. To arrive at a reliable quantification of these processes and parameters, several ingredients must concur, including the quality of the PET scanner, radiotracer, image processing method, and physiological model. In this chapter, we have concentrated on available modelling and identification strategies for interpreting PET images. Focus has been on linear and nonlinear compartmental models to quantitate glucose metabolism, blood flow, and ligand−receptor interaction. The importance of input−output modelling as an aid and guide in model structure selection has also been stressed. Finally, some popular graphical methods have been reviewed in light of their underlying assumptions.

References

[1] Phelps ME, Huang SC, Hoffman EJ, Selin C, Sokoloff L, Kuhl DE. Tomographic measurement of local cerebral glucose metabolic rate in humans with (F-18)2-fluoro-2-deoxy-D-glucose: validation of method. Ann Neurol 1979;6:371−88.

[2] Iida H, Kanno I, Miura S, Murakami M, Takahashi K, Uemura K. Error analysis of a quantitative cerebral blood flow measurement using $H_2{}^{15}O$ autoradiography and positron emission tomography, with respect to the dispersion of the input function. J Cereb Blood Flow Metab 1986;6:536−45.

[3] Kelley DE, Williams KV, Price JC. Insulin regulation of glucose transport and phosphorylation in skeletal muscle assessed by PET. Am J Physiol 1999;277:E361−9.

[4] Kelley DE, Mintun MA, Watkins SC, Simoneau J-A, Jadali F, Fredrickson A, et al. The effect of non-insulin-dependent diabetes mellitus and obesity on glucose transport and phosphorylation in skeletal muscle. J Clin Invest 1996;97:2705−13.

[5] Ruotsalainen U, Raitakari M, Nuutila P, Oikonen V, Sipila H, Teras M, et al. Quantitative blood flow measurement of skeletal muscle using oxygen-15-water and PET. J Nucl Med 1997;38:314−9.

[6] Gambhir S, Schwaiger M, Huang SC, Krivokapich J, Schelbert HR, Nienaber CA, et al. Simple noninvasive quantification method for measuring myocardial glucose utilization in humans employing positron emission tomography and fluorine-18 deoxyglucose. J Nucl Med 1989;30:359−66.

[7] Bergmann SR, Fox KA, Rand AL, McElvany KD, Welch MJ, Markham J, et al. Quantification of regional myocardial blood flow *in vivo* with $H_2{}^{15}O$. Circulation 1984;70:724−33.

[8] Wong DF, Gjedde A, Wagner Jr HN. Quantification of neuroreceptors in the living human brain. I. Irreversible binding of ligands. J Cereb Blood Flow Metab 1986; 6:137−46.

[9] Sokoloff L, Reivich M, Kennedy C, Des-Rosiers MH, Patlak CS, Pettigrew KD, et al. The [^{14}C]deoxyglucose method for the measurement of local cerebral glucose utilization: theory, procedure, and normal values in the conscious and anesthetized albino rat. J Neurochem 1977;28:897−916.

[10] Cobelli C, Carson ER. Introduction to modeling in physiology and medicine. Amsterdam: Academic Press; 2008.

[11] Jacquez JA. Compartmental analysis in biology and medicine. 3rd ed. Ann Arbor, MI: BioMedware; 1996.

[12] Cobelli C, Foster DM, Toffolo G. Tracer kinetics in biomedical research: from data to model. New York, NY: Kluwer Academic Publishers; 2000.

[13] Cunningham V, Jones T. Spectral analysis of dynamic PET studies. J Cereb Blood Flow Metab 1993;13:15−23.

[14] Bertoldo A, Cobelli C. A more general and statistically robust spectral analysis method. J Cereb Blood Flow Metab 1999;19(Suppl. 1):S762.

[15] Meikle SR, Matthews JC, Brock CS, Wells P, Harte RJ, Cunningham VJ, et al. Pharmacokinetic assessment of novel anti-cancer drugs using spectral analysis and positron emission tomography: a feasibility study. Cancer Chemother Pharmacol 1998; 42:183−93.

[16] Meikle SR, Matthews JC, Cunningham VJ, Bailey DL, Livieratos L, Jones T, et al. Parametric image reconstruction using spectral analysis of PET projection data. Phys Med Biol 1998;43:651−66.

[17] Richardson MP, Koepp MJ, Brooks DJ, Fish DR, Duncan JS. Benzodiazepine receptors in focal epilepsy with cortical dysgenesis: an 11C-flumazenil PET study. Ann Neurol 1996;40:188−98.

[18] Turkheimer F, Moresco RM, Lucignani G, Sokoloff L, Fazio F, Schmidt K. The use of spectral analysis to determine regional cerebral glucose utilization with positron emission tomography and [^{18}F]fluorodeoxyglucose: theory, implementation, and optimization procedures. J Cereb Blood Flow Metab 1994;14:406−22.

[19] Alpert NM, Chesler DA, Correia JA, Ackerman RH, Chang JY, Finklestein S, et al. Estimation of the local statistical noise in emission computed tomography. IEEE Trans Med Imag 1982;MI-1:142−6.

[20] Budinger TF, Derenzo SE, Greenberg WL, Gullberg GT, Huesman RH. Quantitative potentials of dynamic emission computed tomography. J Nucl Med 1978;19:309−15.

[21] Budinger TF, Derenzo SE, Gullberg GT, Greenberg WL, Huesman RH. Emission computed assisted tomography with single-photon and positron annihilation emitters. J Comput Assist Tomogr 1977;1:131−45.

[22] Carson RE, Yan Y, Daube-Witherspoon ME, Freedman N, Bacharach SL, Herscovitch P. An approximation formula for the variance of PET region-of-interest values. IEEE Trans Med Imaging 1993;12:240−50.

[23] Huesman RH. A new fast algorithm for the avelutation of regions of interest and statistical uncertainty in computed tomography. Phys Med Biol 1984;29:543−52.

[24] Huesman RH. The effects of a finite number of projection angles and finite laterale sampling of projections on the propagation of statistical errors in transverse section recostruction. Phys Med Biol 1977;22:511−21.

[25] Defrise M, Townsend DW, Deconinck F. Statistical noise in three-dimensional positron emission tomography. Phys Med Biol 1990;35:131−8.

[26] Pajevic S, Daube-Witherspoon ME, Bacharach SL, Carson RE. Noise characteristics of 3-D and 2-D PET images. IEEE Trans Med Imaging 1998;17:9−23.

[27] Mazoyer BM, Huesman RH, Budinger TF, Knittel BL. Dynamic PET data analysis. J Comput Assist Tomogr 1986;10:645−53.

[28] Delforge J, Syrota A, Mazoyer BM. Identifiability analysis and parameter identification of an in vivo ligand-receptor model from PET data. IEEE Trans Biomed Eng 1990; 37:653−61.

[29] Delforge J, Pappata S, Millet P, Samson Y, Bendriem B, Jobert A, et al. Quantification of benzodiazepine receptors in human brain using PET, [^{11}C]flumazenil, and a single-experiment protocol. J Cereb Blood Flow Metab 1995;15:284−300.

[30] Delforge J, Bottlaender M, Loc'h C, Guenther I, Fuseau C, Bendriem B, et al. Quantitation of extrastriatal D2 receptors using a very high-affinity ligand (FLB 457) and the multi-injection approach. J Cereb Blood Flow Metab 1999;19:533−46.

[31] Bertoldo A, Vicini P, Sambuceti G, Lammertsma AA, Parodi O, Cobelli C. Evaluation of compartmental and spectral analysis models of [^{18}F]FDG kinetics for heart and brain studies with PET. IEEE Trans Biomed Eng 1998;45:1429−48.

[32] Blomqvist G, Bergström K, Bergström M, Ehrin E, Eriksson L, Garmelius B, et al. Models for 11C-glucose. In: Greitz T, Ingvar DH, Widen L, editors. The metabolism of the human brain studied with positron emission tomography. New York, NY: Raven Press; 1985. p. 185−94.

[33] Blomqvist G, Stone-Elander S, Halldin C, Roland PE, Widen L, Lindqvist M, et al. Positron emission tomographic measurements of cerebral glucose utilization using [1-^{11}C]D-glucose. J Cereb Blood Flow Metab 1990;10:467−83.

[34] Powers WJ, Dagogo JS, Markham J, Larso KB, Dence CS. Cerebral transport and metabolism of 1-^{11}C-D-glucose during stepped hypoglycemia. Ann Neurol 1995; 38:599−609.

[35] Fanelli CG, Dence CS, Markham J, Videen TO, Paramore DS, Cryer PE, et al. Blood-to-brain glucose transport and cerebral glucose metabolism are not reduced in poorly controlled type 1 diabetes. Diabetes 1998;47:1444−50.

[36] Hasselbalch SG, Knudsen GM, Holm S, Hageman LP, Capaldo B, Paulson OB. Transport of D-glucose and 2-fluorodeoxyglucose across the blood−brain barrier in humans. J Cereb Blood Flow Metab 1996;16:659−66.

[37] Spence AM, Muz M, Graham MM, O'Sullivan F, Krohn KA, Link JM, et al. Glucose metabolism in human malignant gliomas measured quantitatively with PET, 1-[C-11] glucose and FDG: analysis of the FDG lumped constant. J Nucl Med 1998;39:440−8.

[38] Botker HE, Goodwin GW, Holden JE, Doenst T, Gjedde A, Taegtmeyer H. Myocardial glucose uptake measured with fluorodeoxyglucose: a proposed method to account for variable lumped constants. J Nucl Med 1999;40:1186−96.

[39] Krivokapich J, Huang SC, Selin CE, Phelps ME. Fluorodeoxyglucose rate constants, lumped constant, and glucose metabolic rate in rabbit heart. Am J Physiol 1987;252: H777−87.

[40] Utriainen T, Mäkimattilla S, Lovisatti S, Bertoldo A, Bonadonna R, Weintraub S, et al. Lumped constant for [^{14}C]deoxy-D-glucose in human skeletal muscle. Diabetologia 1998;41:A1818.

[41] Kelley DE, Williams KV, Price JC, Goodpaster B. Determination of the lumped constant for [^{18}F]FDG in human skeletal muscle. J Nucl Med 1999;40:1798−804.

[42] Schmidt K, Mies G, Sokoloff L. Model of kinetic behavior of deoxyglucose in heterogeneous tissues in brain: a reinterpretation of the significance of parameters fitted to homogeneous tissue models. J Cereb Blood Flow Metab 1991;11:10−24.

[43] Schmidt K, Lucignani G, Moresco RM, Rizzo G, Gilardi MC, Messa C, et al. Errors introduced by tissue heterogeneity in estimation of local cerebral glucose utilization with current kinetic models of the [^{18}F]fluorodeoxyglucose method. J Cereb Blood Flow Metab 1992;12:823−34.

[44] Patlak CS, Blasberg RG, Fenstermacher JD. Graphical evaluation of blood-to-brain transfer constants from multiple-time uptake data. J Cereb Blood Flow Metab 1983; 3:1−18.

[45] Patlak CS, Blasberg RG. Graphical evaluation of blood-to-brain transfer constants from multiple-time uptake data. Generalizations. J Cereb Blood Flow Metab 1985; 5:584−90.

[46] Kety SS, Schmidt CF. The nitrous oxide method for the quantitative determination of cerebral blood flow in man: theory, procedure, and normal values. J Clin Invest 1948; 27:476−83.

[47] Sokoloff L. Cerebral metabolism and visualization of cerebral activity. In: Greger R, Windhorst U, editors. Comprehensive human physiology. New York, NY: Springer-Verlag; 1996. p. 579−601.

[48] Kety SS. The theory and application of the exchange of inert gas at the lungs and tissues. Pharmacol Rev 1951;3:1−41.

[49] Eichling JO, Raichle MA, Grubb RL, Ter-Pergossian MM. Evidence of the limitations of water as a freelt diffusible tracer in brain of the rhesus monkey. Circ Res 1974; 35:358−64.

[50] Watabe H, Itoh M, Cunningham VJ, Lammertsma AA, Bloomfield PM, Mejia M, et al. Noninvasive quantification of rCBF using positron emission tomography. In: Myers R,

Cunningham V, Bailey D, Jones T, editors. Quantification of brain function using PET. London: Academic Press; 1993. p. 191–5.

[51] Ito H, Okubo Y, Halldin C, Farde L. Mapping of central D2 dopamine receptors in man using [^{11}C]raclopride: PET with anatomic standardization technique. Neuroimage 1999;9:235–42.

[52] Backman L, Robins-Wahlin TB, Lundin A, Ginovart N, Farde L. Cognitive deficits in Huntington's disease are predicted by dopaminergic PET markers and brain volumes. Brain 1997;120:2207–18.

[53] Farde L, Eriksson L, Blomquist G, Halldin C. Kinetic analysis of central [^{11}C]raclopride binding to D2-dopamine receptors studied by PET—a comparison to the equilibrium analysis. J Cereb Blood Flow Metab 1989;9:696–708.

[54] Malizia AL, Cunningham VJ, Bell CJ, Liddle PF, Jones T, Nutt DJ. Decreased brain GABA(A)-benzodiazepine receptor binding in panic disorder: preliminary results from a quantitative PET study. Arch Gen Psychiatry 1998;55:715–20.

[55] Sihver W, Sihve S, Bergstrom M, Murata T, Matsumura K, Onoe H, et al. Methodological aspects for *in vitro* characterization of receptor binding using 11C labeled receptor ligands: a detailed study with the benzodiazepine receptor antagonist [^{11}C]Ro 15-1788. Nucl Med Biol 1997;24:723–31.

[56] Delforge J, Spelle L, Bendriem B, Samson Y, Syrota A. Parametric images of benzodiazepine receptor concentration using a partial-saturation injection. J Cereb Blood Flow Metab 1997;17:343–55.

[57] Le Guludec D, Cohen Solal A, Delforge J, Delahaye N, Syrota A, Merlet P. Increased myocardial muscarinic receptor density in idiopathic dilated cardiomyopathy: an *in vivo* PET study. Circulation 1997;96:3416–22.

[58] Delforge J, Le Guludec D, Syrota A, Bendriem B, Crouzel C, Slama M, et al. Quantification of myocardial muscarinic receptors with PET in humans. J Nucl Med 1993;34:981–91.

[59] Lammertsma AA, Bench CJ, Hume SP, Osman S, Gunn K, Brooks DJ, et al. Comparison of methods for analysis of clinical [^{11}C]raclopride studies. J Cereb Blood Flow Metab 1996;16:42–52.

[60] Lammertsma AA, Hum SP. Simplified reference tissue model for PET receptor studies. Neuroimage 1996;4:153–8.

[61] Logan J, Fowler JS, Volkow ND, Wolf AP, Dewey SL, Schlyer DJ, et al. Graphical analysis of reversible radioligand binding from time-activity measurements applied to [N-^{11}C-methyl]-(-)-cocaine PET studies in human subjects. J Cereb Blood Flow Metab 1990;10:740–7.

[62] Logan J, Fowler JS, Volkow ND, Wang GJ, Ding YS, Alexoff DL. Distribution volume ratios without blood sampling from graphical analysis of PET data. J Cereb Blood Flow Metab 1996;16:834–40.

[63] Hong YT, Fryer TD. Kinetic modelling using basis functions derived from two-tissue compartmental models with a plasma input function: general principle and application to [^{18}F]fluorodeoxyglucose positron emission tomography. Neuroimage 2010;51:164–72.

[64] Rizzo G, Turkheimer FE, Bertoldo A. Multi-scale hierarchical approach for parametric mapping: assessment on multi-compartmental models. Neuroimage 2012;59:2485–93.

[65] Tomasi G, Bertoldo A, Cobelli C. PET parametric imaging improved by global-two-stage method. Ann Biomed Eng 2009;37:419–27.

[66] Tomasi G, Bertoldo A, Cobelli C, Pavese N, Tai YF, Hammers A, et al. Global-two-stage filtering of clinical PET parametric maps: application to [(11)C]-(R)-PK11195. Neuroimage 2011;55:942–53.

[67] Alpert NM, Yuan F. A general method of Bayesian estimation for parametric imaging of the brain. Neuroimage 2009;45:1183−9.

[68] Rizzo G, Turkheimer FE, Keihaninejad S, Bose SK, Hammers A, Bertoldo A. Multi-scale hierarchical generation of PET parametric maps: application and testing on a [^{11}C]DPN study. Neuroimage 2012;59:2485−93.

[69] Zanotti-Fregonara P, Chen K, Liow JS, Fujita M, Innis RB. Image-derived input function for brain PET studies: many challenges and few opportunities. J Cereb Blood Flow Metab 2011;31:1986−98.

[70] Schain M, Benjaminsson S, Varnäs K, Forsberg A, Halldin C, Lansner A, et al. Arterial input function derived from pairwise correlations between PET-image voxels. J Cereb Blood Flow Metab 2013;33:1058−65.

19 Tumor Growth Modelling for Drug Development

Monica Simeoni[a], Paolo Magni[b], Italo Poggesi[c], Giuseppe De Nicolao[b] and Maurizio Rocchetti[d]

[a]Biopharm and ImmunoInflammation, Clinical Pharmacology, Quantitative Sciences, GlaxoSmithKline, Stockley Park West, Uxbridge, Middlesex, UK, [b]Dipartimento di Ingegneria Industriale e dell'Informazione, Università degli Studi di Pavia, Pavia, Italy, [c]Model-Based Drug Development, Janssen R&D, Via Michelangelo Buonarroti, Cologno Monzese(MI), Italy, [d]Independent Consultant, Rho, Milan, Italy

19.1 Introduction

Every new compound appearing on the market is the result of many years of experimental work and testing. The full research and development (R&D) process for a new drug lasts on average 13.7 years, 4.7 of which are dedicated to discovery research and 9 to its clinical development (data from the Pharmaceutical Benchmarking Forum [1]). Among 30 preclinical candidates, only one will succeed in becoming an approved new molecular entity (NME) [1], with an average estimated cost of drug development of $1.137 billion (data from Deloitte LLP and Thompson Reuters [2]).

Over the last three decades, advances in genetics and molecular/cell biology have greatly extended our knowledge of the molecular pathogenesis of cancer. This led to the identification of potential therapeutic targets involved in intracellular signaling processes responsible for malignant transformation and tumor progression. However, despite the high yield of potential new agents through integrative and innovative approaches in drug discovery, the attrition rate of new chemical entities still remains high, with only about 5% of cancer therapeutic or vaccine molecules considered eligible for clinical development that eventually obtain approval in the US and/or in Europe [3−6].

Quantitative model-based drug development (MBDD) has been recognized to have the potential to improve the efficiency of drug development [7]. The acceptance of MBDD from regulatory agencies, industry, and academia has been growing in the past few years [8]. Development of anticancer drugs is certainly a field in which this approach can bring major benefits [9,10].

Modelling Methodology for Physiology and Medicine. DOI: http://dx.doi.org/10.1016/B978-0-12-411557-6.00019-7

19.2   R&D Cycle Time: From Discovery to Launch

A drug discovery program starts because there is a disease or clinical condition characterized by a medical need (e.g. without suitable treatment or with treatments showing low benefit/risk balance). The initial research, often conducted in academia, originates from a hypothesis—e.g. that the inhibition or activation of a protein or pathway will result in a therapeutic effect in a disease state—and generates data in its support. The outcome of this activity is the identification of a target that may require further validation prior to progress into the lead discovery phase in order to justify a drug discovery effort. A good target needs to be efficacious and safe, to meet clinical and commercial needs, and, above all, to be accessible to the putative drug molecule that, upon binding or other forms of interaction, elicits a measurable biological response both *in vitro* and *in vivo*. Target selection and validation are followed by hit identification, where molecules are screened and retested for activity (Figure 19.1). The subsequent step is the lead discovery phase, in which molecules are further screened or modified to produce more potent and selective compounds that possess a likely safe profile and pharmacokinetic (PK) properties adequate to examine their efficacy in the available *in vivo* models. The aim of the lead discovery phase is to find the candidate, a drug-like small molecule, or biological therapeutic compound, which will

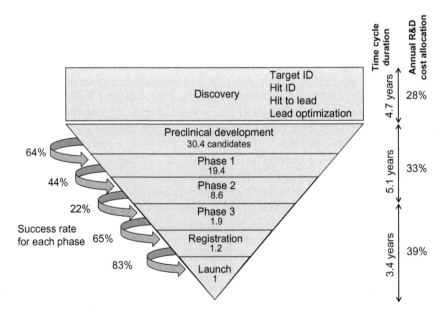

Figure 19.1 Drug development process: attrition rate, time cycle duration, annual R&D cost allocation.
(*Source*: Data from the Pharmaceutical Benchmarking Forum [1] and from Deloitte LLP and Thompson Reuters [2]).

progress into preclinical and clinical development and ultimately be a marketed medicine [11].

A selected candidate will progress first in preclinical development, in which the compound is tested for safety and efficacy in animals, and subsequently in three clinical development phases. In Phase 1, the safety and tolerability of the compound in a restricted population of humans is explored (e.g. up to a few dozen subjects), usually healthy volunteers in other therapeutic areas, but most frequently patients in oncology. Phase II focuses on the effectiveness and safety of the compound in a larger group of patients (e.g. up to a few hundred) with the aim of determining the best dosage and delivery method. Phase III comprises large efficacy confirmatory trials enrolling from several hundred up to a few thousand patients. Registration follows, which is the filing of an application with the health regulatory authority of a country for obtaining approval to market the new medicine. The application provides a description of the medicine manufacturing process along with quality data and trial results in order to demonstrate the safety and effectiveness of the new medicine. In the US, a New Drug Application (NDA) is filed with the US Food and Drug Administration (FDA); in Europe, a Market Authorization Application (MAA) is filed with the European Agency for the Evaluation of Medicinal Products (EMEA). If approval is granted, the new medicine can then be sold for use by patients.

19.3 Preclinical Development in Oncology

The *in vivo* evaluation of the antitumor effect is a fundamental step in the preclinical development of oncology drugs. Human tumors are usually transplanted into animal models (xenografts) with reduced capacity to reject "foreign" tissue, like athymic nude mice. These animals, resulting from the inheritance of a recessive mutation, are hairless and exhibit the congenital absence of thymus with the consequent inhibition of T-lymphocytes that are responsible for the immune reaction of the host. In preclinical development studies, tumor cells from immortalized human cell lines are inoculated into the flank of athymic mice and when the tumors reach a prespecified volume, mice are randomized into control and treated groups. Examples of cell lines are HCT116 (human colon carcinoma), A2780, and H207 (both human ovarian carcinoma). The tumor growth in the different groups is then evaluated by recording the tumor dimensions at selected time points using calipers or other techniques (e.g. imaging). The tumor mass (mg) is calculated as

$$\text{weight} = \rho_{\text{tr}} \, \text{length} \, \text{width}^2$$

approximating the tumor shape with the ellipsoid generated from the rotation of a semi-ellipsis around its larger axis (length) and assuming the tumor density

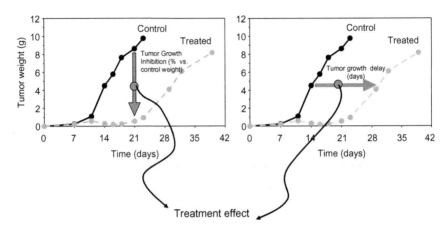

Figure 19.2 Preclinical measures of drug benefit in oncology.

$\rho_{tr} = 1$ mg/mm^3. Experimental results are then summarized using different metrics, such as the ratio of treated to control tumor weight at a prespecified time, or the difference in time (treated-control; tumor delay) to reach a predefined tumor weight (Figure 19.2).

These inhibition metrics are not invariant with respect to the experimental conditions. Only mathematical models that are able to describe tumor growth by dissecting the system-specific properties (i.e. unperturbed tumor progression) from the drug-related effects (described by an exposure-effect [PK−PD] model) can provide compound-specific and experiment-independent model parameters [10].

19.4   A Preclinical Tumor Growth Inhibition Model

Usually a preclinical tumor growth study presents a control arm and one or more treatment arms, each arm composed of five to ten animals [10]. Plasma and, sometimes tumor, drug concentrations are typically measured in ancillary groups or in separate studies. The ideal situation would be to have a mathematical model linking the dosing regimen with the tumor growth dynamic via the plasma concentrations, which is what is normally called a pharmacokinetic/ pharmacodynamic (PK/PD) model (Figure 19.3). So, first the tumor dynamic without intervention needs to be mathematically described (unperturbed model); then the tumor dynamic in the presence of a treatment (perturbed model) can be described. In the following section, we will present and discuss a model proposed in 2004 [12], widely known in the field, even though other tumor growth models are also available.

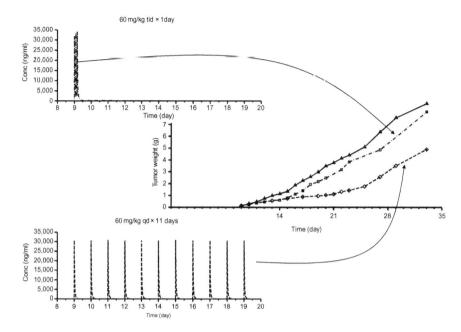

Figure 19.3 PK profiles (dash-dot line: 60 mg/kg TID (i.e. three times per day) for 1 day, dashed line: 60 mg/kg qd (i.e. once per day) for 11 days) and corresponding PD tumor growth (TG). Solid line: TG profile of controls.

19.4.1 The Unperturbed Model

The experimental data collected in the control groups show an exponential-type tumor growth followed by a roughly linear phase. In the literature, tumor masses are reported to achieve an asymptotic weight [13]; however, this phase is not reached in the typical data collected in the preclinical experiments we are considering. For this reason, we adopt a model taking into account only the exponential and linear phases (Figure 19.4).

To get rid of the threshold time t_{th}, when the curve switches from the exponential into the linear phase, it is convenient to let the growth rate of the tumor switch when a threshold tumor weight is reached (this being a biologically relevant, system-related parameter). Calling $w(t)$ the tumor weight, in terms of differential equations, such a model is given by

$$\dot{w}(t) = f_u(w(t))$$
$$w(0) = w_0$$

with:

$$f_u(w) = \begin{cases} \lambda_0 w, & w \leq \tilde{w} \\ \lambda_1, & w > \tilde{w} \end{cases} \tag{19.1}$$

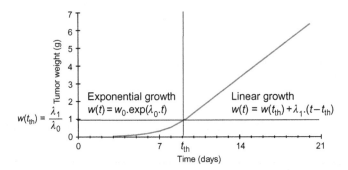

Figure 19.4 Typical tumor growth time profile in untreated mice: exponential and linear phases.

$$\tilde{w} = \frac{\lambda_1}{\lambda_0} \tag{19.2}$$

where w_0 represents the tumor weight at the inoculation time ($t = 0$) and $\lambda_0 > 0$, $\lambda_1 > 0$ are parameters characterizing the rate of exponential and linear growth, respectively. The value $\tilde{w}$ is the threshold weight beyond which the exponential growth turns into linear growth, which may be due to nutrient supply limitations. The continuity of derivative of $f_u(w)$ in $w = \tilde{w}$ follows from Eq. (19.2).

When the parameters w_0, λ_0, and λ_1 have to be estimated from experimental data via nonlinear least squares, for numerical purposes, it is convenient to approximate $f_u(w)$ with the differentiable function:

$$\tilde{f}_u(w) = \frac{\lambda_0 w}{\left[1 + \left(\frac{\lambda_0}{\lambda_1} w\right)^{\Psi}\right]^{1/\Psi}} \tag{19.3}$$

which, for values of Ψ large enough, is a good approximation of f_u.

19.4.2 Perturbed-Growth Model

The perturbed-growth model extends the unperturbed model accounting for the effect of the anticancer drug. This PD model assumes that, after the exposure of tumor cells to the drug, only a fraction retain the ability to proliferate so that the unperturbed tumor growth rate decreases by a factor proportional to the number of proliferating cells and drug concentration. The other cells, which the drug action has rendered nonproliferating, do not die immediately, but only after passing through progressive degrees of damage. This delayed process can be properly described using a transit compartmental system.

Let x_1 indicate the weight of the proliferating cells, x_2, ..., x_n be the weights of the $n - 1$ subpopulations of nonproliferating cells, and w be the weight of the whole tumor. The perturbed-growth model is

$$
\begin{cases}
\dfrac{dx_1(t)}{dt} = f_p(x_1(t), w(t)) - k_2 c(t) x_1(t) \\[2mm]
\dfrac{dx_2(t)}{dt} = k_2\, c(t) x_1(t) - k_1\, x_2(t) \\[2mm]
\dfrac{dx_3(t)}{dt} = k_1\, [x_2(t) - x_3(t)] \\[2mm]
\quad \vdots = \vdots \\[2mm]
\dfrac{dx_n(t)}{dt} = k_1\, [x_{n-1}(t) - x_n(t)] \\[2mm]
w(t) = \displaystyle\sum_{i=1}^{n} x_i(t)
\end{cases}
\tag{19.4}
$$

where

$$
x_1(0) = w_0, \quad x_2(0) = x_3(0) = \cdots = x_n(0) = 0
$$
$$
c(t) = 0, \qquad 0 < t \le t_0
$$

$$
f_p(x_1, w) = \begin{cases}
\lambda_0 x_1, & w \le \dfrac{\lambda_1}{\lambda_0} \\[3mm]
\lambda_1 \dfrac{x_1}{w}, & w > \dfrac{\lambda_1}{\lambda_0}
\end{cases}
\tag{19.5}
$$

Above, $c(t)$ is the plasma drug concentration (that can be described using standard PK models) and t_0 denotes the starting time of the treatment. During the drug treatment, the decrease in tumor cell growth rate is directly proportional to the number of proliferating cells x_1 and the drug concentration $c(t)$ via a proportionality constant k_2 that is a parameter describing the antitumor potency of a drug. The higher the k_2, the more the tumor growth is inhibited (Figure 19.5). This assumption of proportionality was tested in hundreds of experiments successfully described by the model equations (19.4) and (19.5); nevertheless, in some cases, a nonlinear (saturable) relationship has been observed and modelled [13].

The damaged tumor cells stop proliferating and proceed through $\bar{n} = n - 1$ progressive degrees of damage with rate constant $k_1 > 0$, in other words entering a mortality chain (see Figure 19.6 where $k_1 x_n$ represents the weight of damaged cells that die in the unit of time).

It should be noted that, due to the mean residence time $1/k_1$ in the ith transit compartment, the mortality chain introduces an average delay due to the damaging progression that is equal to $\bar{n}/k_1$. If we consider a single damaged cell, this corresponds to assuming that its time-to-death is an Erlang-$\bar{n}$ random variable with probability density function

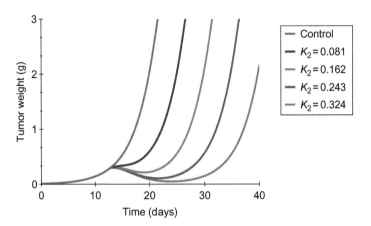

Figure 19.5 Effect of k_2 modulation on simulated tumor growth curves.

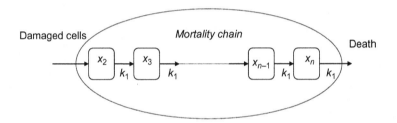

Figure 19.6 Modelling of the damaging process before the cells death.

$$f_s(s) = \begin{cases} \dfrac{k_1^{\bar{n}} s^{\bar{n}-1} e^{-k_1 s}}{(\bar{n}-1)!}, & s \geq 0 \\ 0, & s < 0 \end{cases} \tag{19.6}$$

The higher $\bar{n}$, the more bell-shaped is the distribution function. The smaller k_1, the longer the time before the damaged cells come to death (see Figure 19.7, in which a three-compartment mortality chain was used).

The function f_p takes into account that, for $w > \tilde{w}$, the asymptotic tumor growth rate is slowed down by factors (e.g. nutrient supply limitations) that depend on x_1/w (i.e. the fraction of proliferating cells over the total tumor weight).

It should be noted that from $t = 0$, when the inoculation takes place, to time t_0, when the anticancer treatment starts, the tumor follows an unperturbed growth. In fact, in this time interval, $c(t) = 0$, $x_2(t) = \cdots = x_n(t) = 0$, $w(t) = x_1(t)$, so that

$$f_p(x_1, w) \equiv f_u(w) \quad \text{when } 0 < t \leq t_0$$

As in the case of unperturbed growth, for numerical purposes it is convenient to approximate $f_p(x_1, w)$ with the differentiable function

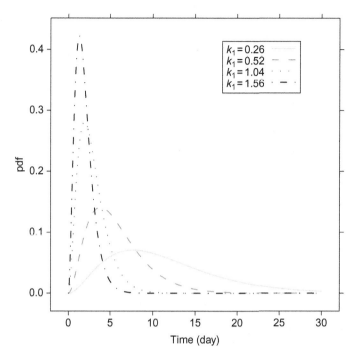

Figure 19.7 Distribution of the time-to-death of tumor cells: effect of k_1 values ($n = 3$). k_1, first-order rate constant of transit.

$$\tilde{f}_p(x_1, w) = \frac{\lambda_0 \, x_1}{\left[1 + \left(\frac{\lambda_0}{\lambda_1} w \right)^{\Psi} \right]^{1/\Psi}} \tag{19.7}$$

which, for large enough values of Ψ, is a good approximation of f_p. Note that Eq. (19.7) differs from Eq. (19.3) in that x_1 appears in the numerator instead of w.

19.5   Mathematical Analysis of the TGI Model

In this section, the dynamic system (Eqs. (19.4) and (19.5)) describing the perturbed tumor growth is analyzed. In particular, we search for the possible equilibrium points of the system when the drug is administered through an infusion yielding a (steady-state) constant concentration $c(t) = \bar{c}$, and we study its stability (both local and global). Then, we obtain the minimum constant concentration $\bar{c}$ that asymptotically guarantees the tumor eradication and define a time efficacy index (TEI) that measures the efficacy of a generic treatment (see Ref. [14]).

19.5.1 Steady-State Analysis: Equilibrium Points

Suppose that the anticancer drug is administered through an infusion yielding a (steady-state) constant concentration $c(t) = \bar{c}$. The equilibrium points of the dynamic system equations (19.4) and (19.5) are the following (see Ref. [14] for details):

1. If $\bar{c} > \lambda_0/k_2$ the only equilibrium point is

$$\bar{x}_1 = \bar{x}_2 = \bar{x}_3 = \cdots = \bar{x}_{n-1} = \bar{x}_n = 0 \Rightarrow \bar{w} = 0$$

2. if $\bar{c} < \lambda_0/k_2$ there are two equilibria:

$$\bar{x}_1 = \bar{x}_2 = \bar{x}_3 = \cdots = \bar{x}_{n-1} = \bar{x}_n = 0 \Rightarrow \bar{w} = 0$$

$$\bar{x}_1 = \frac{\lambda_1}{k_2\bar{c}\left(1 + \frac{(n-1)k_2\bar{c}}{k_1}\right)} \tag{19.8}$$

$$\bar{x}_2 = \bar{x}_3 = \cdots = \bar{x}_{n-1} = \bar{x}_n = \frac{\lambda_1}{k_1 + (n-1)k_2\bar{c}} \Rightarrow \bar{w} = \frac{\lambda_1}{k_2\bar{c}}\tilde{w} \tag{19.9}$$

3. If $\bar{c} = \lambda_0/k_2$ the infinite equilibrium points are

$$0 \leq \bar{x}_1 \leq \frac{\lambda_1}{\lambda_0\left(1 + \frac{(n-1)\lambda_0}{k_1}\right)}$$

$$\bar{x}_2 = \bar{x}_3 = \cdots = \bar{x}_{n-1} = \bar{x}_n = \frac{k_2\bar{c}\bar{x}_1}{k_1} = \frac{\lambda_0}{k_1}\bar{x}_1$$

The locus of equilibrium points for varying values of $\bar{c}$, in the simple (even if not realistic) case with $n = 2$, is shown in Figure 19.8.

19.5.2 Local and Global Stability of the Equilibria

The equilibrium point $\bar{x}_1 = \bar{x}_2 = \bar{x}_3 = \cdots = \bar{x}_{n-1} = \bar{x}_n = 0$ is unstable if $\bar{c} < \lambda_0/k_2$, and globally asymptotically stable if $\bar{c} > \lambda_0/k_2$.

If $\bar{c} < \lambda_0/k_2$, the nonzero equilibrium point, Eqs. (19.8) and (19.9), is asymptotically stable for $n = \{2, 3, 4\}$ (see Ref. [14] for details).

19.5.3 Interpretation of the Steady-State Analysis: Concentration Threshold for Tumor Eradication

From the above analysis of the equilibrium points and their stability, it follows immediately that if the drug is ideally administered via a constant infusion so

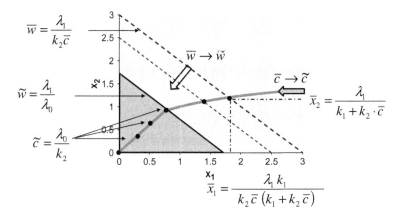

Figure 19.8 The equilibria space in the simple case $n = 2$.

that $c(t) = \bar{c}$ for $t \geq \tilde{t}$, the model predicts different tumor behaviors depending on the drug concentration level $\bar{c}$. More precisely, if $\bar{c} > \lambda_0/k_2$, the model predicts complete tumor eradication starting from any possible initial condition. Conversely, if $\bar{c} < \lambda_0/k_2$, the tumor will not be eradicated, because the origin is no longer a stable equilibrium, and a given tumor weight is asymptotically reached.

Hereafter, the constant $\tilde{c} = \lambda_0/k_2$ will be referred to as concentration threshold for tumor eradication. For illustrative purposes, Figure 19.9 shows the state trajectories predicted by the model for different initial conditions in the simplest case in which the model has only two state variables ($n = 2$) and $\bar{c} < \tilde{c}$. In contrast, Figure 19.10 shows the effect over time of sustaining for a limited time interval a constant concentration either below or above the concentration threshold. In the first case, the tumor weight tends to the positive equilibrium weight, in the latter, to complete tumor eradication. When the treatment is suspended, all of the profiles restart to rise.

Let us assume that a treatment with constant infusion greater than the eradication threshold is started at time 0 (i.e. $c(t) = 0$; $t < 0$; $c(t) = \bar{c}$; $t \geq 0$) and let t_r denote the time when the tumor reaches its maximum weight and the tumor regression begins. It can be demonstrated that if $w(t) < \tilde{w}\ \forall\, t > 0$, then t_r is given by the solution of the following equation:

$$\left[1 - \frac{\lambda_0(k_1+\alpha)^{\bar{n}-1}}{k_1^{n-1}k_2\bar{c}}\right] e^{(k_1+\alpha)\cdot t_r} = 1 + (k_1 + \alpha)\, t_r + \cdots + \frac{(k_1+\alpha)^{n-2}}{(n-2)!} \cdot t_r^{n-2}$$

$$(19.10)$$

where $\alpha = \lambda_0 - k_2\bar{c}$.

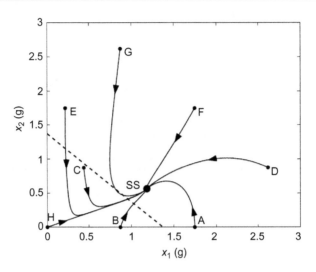

Figure 19.9 Trajectories predicted by the model for different initial conditions with constant infusion concentration $\bar{c}$ less than the threshold $\tilde{c}$ for tumor eradication (Figure 1, [14]). Second order case (i.e. $n = 2$): x_1 and x_2 are the weights of the proliferating and the nonproliferating tumor cells, respectively. Model parameters: $k_1 = 0.615$ days^{-1}, $k_2 = 2.93 \, e^{-4}$ mL/ng/days, $k_0 = 0.369$ day^{-1}, $k_1 = 0.511$ g/day, $w_0 = 0.163$ g. Drug concentrations: $\bar{c} = 1000$ ng/mL, $\tilde{c} = 1260.86$ ng/mL. SS is the equilibrium point: $\bar{x}_1 = 1.182$ g, $\bar{x}_2 = 0.562$ g, $\bar{w} = \bar{x}_1 + \bar{x}_2 = 1.744$ g. The dashed line ($x_1 + x_2 = 1.37$ g) separates the region where tumor growth is exponential from that where growth saturation occurs. Different starting weights: (A) $x_1 = \bar{w}$, $x_2 = 0$, (B) $x_1 = \bar{w}/2$, $x_2 = 0$, (C) $x_1 = \bar{w}/4$, $x_2 = \bar{w}/2$, (D) $x_1 = 1.5 \cdot \bar{w}$, $x_2 = \bar{w}/2$, (E) $x_1 = \bar{w}/8$, $x_2 = \bar{w}$, (F) $x_1 = \bar{w}$, $x_2 = \bar{w}$, (G) $x_1 = \bar{w}/2$, $x_2 = 1.5 \cdot \bar{w}$, (H) $x_1 = \bar{w}/100$, $x_2 = \bar{w}/100$. (*Source*: From Ref. [14].)

Furthermore, the maximum weight (w_{max}) reached by the tumor is

$$
w_{max} = w_0 \left\{ e^{\alpha t_r} \left[1 + \frac{k_2 \bar{c}}{\alpha} \left(1 - \frac{k_1^{n-1}}{(k_1 + \alpha)^{n-1}} \right) \right] \right.
$$
$$
\left. - k_2 \bar{c} \, e^{-k_1 t_r} \sum_{j=0}^{n-2} \frac{k_1^j}{(k_1 + \alpha)^{j+1}} \sum_{i=0}^{n-2-j} (k_1 t_r)^i \right\} \tag{19.11}
$$

which holds if $w_{max} < \tilde{w}$.

From Eq. (19.10), it follows that the delay between the treatment start and tumor regression depends on the parameters k_1, k_2, and λ_0, and the drug concentration $\bar{c}$, but not on the initial weight of the tumor (w_0). Conversely, from Eq. (19.11), the maximum tumor weight depends on the model parameters λ_0, k_1, k_2, and the drug concentration but is directly proportional to the initial weight of the tumor. Consequently, for a given $\bar{c} > \tilde{c}$, there exists a threshold $\hat{w}$ such that, whenever $w_0 < \hat{w}$, the condition $w(t) < \tilde{w}$, $\forall t \geq 0$ is satisfied so that Eqs. (19.10) and (19.11) hold.

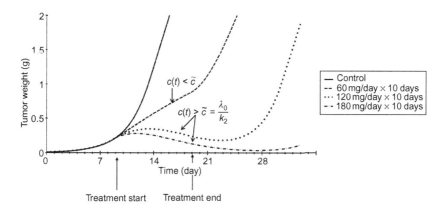

Figure 19.10 Effect of different regimens maintaining drug concentrations above or below the concentration threshold $\tilde{c} = \lambda_0/k_2$; $k_1 = 0.355$ day^{-1}, $k_2 = 3.76$ e^{-4} mL/ng/day, $\lambda_0 = 0.355$ day^{-1}, $\lambda_1 = 0.366$ g/day, $w_0 = 9$ e^{-3} g treatment starting at after 9 days.

19.5.4 Time Efficacy Index

If the treatment has a finite duration, the model predicts that the time course of the tumor weight will eventually follow a straight line with slope λ_1. Figure 19.11 shows that different administration regimens give rise, after a transient, to parallel straight lines; the most leftward being the one related to the controls.

A measure of the antitumor effect of a treatment with finite duration is the amount of delay it causes to the growth of the tumor mass. This can be measured by the time shift of the growth curve of treated animals with respect to the growth curve of the controls (i.e. by the distance between the straight lines).

Let $w_c(t)$ and $w_t(t)$ denote the tumor weight in the control ($c(t) = 0$, $\forall\, t$) and treated cases, respectively, and for a given weight w, let $\tau_c = \tau_c(w)$ and $\tau_t = \tau_t(w)$ be such that $w_c(\tau_c) = w_t(\tau_t) = w$. The TEI is defined as TEI $= \lim_{w \to \infty} [\tau_t - \tau_c]$ (i.e. as the horizontal distance between the asymptotes of the treated and control curves); see Figure 19.11.

Note that in the definition of TEI, the use of limit for w (and therefore t) tending to infinity does not mean that we extend the model validity to an arbitrary large tumor weight. Rather, the limit is needed for determining the distance between the asymptotes observed in the control and treated growth curves.

Consider a finite duration treatment (i.e. $c(t) = 0$ for $t < 0$, and $t > t_f$), and let AUC$_c$ denote the corresponding area under the plasma concentration−time curve (i.e. the integral of $c(t)$), it can be demonstrated (for details see Ref. [14]) that:

$$\text{TEI} \cong \frac{k_2}{\lambda_0}\text{AUC}_c \tag{19.12}$$

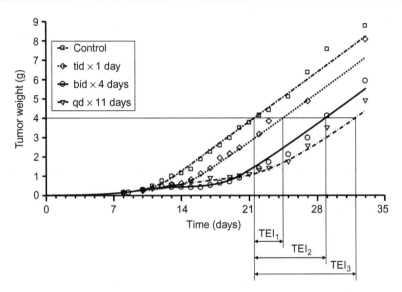

Figure 19.11. Trajectories predicted by the model for different drug schedules: three times daily for 1 day (T1), two times daily for 4 days (T2), and once daily for 11 days (T3). Model parameters are reported in Table 19.1. Dashed line: unperturbed-growth. Continuous lines: perturbed growth. Arrows: time shift of the perturbed-growth curve with respect the unperturbed one. The TEI is the distance between the straight lines. (*Source*: Adapted from Ref. [12].)

Essentially, the TEI depends only on the AUC, and not on the specific shape of the concentration curve $c(t)$. In view of Eq. (19.12), the ratio AUC_c/TEI is approximately equal to the eradication threshold $\tilde{c}$. Then, the threshold concentration, in principle, could be computed only by measuring the TEI and the AUC_c.

From the definition of TEI, it follows that TEI should be evaluated when the tumor weight is infinite. It is clear that measuring the time shift of two experimental curves for a finite weight w provides only an estimate of the TEI.

19.6   Model Identification and its Applications

The tumor growth inhibition (TGI) model just presented has been extensively used to describe and simulate the growth profile of different tumor cell lines, treated with a number of compounds following different administration schedules.

Given that the PK profiles are generally collected only in an ancillary group of animals, the mean profile is generally considered and modeled by a suitable standard PK compartment model, whose parameters are estimated separately from those of the TGI model. Therefore, given the dosing scheme adopted in the tumor growth study and the PK model, the predicted drug concentration profile is computed and provided as an input to the perturbed tumor growth model. In absence of PK collection within the study, the compartment PK parameters can also be retrieved from separate studies, even with different dosing schemes and sampling times, if the assumption that the PK parameters are independent from the study design holds.

Regarding the PD dataset, although individual fitting was also generally successful, it might be convenient, for overcoming a posteriori identification problems, to simultaneously fit the unperturbed and the perturbed models against the average tumor growth data of control and treated arms, respectively. In this way, the control arm drives the identification of w_0, λ_0, and λ_1 with the unperturbed model, so that k_1 and k_2 can be better estimated from the treatment arms.

Both PK and PD model parameters are estimated from experimental data via nonlinear least squares. It is worth noting that a nonlinear mixed effect modelling approach [15] can also be adopted, whose benefit is to evaluate the interindividual variability of the parameters within and across studies (see Chapter 7). Results of this technique are discussed in Refs. [16,17].

A sharp but continuous switch from the exponential to the linear (Eq. (19.1)) was achieved letting $\Psi = 20$ (Eqs. (19.3) and (19.7)). Furthermore, a three-compartment transit model showed the best fitting during the model evaluation, so that the perturbed model was defined as a four-state variables model: the proliferating portion and the three stages of damage, as illustrated in Figure 19.12.

An example of simultaneous fitting is shown in Figure 19.11, with parameters reported in Table 19.1.

The study consists of a control arm and three treatment arms in which Drug A is administered via i.v. from Day 9, at a dose level of 60 mg/kg, every day for 11 days (qd $\times$ 11), twice a day for 4 days (bid $\times$ 4), and three times a day for 1 day (tid $\times$ 1). It can be appreciated that the PD parameters are generally estimated with good precision, which indicates the absence of overparameterization in the TGI model.

Still referring to Figure 19.11, it is easy to verify the validity of Eq. (19.12) for the estimation of the secondary parameter TEI. Since the controls and treated arms share the same estimate of λ_1, the exact TEI is immediately visualized as the horizontal distance between these two curves for a sufficiently large tumor weight, here 4 g. The model showed within-experiment predictive power. For example, referring to the same dataset, estimating the PD parameters from the control and the daily treatment for 11 days (Figure 19.13, upper left panel), it is possible to simulate the remaining two treatment arms (Figure 19.13, upper right panel).

Across-experiment predictive power using different administration modes was also demonstrated. In a first experiment, Drug B was given via i.v. twice a day for 5 days at the dose level of 15 and 30 mg/kg as bolus administrations. The PK profile obtained in a previous experiment is shown in Figure 19.13, inset lower left panel ($V_1 = 1.42$ mL/kg, $k_{10} = 1.17$ h^{-1}, $k_{12} = 0.206$ h^{-1}, $k_{21} = 0.232$ h^{-1}). PD parameters are given in Table 19.2. The observed and model-fitted tumor growth curves are reported in Figure 19.13, lower left panel, showing an excellent agreement ($r^2 = 0.99$). For further development of the drug, it was of interest to test the efficacy of a long-term infusion. On the basis of the λ_0 and k_2 values previously estimated, a threshold concentration for tumor eradication $\tilde{c} \cong 1100$ ng/mL was

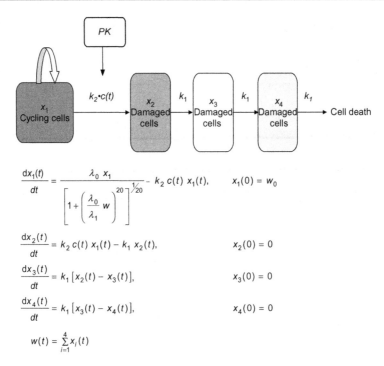

$$\frac{dx_1(t)}{dt} = \frac{\lambda_0 \, x_1}{\left[1 + \left(\frac{\lambda_0}{\lambda_1} w\right)^{20}\right]^{1/20}} - k_2 \, c(t) \, x_1(t), \qquad x_1(0) = w_0$$

$$\frac{dx_2(t)}{dt} = k_2 \, c(t) \, x_1(t) - k_1 \, x_2(t), \qquad x_2(0) = 0$$

$$\frac{dx_3(t)}{dt} = k_1 \, [x_2(t) - x_3(t)], \qquad x_3(0) = 0$$

$$\frac{dx_4(t)}{dt} = k_1 \, [x_3(t) - x_4(t)], \qquad x_4(0) = 0$$

$$w(t) = \sum_{i=1}^{4} x_i(t)$$

Figure 19.12 Scheme of the PK−PD model. k_1, first-order rate constant of transit; k_2, measure of drug potency; $c(t)$, plasma concentration of the anticancer agent.

Table 19.1 PD Parameters Obtained Fitting the Model to Tumor Weight Data in Nude Mice Given i.v. Either the Vehicle or Drug A at a Dose Level of 60 mg/kg

Parameter	Estimate	CV (%)
k_1 (day^{-1})	0.355	20.57
k_2 (ng^{-1} mL day^{-1})	$3.76.10^{-4}$	5.27
λ_0 (day^{-1})	0.355	3.52
λ_1 (g day^{-1})	0.366	3.47
w_0 (g)	0.009	13.08

The parameters refer to the fitting of controls and animals treated every day for 11 days (qd × 11), Twice a day for 4 days (bid × 4), three times a day for 1 day (tid × 1); all treatments started from day 9 [12].

calculated. An infusion experiment was designed to target a steady-state concentration of Drug B of $\cong 2000$ ng/mL (1.8-fold higher than $\tilde{c}$) to observe tumor shrinkage within the time frame of the experiment. On the basis of the plasma clearance obtained in the previously described experiment (1.7 L/h/kg) and on basic PK principles, an infusion rate of 80 mg/kg/day was calculated; the corresponding profile

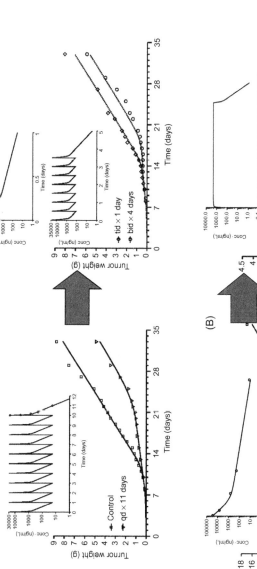
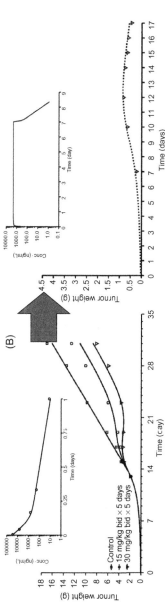

Figure 19.13 Predictive power of the model. Within study (Panel A): estimating the PD parameters from the control and the daily treatment with Drug A for 11 days (left panel), was possible to simulate the remaining two treatment arms (right panel). Across studies (Panel B): In a first TGI experiment, Drug B was given i.v. twice a day for 5 days at the dose level of 15 and 30 mg/kg as bolus administrations. The PK profile obtained in a previous experiment is shown in the inset of the left panel. The observed and model-fitted tumor growth curves are reported in the left panel. On the basis of the λ_0 and k_2 values previously estimated, a threshold concentration for tumor eradication of 1100 ng/mL was calculated. The second TGI experiment was performed at an effective infusion rate of 83.6 mg/kg/day able to maintain an average observed steady-state concentration of 2085 ng/mL. In the right panel, observed and predicted tumor weights are shown.
(*Source*: Adapted from Ref. [12].)

Table 19.2 PD Parameters Obtained Fitting the Model to the Tumor Weight Data in Nude Mice Given i.v. Either the Vehicle or Drug B

Parameters	Bolus		Infusion	
	Estimate	CV^a (%)	Estimate	CV (%)
k_1 (day^{-1})	0.517	16.72	0.615	92.56
k_2 (ng^{-1} mL day^{-1})	2.89×10^{-4}	9.15	2.93×10^{-4}	49.60
λ_0 (day^{-1})	0.309	22.06	0.369	3.51
λ_1 (g day^{-1})	0.796	4.82	0.511	6.18
w_0 (g)	0.034	91.52	0.016	12.31

On the left, the parameters refer to i.v. bolus administrations (15 and 30 mg/kg, twice a day for 5 days from day 13); on the right, the parameters refer to long-term i.v. infusion administrations (83 mg/kg/day, 7-day infusion from day 9)
^{a}CV, coefficient of variation; k_1, first-order rate constant of transit; k_2, measure of drug potency; λ_0, first-order rate constant of tumor growth; λ_1, zero-order rate constant of tumor growth; w_0, tumor weight at the inoculation time. (*Source*: Table from Ref. [12].)

is shown in Figure 19.13, inset lower right panel. The experiment was performed at an effective infusion rate of 83 mg/kg/day; the average observed steady-state concentration was 2085 ng/mL. In Figure 19.13, lower right panel, observed and predicted tumor weights are shown, showing an excellent agreement. As predicted, the observed tumor weights in treated animals indeed showed the start of tumor eradication, confirming the utility of the secondary parameter $\tilde{c}$.

This TGI model was implemented in SimBiology®, a MATLAB® toolbox for modelling, simulating, and analyzing biological systems [18]. Furthermore, a user-friendly software tool, called TGI-Simulator, has been designed to show, also through a 2D graphical animation, the simulated time effect of an anticancer drug [19]. The application is based on a Java graphical user interface (GUI) that includes a self-installing differential equation solver implemented in MATLAB® together with an optimization algorithm that performs model identification via Weighted Least Squares (WLS).

19.7   Combined Administration of Drugs

Anticancer drugs have been shown in many cases to be more effective when given in combination. The rationale for combination chemotherapy is to use drugs that work by different mechanisms of action, thereby decreasing the likelihood that resistant cancer cells will develop. The ideal advantage is that when drugs with different effects are combined, each drug can be used at its optimal dose, without intolerable side effects. Instead of fitting different drug-specific interaction models, Rocchetti et al. [20] developed a model able to predict the response of the tumor to a combination treatment in the null-interaction hypothesis. The null-interaction assumption requires that the death processes triggered by the two drugs are independent of each other.

19.7.1   The Stochastic Interpretation of the TGI Model

It is worthwhile to introduce the stochastic interpretation of the TGI model before extending this concept to the interaction model. As in the above-presented TGI model, it is assumed that the tumor cells are of two types: proliferating cells that can generate new tumor cells by mitotic division and nonproliferating cells that cannot proliferate because they have been damaged by the drug action [21].

Given one proliferating cell, in the time interval $[t, t + dt]$, two types of events can occur:

1. birth (B): the cell duplicates and a new cell is generated;
2. damage (D): the cell becomes nonproliferating and a death-time t_D is assigned to it.

A nonproliferating cell cannot turn back to the proliferating state and when its death-time comes, it leaves the tumor, no longer contributing to the overall tumor weight.

The following probabilities for the events B and D are assumed:

- $p(B) = \lambda_B(w(t))dt + o(dt)$
- $p(D) = \lambda_D(c(t))dt + o(dt)$

where $o(dt)$ is a higher order infinitesimal. The birth function $\lambda_B(w)$ will typically be a function decreasing with the overall tumor mass to allow for effects such as nutrient supply limitations, while the damage function $\lambda_D(c)$, as a rule, will be a function increasing with drug concentration.

Two further assumptions are introduced in the events B and D:

- the probability of two or more events occurring in $[t, t + dt]$ is $o(dt)$.
- the number and type of events occurring in disjointed time intervals are independent.

The above assumptions are typical of (inhomogeneous) Poisson processes. Let us extend these concepts to populations of cells in which

- $N_p(t)$ is the number of proliferating tumor cells at time t;
- $N_{np}(t,s)$ is the number of nonproliferating cells at time t, whose death-time is $t_D \leq t + s$ and with $N_{np}(0, s) = 0$, $\forall s$ as $c(t) = 0$, $t \leq 0$.

It follows that in the interval $[t, t + dt]$:

- the number of newborn cells, N_b, given $N_p(t)$ and $w(t)$, is a Poisson variable with expectation:

$$E[N_b(t)dt|\{N_p(t), w(t)\}] = \lambda_B(w(t))N_p(t)dt \quad \text{and}$$

- the number of damaged cells, N_d, given $N_p(t)$, is a Poisson variable with expectation:

$$E[N_d(t)dt|N_p(t)] = \lambda_D(c(t))N_p(t)dt$$

Let us now define the mass of proliferating cells by $x_p(t) := N_p(t)/N_c$, where N_c is the number of tumor cells per mass unit (around $10^9/g$), and let us remember that $x_p(0) = w(0)$, as $N_{np}(0, s) = 0$, $\forall s$. Recalling that the variance of a Poisson variable is equal to its mean, the conditional expectation and variance of $x_p(t)$ evolve according to

$$E\left[x_p(t+dt)|\{N_p(t),w(t)\}\right] = \frac{N_p(t)}{N_c} + \frac{N_p(t)}{N_c}[\lambda_B(w(t)) - \lambda_D(c(t))]dt$$

$$\text{Var}\left[x_p(t+dt)|\{N_p(t),w(t)\}\right] = \frac{N_p(t)}{N_c^2}[\lambda_B(w(t)) + \lambda_D(c(t))]dt$$

and it is possible to demonstrate [21] that the expectation $\bar{x}_p(t): \triangleq E[x_p(t)|w(0)]$ satisfies the equation:

$$\dot{\bar{x}}_p(t) = [\lambda_B(w(t)) - \lambda_D(c(t))]x_p(t)$$

Regarding the number $N_{np}(t,s)$ of nonproliferating cells, the following equation holds:

$$N_{np}(t+dt,s)|\{N_p(t),N_{np}(t,\cdot)\} = N_{np}(t,s+dt) - N_{np}(t,dt) + (\tilde{N}_d(t+dt,s)|N_p(t))dt$$

where $\tilde{N}_d(t+dt,s)$ is the number of proliferating cells damaged at time $t+dt$ that are assigned a time-to-death not greater than s, $N_{np}(t,dt)$ is the number of cells that die in $[t, t+dt]$, whereas $N_{np}(t, s+dt)$ takes into account the "aging" of the nonproliferating cells. Its expectation is

$$E[N_{np}(t+dt,s)|\{N_p(t),N_{np}(t,\cdot)\}] = N_{np}(t,s+dt) - N_{np}(t,dt)$$
$$+ F_s(s)\lambda_D(c(t))N_p(t)dt \qquad (19.13)$$

where $F_s(s)\lambda_D(c(t))N_p(t)dt$ is the expectation of the Poisson variable $(\tilde{N}_d(t+dt,s)|N_p(t))dt$ and $F_s(s) = \int_0^s f_s(\xi)d\xi$ is by definition the probability that a generic damaged cell is assigned a time-to-death not greater than s.

Let us define $y(t,s) \triangleq 1/N_c \; \partial N_{np}(t,s)/\partial s$ and $y(\bar{t},s) \triangleq E[y(t,s)]$. From Eq. (19.13), it follows:

$$\frac{\partial \bar{y}(t,s)}{\partial t} = \frac{\partial \bar{y}(t,s)}{\partial s} + f_s(s)\lambda_D(c(t))x_p(t)$$

Given that $x_p(t)$ and $y(t, s)$ can be approximated by their expectations, as the coefficients of variation can be proven to be close to zero, after some passages the cell population model can then be summarized as follows:

$$\dot{x}_p(t) = [\lambda_B(w(t)) - \lambda_D(c(t))]x_p(t) \qquad (19.14)$$

$$\frac{\partial \bar{y}(t,s)}{\partial t} = \frac{\partial \bar{y}(t,s)}{\partial s} + f_s(s)\lambda_D(c(t))x_p(t)$$

$$w(t) = x_p(t) + \int_0^\infty y(t,s)\,ds$$

with initial conditions $x_p(0) = w(0)$ and $y(0, s) = 0 \ \forall s$, and having defined:

- the birth function $\lambda_B(w)$;
- the damage function $\lambda_D(c)$;
- the time-to-death probability density function $f_s(s)$.

This model is an infinite-dimensional system since it includes a partial differential equation. Special cases can be determined choosing an appropriate time-to-death probability density function. If we select the Erlang-n one, similarly to Eq. (19.6), the model can then be reformulated splitting the nonproliferating cells into n classes. The ith class, $i = 1, \ldots, n$, contains the nonproliferating cells that have undergone i stages of damage. For a nonproliferating cell, only one event is possible in $[t, t + dt]$:

- transition (T): if $i < n$, the cell passes from damage stage i to stage $i + 1$; if $i = n$, the cell dies.

Since the residence time in the ith class is exponentially distributed, the transition between one class and the next one is a Poisson event with intensity λ_T. Therefore, $p(T) = \lambda_T\,dt + o(dt)$.

Note that in this model version, when a proliferating cell is damaged, it is not assigned a death-time but just passes to the first stage of damage.

The model will then be defined by the following equations:

$$\dot{x}_p(t) = [\lambda_B(w(t)) - \lambda_D(c(t))]x_p(t)$$
$$\dot{x}_1(t) = \lambda_D(c(t))x_p(t) - \lambda_T x_1(t)$$
$$\dot{x}_2(t) = \lambda_T(x_1(t) - x_2(t))$$
$$\vdots = \vdots$$
$$\dot{x}_n(t) = \lambda_T(x_{n-1}(t) - x_n(t))$$
$$w(t) = x_p(t) + \sum_{i=1}^n x_i$$

and assuming:

- $\lambda_B(w) = \begin{cases} \lambda_0, \ \tilde{w}, \ w \le \tilde{w} \\ \lambda_0 \dfrac{\tilde{w}}{w}, \ w > \tilde{w} \end{cases}$
- $\lambda_D(c) = k_2 c$
- $\lambda_T = k_1$
- $n = 3$

we obtain the TGI in Eqs. (19.4) and (19.5).

19.7.2 The "No-Interaction" Model

In a combination regimen with k coadministered compounds, we define:

- the birth function $\lambda_B(w)$;
- the damage functions $\lambda_{D_i}(c_i)$, $i = 1, \ldots, k$;
- the time-to-death probability density functions $f_{si}(s_i)$, $i = 1, \ldots, k$.

Under the hypothesis of no interaction between drug effects, the total number of damaged proliferating cells is the sum of independent Poisson variables, which is a Poisson variable with intensity equal to the sum of all of the intensities, $N_p(t) \cdot \sum_{i=1}^{k} \lambda_{D_i}(c_i(t))$ and Eq. (19.14) becomes

$$\dot{x}_p(t) = \left[\lambda_B(w(t)) - \sum_{i=1}^{k} \lambda_{D_i}(c_i(t)) \right] x_p(t)$$

(see Ref. [22]).

In case $f_{si}(s_i)$ are Erlang-n_i time-to-death probability density functions, the time-to-death can be viewed as the time that occurs between $n_i + 1$ consecutive Poisson events with rate denoted by λ_{T_i}. Just as for the single-agent TGI model, the tumor cells damaged by one of the drugs, when exposed again to the same drug, do not further accelerate their death process. On the other hand, it cannot be excluded that a cell already damaged by one of the drugs could undergo a further damage due to the action of another drug of the combination.

In the case of $k = 2$, the finite dimensional system equations will be

$$
\begin{aligned}
\dot{x}_p(t) &= \left[\lambda_B(w(t)) - \sum_{i=1}^{2} \lambda_{D_i}(c_i(t)) \right] x_p(t) \\
\dot{x}_{10}(t) &= \lambda_{D_1}(c_1(t))x_p(t) - [\lambda_{D_2}(c_2(t)) + \lambda_{T_1}]x_{1,0}(t) \\
\dot{x}_{01}(t) &= \lambda_{D_2}(c_2(t))x_p(t) - [\lambda_{D_1}(c_1(t)) + \lambda_{T_2}]x_{0,1}(t) \\
\dot{x}_{l0}(t) &= \lambda_{T_1} x_{l-1,0}(t) - [\lambda_{D_2}(c_2(t)) + \lambda_{T_1}]x_{l,0}(t), \quad 2 \le l \le n_1 \\
\dot{x}_{0m}(t) &= \lambda_{T_2} x_{0,m-1}(t) - [\lambda_{D_1}(c_1(t)) + \lambda_{T_2}]x_{0,m}(t), \quad 2 \le m \le n_2 \\
\dot{x}_{11}(t) &= \lambda_{D_1}(c_1(t))x_{0,1}(t) + \lambda_{D_2}(c_2(t))x_{1,0}(t) - \sum_{i=1}^{2}\lambda_{T_i} x_{1,1}(t) \\
\dot{x}_{l1}(t) &= \lambda_{T_1}x_{l-1,1}(t) + \lambda_{D_2}(c_2(t))x_{l,0}(t) - \sum_{i=1}^{2}\lambda_{T_i} x_{l,1}(t), \quad 2 \le l \le n_1 \\
\dot{x}_{1m}(t) &= \lambda_{T_2} x_{1,m-1}(t) + \lambda_{D_1}(c_1(t))x_{0,m}(t) - \sum_{i=1}^{2}\lambda_{T_i} x_{1,m}(t), \quad 2 \le m \le n_2 \\
\dot{x}_{lm}(t) &= \lambda_{T_1} x_{l-1,m}(t) + \lambda_{T_2} x_{l,m-1}(t) - \sum_{i=1}^{2}\lambda_{T_i} x_{l,m}(t), \quad 2 \le l \le n_1, 2 \le m \le n_2
\end{aligned}
$$

(19.15)

The model is illustrated in Figure 19.14, where $n_i = 3$ $\forall i$.

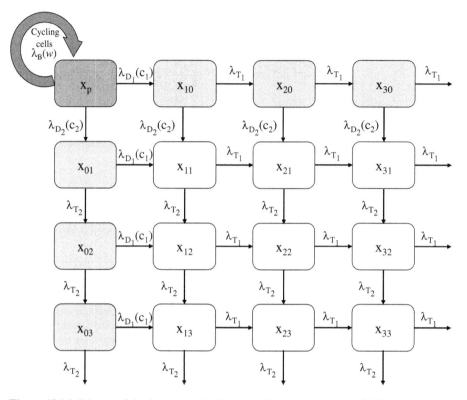

Figure 19.14 Scheme of the "no-interaction" model of two active agents. Light gray compartments represent the states of damage resulting from the action of one single drug, white compartments of both drugs.

The tumor weight being

$$w(t) = x_p(t) + \sum_{l=0}^{n_1} \sum_{m=0}^{n_2} x_{l,m}(t)$$

The model was defined as TGI "no-interaction" model [20] with the following assumptions:

- $\lambda_B(w) = \begin{cases} \lambda_0, \ \tilde{w}, \ w \leq \tilde{w} \\ \lambda_0 \dfrac{\tilde{w}}{w} = \dfrac{\lambda_1}{w}, \ w > \tilde{w} \end{cases}$

- $\lambda_{D_i}(c_i) = k_{2i} \cdot c_{1i}$
- $\lambda_{T_i} = k_{1i}$
- $k = 2$ and $n_i = 3 \ \forall i$

An example of the fitting of the data and estimated parameters is presented in Figure 19.15.

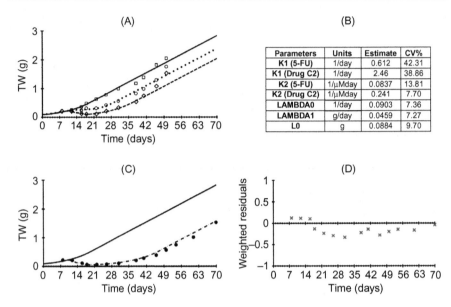

Figure 19.15 An example of fitting of the combination data. (A) Observed and model-fitted tumor growth curves obtained in HT29 tumor-bearing mice given either the vehicle (□) or 5-FU 50 mg/kg i.v. q4dx3 days treatment (○) or Drug C2 60 mg/kg orally given on days 10, 11, 12, 14, 15, 16 (◊), each administered as a single agent. (B) PD parameters obtained simultaneously fitting the TGI model to the tumor weight data obtained after administration of the vehicle and the single agents. (C) Observed tumor growth data obtained in nude mice given 5-FU 50 mg/kg i.v. q4dx3 days treatment in combination with Drug C2 60 mg/kg o.s. repeated doses treatment (•) overlapped to the predicted tumor growth curve obtained applying the additive TGIadd model with parameters previously estimated. (D) Weighted residuals (i.e. (observed-estimated)/observed) between observed data in the combination arm and the predicted curve. The hypothesis of additivity cannot be rejected (p-value $\cong 1$). (*Source*: Figure from Ref. [20].)

It is worth noting that the hypothesis of no interaction between drug actions implies, under reasonable hypothesis, the additivity of the TEI:

$$\text{TEI}_{\text{add}} \cong \frac{\text{AUC}_{c_1}}{\lambda_0} + \frac{\text{AUC}_{c_2}}{\lambda_0}$$

The TGI$_{\text{add}}$ model simulations can be compared with the experimental data of the combination therapy assessing the validity of zero-interaction hypothesis by a suitable statistical test. In summary, the model is able to predict the TGI in case the "additivity" of the combined drug effects holds and departures from its predictions allow for the establishment of the synergistic or antagonistic behaviors, depending on the effect being larger or smaller than predicted.

19.7.3 The Interaction Model

In order to be able not only to test, but also to fit and predict synergistic and antagonistic TGI profiles, Terranova et al. [23] modified the system equations of the "no-interaction model," introducing an interaction term on the proliferating cell compartment in Eq. (19.15):

$$\dot{x}_p(t) = [\lambda_B(w(t)) - \sum_{i=1}^{2} \lambda_{D_i}(c_i(t)) - \gamma c_1(t)c_2(t)]x_p(t)$$

$$\dot{x}_{11}(t) = \lambda_{D_1}(c_1(t))x_{0,1}(t) + \lambda_{D_2}(c_2(t))x_{1,0}(t) - \sum_{i=1}^{2} \lambda_{T_i}x_{1,1}(t) + \gamma c_1(t)c_2(t)x_p(t)$$

If the value of γ is higher than, lower than, or close to zero, the interaction of drug effects has a synergistic, antagonistic, or additive nature, respectively. Although γ cannot be used directly to compare different combination treatments, because its value is not a pure measure of strength of the drug effect interaction, but depends on the potencies of the considered drugs, nevertheless the model can find its best application in the simulation of different regimen of the same coadministered drugs.

Moreover, starting from the demonstration that in the interaction model, under reasonable hypothesis, the TEI is

$$TEI_{inter} \cong \frac{AUC_{c_1}}{\lambda_0} + \frac{AUC_{c_2}}{\lambda_0} + \gamma \frac{AUC_{c_1 c_2}}{\lambda_0}$$

it is possible to derive two combination indexes:

- the synergistic $\frac{TEI_{inter} - TEI_{add}}{TEI_{inter}}$ (%)
- the antagonistic $\frac{TEI_{add} - TEI_{inter}}{TEI_{add}}$ (%)

that provide a quantitative measure of the strength of the interaction, useful for comparing and ranking different combination treatments.

19.8 Model-Based Clinical Dose Prediction

Being able to predict the activity of new compounds in humans from preclinical data can substantially reduce the number of failures, especially in oncology. Therefore, the correlations between the "active clinical doses" and the drug-specific potency-related TGI parameters were considered for a number of different drugs used in clinical practice. For this purpose, known anticancer drugs with different mechanisms of action, including topoisomerase inhibitors, antimicrotubule assembly inhibitors, antimetabolites, and alkylating agents, were considered.

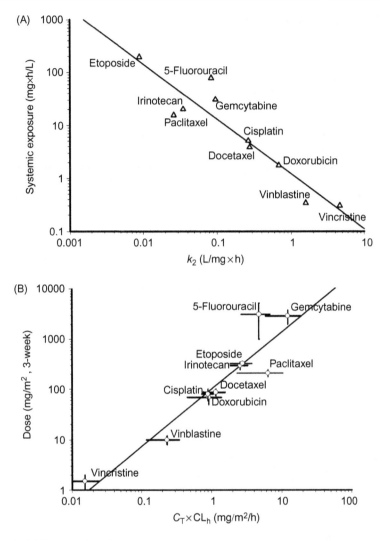

Figure 19.16 Panel A: Relationship between the systemic exposure obtained at the midpoint of the dose range used in the clinics (cumulative doses given in 3-week cycles) and k_2 (potency parameter) estimated in animals. Regression performed on log–log scale: intercept = 0.0835, slope = -1.03, $r = -0.927$ [24]. Panel B: Relationship between the doses used in the clinics (cumulative doses given in 3-week cycles) and $C_T \times CL_h$ (with C_T model-derived threshold plasma concentration required to block the growth of the tumor and CL_h human plasma clearance). Regression (based on the midpoint of dose range) performed on log–log scale: intercept = 2.01, slope = 1.14, $r = 0.939$. Vertical error bars represent the dose range reported in the literature. Horizontal error bars defined based on the standard deviation of CL_h. (*Source*: Adapted from Ref. [24].)

Anticancer drugs are given to patients at different dosages and administration schedules, according to individual clinical cases. Due to this, the lowest and highest dose levels commonly encountered in the clinical literature (see Ref. [24] and references therein), expressed in terms of cumulative doses given over a 3-week period, were considered as "active clinical doses" for the purpose of this analysis. The 3-week period was chosen because it is among the most frequently adopted cycle duration for the majority of these compounds.

The results of this analysis showed a strong correlation between the preclinical potency parameter, k_2, and the clinical exposure over a 3-week period ($r = -0.927$), Figure 19.16 panel A, as well as between the clinical dose and the preclinical concentration threshold multiplied by the clearance in humans ($r = 0.936$), Figure 19.16 panel B [24], the latter being more easily suited for estimating the exposure expressed in terms of dose. Remembering that the time interval is fixed, $C_T = \lambda_0/k_2$, and noticing that λ_0, which describes the free exponential tumor growth, has low variability across studies, the two results are clearly linked.

New oncology agents which, in correspondence with their estimated potency parameter C_T multiplied by their clearance in human, cannot be administered at clinical doses approaching or exceeding those defined by the regression line in Figure 19.16 panel B (e.g. for toxicity reasons) are unlikely to achieve technical and regulatory success, so that the continuation of their clinical development should be considered with reservation, acknowledging the risk of failure.

19.9 Conclusions

In the current chapter, the various steps for the development of a TGI model have been illustrated. Although it is not the purpose of this model to describe in fine detail the biological mechanisms of tumor growth, it is nevertheless possible to associate a macroscopic biological meaning to its parameters. In turn, this makes it possible to translate into biological terms some useful mathematical properties of the model (e.g. TEI). It has also been shown that the model can be extended to describe other cases; one example is the combination-therapy regimen. This flexibility has contributed to the robustness of the model, which has been successfully applied in many preclinical programs through the years.

One of the major challenges in drug development is the ability to transfer the knowledge acquired in preclinical studies into the clinical setting. Indeed, the identification via modelling tools of preclinical parameters that can be meaningfully linked to clinical endpoints could have a great impact on reducing the attrition rate in drug development. This modelling project has clearly demonstrated the added value of a model-based approach: optimizing the design of preclinical experimentation (substantially decreasing the number of animals used) and streamlining the early phases of clinical development.

References

[1] http://www.kmrgroup.com/PressReleases/2012_08_08%20KMR%20PBF%20Success%20Rate%20&%20Cycle%20Time%20Press%20Release.pdf

[2] Deloitte. Measuring the return from pharmaceutical innovation 2012. Is R&D earning its investment? http://www.deloitte.com/assets/Dcom-Ireland/Local%20Assets/Documents/Life%20sciences/2012%20RD%20Measuring%20the%20Return%20from%20Innovation.pdf

[3] Kola I, Landis J. Can the pharmaceutical industry reduce attrition rates? Nature Rev Drug Disc 2004;3:711−5.

[4] Raichert JM, Wenger JB. Development trends for new cancer therapeutics and vaccines. Drug Discovery Today 2008;13(1/2):30−7.

[5] Zhou Q, Gallo JM. The pharmacokinetic/pharmacodynamic pipeline: translating anticancer drug. pharmacology to the clinic. AAPS J 2011;13(1):111−20.

[6] Bernard A, Kimko H, Mital D, Poggesi I. Mathematical modelling of tumor growth and tumor growth inhibition in oncology drug development. Expert Opin Drug Metabol Toxicol 2012;8(9):1057−69.

[7] Lalonde RL, Kowalski KG, Hutmacher MM, Ewy W, Nichols DJ, Milligan PA, et al. Model-based drug development. Clin Pharmacol Ther 2007;82:21−32.

[8] Zhang L, Pfister M, Meibohm B. Concepts and challenges in quantitative pharmacology and model-based drug development. AAPS J 2008;10(4):552−9.

[9] Keizer RJ, Schellens JHM, Beijnen JH, Huitema ADR. Pharmacodynamic biomarkers in model-based drug development in oncology. Current Clin Pharmacol 2011;6:30−40.

[10] Simeoni M, De Nicolao G, Magni P, Rocchetti M, Poggesi I. Modelling of human tumor xenografts and dose rationale in oncology. Drug Discovery Today: Technol 2013;10(3):e365−72.

[11] Hughes JP, Rees S, Kalindjian SB, Philpott KL. Principles of early drug discovery. Brit J Pharmacol 2011;162:1239−49.

[12] Simeoni M, Magni P, Cammia C, De Nicolao G, Croci V, Pesenti E, et al. Predictive pharmacokinetic−pharmacodynamic modelling of tumor growth kinetics in xenograft models after administration of anticancer agents. Cancer Res 2004;64:1094−101.

[13] Marusic M, Bajzert Z, Freyert JP, Vuk-Pavlovic S. Analysis of growth of multicellular tumor spheroids by mathematical models. Cell Prolif 1994;21:13−94.

[14] Magni P, Simeoni M, Poggesi I, Rocchetti M, De Nicolao G. A mathematical model to study the effects of drugs administration on tumor growth dynamics. Math Biosci 2006;200:127−51.

[15] Sheiner LB, Rosenberg B, Marathe VV. Estimation of population characteristics of pharmacokinetic parameters from routine clinical data. J Pharmacokin Pharmacodyn 1977;5:445−7.

[16] Poggesi I, Simeoni M , Germani M, De Nicolao G, Rocchetti M. Population modelling of tumor growth in untreated xenografted mice. Abstracts of the annual meeting of the population approach group in Europe, Uppsala PAGE 13, 2004; Abstr 535 [http://www.page-meeting.org/?abstract = 535]

[17] Simeoni M, Poggesi I, Germani M, De Nicolao G, Rocchetti M. Population modelling of tumor growth inhibition *in vivo*: application to anticancer drug development. Abstracts of the annual meeting of the population approach group in Europe, Uppsala page, 2004; Abstr 503 [http://www.page-meeting.org/?abstract = 503]

[18] http://www.mathworks.com/discovery/supporting-docs/pkpd-model-of-tumor-growth-kinetics-under-therapy.pdf

[19] Terranova N, Magni P. TGI-Simulator: a visual tool to support the preclinical phase of the drug discovery process by assessing *in silico* the effect of an anticancer drug. Comput Meth Prog Biomed 2012;105:162−74.

[20] Rocchetti M, Del Bene F, Germani M, Fiorentini F, Poggesi I, Pesenti E, et al. Testing additivity of anticancer agents in pre-clinical studies: a PK/PD modelling approach. Eur J Cancer 2009;45:3336−46.

[21] Magni P, Germani M, De Nicolao G, Bianchini G, Simeoni M, Poggesi I, et al. A minimal model of tumor growth inhibition. IEEE Trans Biomed Eng 2008;55(12): 2683−90.

[22] Magni P, Terranova N, Del Bene F, Germani M, De Nicolao G. A minimal model of tumor growth inhibition in combination regimens under the hypothesis of no interaction between drugs. IEEE Trans Biomed Eng 2012;59:2161−70.

[23] Terranova N, Germani M, Del Bene F, Magni P. A predictive pharmacokinetic−pharmacodynamic model of tumor growth kinetics in xenograft mice after administration of anticancer agents given in combination. Cancer Chemother Pharmacol 2013;72 (2):471−82.

[24] Rocchetti M., Simeoni M., Pesenti E., De Nicolao G., Poggesi I. Predicting the active doses in humans from animal studies: a novel approach in oncology. Eur J Cancer 2007;43:1862−1868.

20 Computational Modelling of Cardiac Biomechanics

Emiliano Votta and Alberto Redaelli

Dipartimento di Elettronica, Informazione e Bioingegneria (DEIB),
Politecnico di Milano, Milan, Italy

20.1 Introduction

The socioeconomic burden of cardiovascular diseases (CVDs) in the Western world is dramatic: CVDs are the largest single cause of mortality in the European Union and cause one of every three deaths in the United States with associated direct and indirect costs far above those of any other major disease [1]. Among CVDs, the subclass of cardiac pathologies plays a relevant role: in the United States, coronary heart disease and heart failure accounts for about 1/6 and 1/9 of every death, respectively, while heart valve diseases affect approximately 2.5% of the population. For some cardiac diseases, and in particular for heart valve diseases and heart failure, the best therapeutic solution is still an open issue despite continuous improvements and the introduction of novel solutions, such as minimally invasive surgery and percutaneous heart valve repair/replacement. The optimization of these novel approaches, as well as the development of new ones, needs advanced tools for the analysis of heart chambers and valves, as well as for the quantification of the effects of different therapeutic options. Among these, a relevant role is played by the combination of image processing and numerical modelling to realistically simulate clinically relevant scenarios.

This chapter provides an overview of the modelling approaches available for assessing *in vivo* ventricular and heart valve biomechanics, as well as for the simulation of surgical and medical treatments to support clinical training and planning. The focus of the overview is on those methods that are currently exploitable for clinical purposes and on those that may become exploitable in the short term. To fit with this aim, these approaches should lead to models satisfying at least four requirements: (i) they must include patient-specific anatomical detail, which can be achieved by integrating biomechanical modelling with image processing; (ii) they must be flexible, by making models parametric and by allowing for an easy tuning of parameters on a patient-specific basis; (iii) they must perform computationally, so as to provide nearly real-time analyses to be used to support clinical planning; and (iv) they must guarantee ease of use for clinicians, possibly through a user-

Modelling Methodology for Physiology and Medicine. DOI: http://dx.doi.org/10.1016/B978-0-12-411557-6.00020-3

friendly graphical interface complying with the conventions and standards of clinicians.

As will be evident from the next sections, none of the current approaches matches all of the mentioned requirements, although in some cases a good tradeoff is reached between model sophistication on the one hand and flexibility and computational performance on the other hand, so that clinical application appears within reach.

20.2   Modelling of Ventricular Biomechanics

Several models have been proposed to describe the mechanics of the ventricular chambers, ranging from very simple models based on the Young—Laplace equation first formulated in 1806, to very complex image-based multiscale and multiphysics models such as the ones recently developed in the framework of the human Physiome Project (http://physiomeproject.org/) and the euHeart project (http://www.euheart.eu/). In this section, three different modelling approaches with increasing levels of sophistication will be exemplified referring to the scientific literature: models based on simple ventricle geometry representations, models extracting accurate stress—strain data from 4D imaging, and image-based multiphysics models. For each approach, pros and cons will be highlighted from the implementation standpoint.

20.3   Models Assessing Ventricular Global Function

This approach dates back to the early 1970s. It is based on functional or simple geometrical models that are able to capture the overall performance of the ventricle using a limited number of equilibrium equations and can effectively depict the interaction between the heart and the vascular bed under different working conditions, providing quantitative data in almost real time. Regarding their personalization and ability to mimic the behavior of a specific heart, these models offer one advantage, consisting in the need for only a limited number of input parameters. However, at the same time such personalization is limited in several ways; in particular, by the fact that these models replicate global ventricular function and any tuning of the model's parameters affects homogeneously the entire model, due to the axial symmetry assumption. As a consequence, local impairments, such as localized ischemia, necrosis, or aneurysm, cannot be simulated, nor can their treatments (e.g., through cardiac resynchronization therapy).

The first model here presented is the time-varying elastance model proposed by Suga and coworkers [2,3], which is a purely functional model with no geometric assumption and is based on the elastance function $E(t)$, defined by

$$E(t) = \frac{P(t)}{V(t) - V_0} \qquad (20.1)$$

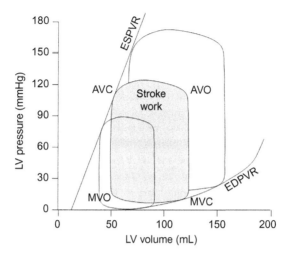

Figure 20.1 P—V loop diagram. Each loop represents a cardiac cycle under different preload and afterload conditions. The area included in the loop is the stroke work. The EDPVR collects the end-diastolic pressure—volume points obtained at different filling conditions. Depending on the afterload, the aortic valve opens and blood is ejected from the ventricle. The slope of the ESPVR corresponds to the maximum value of the time-varying elastance function. AVO, aortic valve opening; AVC, aortic valve closure; MVO, mitral valve opening; MVC, mitral valve closure.

where $P(t)$ is the intraventricular pressure and $V(t)$ and V_0 are the current- and end-diastolic ventricular volumes. The elastance function sums the active and passive properties of the myocardium where the active contribution is bell shaped, to mimic myocardial fiber activation in the systolic phase. The elastance model has to be coupled to a lumped parameter model of the circulation, which provides the closure of the mathematical problem. In the most simple configuration, a Resistor—Inductor—Capacitor (RLC) model already provides a realistic ventricular afterload; by adding a further block (either a capacitor or a RCL block), the atrium behavior can also be included and a closed-loop comprehensive of the ventricular preload and afterload is obtained. This very simple model has been used to characterize the ventricular behavior in terms of the so-called pressure—volume (P—V) loop and energetics (Figure 20.1). The P—V loop represents the intraventricular pressure and volume changes during the cardiac cycle as a function of the preload and afterload; by changing the filling pressure of the ventricle, the end-diastolic pressure—volume relation (EDPVR) is obtained, which depends on the passive properties of the myocardium; the end-systolic pressure—volume relation (ESPVR) is obtained by connecting the end-ejection points. The ESPVR curve is approximately linear and its slope corresponds to the maximum of the elastance function. The area included in the loop represents the stroke work (i.e., the net energy produced in the cardiac cycle).

As previously mentioned, even this elemental model can provide an effective description of a patient-specific heart, at least from the phenomenological point of view, although limited to cardiac global function representation. In particular, the magnitude and shape of the elastance curve can be related to heart rate and contractility; thus, it is possible to simulate heart impairment by modifying the elastance function. The passive elastance behavior can be changed to simulate tissue stiffening due to fibrosis.

Alternatively, more sophisticated yet simple geometrical models of the LV are available, which allow one to simulate heart behavior in real time; the ventricle is considered as a thick-walled axial—symmetrical structure with myofibers varying their orientation across the wall thickness (Figure 20.2). Ventricular contraction and relaxation during the cardiac cycle is determined by the mechanical equilibrium, governed by the Young—Laplace equation, between intracavitary blood pressure and wall stresses, which are modeled as the sum of an active and a passive contribution, related to the active contraction of myofibers and the passive stretch of the extracellular matrix, respectively. With this kind of model, on the basis of the complexity of the fiber model, it is hence possible to simulate the patient-specific ventricle behavior from the geometrical point of view. It is also possible to simulate its inotropic status, for instance by accounting for myocardial fibrosis, which impairs both the passive behavior, by stiffening the myocardium, and the active behavior, by decreasing the active stress. In the following, we will focus on two examples.

Beyar and Sidemann [4] proposed an LV model based on one single equilibrium equation. The left ventricle is described as a prolate ellipsoid with a fixed major to minor axis ratio. The wall is assumed to be incompressible and with uniform thickness; in its reference configuration, it is divided into n concentric layers, each one characterized by a fiber inclination angle calculated from histological observations. Under the hypothesis of wall incompressibility, the equilibrium condition at the equator in the circumferential direction for a generic myocardial layer at a generic instant is given by

$$\Delta P_{im} = \sum_k \Delta P_{im,k} = \sum_k \sigma_{fk} \left(\frac{\cos^2 \alpha_k}{r_k} + \frac{\sin^2 \alpha_k}{(r_k - a + b)^2 / r_k} \right) (r_{k+1} - r_k) \qquad (20.2)$$

where ΔP_{im} is the intraventricular pressure increment, a and b are the prolate ellipsoid major and minor axes, α is the local fiber inclination, and σ_f is the local fiber stress, whose passive and active components must be defined as functions of time and fiber contractility state, preload, and afterload.

In order to complete the model and compute the LV P—V loop, one must define the LV reference: geometrical configuration must be set (i.e., wall thickness, major and minor axes, and fiber angle distribution across the wall), as well as an afterload, typically consisting of an RLC lumped parameter network relating the ventricular volume changes to the intraventricular pressure.

Arts et al. [5] adopted a similar approach as compared to Beyar and Sidemann [4], but assumed a different geometrical model; they described the ventricle as an incompressible thick-walled cylinder, divided into layers which, in the reference, end-diastolic

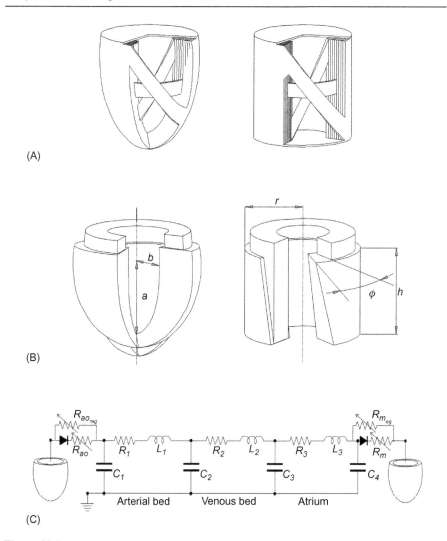

Figure 20.2 Thick-walled models of the ventricle. (A) The schematics of the Beyar and Sideman and Arts and Reneman models are shown; in both cases the models account for the myofibers varying their orientation through the wall. (B) Wall equilibrium is calculated using one single equation and three equations, respectively; indeed, considering the prolate spheroid, the ventricle the major and minor axis have a constant ratio so that the unknowns reduce to one. (C) To calculate the relationship between the ventricular volume variation and the ventricular pressure, the ventricular models have to be connected to a lumped parameter model of the circulation. In the presented case, the capacitor C_1 represents the aorta compliance and the three RCL blocks represent the arterial venous and atrial districts, respectively. The model also includes the transvalvular nonlinear resistances both in the forward and backward (regurgitant) directions. Forward losses are usually negligible except in the case of aortic or mitral stenosis. Backward resistance is usually very high, and backward flow is negligible, unless valve regurgitation has to be simulated.

configuration have equal thickness. Myocardial fibers are arranged like helixes with an angle varying from layer to layer consistent with anatomical observations. LV internal radius, height, and twisting can be computed solving the cylinder equilibrium in the circumferential and axial directions and with respect to torque (Eqs. (20.3)–(20.5), respectively):

$$\Delta P_{imk} = \sigma_{fk} \frac{\cos^2 \alpha_k}{r_k}(r_{k+1} - r_k) \tag{20.3}$$

$$\sum_{k=1}^{n}(\sigma_k \cdot \sin^2 \alpha_k)(r_{k+1}^2 - r_k^2) + R_i^2 \cdot P_v = 0 \tag{20.4}$$

$$\sum_{k=1}^{n}(\sigma_{fk} \cdot \cos \alpha_k \cdot \sin \alpha_k)\pi(r_{k+1}^2 - r_k^2)(r_{k+1} + r_k)/2 = 0 \tag{20.5}$$

Then, the LV wall external radius is obtained through the incompressibility condition.

As for the ellipsoidal model, the knowledge of the fiber mechanical performance drives the myocardium contraction and hence the LV geometry modifications through time, and the assignment of the proper afterload, used to mimic the interaction with the systemic circulation, allows for the coupling between LV volume variations and LV pressure.

20.4 Image-Based Assessment of Ventricular Biomechanics

This section is focused on those models specifically developed for clinical purposes (i.e., whose aim is to describe the behavior of the heart of a specific patient). This approach exploits the information provided by 4D imaging through biomechanical methods to rapidly extract myocardial regional 3D stress and strain data. Thus, these models represent an advancement with respect to existing software for image processing, such as Syngo™ by Siemens and Qlab™ by Philips, which allows for performing quantitative analysis from imaging in order to extract ventricle mass and ejection fraction, wall thickening, and regional motion (e.g., by computing the regional fractional area change). Also, computed stress and strain data can feed biomechanical models to predict the effects of local surgical or therapeutic interventions.

In the following paragraphs, a brief overview will be provided of the standard workflow needed to attain data from imaging; moreover, two different methods will be presented to accurately quantify strains and stresses, respectively.

20.4.1   4D Imaging and Image Processing

With respect to state-of-the-art cardiac imaging, such as magnetic resonance (MR), Echo Doppler, and computer tomography (CT), cardiac MR imaging (cMRI) is usually considered the reference standard for assessing myocardial function, in that it is characterized by low invasiveness, good spatial and temporal resolution, and good contrast between ventricular myocardium and the surrounding tissues. Moreover, different acquisition sequences can be implemented to obtain complementary information: cine sequences allow for reconstructing ventricular anatomy and motion, tagged cMRI can yield regional 2D strains, late gadolinium enhanced sequences provide information on local tissue necrosis/fibrosis, diffusion tensor cMRI can be used to assess myocardial fiber architecture, and phase contract cMRI provides the blood velocity field. However, in clinics only cine-cMRI and late gadolinium enhanced acquisitions are routinely adopted; the other sequences are not usually used outside of academic centers, due to their complex implementation and lengthy times for analysis. In the following, we will hence focus on the extraction of myocardial strains from standard cine-cMRI. The input for these methods consists in a stack of short-axis images acquired throughout the cardiac cycle (Figure 20.3A), possibly integrated with acquisitions on long-axis rotational planes. Endocardial end epicardial contours are identified on the acquired planes (Figure 20.3B) and, if long-axis images are available, coregistration between short- and long-axis data is performed to compensate for possible misalignments due to chest motion during image acquisition. The reconstructed contours are then interpolated with discretized 3D surfaces corresponding to the endocardium and epicardium (Figure 20.3C), which can be regularized through smoothing algorithms. Alternatively, regular endocardial and epicardial surfaces can be obtained through morphing of predetermined 3D templates of the ventricular wall on the detected contours [6,7].

20.4.2   Time-Dependent 3D Strain Estimation

This section is dedicated to methods based on nearest neighbor algorithms. According to these methods, the regional 3D strains experienced by the ventricular wall can be computed through three steps.

First, in the end-diastolic frame, assumed as the reference configuration, endocardial and epicardial surfaces are divided into 16−18 sections, consistently the commonly adopted bull's eye representation (Figure 20.3D); on the boundaries of each sector, reference points are identified and characterized in terms of spatial position and local curvature [8,9].

Second, the characterized reference points are tracked throughout the subsequent time points by means of a nearest neighbor search. Under the assumption that local changes in position and geometry occur gradually through time, and that their entity must be small in a brief time frame, the nearest neighbor of each reference point P_{ti} at time t_i is identified as the vertex P_{ti+1} on the corresponding triangulated surface at time t_{i+1} that minimizes the change in local principal curvature [8], or the linear combination of changes in curvature, changes in position, and a third term accounting for twisting and untwisting of the ventricle during the cardiac cycle [9]. The nearest neighbor search can be made symmetric and hence more robust; after identifying P_{ti+1}, its nearest neighbor is searched for backwards

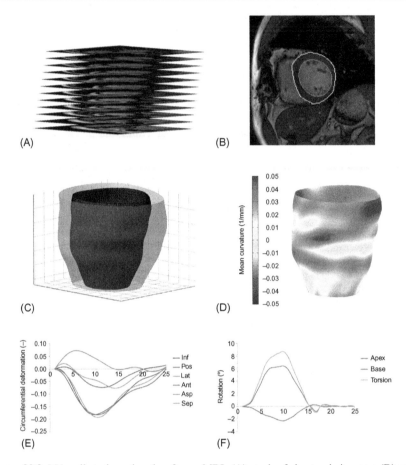

Figure 20.3 LV wall strain estimation from cMRI: (A) stack of short-axis images; (B) endo- and epicardial contours tracing on a short-axis plane; (C) 3D endo- and epicardial surfaces of endocardium (blue, with superimposed mesh) and epicardium (light blue) obtained through interpolation of 2-D contours and smoothing; (D) representative contour plot of mean curvature values on the endocardial surface; (E) circumferential wall strains in the LV basal region of a patient with localized dyskinesia; (F) LV rotation at the apex (blue) and at the base (red). Torsion (green) is defined as the difference between apical and basal rotations. (For interpretation of the references to color in this figure legend, the reader is referred to the web version of this book.)

(i.e., on the triangulated surface at time t_i). Only if this further operation retrieves the original point P_{ti} is the search considered successful [10]. In any case, the nearest neighbor search provides the time-dependent displacements of the reference points throughout the cardiac cycle.

Third, each one of the 16–18 sectors of the ventricular wall is treated as a hexahedral finite element (FE), and the reference points on its edges coincide with the nodes of the FE, whose displacements have been computed. Accordingly, these can be interpolated to obtain the continuous displacement field within each element/ wall sector, whose derivatives provide the 3D regional strains. Since myocardial fiber systolic shortening can reach 20−25%, the finite strain theory should be adopted to compute strains; the deformation gradient tensor F is hence obtained through the spatial derivative of the displacement field u:

$$F = \nabla u + I \tag{20.6}$$

where I is the unit rank-2 tensor. The Green−Lagrange strain tensor E can hence be computed as

$$E = \frac{1}{2}(F^{\mathrm{T}}F - I) \tag{20.7}$$

Moreover, from the displacements of the most basal and most apical reference points, it is straightforward to compute basal and apical wall rotation with respect to a reference position, and hence to obtain ventricular torsion.

This approach just described was successfully applied to quantify regional 3D strains and torsion from conventional cine short-axis cMRI images (Figure 20.3E and F), in a physiological human ventricle and pathological human ventricles affected by increasingly severe ischemic disease [9]. Computations well agreed with previous strain and torsion measurements obtained through the processing of more complex and less common medical imaging techniques, such as Doppler strain rate echo cardiography and three-dimensional tagged magnetic resonance imaging. Also, it is worth stressing that this approach yields the full strain tensor E, which means that the 3D strain field can be computed. As compared to the in-plane strain data that can be extracted from tagged MRI or echocardiographic images via spackle tracking, which are inherently 2D, this information is more exhaustive and consistent with the actual deformation modes experienced by the ventricle during the cardiac cycle (i.e., torsion superimposed to circumferential and axial shortening during systole and untwisting combined with circumferential and axial re-elongation during diastole).

20.4.3 Contractility Estimation

In the previous section, we have shown how regional 3D strains can be computed and deeper insight into myocardial function can be obtained as compared to that provided by mere motion assessment from 2D clinical images. The computation of 3D strains is hence a valid tool for assessing the presence of contractility impairment; however, this does not provide a direct link to myocardial properties. To obtain this further information, one must formulate a constitutive model (i.e., a stress−strain relationship) for the myocardium. If the constitutive parameters (i.e., the parameters of the constitutive model) have a clear physical meaning, one can identify them to gain insight into the cause of regional impairments. Also, one can simulate the effects of possible interventions by changing the constitutive parameters these directly affect, thus predicting their success or failure.

A noninvasive method for estimating regional myocardial constitutive parameters is inverse FE modelling. Based on *in vivo* imaging, anatomic and kinematic
boundary conditions of the ventricle can be defined and displacements/strains can
be measured. An FE model of the ventricle can be implemented based on the
reconstructed anatomy and kinematic constraints, and completed with intraventricular pressure loads and with a constitutive law for myocardial tissue accounting for
its anisotropy and active contraction. Simulations of ventricular function can be run
iteratively to explore the parameters' space and to identify the set of constitutive
parameters that minimizes the errors in the computation of regional displacements/
strains with respect to measured data [6,11].

This approach is well exemplified by the work by Wenk et al. [11], who have
induced localized myocardial infarction in an adult sheep and have acquired a series
of orthogonal short- and long-axis tagged-cMRI images at 14 weeks postmyocardial
infarction. During cMRI acquisitions, intraventricular pressure has been measured
with a nonferromagnetic transducer-tipped pressure catheter. Through image segmentation, ventricular geometry is reconstructed and discretized at early diastole,
considered to be the reference configuration since pressure is at its minimum at this
time point, and time-dependent kinematic boundary conditions are derived throughout systole. In the discretization process, three layers of elements are defined across
the wall thickness, each with a different fiber orientation. Also, three myocardial
regions are defined consistently with clinical images: healthy region, infarct region,
and transition region. The endocardial surface is loaded by the measured time-
dependent pressure. The mathematical problem is closed by formulating a nonlinear, nearly incompressible, transversely isotropic elastic constitutive model of
passive myocardium, represented by the following strain energy function W:

$$W = k_1(e^Q - 1) \tag{20.8}$$

$$Q = k_2 E_{ff}^2 + k_3(E_{ss}^2 + E_{nn}^2 + 2E_{sn}^2) + 2k_4(E_{fs}E_{sf} + E_{fn}E_{nf}) \tag{20.9}$$

where E_{ij} are the components of the Green−Lagrange strain tensor, k_1-k_4 are four
constitutive parameters, which are different for the three myocardial regions, and
subscripts f, n, s correspond to three local orthonormal axes, f being oriented as the
myocardial fiber and n running through the wall thickness.

Since the myocardium is assumed to be nearly incompressible, the strain energy
function W can be decomposed into two terms, one accounting for the volumetric
strains $(U(J))$ and one accounting for the deviatoric stress associated with changes
in shape $(\tilde{W}(\tilde{C}))$:

$$W = U(J) + \tilde{W}(\tilde{C}) \tag{20.10}$$

The passive component of the local stress tensor is calculated deriving $\tilde{W}$ with
respect to the strain tensor and is added to an active component T_a to obtain the
second Piola−Kirchoff stress S in the form:

$$S = pJC^{-1} + 2J^{-2/3}\text{Dev}\left(\frac{\partial \tilde{W}}{\partial C}\right) + T_a(t, \text{Ca}_0, l, T_{\max}) \tag{20.11}$$

where C is the right Cauchy strain tensor and is equal to $2E + I$, $J = \det(F)$, p is a Lagrangian multiplier, and T_a depends on time (t), intracellular calcium concentration (Ca_0), sarcomere length (l), and maximal isometric sarcomere tension (T_{max}) through a bell-shaped time function [12]. Finally, the deviatoric operator Dev is defined as

$$\text{Dev}(\cdot) = (\cdot) - \frac{1}{3}([\cdot]:C)C^{-1} \qquad (20.12)$$

Ventricular systolic contraction is iteratively simulated and the strain tensor E is computed for each FE. At each iteration, the projections of E on the acquired short-axis cMRI planes are considered in a subset of elements and compared to the corresponding cMRI-based strain measurements by computing the mean square error. The myocardial constitutive parameters are modified through a systematic search methodology to automatically explore the parameter space and find an optimum design.

This approach proves successful in identifying the material properties of healthy and ischemic myocardium. However, from an application standpoint it has two limitations: first, it requires the use of tagged-cMRI images to obtain the *in vivo* measurements used as terms of comparison for the strains computed by the FE model; second, the identification of myocardial parameters is time-consuming (the authors have found it to take about 4.5 h on a 4-processor workstation), although it could be sped up by further exploitation of parallel computing.

20.5 Multiphysics Patient-Specific Models of the Left Ventricle

The previous two classes of ventricular models, even if characterized by different degrees of complexity, all focus on myocardial mechanics, in that their ultimate goal is the characterization of P–V relations or of regional 3D wall strains and stresses. In contrast, this third and last class of model measures the exhaustive patient-specific modelling of the cascade of multiphysics phenomena that leads from electrical pacing to myocardial contraction and to blood ejection; myocardial wall mechanics can be coupled with electrical activation within an electromechanical (EM) model, and with blood fluid dynamics within a fluid–structure interaction (FSI) model. Of note, comprehensive models including all three of the types of physics mentioned can be implemented.

The potential of this approach consists in the possibility of accounting for the actual links between the different aspects of ventricular function [13], and hence analyzing the real mechanisms underpinning the effects of alterations of any of its aspects (Figure 20.4). For example, EM models can be used to quantify the effects of impairments in electrical pacing and conduction on the heart pumping function or of therapies aimed at treating them (e.g., CRT and ablation). FSI models can elucidate the relationship between the properties of the myocardial wall (e.g., diastolic stiffness and regional

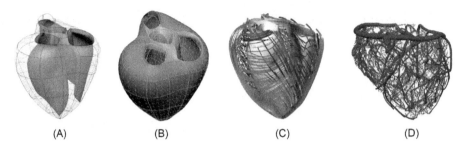

(A)   (B)   (C)   (D)

Figure 20.4 EM heart model developed at the Auckland Bioengineering Institute: (A) FE surfaces fitted to measurements from the left and right ventricles; (B) 3D FE model of the heart; (C) epicardial and midwall fiber directions; and (D) integrated coronary artery model. *Source*: Reproduced from Ref. [13] with permission of Elsevier Science.

contractility) and blood fluid dynamics (e.g., fluid dynamic patterns during ventricular filling and energetics of ventricular systolic ejection). Still, these models have some major limitations with respect to their clinical application; first, the computational expense associated to their simulation in many cases is too high to be compatible with the timing of clinical practice. Second, they require a large amount of input data, which in some cases needs to be derived from uncommon clinical imaging modalities through time-consuming image-processing techniques, and in some cases cannot actually be defined on a patient-specific basis. In particular, as compared to the simpler image-based models described in Section 2.2, they require the combination of detailed anatomical data with extra information on the architecture of heart fibers, which is needed for the definition of myocardial mechanical and electrical conduction properties.

20.5.1 Modelling of Cardiac Tissue Electrical Properties and EM Coupling

Myocardial active contraction occurs along the myocardial fiber axis; also, the propagation of the action potential triggering sarcomere contraction is about three times faster along myocardial fibers as compared to the other directions. Hence, in the context of EM models, a pivotal role is played by the description of *in vivo* cardiac fiber architecture. The latter can be assessed through MR diffusion tensor imaging. The diffusion tensor is a rank-2 tensor, whose components correspond to the direction-dependent diffusivities of a tissue. These can be measured by encoding the additional signal loss associated with the translational motion of water molecules in the presence of spatially varying magnetic fields, along with the signal associated with the spin relaxation from their random rotational motion. With regards to myocardial tissue, this technique can yield information on the layered structure of the LV wall and on fiber direction within each layer. The direction of fastest diffusion (i.e., the primary diffusion tensor eigenvector) identifies the local myocardial fiber orientation; the plane spanned by the directions of the two fastest diffusion dimensions (i.e., the primary and secondary eigenvectors) can be taken as the plane locally tangent to the fiber layer. The direction of slowest diffusion (i.e., the tertiary eigenvector) can be taken as the axis normal to the layer [14]. However, this imaging modality has important limitations when used to assess *in vivo* myocardial fiber architecture: to entirely define the diffusion tensor, at least seven scans are needed, leading to time-consuming acquisitions. Also, the translational motion of water molecules associated with diffusion is 3 orders of magnitude smaller than the bulk motion of the heart, resulting in inaccurate signal and low signal-to-noise ratio, so that sophisticated offline postprocessing techniques are needed to retrieve reliable information [15].

When the diffusion tensor D is defined, it can be used in the differential equations governing the time evolution of the sarcomere transmembrane potential V at a given point in space. In order to limit the number of parameters involved in the equations while obtaining accurate results, a phenomenological model may be used in the following form [16]:

$$\frac{\partial V}{\partial t} = \nabla \cdot (D\nabla V) + \frac{zV^2(1 - V)}{\tau_{\text{in}}} - \frac{V}{\tau_{\text{out}}} + J_{\text{stim}}(t) \tag{20.13}$$

The first term in the right-hand side of the equation is the diffusion term depending on D; in a physiological myocardium it can be set to $D = d \cdot \text{diag}(1, 1/r^2, 1/r^2)$, where d is a scalar parameter and $r = 2.5$, so to account for the faster diffusion along fiber axis (D_{11}) as compared to the other two local orthogonal directions (D_{22} and D_{33}). D can also be modified to simulate pathological or postprocedure conditions; for instance, ablation can be simulated by setting $D_{11} = D_{22} = D_{33} = 0$ in the treated region. The second term in the right-hand side of the equation is associated with the transmembrane ion currents leading to a V increase (Na^+ and Ca^{2+}), characterized by the time constant τ_{in}. These are gated by an auxiliary variable z, which ranges between 0 (closed ion channels) and 1 (open ion channels) according to the following dynamics:

$$\frac{\partial z}{\partial t} = \begin{cases} \dfrac{1 - z}{\tau_{\text{open}}} & \text{if } V < V_{\text{gate}} \\[2ex] \dfrac{-z}{\tau_{\text{close}}} & \text{if } V > V_{\text{gate}} \end{cases} \tag{20.14}$$

where τ_{open} and τ_{close} are the time constants of channels opening and closure and V_{gate} is the changeover voltage. The third term is associated with the nongated transmembrane ion currents leading to a V decrease (K^+) characterized by the time constant τ_{out}, while the fourth term represents the stimulation current starting the diffusion process and representing the physiological or the artificial pacing. $J_{\text{stim}}(t)$ can hence be locally tuned to simulate CRT procedures [17].

The transmembrane potential V can be used as a control variable in the definition of sarcomere active tension T_a to obtain a coupled EM model. This approach was recently applied by Talbot et al. [16], who proposed an interactive image-based biventricular model aimed at rehearsal and training in cardiac catheter ablation, and potentially for the planning of cardiac resynchronization therapy.

20.5.2 Modelling of Blood Flow and FSI Coupling

Intraventricular blood flow is the fluid dynamic counterpart of myocardium structural modelling and an important aspect of the assessment of cardiac function. The equations governing fluid dynamics explain how vortex dynamics depend on patient-specific anatomical features, such as cavity shape and cavity shape modification, and on flow features, such as inflow velocity profile. Up to now, however, few models have considered this aspect due to the complexity and the difficulty of the problem, which makes this issue an ongoing challenge.

From the fluid mechanics standpoint, the deformation of the fluid domain is pivotal in generating the intraventricular flow pattern. Wall motion determines a coherent blood flow at the wall−blood interface; the motion of blood determines blood pressure gradients, which impacts wall motion. This two-way coupling is extremely delicate and requires that both velocities v and Cauchy stress tensors σ coincide at the boundary

$$v_f - v_s = 0 \quad \sigma_f \cdot n_f + \sigma_s \cdot n_s = 0 \tag{20.15}$$

where n is the unit vector normal to the interface, and subscripts s and f refer to wall and blood, respectively.

One of the first simulations of ventricular blood flow was the seminal work conducted by Peskin using the immersed boundary method [18,19], which, however, suffers from a lack of precise conservation near fibers (allowing blood flow across the heart wall). Alternatively, the arbitrary Lagrangian−Eulerian approach can be used, which allows for a strong FSI coupling. One of the first attempts in this direction was presented by Redaelli and Montevecchi [20], who analyzed the effects of coupled structure and flow mechanics on intraventricular pressure gradient generation. About one decade later, fully 3D, realistic, models adopting the arbitrary Lagrangian-Eulerian (ALE) method and based on MR imaging have become available [21,22]. These works have shown that the blood enters the ventricle, changes its direction to form a vortex ring, and is finally squeezed out from the ventricle according to a dynamic that is strictly influenced by ventricular chamber shape and contraction. Alterations in the vortex dynamics associated with wall local hypo- or dyskinesia, as well as with the loss of the funnel-like shape of the outflow tract, can be related to ventricular impairment and dysfunction and can prospectively be used as early detectors of disease onset [23,24]. Moreover, FSI models can be effectively used to assess surgical procedures for the repair of the ischemic heart [22].

20.6 3D Patient-Specific Heart Valve Modelling: Early Approaches

The development of patient-specific models of cardiac valves is quite a recent practice due to the complexity of valve apparatuses and their 3D visualization. Nonetheless, in the case of heart valves there are also simple analytical tools that have been used to characterize their performance from either the fluid dynamics or the structural point of view.

The simplified Bernoulli's equation is routinely used to assess pressure gradients across the valve in conjunction with Doppler velocimetry [25,26]. It is based on the application of the energy conservation equation, neglecting the inertia and assuming that the kinetic energy is completely dissipated. When the pressure is expressed in mmHg, it has the form:

$$\Delta P_{12} = 4\left(v_2^2 - v_1^2\right) \tag{20.16}$$

where subscripts 1 and 2 indicate two locations along the blood streamline. The simplified Bernoulli formula is used to calculate the degree of stenosis in a vessel or through a valve and overestimates the effective pressure losses, since it neglects the pressure recovery downstream from the obstacle.

It must be considered that this is only one form (the simplest one) to calculate the pressure losses due to a local area change also called shock losses. A complete review of the formula describing the shock losses of flow through orifices can be found in Ref. [27].

Bernoulli's equation is also used in lumped parameter models of the peripheral circulation, representing the entire ventricle or of the entire heart, where the role played by heart valves mainly consists in defining the different phases of the cardiac cycle through their opening and closure [16]. In this case, the aortic and mitral valves are modeled as electric diodes in series with a nonlinear resistance. The resistance value is in general negligible, but becomes relevant when the valve is stenotic. In the case of incompetent valves, the additional (nonlinear) resistance is put in parallel to simulate the regurgitant blood flow.

Finally, a rough estimation of the stresses in the leaflets can be attained through the Young–Laplace formula provided the main curvature radii of the leaflets is measured with Echo or other imaging techniques [28].

20.7   3D Patient-Specific Heart Valve Modelling: Recent Advances

Despite being integrated into heart chambers, with few exceptions [29], heart valves are usually modeled separately, accounting for the interaction with the surrounding myocardial wall through proper boundary conditions. This approach is due to the complexity of valvular structures, and to the fact that it can be extremely challenging to obtain the desired level of detail in the valve and in the surrounding chambers at the same time. For this reason it is worth reviewing current advances in heart valve patient-specific modelling separately, with an emphasis on the left heart valve (i.e., the mitral valve (MV)) and on the aortic root (AR), whose pathologies are most frequently affected by pathologies requiring surgical treatment. In the following section, the currently most advanced approaches in MV and AR patient-specific numerical modelling are overviewed.

20.7.1 Image-Based Reconstruction of Heart Valve Morphology and Kinematics

The MV and the AR are complex apparatuses. The MV is the left atrioventricular valve and is composed of four principal substructures: the mitral annulus (MA), the anterior and posterior leaflets, the chordae tendineae, and two papillary muscles (PMs). The MA and the PMs act dynamically and synergically due to their continuity with the LV myocardium. In systole, as LV pressure increases forcing the leaflets' closure, the MA shrinks and bends, while the PMs contract. This tightens the chordae tendineae, thus regulating the leaflets' dynamics and preventing both leaflet prolapse and lack of coaptation. The AR consists of three asymmetrical units, each one consisting of one of the three leaflets of the aortic valve (AV) and of the corresponding the sinus of Valsalva. The three units are separated by three interleaflet triangles that connect the AV to the aorto-ventricular junction. As in the MV, in the aortic root, the dynamic interplay between the different substructures is crucial to its function.

The morphologically realistic reconstruction of these complex structures in their stress-free configuration is the first step to creating a numerical model. Different image modalities can be exploited for this aim: real-time 3D echocardiography (RT3DE) [30−35], cMRI [29,36−38], or multislice computed tomography (MSCT) [39−42]. Depending on the image modality and on the adopted image processing approach, different levels of detail and automation can be obtained. Regarding detail, current processing techniques allow for reconstructing even fine details such as mitral leaflet thickness from transesophageal RT3DE imaging [34]. Yet, only MSCT allows for capturing fine details, such as leaflet profiles in AR reconstruction or chordae distribution, origins of the PM tips, or insertions into the leaflets in MV reconstruction [39]. Regarding automation, cMRI images are typically manually segmented [29,36−38], while for RT3DE imaging datasets, algorithms for the semiautomated partial [30,31] or complete [34] reconstruction of the valve are available, as well as for its fully automated geometrical modelling [33,43−45]. Also, when images are acquired throughout the cardiac cycle, the motion of valvular substructures acting as interface between the valve itself and the surrounding myocardium can be quantitatively reconstructed. Algorithms are available for the semiautomated tracking of mitral and aortic annuli [46,47], and for analyzing their coupling [48,49]. The motion of MV PMs can be obtained from their manual and frame-by-frame identification [38]. This extra information can be included in numerical analyses as kinematic boundary conditions to account for the dynamic contraction of these valvular substructures, as well as to realistically reproduce pathological conditions (e.g., PM dislocation and annular dilatation).

20.7.2 Modelling of Tissue Mechanical Properties

In order to implement patient-specific heart valve models, it would be necessary to identify tissues' patient-specific mechanical properties, but their measurement is currently infeasible *in vivo*. Indirect estimations were recently performed for the anterior mitral leaflet (AML) only in sheep, highlighting that its *in vivo* properties

differ greatly from *ex vivo* ones due to AML active contraction [50,51] and are pivotal in determining its time-dependent configuration [52,53]. Still, that procedure
can't be adopted in humans, since it involves the identification of mechanical parameters through inverse FE modelling by fitting the time-dependent position of radi-
opaque markers surgically implanted into the AML.

For this reason, tissues' *in vitro* stress−strain response is normally adopted, thus
describing mechanical properties not on a patient-specific basis, but rather on a
condition-specific basis (i.e., healthy tissue, calcified tissue, etc.). Valve leaflets
and aortic wall tissues exhibit a heterogeneous, incompressible, nonlinear elastic,
and anisotropic behavior. Accordingly, the latter is modeled through hyperelasticity
(i.e., the same theoretical framework adopted for modelling myocardial passive
mechanical properties). The simplest tissue to be modeled is the MV chordae tendineae, which is normally assumed to be isotropic hyperelastic or even linear elastic.
Mitral and aortic leaflets, as well as the sinus of Valsalva, are commonly described
as orthotropic or transversely isotropic, owing to the preferentially oriented collagen fibers embedded in their extracellular matrices. To this aim, two main
approaches are available: the most accurate one accounts for collagen fibers' splay
by defining a probability density function of the fiber direction (e.g., a Gaussian
function of the fiber angle θ with respect to a reference direction) so that the
Piola−Kirchoff stress tensor S can be expressed as

$$S + pC^{-1} = \int_{-\pi/2}^{\pi/2} S_\mathrm{f}(\theta, C)R(\vartheta)a_0 \otimes a_0(\theta)\mathrm{d}\theta \tag{20.17}$$

where C is the right Cauchy−Green strain tensor, p is a Lagrangian multiplier,
$S_\mathrm{f}(\theta,C)$ is a function describing collagen fibers' stress−strain relation, $R(\theta)$ is the
probability density function, and a_0 is the unitary vector identifying fibers' direction in the stress-free configuration. The main drawback of this approach consists
in the computational expense associated with the numerical computation of the
integral. The alternative and more efficient method consists in assuming that collagen fibers are perfectly aligned along a given direction a_0 and in accounting for
anisotropy by including in the strain energy potential W a term that depends on the
pseudoinvariant $I_4 = a \otimes a$, with t $a = F \cdot a_0$:

$$W = k_0\{e^{[k_1(I_1-3)^2+k_2(I_4-1)^2]} - 1\} \tag{20.18}$$

where k_0, k_1, and k_2 are the constitutive parameters and $I_1 = tr(C)$. Similar to
Eq. (20.11), the second Piola−Kirchoff stress tensor S can be expressed as

$$S = pJC^{-1} + 2J^{-2/3}\mathrm{Dev}\left(\frac{\partial \tilde{W}}{\partial C}\right) \tag{20.19}$$

The main drawback of this second approach is the lower accuracy in computing
stresses associated with shear deformations, which can be relevant in some regions
of the leaflets.

20.7.3   Heart Valve–Blood Interaction

During the cardiac cycle, heart valves open and close due to their interaction with blood. The most accurate way to account for it is FSI modelling. However, the latter is even more challenging than ventricle FSI modelling, due to the abrupt nature of valve transient closure and opening. It is also due to leaflets' coaptation, which implies the use of very fine fluid meshes and of small time steps in the numerical solution of the problem, thus leading to excessive computational expense. As a

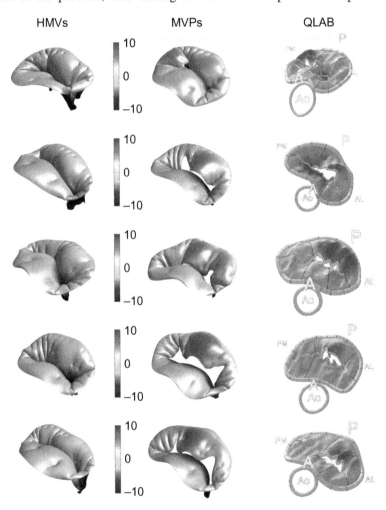

HMVs   MVPs   QLAB

Figure 20.5 Computed MV configuration at peak systole for five healthy subjects (left column) and five patients with mitral prolapse (middle column); color-coded contour maps represent the signed distance (mm) along the axial direction from the insertion on the annulus. For PMVs, the comparison with ground truth data reconstructed from TE-RT3DE via Q-Lab is provided (right column).

Source: Reproduced from Ref. [31] with permission of Elsevier Science.

result, FSI modelling is commonly adopted only in idealized models characterized by simple geometries and nonphysiological parameter ranges (e.g., blood bulk modulus and Reynolds number), although Chandran and Vigmostad lately proposed the preliminary results of their in-home FSI algorithm allowing for patient-specific simulations with physiologic Reynolds numbers, realistic material properties, and highly resolved grids [35]. With this exception, in advanced patient-specific models it is usual to account for blood pressure simply by applying time-dependent distributed pressure loads on the leaflets surfaces.

20.7.4   Clinically Relevant Applications

In the last five years, heart valve patient-specific modelling has evolved from pilot studies testing the possibility of effectively integrating image processing with FE

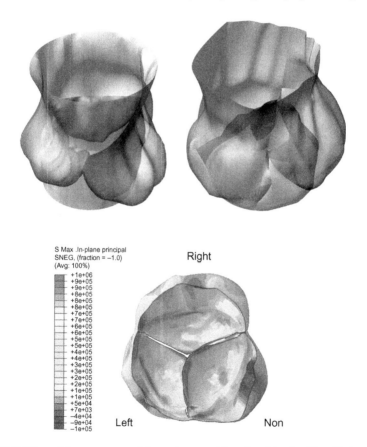

Figure 20.6 3D models reconstructed from ultrasound imaging for a healthy AR (top left) and in case of bicuspid aortic valve (top right). An example of computed stress distribution is provided (bottom).
Source: Adapted from Ref. [35] with permission of Elsevier Science.

modelling [30] to the development of tools that exploit the information from routinely adopted imaging modalities (mostly RT3DE) and that can be systematically adopted to simulate valve function on a patient-specific basis.

These tools can be applied to analyze in detail the biomechanics of healthy and pathological valves (Figure 20.5). We recently proposed a robust method for simulating MV *in vivo* systolic function in healthy subjects and in patients affected by MV prolapse, showing how it is possible to capture the specific functional defects of prolapsed valves and to quantify the associated alterations in tissue tensions, although the clinical application of the method is currently limited by the time expense of FE simulations, which is approximately 6 h [31]. This limit can be reduced to <1 h through the use of alternative numerical methods, such as corotated FEs, and through the nearly complete automation of the pipeline from image processing to simulations, as eloquently shown by the recent work by Mansi et al. [33]. Also, Chandran and Vigmostad [35] developed a method for the biomechanical analysis of the AR from RT3DE imaging (Figure 20.6) and applied it to the analysis of healthy subjects and patients affected by different types of bicuspid aortic valve.

Moreover, patient-specific modelling is now starting to also be tested as a tool to predict the effects of heart valve surgical repair, as in the case of mitral annuloplasty [38] and mitral prolapse correction by means of the percutaneous MitraClip [33] approach (Figure 20.7) and of percutaneous aortic valve replacement [40].

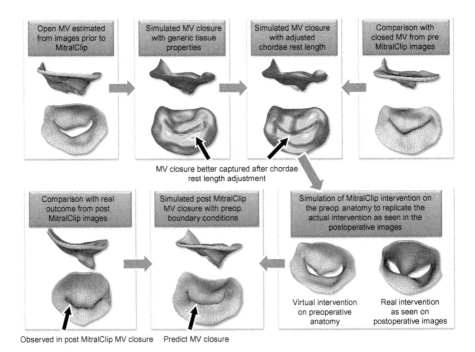

Figure 20.7 Patient-specific simulation of MitralClip intervention on an MV; prediction from the model is qualitatively consistent with the true postoperative outcome.
Source: Reproduced from Ref. [33] with permission of Elsevier Science.

20.8 Conclusion

Currently, different computational modelling techniques are available to quantitatively analyze different aspects of cardiac mechanics. These span different levels of complexity; increasing their complexity leads to deeper insight into the modeled system, but also to greater computational expenses and difficulties in translating the adopted technique into a clinically usable tool. However, the most recent advances in LV and heart valve patient-specific modelling show that even for the most sophisticated approaches, the gap between their use in basic research and their clinical application is narrowing.

References

[1] Go AS, Mozaffarian D, Roger VL, Benjamin EJ, Berry JD, Borden WB, , et al. on behalf of the American Heart Association Statistics Committee and Stroke Statistics Subcommittee Heart disease and stroke statistics—2013 update: a report from the American Heart Association. Circulation 2013;127(1):e6−e245.

[2] Suga H, Sagawa K, Shoukas AA. Load independence of the instantaneous pressure−volume ratio of the canine left ventricle and effects of epinephrine and heart rate on the ratio. Circ Res 1973;32(3):314−22.

[3] Suga H, Sagawa K, Kostiuk DP. Controls of ventricular contractility assessed by pressure−volume ratio, E_{max}. Cardiovasc Res 1976;10(5):582−92.

[4] Beyar R, Sideman S. A computer study of the left ventricular performance based on fibre structure, sarcomere dynamics, and transmural electrical propagation velocity. Circ Res 1984;55(3):358−75.

[5] Arts T, Veenstra PC, Reneman RS. A model of the mechanics of the left ventricle. Ann Biomed Eng 1979;7(3−4):299−318.

[6] Linte CA, Wierzbicki M, Peters TM, Samani A. Towards a biomechanics-based technique for assessing myocardial contractility: an inverse problem approach. Comput Methods Biomech Biomed Eng 2008;11(3):243−55.

[7] de Vecchi A, Nordsletten DA, Razavi R, Greil G, Smith NP. Patient-specific fluid structure ventricular modelling for integrated cardiac care. Med Biol Eng Comput 2013; Available from: http://dx.doi.org/10.1007/s11517-012-1030-5 [Epub ahead of print].

[8] Shi P, Sinusas AJ, Constable RT, Ritman E, Duncan JS. Point-tracked quantitative analysis of left ventricular surface motion from 3-D image sequences. IEEE Trans Med Imaging 2000;19(1):36−50.

[9] Conti CA, Votta E, Corsi C, De Marchi D, Tarroni G, Stevanella M, et al. Left ventricular modelling: a quantitative functional assessment tool based on CMR imaging. Interface Focus 2011;1(3):384−95.

[10] Papademetris X, Sinusas AJ, Dione DP, Constable RT, Duncan JS. Estimation of 3-D left ventricular deformation from medical images using biomechanical models. IEEE Trans Med Imaging 2002;21(7):786−800.

[11] Wenk JF, Sun K, Zhang Z, Soleimani M, Ge L, Saloner D, et al. Regional left ventricular myocardial contractility and stress in a finite element model of posterobasal myocardial infarction. J Biomech Eng 2011;133(4): [manuscript 044501].

[12] Tözeren A. Continuum rheology of muscle contraction and its application to cardiac contractility. Biophys J 1985;47(3):303−9.

[13] Hunter P, Smith N, Fernandez J, Tawhai M. Integration from proteins to organs: the IUPS Physiome Project. Mech Ageing Dev 2005;126(1):187−92.

[14] Guccione JM, Kassab GS, Ratcliffe MB, editors. Computational cardiovascular mechanics—modelling and applications in heart failure. New York, NY: Springer; 2010.

[15] Toussaint N, Stoeck CT, Sermesant M, Schaeffter T, Kozerke S, Batchelor PG. *In vivo* human cardiac fibre architecture estimation using shape-based diffusion tensor processing. Med Image Anal 2013; Available from: http://dx.doi.org/10.1016/j.media.2013.02.008 [Epub ahead of print].

[16] Talbot H, Marchesseau S, Duriez C, Sermesant M, Cotin S, Delingette H. Towards an interactive electromechanical model of the heart. Interface Focus 2013;3(2): [manuscript 20120091].

[17] Mitchell C, Schaeffer D. A two-current model for the dynamics of cardiac membrane. B Math Biol 2003;65(5):767−93.

[18] Peskin CS. Numerical analysis of blood flow in the heart. J Comp Phys 1977;25 (3):220−52.

[19] Peskin CS, McQueen DM. A three-dimensional computational method for blood flow in the heart. J Comp Phys 1989;81(2):372−405.

[20] Redaelli A, Montevecchi FM. Computational evaluation of intraventricular pressure gradients based on a fluid−structure approach. J Biomech Eng 1996;118(4):529−37.

[21] Watanabe H, Sugiura S, Kafuku H, Hisada T. Multiphysics simulation of left ventricular filling dynamics using fluid−structure interaction finite element method. Biophys J 2004;87(3):2074−85.

[22] Tang D, Yang C, Geva T, Gaudette G, del Nido PJ. Multi-physics MRI-based two-layer fluid−structure interaction anisotropic models of human right and left ventricles with different patch materials: cardiac function assessment and mechanical stress analysis. Comput Struct 2011;89(11−12):1059−68.

[23] Dimasi A, Cattarinuzzi E, Stevanella M, Conti CA, Votta E, Maffessanti F, et al. Influence of mitral valve anterior leaflet *in vivo* shape on left ventricular ejection. Cardiovasc Eng Technol 2012;3(4):388−401.

[24] Sugiura S, Washio T, Hatano A, Okada J, Watanabe H, Hisada T. Multi-scale simulations of cardiac electrophysiology and mechanics using the University of Tokyo heart simulator. Prog Biophys Mol Biol 2012;110(2−3):380−9.

[25] Holen J, Aaslid R, Landmark K, Simonsen S. Determination of pressure gradient in mitral stenosis with a non-invasive ultrasound Doppler technique. Acta Med Scand 1976;199(6):455−60.

[26] Yoganathan AP, Cape EG, Sung HW, Williams FP, Jimoh A. Review of hydrodynamic principles for the cardiologist: applications to the study of blood flow and jets by imaging techniques. J Am Coll Cardiol 1988;12(5):1344−53.

[27] Idelchik IE, Ginevskiy AS, Kolesnikov AV, editors. Handbook of hydraulic resistance. 4th ed. New York, NY: Begell House; 2007.

[28] Thubrikar MJ, Nolan SP, Aouad J, Deck JD. Stress sharing between the sinus and leaflets of canine aortic valve. Ann Thorac Surg 1986;42(4):434−40.

[29] Wenk JF, Zhang Z, Cheng G, Malhotra D, Acevedo-Bolton G, Burger M, et al. First finite element model of the left ventricle with mitral valve: insights into ischemic mitral regurgitation. Ann Thorac Surg 2010;89(5):1546−53.

[30] Votta E, Caiani E, Veronesi F, Soncini M, Montevecchi FM, Redaelli A. Mitral valve finite-element modelling from ultrasound data: a pilot study for a new approach to

understand mitral function and clinical scenarios. Philos Trans A Math Phys Eng Sci 2008;366(1879):3411−34.

[31] Votta E, Le TB, Stevanella M, Fusini L, Caiani EG, Redaelli A, et al. Toward patient-specific simulations of cardiac valves: state-of-the-art and future directions. J Biomech 2013;46(2):217−28.

[32] Xu C, Brinster CJ, Jassar AS, Vergnat M, Eperjesi TJ, Gorman RC, et al. A novel approach to *in vivo* mitral valve stress analysis. Am J Physiol Heart Circ Physiol 2010;299(6):H1790−4.

[33] Mansi T, Voigt I, Georgescu B, Zheng X, Mengue EA, Hackl M, et al. An integrated framework for finite-element modelling of mitral valve biomechanics from medical images: application to MitralClip intervention planning. Med Image Anal 2012;16 (7):1330−46.

[34] Pouch AM, Xu C, Yushkevich PA, Jassar AS, Vergnat M, Gorman III JH, et al. Semi-automated mitral valve morphometry and computational stress analysis using 3D ultrasound. J Biomech 2012;45(5):903−7.

[35] Chandran KB, Vigmostad SC. Patient-specific bicuspid valve dynamics: overview of methods and challenges. J Biomech 2013;46(2):208−16.

[36] Conti CA, Della Corte A, Votta E, Del Viscovo L, Bancone C, De Santo LS, et al. Biomechanical implications of the congenital bicuspid aortic valve: a finite element study of aortic root function from *in vivo* data. J Thorac Cardiovasc Surg 2010;140 (4):890−6 896.e1−2.

[37] Conti CA, Votta E, Della Corte A, Del Viscovo L, Bancone C, Cotrufo M, et al. Dynamic finite element analysis of the aortic root from MRI-derived parameters. Med Eng Phys 2010;32(2):212−21.

[38] Stevanella M, Maffessanti F, Conti CA, Votta E, Arnoldi A, Lombardi M, et al. Mitral valve patient-specific finite element modelling from cardiac MRI: application to an annuloplasty procedure. Cardiovasc Eng Technol 2011;2(2):66−76.

[39] Wang Q, Sun W. Finite element modelling of mitral valve dynamic deformation using patient-specific multi-slices computed tomography scans. Ann Biomed Eng 2013;41 (1):142−53.

[40] Capelli C, Bosi GM, Cerri E, Nordmeyer J, Odenwald T, Bonhoeffer P, et al. Patient-specific simulations of transcatheter aortic valve stent implantation. Med Biol Eng Comput 2012;50(2):183−92.

[41] Wang Q, Sirois E, Sun W. Patient-specific modelling of biomechanical interaction in transcatheter aortic valve deployment. J Biomech 2012;45(11):1965−71.

[42] Sirois E, Wang Q, Sun W. Fluid simulation of a transcatheter aortic valve deployment into a patient-specific aortic root. Cardiovasc Eng Technol 2011;2(3):186−95.

[43] Jassar AS, Brinster CJ, Vergnat M, Robb JD, Eperjesi TJ, Pouch AM, et al. Quantitative mitral valve modelling using real-time three-dimensional echocardiography: technique and repeatability. Ann Thorac Surg 2011;91(1):165−71.

[44] Vergnat M, Jassar AS, Jackson BM, Ryan LP, Eperjesi TJ, Pouch AM, et al. Ischemic mitral regurgitation: a quantitative three-dimensional echocardiographic analysis. Ann Thorac Surg 2011;91(1):157−64.

[45] Voigt I, Mansi T, Ionasec RI, Mengue EA, Houle H, Georgescu B, et al. Robust physically-constrained modelling of the mitral valve and subvalvular apparatus. Med Image Comput Comput Assist Interv 2011;14(Pt 3):504−11.

[46] Veronesi F, Corsi C, Sugeng L, Caiani EG, Weinert L, Mor-Avi V, et al. Quantification of mitral apparatus dynamics in functional and ischemic mitral

regurgitation using real-time 3-dimensional echocardiography. J Am Soc Echocardiogr 2008;21(4):347−54.

[47] Schneider RJ, Perrin DP, Vasilyev NV, Marx GR, del Nido PJ, Howe RD. Mitral annulus segmentation from four-dimensional ultrasound using a valve state predictor and constrained optical flow. Med Image Anal 2012;16(2):497−504.

[48] Ionasec RI, Voigt I, Georgescu B, Wang Y, Houle H, Hornegger J, et al. Personalized modelling and assessment of the aortic-mitral coupling from 4D TEE and CT. Med Image Comput Comput Assist Interv 2009;12(Pt 2):767−75.

[49] Veronesi F, Corsi C, Sugeng L, Mor-Avi V, Caiani EG, Weinert L, et al. A study of functional anatomy of aortic-mitral valve coupling using 3D matrix transesophageal echocardiography. Circ Cardiovasc Imaging 2009;2(1):24−31.

[50] Krishnamurthy G, Itoh A, Swanson JC, Bothe W, Karlsson M, Kuhl E, et al. Regional stiffening of the mitral valve anterior leaflet in the beating ovine heart. J Biomech 2009;42(16):2697−701.

[51] Krishnamurthy G, Itoh A, Swanson JC, Miller DC, Ingels Jr. NB. Transient stiffening of mitral valve leaflets in the beating heart. Am J Physiol Heart Circ Physiol 2010;298 (6):H2221−5.

[52] Kvitting JP, Bothe W, Goktepe S, Rausch MK, Swanson JC, Kuhl E, et al. Anterior mitral leaflet curvature during the cardiac cycle in the normal ovine heart. Circulation 2010;122(17):1683−9.

[53] Stevanella M, Krishnamurthy G, Votta E, Swanson JC, Redaelli A, Ingels NB. Mitral leaflet modelling: Importance of *in vivo* shape and material properties. J Biomech 2011;44(12):2229−35.

21 Downstream from the Heart Left Ventricle: Aortic Impedance Interpretation by Lumped and Tube-Load Models

Roberto Burattini

Previously at the Department of Information Engineering, Polytechnic University of Marche, Italy; Department of Veterinary and Comparative Anatomy, Pharmacology and Physiology, Washington State University, Pullman, Washington, USA.

21.1 Introduction

The systemic arterial system (SAS) is a branching network of elastic tubes with decreasing diameter and increasing stiffness toward the periphery. Because the action of the heart is intermittent, the SAS accepts pulsatile flow from the left ventricle (LV) of the heart and passes it on in an almost steady stream into the arterioles. The SAS thus has the primary functions of providing steady flow of blood to bodily organs and tissues (its conduit function) and damping out blood flow wave fluctuations which are consequences of intermittent ventricular ejection (its cushioning function). These functions cannot be separated anatomically, but can be separated in mathematical descriptions of pressure–flow relationships. Because the pressure and flow waves comprise mean (steady) and pulsatile components, the conduit function can be described in terms of mean pressure and mean flow, while information on the cushioning function resides in pulsatile components [1–3].

Vessel resistance to blood flow is the physical property that characterizes the conduit function. Since Poiseuille's law [4] states that hydraulic resistance is inversely related to the fourth power of vessel radius, it is easily recognized that the major part of resistance to blood flow takes place in the smallest peripheral vessels. When all individual resistances of proximal (small resistance) and distal (high resistance) vessels are properly composed, by taking into account their series and parallel connections, the resistance of the entire systemic vascular bed is obtained. This is the overall resistance the LV faces and is called total peripheral resistance, R_p. Denoting Q'_{ao} and P'_{ao}, respectively, as the mean, over the heart cycle, of flow and pressure waves in the ascending aorta, and p'_{vc}

Modelling Methodology for Physiology and Medicine. DOI: http://dx.doi.org/10.1016/B978-0-12-411557-6.00021-5

as the mean, over the heart cycle, of central venous pressure, the R_p is quantified by
the following equation [2,3]:

$$R_p = \frac{P'_{ao} - P'_{vc}}{Q'_{ao}} \tag{21.1}$$

Because P'_{vc} is negligible compared to P'_{ao}, and considering that Q'_{ao} (equal to
the product of stroke volume $\times$ heart rate) is commonly referred to as
cardiac output, CO, the following equation is used for practical purposes [3,5]:

$$R_p = \frac{P'_{ao}}{Q'_{ao}} = \frac{P'_{ao}}{CO} \tag{21.2}$$

Total peripheral resistance quantified by Eqs. (21.1) or (21.2) constitutes a simple
lumped-parameter model of LV load. Based on a simplistic approach, the Poiseuillian
model of the arterial system, which was formerly used by clinicians, yielded poor comprehensive benefit as to how the pulsatile flow wave generated by left ventricular contraction is converted into the sustained aortic pressure wave that is essential for normal
organ perfusion [5,6].

By the end of the twentieth century, high systolic pressure and wide pulse pressure
(systolic minus diastolic) have emerged as important risk factors for cardiovascular disease. These observations have shifted interest to the main determinants of increased systolic blood pressure and pulse pressure, such as arterial stiffness and amplitude and the
timing of wave reflections [3,5,7,8]. To unmask these factors, the vascular input impedance in the ascending aorta is of particular significance. Not only does this describe the
relationship between pressure and flow at the root of the aorta, but can also be taken to
represent the hydraulic load presented by the systemic circulation to the LV [1–3,9].

Mathematical models are needed to interpret impedance patterns in terms of systemic
resistive and reactive properties, and in terms of wave propagation and reflection phenomena. Among the wide variety of reported models, this chapter focuses on models of
reduced complexity that are characterized by a low number of parameters, thus being
suitable for allowing quantitative assessment of LV load properties and their roles in
determining pressure wave contour on an individual basis [10]. These models are
divided into two classes. The class of lumped-parameter models, most of which are
derived from the Windkessel [1–3,5,6,9–18], and the class of partially distributed-
parameter models, generally referred to as tube-load models [1–3,10,13,19–28]. This
subdivision is relevant in that lumped-parameter models assume infinite pulse wave
velocity and, therefore, cannot reproduce wave propagation and reflection phenomena.
Reproduction of these phenomena is made possible by tube-load models.

21.2 Lumped-Parameter Models

21.2.1 Two- and Three-Element Models

The main purpose for which lumped-parameter arterial models have been developed is for gaining insight into the overall elastic arterial properties that permit a

damping mechanism through which the cyclic blood flow coming from the heart is changed into an almost steady flow at the arteriolar (resistance) level. A capacitive element is used as marker of the overall elastic arterial properties expressed in the quantitative terms of compliance, C, defined as the ratio of volume change, ΔV, and the resulting pressure change, ΔP[2,3,5]:

$$C = \frac{\Delta V}{\Delta P} \tag{21.3}$$

Because experimental techniques for direct measurement of overall compliance are difficult to implement, model-based methods are needed [5]. Most of these rely on a conceptual basis that originates from the studies of Hales [29], who was the first to measure arterial pressure in an animal, to document the response of arterial pressure to blood loss, to formulate the concept of peripheral resistance, and to show in an elegantly simple experiment that the greatest resistance to blood flow resides in the tiny blood vessels that are invisible to the naked eye. Hales also linked the arterial system to the inverted, air-filled dome of the contemporary fire engine that smoothed the oscillations due to intermittent pumping so that flow of water through the fire hose nozzle was almost perfectly continuous [1,3]. Hales saw the arterial system as an equivalent elastic chamber and the arterioles as an equivalent resistance to the intermittent pumping of the heart (Figure 21.1), so that flow accepted from the LV was converted into a virtually steady flow through the tissues. Frank [11] and his school of thought pioneered putting Hales' idea into mathematical form. In the German translation of *Haemastaticks*, the fire engine conversion chamber was referred to as *Windkessel*, a term that has stuck as the description of the functional cushioning role of the arterial system [3]. The electrical analogue of the Windkessel consists of a capacitor, representing total arterial compliance, connected in parallel with a resistor representing total peripheral resistance (W2 model, Figure 21.1).

Shortcomings of the Windkessel were noticed but not quantified until measurements of aortic flow were made possible and arterial system input impedance data could be computed by Fourier analysis of pressure and flow signals simultaneously measured in the ascending aorta [1–3,5,9]. Impedance data were then compared with the W2 model input impedance, which has the features of a one-pole frequency response, $Z_p(j\omega)$, [10,30]:

$$Z_p(j\omega) = \frac{G}{1 + j\omega \cdot \tau} \tag{21.4}$$

where ω is angular frequency and j the imaginary operator. The time constant, τ, and the gain, G, are purely phenomenological parameters that are related to the physical model parameters, R_p and C_{W2}, by the following equations:

$$\tau = R_p \cdot C_{W2} \tag{21.5}$$

$$G = R_p \tag{21.6}$$

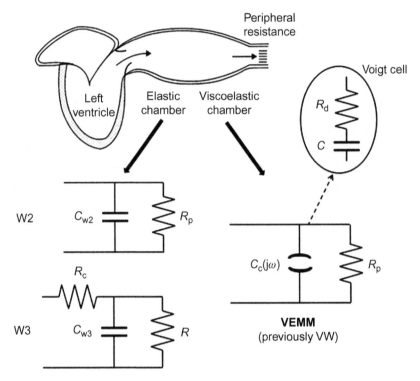

Figure 21.1 Schematic representation of the systemic arterial load in lumped terms. The arterial tree operates as a reservoir that accepts pulsatile flow from the LV of the heart and converts this into an almost steady flow through the smallest high-resistance vessels. The reservoir can be conceived to be elastic or viscoelastic. From pure elasticity hypothesis Frank's Windkessel (W2) is derived. Its electrical analogue consists of a capacitor, C_{w2}, representing total systemic arterial compliance, in parallel to a resistor, R_p, representing total peripheral resistance. Addition of a series resistor, R_c, supposed to represent aortic characteristic impedance, yields the W3 model. The capacitor C_{W3} represents total arterial compliance. Total peripheral resistance is dissected into the sum of R_c and a peripheral resistance, R. From viscoelasticity hypothesis the VEMM is derived. This consists of total peripheral resistance, R_p, in parallel with a complex and frequency-dependent compliance, $C_c(j\omega)$, represented by a Voigt cell. The resistor R_d accounts for the viscous response of the arterial wall, while the capacitor C represents overall static compliance. This model was previously denominated VW (Section 21.2.1.1).

To discuss W2 limitations in data modelling, typical results from W2 behavior analysis in time and frequency domains are displayed in Figure 21.2. Solid lines in Figure 21.2A and B represent the wave shapes of flow and pressure, respectively, simultaneously measured in the ascending aorta of an anesthetized dog, in steady state. The measured flow wave was input into the W2 model and model-predicted pressure was fitted to measured pressure by iteratively changing the values of

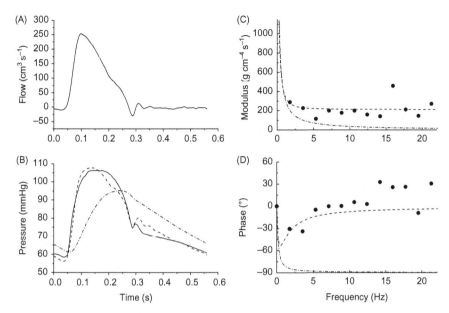

Figure 21.2 (A) Flow measured in the ascending aorta of a dog in basal state (solid line). (B) Corresponding, measured pressure wave (solid line) is superimposed with the best fitting pressure waves produced by the W2 model (dash−dot line) and by both the W3 and the VEMM (dashed line). (C and D) Data (closed circles) of aortic input impedance modulus and phase angle, respectively, as obtained from Fourier analysis of measured pressure and flow waves, are superimposed with corresponding impedance patterns provided by the W2 (dash−dot line) and by both the W3 and the VEMM (dashed line). Zero-frequency impedance modulus (not displayed) is 1967 g/(cm^4 s).

model parameters until the difference between the two pressure waves (as measured by the sum of squared errors between corresponding samples) was minimized [10,30]. The optimal estimates of model parameters were then associated to the best fitting pressure wave (dash−dot line in Figure 21.2B). Model-predicted impedance modulus (dash−dot line in Figure 21.2C) and phase angle (dash−dot line in Figure 21.2D) were computed from estimated parameters, and then superimposed on impedance data (closed circles) obtained from Fourier analysis of measured pressure and flow waves. The limited ability of the Windkessel output to approximate the experimental pressure wave (dash−dot vs. solid line in Figure 21.2B) is linked to the poor description of impedance patterns provided by the model (dash−dot line vs. closed circles in Figure 21.2C and D). At zero frequency, the modulus of the W2-predicted impedance equals R_p and, with increasing frequency, reduces to negligible values, while impedance data (in most cases) settle about a plateau value of 5−10% of R_p (Figure 21.2C). The W2-predicted impedance phase angle is zero at zero frequency and reaches −90° with increasing frequency, thus significantly deviating from experimental data, which hover around zero for high frequencies (Figure 21.2D).

A marker of the rather constant level of aortic impedance modulus at higher frequencies was originally estimated by averaging impedance data over an appropriate frequency range, though with some degree of arbitrariness [2,3,9]. This marker was found to be equal to the characteristic impedance of the proximal aorta and led to the addition of aortic characteristic impedance, R_c, in series at the input of Frank's model, thus giving rise to the three-element (W3) structure shown in Figure 21.1 [5,9,12]. The input impedance of this model structure has the generalized form of a gain constant multiplied by the ratio between one zero and one pole, $Z_{zp}(j\omega)$ [30]:

$$Z_{zp}(j\omega) = G \cdot \frac{1 + j\omega \cdot \tau_n}{1 + j\omega \cdot \tau_d} \tag{21.7}$$

The τ_n and τ_d time constants and the gain, G, are purely phenomenological parameters that are related to the physical parameters, R_c, R, and C_{W3}, by the following equations:

$$\tau_n = \frac{R \cdot R_c}{R + R_c} \cdot C_{W3} \tag{21.8}$$

$$\tau_d = R \cdot C_{W3} \tag{21.9}$$

$$G = R_c + R \tag{21.10}$$

The constraint that the impedance equals R_p at zero frequency implies the following additional equality:

$$R_p = R_c + R \tag{21.11}$$

Once the W3 is fed with measured aortic flow and its pressure output is fitted to measured pressure for parameter estimation and impedance calculation, the model-related dashed lines reproduced in Figure 21.2B−D are typically obtained. The improvement in data approximation, over the W2, is evident after comparison of dashed and dash−dot lines, both in the time and the frequency domains.

Exactly the same behavior as that of the W3, in terms of data modelling, is obtained from a model structure that, different from Frank's Windkessel, is supposed to account for viscoelasticity, rather than pure elasticity [6,15,30,31]. In this alternative model, which will henceforth be referred to as the viscoelastic minimal model (VEMM), a complex and frequency-dependent compliance, $C_c(j\omega)$, rather than a constant compliance, is placed in parallel with total peripheral resistance, R_p (Figure 21.1). The Voigt cell, the electrical analogue that is represented by a resistor, R_d, in series with a capacitor, is assumed as a minimal model capable of roughly describing the creep that is the gradual extension observed in the arterial wall material in response to the step change of tension [2,3,32−34]. Under this assumption, the VEMM assumes the configuration of an alternate three-element model characterized by the following expression for complex compliance [15,30]:

$$C_c(j\omega) = C \cdot \frac{1}{1 + j\omega \cdot C \cdot R_d} \tag{21.12}$$

Parallel connection of $C_c(j\omega)$ with R_p yields the generalized form for arterial input impedance as expressed by Eq. (21.7). Different from the W3, however, the VEMM is characterized by the following relations between phenomenological and physical parameters [15,30]:

$$\tau_n = R_d \cdot C \tag{21.13}$$

$$\tau_d = (R_p + R_d) \cdot C \tag{21.14}$$

$$G = R_p \tag{21.15}$$

The pressure wave and the impedance patterns obtained from the VEMM, after fitting to data, are indistinguishable from those produced by the W3 (dashed lines in Figure 21.2B−D). With increasing frequency, the impedance modulus of the VEMM tends to a constant value equal to the parallel of R_d and R_p. The result is an equivalent resistance equal to the R_c of the W3:

$$R_c = \frac{R_p \cdot R_d}{R_p + R_d} \tag{21.16}$$

Compared to the W3, the VEMM offers a stimulating alternate paradigm for physiological interpretation of aortic input impedance [15,30] that helps clarify why, as noticed by Westerhof et al. [5], Frank's model explains aortic pressure decay in diastole, but falls short in systole. Rebuttal of the VEMM, based on the dogmatic statement that "this model is not an improvement of Frank's two-element model," [5] is indeed questionable.

21.2.1.1 How the VEMM Improves Frank's Model

Frank's model [11] is based on the characteristics of a fire engine's air-filled dome (*Windkessel*). To be consistent, the electrical analogue (W2 in Figure 21.1) incorporates a purely capacitive element with the underlying assumption that the overall arterial system behaves as a purely elastic reservoir. After Bergel [33,34], however, it is generally recognized [2,3,32,35,36] that, although the response of the arterial tree to relatively slow changes in blood pressure is determined by its static elastic properties [33], the rapid pressure changes occurring at each heartbeat will result in rather different behavior due to the viscoelastic properties of the arterial wall [34]. Indeed, as a viscoelastic material, the arterial wall shows a mechanical response that depends both on the force applied (elastic response) and on the time it acts (viscous response) [34]. The viscous component of the arterial wall motion gains importance toward the periphery of the arterial system and should logically be involved in a model of aortic input impedance, which, by definition, should account for the properties of the overall arterial system, rather than the properties of the proximal part of it. Consistently, the VEMM constitutes a conceptual improvement over Frank's Windkessel in that arterial viscoelasticity, rather than pure elasticity, is incorporated [15,30,31], such that the former model reduces to the latter when the arterial viscous response is disregarded (R_d is canceled).

A formal objection may arise as to the notation of the *viscoelastic Windkessel* (VW) used previously [15,18,27,30] for the viscoelastic variant of Frank's Windkessel. The VEMM notation, introduced here, appears more appropriate to avoid conceptual misunderstanding. The hydraulic image associated with this model should indeed involve a viscoelastic reservoir, rather than an air chamber (*Windkessel*), connected to an output resistance.

The static compliance, C_{W2}, incorporated in the W2, helps describe the decay (approximated by an exponentially decreasing line) of aortic pressure in diastole since the static elastic properties of the arterial wall prevail. Due to the absence of an element capable of accounting for the viscous arterial response, however, the W2 falls short in systole, during which a faster deformation of the arterial wall occurs in response to pressure. The viscoelastic model resolves this shortcoming by incorporating the resistor R_d in series with compliance. A proof is found in that the results obtained from fitting the output pressures of both the W2 and the VEMM to the aortic pressure pulse measured in several dogs under basal, vasodilated, and vasoconstricted conditions, showed similar values of static compliance, C_{W2}, and C, respectively, as displayed in the scattergram of Figure 21.3A[15]. In spite of this similarity, a much better approximation of the pressure contour was produced by the VEMM [15]. This improvement is to be ascribed to the R_d-C related viscoelastic response that allows the model output to follow the rapid pressure change in systole, as well as its slower decrease in diastole (Figure 21.2B).

A further controversial point brought about by Frank's Windkessel comes from the finding that C_{pp} estimates of compliance obtained after limiting the pressure fit to pulse pressure (systolic minus diastolic), as prescribed by the pulse pressure method (PPM [37]), are lower than the corresponding C_{W2} estimates obtained from fitting the W2 output to the entire pressure pulse [15,38]. Establishing which is better between the compliance estimates provided by these two different estimation procedures is an issue difficult to address in the absence of a "gold standard" [15,30,38]. It can be noticed, however, that the PPM captures information focused on systolic arterial wall dynamics, which is strongly affected by the viscous, beyond elastic, response [34]. Thus, the lower C_{pp} estimates are likely to be markers of dynamic (rather than static, C_{W2}) compliance. A proof is found in that, in all cases considered in [15], values of the modulus of dynamic compliance, $|C_c(\omega_h)|$, calculated at individual heart pulsation, $\omega_h = 2\pi/T$, by inputting the C and R_d estimates obtained from the VEMM (after fitting to the entire pressure wave) into Eq. (21.12), are well correlated with C_{pp} estimates obtained from the pulse pressure method (Figure 21.3B).

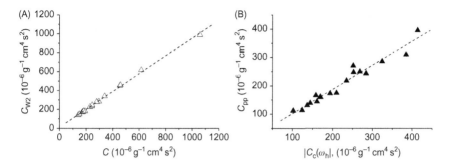

Figure 21.3 (A) C_{W2} individual compliance estimates obtained from four dogs after fitting the W2-predicted pressure to the pressure wave measured in the ascending aorta, under a variety of hemodynamic states, are plotted versus the corresponding estimates of static compliance, C, obtained from the VEMM after the same fitting procedure. Dashed regression line: $C_{W2} = 0.94.C + 11.9$; $\rho = 0.999$. (B) C_{pp} individual compliance estimates obtained from the W2 by a fitting procedure focused on pulse pressure (systolic minus diastolic) are plotted versus the corresponding moduli of dynamic compliance, $|C_c(\omega_h)|$, calculated (Eq. (21.12)) at the heart-rate pulsation, ω_h, from the VEMM parameters estimated by fitting the model output to the entire pressure wave. Dashed regression line: $C_{pp} = 0.85 \cdot |C_c(\omega_h)| + 16.6$; $\rho = 0.975$.
Source: After Burattini and Natalucci [15].

21.2.1.2 The Visco-Elastic Minimal Model Versus the Three-Element Windkessel

Improvement of Frank's model incorporated in the W3 structure by addition of the series resistance, R_c, causes the dissection of total peripheral resistance, R_p, into the two R and R_c components, thus implying that the arterial system stores blood during systole with the LV facing the entire peripheral resistance, whereas only the R part of it is involved during blood delivery to the peripheral tissues in diastole. During systole the R_c causes a pressure drop that makes the pressure applied to the capacitor C_{W3} (i.e., to the wall of the elastic equivalent reservoir) lower than the pressure generated by the heart [15]. Moreover, the R_c dissipates energy [39]. This characteristic conflicts with the fact that no energy dissipation is associated with the characteristic impedance (a real number) of a lossless transmission tube (Section 21.3). Interpretation in terms of wave propagation sounds, indeed, read too much into the nature of the lumped Windkessel theory.

The VEMM incorporates the notion that aortic pressure is applied to both the dynamic (complex) compliance and the total peripheral resistance during the entire cardiac cycle as it is logically expected in a congruent lumped representation of the arterial system. Different from the low resistance, R_c, of the W3, the low resistance, R_d, of the VEMM is detached from R_p and does not interfere in the relation between mean aortic pressure and cardiac output. Rather, it plays a role in modulating the arterial wall response to pulsatility as was shown previously for the carotid circulation [40] and the terminal aortic circulation [41].

When information from the full pressure and flow waves is used for parameter estimation, the VEMM produces C estimates that are similar to the C_{W2} estimates obtained from Frank's model (Figure 21.3A) and explains that both C and C_{W2} represent static compliance. As noticed in Section 21.2.1.1, static compliance alone is insufficient to reproduce systolic pressure from flow, unless the viscous component, R_d, of the arterial wall motion [34] is accounted for [15,30]. Compared to C and C_{W2} estimates, which are similar (Figure 21.3A), overestimation of static compliance is provided by the C_{W3} element of the W3 model [15]. It is likely that the C_{W3} values estimated by fitting to the entire pressure wave are affected by the fact that, during systole, the capacitor of the W3 model charges (elastic energy storage) in the presence of energy dissipation in the resistor R_c, while the capacitor discharges in diastole through the resistor R that has a lower value than total peripheral resistance, R_p. Nevertheless, with its optimized parameters, the W3 produces the same pressure fit as the VEMM (Figure 21.2B). The point is that the R_c appears to behave in the W3 as a viscous element (it dissipates energy) necessary for improving the reproduction of the rapid pressure response in systole, over that obtained from the W2. Unfortunately, the topological location of such a necessary R_c viscous element in series at the W2 entrance causes its action to disappear in diastole, where, however, the viscous response is less effective.

21.2.2   Four-Element Models

It has been reported [16,38] that the W3 model can produce realistic impedance spectra and aortic pressures and flows, but only with values for C_{W3} and R_c that overestimate and underestimate, respectively, the vascular properties they are supposed to represent. To improve the W3 model behavior and avoid dissection of total peripheral resistance R_p into the sums of R and R_c, a four-element model, previously proposed by Burattini and Gnudi [13] was, then, reconsidered [16]. The fourth element is an inductance, L, connected in parallel with the resistor R_c (W4P model, Figure 21.4). Contradictory conclusions have been reported as to the meaning of the L term. Burattini and Gnudi [13] could not find a precise physical interpretation. Thus, the L was considered a further degree of freedom that allows improvement of data fit. Campbell et al. [42] used the W4P model and made the observation that R_c and L combine to determine the characteristic impedance and do not, in themselves, represent physiologic entities. Stergiopulos et al. [16] attributed to L the physiological meaning of total arterial inertance. With the aid of generalized sensitivity function analysis [43], a recent study by Burattini and Bini [44] yielded the conclusion that the L term of the W4P is unsuitable for representing the inertial properties of blood motion.

Unreliability of arterial inertance representation can explain why the W4P fell short when applied by Sharp et al. [17] to interpret pressure−flow relationships in infants and children, who show aortic impedance data with much stronger inertial character than adults. Better impedance approximation was obtained by these authors from an alternative four-element model derived from the W3 by

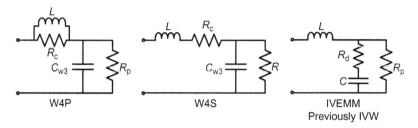

Figure 21.4 Electrical analogues of four-element models of the systemic arterial circulation. The W4P and W4S structures are derived from W3 by connecting an inductance, L, in parallel or in series to R_c, respectively. The IVEMM is derived from the VEMM by adding an inductance, L, in series at the input. This model was previously denominated IVW (Section 21.2.2). In all three model structures the L term is supposed to represent inertial properties of blood motion. See Figure 21.1 for meaning of the other elements.

connecting L in series with R_c (W4S model, Figure 21.4). From W4S parameter estimates, Sharp et al. [17] inferred that significant changes are in progress during vascular development from infancy to adulthood. In particular, a significant difference was observed in the development rate of R_c and R_p, such that a rapid reduction characterized the R_c/R_p ratio during early development. Because R_c is part of R_p, this rapid reduction could not be explained. In an attempt to resolve this shortcoming by means of a model accounting for the viscous, beyond elastic, arterial response disregarded by Sharp et al. [17], Burattini and Di Salvia [18] applied one other four-element structure derived from the VEMM by connecting L in series at the input (inertance viscoelastic minimal model, IVEMM, Figure 21.4). Consistently with the reasons set out in Section 21.2.1.1, the IVEMM denomination is introduced here to replace the IVW (inertance viscoelastic Windkessel) denomination used in previous works [18,41,44]. Because the input impedance of the IVEMM has the same generalized expression as that of the W4S (i.e., the ratio of two zeros to one pole), the two models produce identical data fit [18]. IVEMM parameter estimates showed a difference in the development rate of R_d and R_p similar to that found for R_c and R_p in the W4S. Because R_d is detached from R_p and is supposed to account for the viscous response of the arterial wall, the observed rapid reduction in the R_d/R_p ratio during early development was ascribed [18] to the process of continuous adaptation of viscous, as well as elastic, vessel properties to changing hemodynamic load [45–47].

Generalized sensitivity function analysis [43] applied by Burattini and Bini [44] for assessment of the physiological meaning of low resistance and inductance terms of the four-element models of Figure 21.4, confirmed that R_d is likely to account for viscous losses of arterial wall motion, while the meaning of aortic characteristic impedance ascribed to R_c is questionable. Different from the W4P, in both the W4S and IVEMM structures, the L term connected in series is likely to portray information on inertial properties of blood motion.

21.3 Tube-Load Models

Lumped-parameter models cannot account for wave propagation and reflection phenomena, which take place inside the arterial tree and play a key role in shaping pressure and flow pulse contour from the ejecting ventricle. Identification of source, magnitude, and timing of reflected waves is desirable to characterize their link with mechanical and topological properties of the arterial system and thus better appreciate the means by which the arterial load can exert control and modulate LV function. To account for wave reflection, distributed-parameter models are needed. So-called tube-load models have been set-up in the attempt to meet a suitable compromise between reduced model complexity for mathematical tractability and ability to capture essential features of aortic pressure–flow relationships [1–3,10,13,19–28,30]. One of these models, constituted by a uniform, lossless elastic tube terminating in complex and frequency-dependent load, has been shown to be able to describe some experimental aortic pressure–flow relationships well [10,13,19,20]. However, a shortcoming was found in that it cannot describe a condition in which there is a prominent diastolic oscillation in the aortic pressure waveform [10,21]. More elaborate distributed models are indeed required to capture this event.

The asymmetric T-tube model, championed by O'Rourke and co-workers [1,3], has exerted considerable appeal. In this model, two loss-free transmission tubes (transmission lines in the electrical analogy) connected in parallel represent the two parallel vascular beds originating from the ascending aorta. These are: the vascular bed supplying the upper part of the body, with the summation of arterial terminations relatively close, and the vascular bed supplying the lower part of the body, with the summed arterial terminations much farther away. This representation has been used primarily as a conceptual model to explain aortic pressure contour features as they relate to reflections arising from two functionally discrete reflection sites. Nevertheless, direct application of the model to arterial pressure and flow data for characterization of the arterial system via quantification of specific arterial properties has been limited by the observation that, under the assumption of a purely resistive load at the termination of head-end and body-end transmission tubes, reliable reproduction of aortic impedance patterns is not possible [2,3]. Burattini and Campbell [21] found a solution to this shortcoming by conceiving a modified T-tube model configuration (Figure 21.5) that consists of two parallel, loss-free transmission tubes terminating in first-order low-pass filter loads, which lump the random branching of the arterial system downward and serve to uncouple the LV from peripheral resistances [21–27].

Each one of the two transmission tubes is characterized by the following parameters (the i subscript equals h or b depending on whether reference is made to head-end or body-end circulation, respectively): tube inertance, ℓ_i, and tube compliance, c_i, per unit length; tube length, d_i; tube characteristic impedance, Z_{ci}; and one-way wave transit time, τ_i, across each tube. Each lumped terminal load is characterized by an input impedance, $Z_{Li}(j\omega)$, which is given the form of Eq. (21.7)[24]:

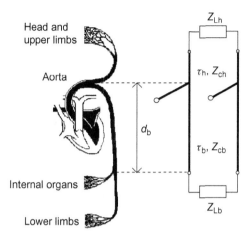

Figure 21.5 Modified asymmetric T-tube model of the systemic arterial circulation. The short and long lossless, uniform transmission tubes connected in parallel represent the portions of the systemic circulation directed towards the upper (head and upper limbs) and lower (internal organs and lower limbs) parts of the body, respectively. Both transmission paths terminate in complex and frequency dependent, lumped-parameter loads with input impedances (Z_{Lh} and Z_{Lb}, respectively) described by Eq. (21.17). Z_{ch} and Z_{cb}, head-end and body-end tube characteristic impedance, respectively; τ_h and τ_b, head-end and body-end, one-way wave transit time, respectively; d_b, distance to the body-end reflection site.

$$Z_{Li}(j\omega) = R_{pi} \cdot \frac{1 + j\omega.\tau_{ni}}{1 + j\omega.\tau_{di}}; \quad i = h, b \tag{21.17}$$

where R_{pi} represents peripheral resistance, while τ_{ni} and τ_{di} are time constants. The constraint:

$$\tau_{di} = R_{pi} \cdot \frac{\tau_{ni}}{Z_{ci}}; \quad i = h, b \tag{21.18}$$

is added to formalize the assumption that with increasing frequency $Z_{Li}(j\omega)$ approximates the tube characteristic impedance, $Z_{ci} = \sqrt{\ell_i/c_i}$. This implies that negligible wave reflection occurs at higher frequencies.

After designating $\Gamma_{Li}(j\omega)$ as the reflection coefficient at the junction between tube and terminal load:

$$\Gamma_{Li}(j\omega) = \frac{Z_{Li}(j\omega) - Z_{ci}}{Z_{Li}(j\omega) + Z_{ci}}; \quad i = h, b \tag{21.19}$$

the ascending aortic input impedance is determined by the parallel of the head-end ($i = h$) and the body-end ($i = b$) input impedances, $Z_i(j\omega)$, which assume the following expression:

$$Z_i(j\omega) = Z_{ci} \cdot \frac{1 + \Gamma_{Li}(j\omega) \cdot \exp(-j\omega \cdot 2 \cdot \tau_i)}{1 - \Gamma_{Li}(j\omega) \cdot \exp(-j\omega \cdot 2 \cdot \tau_i)}; \quad i = h, b \qquad (21.20)$$

According to Eqs. (21.17)–(21.20), the model parameters to be estimated from an input–output experiment are: Z_{ci}, τ_i, τ_{ni}, and R_{pi}, $i = h$, b. The τ_{di} time constants are computed from τ_{ni}, Z_{ci}, and R_{pi} by Eq. (21.18). Tube compliance, $c_i \cdot d_i$, and tube inertance, $\ell_i \cdot d_i$, are given by the following equations:

$$c_i \cdot d_i = \tau_i / Z_{ci}; \quad i = h, b \qquad (21.21)$$

$$\ell_i \cdot d_i = \tau_i \cdot Z_{ci}; \quad i = h, b \qquad (21.22)$$

The optimal set of measurements required for model parameter estimation was assessed after analysis of data fit and accuracy of model parameter estimates [21,22]. It consists of simultaneous measurements of pressure and flow in the ascending aorta and flow in the upper descending aorta. From these measurements, the peripheral resistance values of R_{ph} and R_{pb} are calculated so that the six parameters, Z_{ch}, Z_{cb}, τ_h, τ_b, τ_{nh}, and τ_{nb}, remain to be estimated from fitting to measured flow waves the flow waves predicted by the model using measured pressure as input [22–25,27]. After parameter estimation, tube length, d_i, can be derived from wave transit time, τ_i, across the tube, if a measurement of pulse wave velocity, ν_i, is available. Both d_h and d_b are supposed to represent the distance to a functionally discrete, effective reflection site as seen from the ascending aorta toward the head-end and body-end arms of systemic arterial circulation, respectively. At the level of each effective reflection site, arterial reflections from peripheral sites are supposed to compose one wave traveling backward, toward the LV [1–3,10,21–28]. The denomination of *effective length* of the related arterial circulation is then used for both d_h and d_b[19,20].

To quantify the overall resistive and capacitive properties of the head-end, Z_{Lh}, and body-end, Z_{Lb}, terminal loads from the estimates of R_{pi}, τ_{ni}, and τ_{di}, $i = h$, b, the model structure of either the three-element Windkessel (W3) or the VEMM can be assumed as discussed in Section 21.2.1. The VEMM has been the preferred structural representation of these terminal loads [21–27].

Because the head-end and body-end transmission paths terminate in complex and frequency-dependent loads, the related reflection coefficients are complex and frequency dependent (Eq. (21.19)). As a consequence, the effective lengths, d_h and d_b, can be estimated according to the following equation [19,20]:

$$d_i = \frac{\nu_i}{4 \cdot f_{oi}} \left(1 + \frac{\vartheta_i(f_{oi})}{\pi} \right); \quad i = h, b \qquad (21.23)$$

where ν_i is the pulse wave velocity along the head-end (i equals h) or body-end (i equals b) transmission tube, f_{oi} is the frequency at the first zero-crossing of the (head-end or body-end) impedance phase angle, and $\vartheta_i(f_{oi})$ is the phase of the load reflection coefficient at f_{oi}. This phase affects the timing of reflected waves. When an elastic and frictionless tube is loaded with a resistor, the $\vartheta_i(f_{oi})$ equals zero and Eq. (21.23) reduces to the one-quarter wavelength formula [2,3,10,19,20]. Because the ratio of Eq. (21.23) to the one-quarter wavelength formula is equal to $1 + \vartheta_i(f_{oi})/\pi$ and the phase of load reflection coefficient is negative, reported effective length values computed by the one-quarter wavelength formula [3] are overestimated with respect to the values derived from tube-load models terminating in complex loads [2,20,24,30].

Problems of identifiability arise from the fact that the interpretation of arterial impedance by uniform tube models leaves room for an infinite number of exact solutions for tube length and terminal-load impedance [48]. Once an identifiable structure is specified for the terminal-load, as in the modified T-tube model, an infinite number of equivalent solutions exists for the wave transit time, τ_i, owing to the presence of the $\exp(-j\omega \cdot 2.\tau_i)$ periodic function in the input impedance expression of Eq. (21.20). This holds also for the effective lengths, d_i, which is obtained from the product between wave transit time and pulse wave velocity $(d_i = \tau_i \cdot \nu_i,\ i = h,\ b)$. Infinite solutions enhance a *problem of determinacy*[49], which can be resolved by demonstrating, with aid of model-independent information, that only one of these solutions is acceptable. On this basis, by focusing on the body-end arm of the modified T-tube model, it has been demonstrated [26,30] that only the selection of the lowest value, among the infinite possible estimates of the wave transit time (τ_b), is compatible with the physical and topological properties of the descending aortic circulation. This τ_b value is indeed the only one that yields a distance, d_b, to the lower body effective reflection site compatible with the requirement for this distance to not be greater than body dimension. The same conclusion is inferred for head-end circulation. Once a constraint is assumed to address a fitting procedure to the unique acceptable τ_h and τ_b estimates, all other T-tube model parameters are uniquely identifiable. Especially, from τ_b estimates and pulse wave velocity measurements via the foot-to-foot detection technique, d_b values were found that, compared to descending aorta anatomy, located the body-end effective reflection site, as seen from the ascending aorta, in the abdominal region, where major branching occurs (Figure 21.5) [21,24,26,27,30].

Although reduction of model complexity has been accomplished by neglecting arterial elastic and geometric tapering, and multi-level branching [2,3,28,50], the modified T-tube model with parameter estimation techniques based on central pressure and flow measurements has been shown to be able to explain wave reflection effects on pressure and flow wave features and to track their changes in the presence of mechanical and pharmacological interventions [21−23,25]. Emphasis was also placed on the model's ability to discriminate between proximal and distal mechanical properties of descending aortic circulation [24−26]. It was also shown how the two arterial effective reflection sites can sometimes appear as one to the heart, such that a single tube terminating in complex load can suffice in approximating the central aortic pressure waveform [23].

From the modified, asymmetric T-tube model, multiple-path models have been derived, which use alternative parameter estimation methods based on pressure measurements to monitor wave reflection, pulse transit time, and central aortic pressure [28,51−53]. With these methods, mostly characterized by phenomenological parameters, the difficult task of involving flow measurements is avoided, but information on impedance as it relates to mechanical arterial properties, compliance above all, is missed. Advantages are that the order of pole-zero impedance formulation for tube terminal load is determined from measured waveforms instead of being set *a priori* to one pole and one zero, and the estimation is achieved through a convenient and reliable closed-form solution, rather than a difficult and imperfect numerical search [53]. In conceptual terms, parameter estimation from peripheral pressure measurements moves the observation point from the ascending aorta to peripheral measurement sites, such that a transmission tube represents the wave propagation path from the central aorta to the peripheral artery where the observation point (pressure measurement) is located, whereas the load, representing the composition of arterial reflection from distal sites, is placed downward. Tube-load model parameter estimation techniques for monitoring arterial hemodynamics have been elegantly reviewed by Zhang et al. [28].

Although T-tube model derived techniques have been applied to various animal species, their applicability to human cardiovascular physiology is not yet established. Nevertheless, evidence has been reported that the modified T-tube model is sufficiently versatile and general in its representation of mammalian arterial systems that it is able to discriminate between substantial differences in aortic impedance and wave reflection features as they relate to body size and location of the heart along the body's axis of mammals of decidedly different body sizes and shapes [27]. These aspects are treated in the following section.

21.3.1 Effects of Body Size and Shape on Aortic Input Impedance

The modified T-tube model of Figure 21.5 was applied to a group ($n = 5$) of dogs (26.5 ± 1.0 kg body weight) and a group ($n = 5$) of ferrets (1.6 ± 0.1 kg). The model was fed with pressure measured in the ascending aorta. Parameter estimation was then accomplished by fitting model-predicted flows to flow waves measured in the ascending aorta and upper descending aorta. Almost perfect data fits were obtained. From estimated parameters, model impedance patterns were then computed in individual cases and compared with impedance data obtained from Fourier analysis of pressure and flow measured in the ascending aorta [27]. The average features are displayed in Figure 21.6. In the dog, the experimental aortic impedance modulus (closed circles in Figure 21.6A) shows two distinct minima in the frequency range from 0 to 10 Hz. These are captured by the model-predicted aortic input impedance (solid line). This results from the parallel of head-end and body-end input impedances and, based on model predictions, the first minimum in modulus corresponds to the first minimum of body-end impedance modulus (dash−dot line), whereas the second minimum corresponds to the first minimum of head-end impedance modulus (dashed line). Each minimum correlates with the corresponding first zero-crossing of the body-end and head-end impedance phase angle (dash−dot and dashed lines, respectively, in Figure 21.6B). Different from the dog, in the experimental aortic impedance modulus of the ferret only one broad,

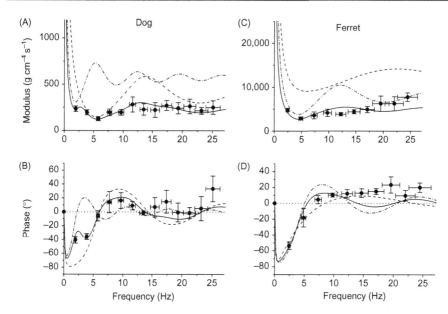

Figure 21.6 Average (± SE) impedance data (solid circles with ± SE bars) computed over five dogs (A and B) and five ferrets (C and D), after Fourier analysis of pressure and flow waves measured in the ascending aorta, are superimposed with predictions of input impedance of the ascending aorta (solid line), and the head-end (dashed line) and body-end (dash−dot line) portions of systemic arterial circulation as provided by the modified T-tube model filled with average parameter estimates (table 2 in Ref. [27]) over the five dogs and the five ferrets, respectively. Dotted horizontal line in the phase graph (B and D) is zero line. *Source*: After Burattini and Campbell [27].

indistinct minimum is identified in the low-frequency range (closed circles in Figure 21.6C). The model is able to capture this different feature (solid line) and explains that it is due to different configurations of body-end (dash−dot line) and head-end (dashed line) impedance moduli, with frequency values at which their first minimum occurs being closer than in the dog. Nevertheless, the frequency values at the first zero-crossing of body-end (dash−dot line) and head-end (dashed line) impedance phase angle are clearly distinct in the ferret as in the dog (compare Figure 21.6B and D). According to Eq. (21.23), transmission path lengths are inversely related to the frequency at which the first zero-crossing of related input impedance phase angle occurs.

From model-provided estimates of one-way wave transit time across the head-end and body-end transmission path (τ_h and τ_b, respectively), and information on arterial pulse wave velocity, estimation of head-end and body-end effective lengths was accomplished. These were about 12 and 30 cm, respectively, for the dog and 6.5 and 13 cm, respectively, for the ferret [27]. Because pulse wave velocity is similar in different mammalian species [3], the τ_h/τ_b ratio yields the information on the ratio between the distances to the two effective reflection sites as seen from the

LV in the two species. Thus, from the average τ_h/τ_b values of 0.39 ± 0.04 and 0.51 ± 0.05 estimated for the dog and the ferret, respectively, it was inferred that in the latter species the heart is located more centrally than in the former species. Measurements of the ratio between aortic valve to mandible distance (AV−M) and aortic valve to base of tail distance (AV−T) were found to be 0.54 ± 0.02 in the dog and 0.71 ± 0.01 in the ferret. Thus, model-predicted head-end and body-end transmission paths supplied by the ascending aorta were found to be consistent with externally measured long-axis body morphometry parameters with respect to the heart location. Differences in body size with concomitant differences in arterial system size appeared to be reflected clearly by the differences in all model-based estimates of arterial parameters (table 2 in Ref. [27]). Especially, the ratio between total compliance of the head-end circulation and total compliance of the body-end circulation provided by the model was also consistent with externally measured long-axis body morphometry ratio, (AV−M)/(AV−T), both in the dog and in the ferret [27].

Differences in body morphometry and arterial properties between the dog and the ferret yield differences in the features of ascending aortic impedance patterns that are clearly visible in the modulus (Figure 21.6A and C). Analysis of these differences provides an interesting spin-off for interpretation of aortic impedance data reported for humans. The input impedance modulus of the systemic circulation shows in adult humans [2,3,54] a broad minimum in the frequency range from 0 to 10 Hz, rather than two distinct minima as observed in large dogs (Figure 21.6A), and thus resembles the ascending aortic impedance modulus seen in the ferret (Figure 21.6B). The similarity of ascending aortic impedance patterns in humans and ferrets is likely due to the fact that these two species have a heart that is more centrally located, with respect to head-end and body-end arterial effective reflection sites, than is the case with the dog. Indeed, as reported in Ref. [3], rough estimates of head-end and body-end effective lengths by the one-quarter wavelength formula yielded a d_h/d_b ratio of 0.71 for the human and 0.53 for the dog. These numbers indicate that the heart in humans is more centrally located than in dogs, just as was found in Ref. [27] that in ferrets the heart is more centrally located than in dogs. Consequently, it is likely that in humans the low-frequency minimum in the ascending aortic impedance modulus is a result of closeness between the first minima in the input impedances of the head-end and body-end sections of modified T-tube model as shown for the ferret in Figure 21.6C.

21.3.2   Effects of Wave Reflections on Aortic Pressure Wave Shape

In the time domain, the modified T-tube model allows quantification of timing and shape of head-end and body-end reflected waves and assessment of their roles in shaping the pressure waves as seen in the ascending aorta. The head-end reflected wave travels across a relatively short pathway so that it gets quickly to the ascending aorta and its crest affects systolic pressure. Under normal conditions, the relatively longer body-end transmission path causes the crest of the body-end reflected wave to occur in diastole, thus producing a boost (augmentation) of diastolic

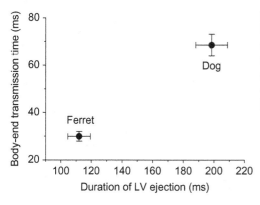

Figure 21.7 Plot of average ($\pm$SE) body-end wave transmission time (τ_b), over five ferrets and five dogs versus average ($\pm$SE) duration (T_s) of LV ejection.
Source: After Burattini and Campbell [27].

pressure that aids coronary perfusion, while the wave trough (generally negative) contributes to attenuate systolic pressure, thus aiding the pumping function of the heart [21–23,26–28]. This favorable correspondence between cardiac performance and timing of body-end reflected wave, extensively discussed by O'Rourke and co-workers [1–3], is irrespective of body size because, as displayed in Figure 21.7 for the ferret in comparison to the dog, the duration, T_s, of LV ejection changes in an appropriate fashion with body size and related body-end transit time, τ_b.

21.4 Conclusion

The input impedance of the ascending aorta is a powerful tool for characterizing the properties of the whole systemic circulation by aid of interpretative models. Model structures characterized by a number of parameters sufficiently low to be estimated from aortic pressure and flow measurements, on an individual basis, have been discussed in this chapter. Distinction has been made between lumped- and distributed-parameter models by focusing on model purpose, underlying hypotheses, and their implications in impedance interpretation. The Windkessel is the simplest model with lumped capacitance and resistance elements. Historical link of this capacitance with the air-filled dome (Windkessel) of a fire engine is suggestive, but has exerted an almost hypnotic influence on the development of improved three- and four-element model structures. These structures, to be consistent with the underlying hydraulic image, account for static elastic arterial properties, although the arterial wall is viscoelastic. A minimal lumped-model configuration (here redenominated VEMM) that incorporates viscoelasticity yields an improvement over Frank's Windkessel that deserves consideration. Comparative analysis of elastic and viscoelastic lumped models in terms of data modelling and physiological meaning of parameters improves interpretation of aortic input impedance, thus resolving contradictions in Windkessel-based evaluations of overall arterial compliance.

Models with distributed parameters are needed to account for wave propagation and reflection phenomena, which take place inside the arterial tree and play a key role in shaping arterial pressure and flow waves. The modified, asymmetric T-tube model, characterized by two uniform transmission tubes connected in parallel and terminating in lumped-parameter loads, constitutes a suitable compromise between reduced model complexity for mathematical tractability and the ability to capture essential features of the aortic input impedance. Although simplification is accomplished by neglecting arterial elastic and geometric tapering and multi-level branching, this model constitutes a suitable tool for estimating physiologically important properties of the arterial system, for predicting pressure and flow events at the aortic entrance and at various locations along the aorta, and for interpreting the composition of aortic pressure and flow waves in terms of forward and reflected wave components. The modified T-tube model has been shown to be sufficiently versatile and general in its representation of mammalian arterial system that it is able to discriminate between substantial differences in arterial properties and wave reflection effects related to body size and location of the heart along the body's axis of mammals of different sizes and shapes. This model is indeed currently used as a suitable platform for development of parameter estimation techniques that allow improved monitoring of arterial hemodynamics.

References

[1] O'Rourke MF. Principles of arterial hemodynamics. In: Ter Keurs HEDJ, Tyberg JV, editors. Mechanics of the circulation. Dordrecht, NL: Martinus Nijhoff Publishers; 1987. p. 233–59.

[2] Milnor WR. Hemodynamics. 2nd ed. Baltimore, MD: Williams & Wilkins; 1989.

[3] Nichols WW, O'Rourke MF, Vlachopoulos C. McDonald's blood flow in arteries: theoretical, experimental and clinical principles. 6th ed. London: Hodder Arnold; 2011.

[4] Poiseuille JLM. Recherhes expérimentales sur le mouvement des liquides dans les tubes de très-petits diamètres. In: Mémories presentés par divers savants à l'Académie Royale des Sciences de l'Institut de France, Tome IX Paris; 1846. p. 433–544.

[5] Westerhof N, Lankhaar JW, Westerhof BE. The arterial Windkessel. Med Biol Eng Comput 2009;47:131–41.

[6] Marcus RH, Korcarz C, McCray G, Neumann A, Murphy M, Borow K, et al. Noninvasive method for determination of arterial compliance using Doppler echocardiography and subclavian pulse tracings. Circulation 1994;89:2688–99.

[7] Franklin SS, Lopez VA, Wong ND, Mitchel GF, Larson MG, Vasan RS, et al. Single versus combined blood pressure components and risk of cardiovascular disease: the Framingham Heart Study. Circulation 2009;119:243–50.

[8] Mitchell GF. Arterial stiffness and wave reflection: biomarkers of cardiovascular risk. Artery Res 2009;3(2):56–64.

[9] Westerhof N, Sipkema P, Elzinga G, Murgo JP, Giolma JP. Arterial impedance. In: Hwang NHC, Gross DR, Patel DJ, editors. Quantitative cardiovascular studies: clinical and research applications of engineering principles. Baltimore, MD: University Park Press; 1979. p. 111–50.

[10] Burattini R. Reduced models of the systemic arterial circulation. In: Westerhof N, Gross DR, editors. Vascular dynamics: physiological perspectives. New York, NY: Plenum; 1989. p. 69–85.

[11] Frank O. Die grundform des arteriellen pulses. Erste abhandlung. Mathematische analyse. Z Biol 1899;37:483–526.

[12] Westerhof N, Elzinga G, Sipkema P. An artificial arterial system for pumping hearts. J Appl Physiol 1971;31:776–81.

[13] Burattini R, Gnudi G. Computer identification of models for the arterial tree input impedance: comparison between two new simple models and first experimental results. Med Biol Eng Comput 1982;20:134–44.

[14] Yoshigi M, Keller BB. Characterization of embryonic aortic impedance with lumped parameter models. Am J Physiol Heart Circ Physiol 1997;273:H19–27.

[15] Burattini R, Natalucci S. Complex and frequency-dependent compliance of viscoelastic Windkessel resolves contradictions in elastic Windkessels. Med Eng Phys 1998;20:502–14.

[16] Stergiopulos N, Westerhof BE, Westerhof N. Total arterial inertance as the fourth element of the Windkessel model. Am J Physiol 1999;276:H81–8.

[17] Sharp MK, Pantalos GM, Minich L, Tani LY, McGough EC, Hawkins JA. Aortic input impedance in infants and children. J Appl Physiol 2000;88:2227–39.

[18] Burattini R, Di Salvia PO. Development of systemic arterial mechanical properties from infancy to adulthood interpreted by four-element Windkessel models. J Appl Physiol 2007;103:66–79.

[19] Sipkema P, Westerhof N. Effective length of the arterial system. Ann Biomed Eng 1975;3:296–307.

[20] Burattini R, Di Carlo S. Effective length of the arterial circulation determined in the dog by aid of a model of the systemic input impedance. IEEE Trans Biomed Eng 1988;35:53–61.

[21] Burattini R, Campbell KB. Modified asymmetric T-tube model to infer arterial wave reflection at the aortic root. IEEE Trans Biomed Eng 1989;36:805–14.

[22] Campbell KB, Burattini R, Bell DL, Kirkpatrick RD, Knowlen GG. Time-domain formulation of asymmetric T-tube model of arterial system. Am J Physiol 1990;258: H1761–74 [Heart Circ Physiol 27].

[23] Burattini R, Knowlen GG, Campbell KB. Two arterial effective reflecting sites may appear as one to the heart. Circ Res 1991;68:85–99.

[24] Burattini R, Campbell KB. Effective distributed compliance of the canine descending aorta estimated by modified T-tube model. Am J Physiol 1993;264:H1977–87 [Heart Circ Physiol 33].

[25] Shroff SG, Berger DS, Korcarz C, Lang RM, Marcus RH, Miller DE. Physiological relevance of T-tube model parameters with emphasis on arterial compliances. Am J Physiol Heart Circ Physiol 1995;269:H365–74.

[26] Burattini R, Campbell KB. Physiological relevance of uniform elastic tube-models to infer descending aortic wave reflection: a problem of identifiability. Ann Biomed Eng 2000;28:512–23.

[27] Burattini R, Campbell KB. Comparative analysis of aortic impedance and wave reflection in ferrets and dogs. Am J Physiol Heart Circ Physiol 2002;282:H244–55 [Corrigenda: Am J Physiol Heart Circ Physiol 2002;283(3)].

[28] Zhang G, Hahn J-O, Mukkamala R. Tube-load model parameter estimation for monitoring arterial hemodynamics. Front Physiol 2011;2:1–18.

[29] Hales S. Statical essays: containing haemastatics, vol. 2. London: Innys & Manby; 1733 [Reprinted: History of Medicine Series. Library of the New York Academy of Medicine. New York, NY: Hafner Publishing; 1964, 22].

[30] Burattini R. Identification and physiological interpretation of aortic impedance in modelling. In: Carson E, Cobelli C, editors. Modelling methodology for physiology and medicine. San Diego, CA: Academic Press; 2001. p. 213−52.

[31] Canty Jr JM, Klocke FJ, Mates RE. Pressure and tone dependence of coronary diastolic input impedance and capacitance. Am J Physiol 1985;248:H700−11.

[32] Westerhof N, Noordergraaf A. Arterial visco-elasticity: a generalized model. J Biomech 1970;3:357−79.

[33] Bergel DH. The static elastic properties of the arterial wall. J Physiol 1961;156:445−57.

[34] Bergel DH. The dynamic elastic properties of the arterial wall. J Physiol 1961;156:458−69.

[35] Apter JT, Rabinowitz M, Cummings DH. Correlation of visco-elastic properties of large arteries with microscopic structure I, II, and III. Circ Res 1966;19:104−21.

[36] Learoyd B, Taylor MG. Alterations with age in the visco-elastic properties of human arterial walls. Circ Res 1966;18:278−92.

[37] Stergiopulos N, Meister J-J, Westerhof N. Simple and accurate way for estimating total and segmental arterial compliance: the pulse pressure method. Ann Biomed Eng 1994;22:392−7.

[38] Stergiopulos N, Meister JJ, Westerhof N. Evaluation of methods for estimation of total arterial compliance. Am J Physiol Heart Circ Physiol 1995;268:H1540−8.

[39] Burattini R, Campbell KB. Assessment of aortic pressure power components and their link to overall elastic and resistive arterial properties. Med Biol Eng Comput 1999;37:366−76.

[40] Burattini R, Montanari L, Mulligan LJ, Cannon MS, Gross DR. Evaluation of hypercholesterol diet-induced changes in visco-elastic properties of carotid circulation in pigs. Am J Physiol 1992;263:H1919−26 [Heart Circ Physiol 32].

[41] Burattini R, Natalucci S, Campbell KB. Visco-elasticity modulates resonance in the terminal aortic circulation. Med Eng Phys 1999;21:175−85.

[42] Campbell KB, Ringo JA, Neti C, Alexander JE. Informational analysis of left-ventricle/systemic-arterial interaction. Ann Biomed Eng 1984;12:209−31.

[43] Thomaseth K, Cobelli C. Generalized sensitivity functions in physiological system identification. Ann Biomed Eng 1999;27:607−16.

[44] Burattini R, Bini S. Physiological interpretation of inductance and low-resistance terms in four-element Windkessel models: assessment by generalized sensitivity function analysis. Med Eng Phys 2011;33:739−54.

[45] Bendeck MP, Langille BL. Rapid accumulation of elastin and collagen in the aortas of sheep in the immediate perinatal period. Circ Res 1991;69:1165−9.

[46] Bendeck MP, Keeley FW, Langille BL. Perinatal accumulation of arterial wall constituents: relation to hemodynamic changes at birth. Am J Physiol Heart Circ Physiol 1994;267:H2268−79.

[47] Safar ME, Boudier HS. Vascular development, pulse pressure, and the mechanisms of hypertension. Hypertension 2005;46:205−9.

[48] Campbell KB, Lee LC, Frasch HF, Noordergraaf A. Pulse reflection sites and effective length of the arterial system. Am J Physiol 1989;256:H1684−9 [Heart Circ Physiol 25].

[49] Brown RF, Godfrey KR. Problem of determinacy in compartmental modeling with application to bilirubin kinetics. Math Biosci 1978;40:205−24.

[50] Westerhof BE, Westerhof N. Magnitude and return time of the reflected wave: the effects of large artery stiffness and aortic geometry. J Hypertens 2012;30:932−9.

[51] Swamy G, Xu D, Olivier NB, Mukkamala R. An adaptive transfer function for deriving the aortic pressure waveform from a peripheral artery pressure waveform. Am J Physiol Heart Circ Physiol 2009;297:1956−63.

[52] Hahn J-O, Reisner AT, Asada HH. Estimation of pulse transit time using two diametric blood pressure waveform measurements. Med Eng Phys 2010;32:753−9.

[53] Swamy G, Olivier NB, Mukkamala R. Calculation of forward and backward arterial waves by analysis of two pressure waveforms. IEEE Trans Biomed Eng 2010;57:2833−9.

[54] Nichols WW, Conti CR, Walker WE, Milnor WR. Input impedance of the systemic circulation in man. Circ Res 1977;40:451−8.

22 Finite Element Modelling in Musculoskeletal Biomechanics

Zimi Sawacha[a] and Bernhard Schrefler[b]

[a]Department of Information Engineering, University of Padova, Padova, Italy, [b]Department of Civil, Environmental and Architectural Engineering, University of Padova, Padova, Italy

22.1 Introduction

Many researchers have pointed out that biomechanical factors play an important role in the etiology, treatment, and prevention of musculoskeletal disorders. Hence it is essential to understand the biomechanics associated with both normal and pathologic functioning of the musculoskeletal apparatus. The human musculoskeletal system is a complex mechanism. The principal function of this apparatus is the transmission of mechanical loads, both in terms of tissues and organs. The three-dimensional (3D) kinematics of the joints in dynamic conditions is evaluated through noninvasive methodologies using stereophotogrammetry, force plates, surface electromyography systems, along with advanced musculoskeletal modelling software and image segmentation methods. The measured variables (kinematics, ground reaction forces, muscle activation patterns, and forces) allow the simulation of the whole musculoskeletal system and the determination of the forces acting locally on each joint. On the other hand, information on the internal stress and strain of each joint is essential, however, direct measurement of those parameters is difficult, while this important information can be obtained through a comprehensive computational model. The finite element method can be an adjunct to an experimental approach to predict the load distribution between different structures, offering additional information, such as the internal stress and strain of a musculoskeletal joint. The local mechanical response of musculoskeletal joints is then studied through finite element computer simulations from the scale of the whole body down to the organ level.

Computational procedures, based on the finite element method, have the capability of modelling structures with irregular geometry and complex material properties, and the ease of simulating complicated boundary and loading conditions in both static and dynamic analyses.

Modelling Methodology for Physiology and Medicine. DOI: http://dx.doi.org/10.1016/B978-0-12-411557-6.00022-7

Numerical approximation of the solutions to boundary value problems, by means of finite element analysis (FEA), has proven to be of significant benefit to the field of musculoskeletal biomechanics.

This chapter outlines the conceptual basis of FEA and discusses the technical considerations involved in modelling musculoskeletal structures. The modelling process is described, including defining the geometry and the mesh, applying the boundary condition and the material properties, validating the model, and interpreting the results. A case study is presented from diabetic foot pathology application, involving the development of foot finite element models.

22.2 Background

The finite element method is now the most-used discretization method in applied sciences and mechanics. Proof of this is the sheer number of books that have been written or are still appearing and the large number of general purpose codes based on this method. They address all fields of solid and fluid mechanics, heat transfer, electromagnetics, computational chemistry and physics, and, now in increasing number, also interaction problems between these fields. This has made simulations possible that in the recent past were simply unthinkable.

22.2.1 The Procedure of Discretization by the Finite Element Method

The general procedures of the finite element discretization of equations are described in detail in various texts. Here we shall use the notation and methodology introduced by Zienkiewicz et al. [1], which is the most recent (sixth) edition of the first text for the finite element method published in 1967.

In its application to problems in biomechanics, we shall typically be solving partial differential equations that can be written as

$$A\ddot{\Phi} + B\dot{\Phi} + L(\Phi) = 0 \tag{22.1}$$

where A and B are matrices of constants and L is an operator involving spatial differentials, such as $\partial/\partial x$ and $\partial/\partial y$, which can be, and frequently are, nonlinear.

The dot notation implies time differentiation so that

$$\frac{\partial \Phi}{\partial t} = \dot{\Phi}, \quad \frac{\partial^2 \Phi}{\partial t^2} = \ddot{\Phi} \tag{22.2}$$

In all of the above, Φ is a vector of dependent variables (representing the displacements $\mathbf{u}$ and the pressure p).

The finite element solution of the problem will always proceed in the following pattern:

1. The unknown functions Φ are "discretized" or approximated by a finite set of parameters $\overline{\Phi}_k$ and shape function N_k, which are specified in spatial dimensions. Thus we shall write

$$\Phi \cong \Phi^h = \sum_{k=1}^{n} N_k \overline{\Phi}_k \tag{22.3}$$

where the shape functions are specified in terms of the spatial coordinates, i.e.,

$$N_k = N_k(x, y, z) \tag{22.4a}$$

or $N_k = N_k(\mathbf{x})$ and

$$\overline{\Phi}_i \equiv \overline{\Phi}_i(t) \tag{22.4b}$$

are usually the values of the unknown function at some discrete spatial points called nodes that remain as variables in time.

2. Inserting the value of the approximating function $\hat{\Phi}$ into the differential equations, we obtain a *residual* which is not identically equal to zero but for which we can write a set of weighted residual equations in the form

$$\int_{\Omega} \mathbf{W}_j^{\mathrm{T}}(A\ddot{\Phi}^h + B\dot{\Phi}^h + L(\Phi^h))\mathrm{d}\Omega = \mathbf{0} \tag{22.5}$$

which on integration will always reduce to a form

$$M\ddot{\overline{\Phi}} + C\dot{\overline{\Phi}} + P(\overline{\Phi}) = \mathbf{0} \tag{22.6}$$

where M, C, and P are matrices or vectors corresponding in size to the full set of numerical parameters $\overline{\Phi}_k$. A very suitable choice for the weighting function W_j is to take them being the same as the shape function $\mathbf{N}_j$:

$$W_j = N_j \tag{22.7}$$

Indeed, this choice is optimal for accuracy in so-called self-adjoint equations as shown in the basic texts and is known as the Galerkin process.

If time variation occurs (i.e., if the parameters $\overline{\Phi}_k$ are time dependent), Eq. (22.6) that is now an *ordinary differential equation* requires solution in the time domain. This can be, once again, achieved by discretization in time and use of finite elements, although there are many alternative approximations (such as the use of finite differences or other integration schemes).

Usually, the parameters $\overline{\Phi}$ represent simply the values of Φ^h at specified points—called *nodes*—and the shape functions are derived on a polynomial basis by interpolating between the nodal values for elements into which the space is assumed to be divided.

22.3   Finite Element Modelling in Biomechanics

Biomechanics did not become a clearly defined and active research field until the mid-1960s.

Three major events were necessary for the evolution of this field [2]:

1. the development of the nonlinear field theory of mechanics during the 1950s and 1960s;
2. the development of the finite element method in the 1950s; and, subsequently,
3. the rapid advances in computer technology.

Hence, clearly the finite element method has an important place in the advent of computational biomechanics. In particular bone, cell and tissue mechanics have received great attention [2–4]. A few fields in which FEA has been applied successfully are brain injuries [5], dental implants [6], multiscale modelling of the heart [7], lifetime assessments of stents [8], multiphysics modelling of blood flow and drug transport [9], intervertebral disk modelling [10], kinetics and kinematics of articulations [11], pathologic foot mechanics and blood transport [12,13], and tumor growth modelling [14,15]. This is just a short list of recent papers dealing with FEA in biomechanics and is far from being exhaustive. The interested reader is referred to the books listed in the references and the journals where these papers appeared. In general, the mechanics of biological structures at the molecular, cellular, and tissue levels is a multidisciplinary area of research that is expanding rapidly and brings together researchers in biology, medicine, engineering, physics, chemistry, material science, and applied mathematics, and in which numerical modelling plays an important role [3].

22.4   The Modelling Process

Finite element modelling generally involves four stages. When first applied to biomechanics [16,17], it involved five stages, if we consider that a first stage to develop special coding in order to address the specific computational problem was necessary. This is still ongoing with the development of better performing procedures, such as the isogeometric analysis concept [18], but nowadays many of the necessary features are already incorporated into the major program packages available on the market (e.g., ABAQUS, ANSYS, NASTRAN, MARC, ADINA, COMSOL, CASTEM, and LUSAS). However, many established research groups in the finite element community have developed their own research codes for advanced problems that are not adequately covered by the above general purpose codes. Examples are the isogeometric analysis code [18,19], COMES-GEO [20], Pandas [21], Swandyne [22], and FEAP [23], just to mention a few. These allow advances in computational biomechanics to push even further.

The aforementioned first phase is now known as *preprocessing* and involves defining the mesh geometry, specifying the material property distributions, and designating the loading conditions. During the second phase, the computation of the finite element solution is carried out (this phase involves the choice of numerical algorithm and choice of convergence criteria). The third phase is generally known as *postprocessing* and it includes the extraction of meaningful parameters and variables of interest from the finite element simulation (generally outputs are calculated from raw simulation results). The last phase is represented by the validation procedure and the interpretation of the results.

Validation of the models entails that models should be verified and validated by means of combining computational and experimental protocols [24].

22.4.1 Geometry: Image Segmentation

Regarding geometry, we should differentiate the case of 3D and two-dimensional (2D) models. 2D models are generally adopted when the object of interest can be modeled considering approximately both geometrical and material symmetry [25] about a single axis (axisymmetric simplification), in the case of 2D plane stress simulations (e.g., influence of surrounding tissues on biomechanics of the aortic wall [26]). 2D simulation is also often used as a preliminary stage of development of a 3D computational model. Otherwise, nowadays 3D models are the state of the art of FEA in biomechanics applications.

In order to define the geometry of the model, digital images are generally preferred as obtained by computed tomography (CT) and magnetic resonance imaging (MRI). The latter is favored as it represents an accurate, noninvasive method for acquiring musculoskeletal anatomy in living subjects. MRI is used to develop subject-specific biomechanical models, either of normal or of pathologic subjects (e.g., preoperative implant of orthesis studies, development of specific orthotic devices). When subject-specific geometries were not available, some researchers developed different scaling methods that enable the use of MRI/CT scan data from other subjects in combination with a few subject-specific parameters. Murray et al. [27] developed a scaling method that scales the shapes of the bones and the origin and insertion sites of the muscles based on both the height and the weight of the subject. A similar method was developed by Arnold et al. [28] that can estimate the lengths and moment arms of the hamstrings and psoas muscles in children with cerebral palsy. In this context also, the methodology developed by Fernandez and Pandy [29] should be mentioned, which is known as the free-form deformation and is used to customize specific features of bone identified from CT and MRI. Obviously, subject-specific digitized images are always preferable for purposes of accuracy in the reconstruction of biological geometries. In the case of prosthetic implants, Computer-aided engineering (CAE) files, derived from the manufacturer, are the best source geometries. When possible, the finite element mesh can be coupled with the source geometry for simplifying result interpretation. Nowadays, several software packages have been developed for image segmentation purposes. Some can directly generate the mesh to run the finite element model simulation (e.g., Besier et al. [30] used this procedure to generate patellofemoral finite element models based on segmented MRI images), some others may be used to segment tissue geometry automatically and to provide the surface STL format (which is

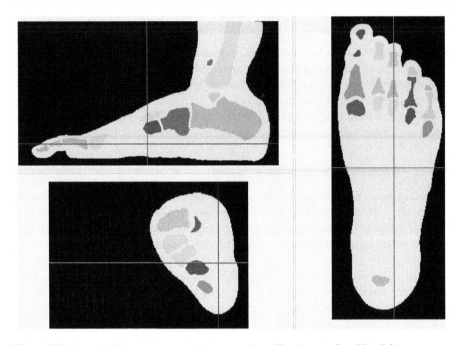

Figure 22.1 Foot MRI image segmentation procedure (Simpleware-ScanIP v.5.0).

a computer-aided design format) to be imported either into a mesh generator specific software or into the software for running the complete finite element simulation. The image segmentation procedure is one of the most time-consuming procedures within the finite element model analysis, therefore it can only be applied to a small number of subjects. A suitable approach to developing an accurate finite element model of the musculoskeletal apparatus should be a combination of a subject-specific model and scaling methods. This may be accomplished by applying automatic segmentation of MRI images to the region of interest (e.g., the contact surfaces where there is the maximum stress concentration) and then scaling the other bones from a generic finite element mesh through one of the available methods, thus reducing the cost of model development.

An example of image segmentation procedure can be found in Figure 22.1 in the case of MRI foot images. In this case, images were acquired with 1.5 T devices (Philips Achieva and Siemens Avanto), spacing between slides of 0.6−0.7 mm, with a slice thickness of 1.2−1.5 mm.

22.4.2 Mesh Definition

The next step after image segmentation (i.e., the construction of the object geometry from 3D scans) is to convert the object into a 3D volumetric mesh with a smooth surface that is as close as possible to the actual domain. The accurate surface definition of an object depends heavily on the segmentation algorithm used, while the mesh quality influences the quality of the FEA. The spatial discretization should be of the right size, quality, and shape for obtaining an accurate solution [31]. For traditional engineering applications, including implants in which object boundaries may be described, analytically or piecewise analytically powerful mesh generating methods exist [32,33]. This is not the case in patient-specific biomedical geometries where surfaces are not well defined and analytical description is not easy. The main difference between domain discretization of a well-defined object and patient-specific geometry is therefore in building the surface mesh [31]. Methods currently applied for patient-specific surface mesh generation are: (i) use of standard NURBS (nonuniform rational B-splines) [18] as in engineering application; (ii) regridding methods [34], starting from an initial surface mesh that may also not be of optimal quality; (iii) mapping methods where the surface is mapped on a plane [35] that is triangulated, producing a high quality 2D mesh (the 2D mesh is then mapped back to the boundary of the biomedical geometry); and (iv) mesh-cosmetic operations converting an initial consistent mesh into a valid mesh for computation [31]. Once a proper surface mesh is obtained, the rest of the domain can be discretized as in standard engineering problems. For very complicated object geometry, the reader is referred to Ref. [31].

22.4.3   Boundary Condition (Loads and Load Application)

In order to simulate the physiological loading on the musculoskeletal system, information on the ground reaction forces and the specific joint (or limb of interest) position, which can be measured from the force of plantar pressure plates and the stereophotogrammetric system, should first be considered. Joint kinematics, ground reaction forces, and surface electromyography data can be recorded simultaneously during a gait analysis session. The musculoskeletal joints' positions should be simulated during different phases of the gait cycle (e.g., for a subject of 80 kg during standing balance a force of approximately 400 N should be applied on each foot) [29]. For the sake of accuracy, 3D joint motion should also be recorded simultaneously using bi-plane X-ray fluoroscopy. The methodologies of Banks and Hodge [36] can be adopted, which entails obtaining the 3D joint motion by projecting the solid bone models obtained from MRI onto the fluoroscopic images. It is an iterative procedure of 3D alignment based on shape matching; then the positions of the bony segments can be tracked with the projections of the bones in the fluoroscopic images [36,37]. It should be decided whether the simulation is run by considering displacements or constraint (kinematics), or loads (kinetics). In this phase of the finite element modelling procedure, it is important to define which region of the model is affected by: the loading condition applied, the load magnitude and its direction, and how it is related to the physiological condition that is the object of the simulation [38]. In this context, in the case of nonlinear static analysis, the time history of the boundary condition could also be considered. Whenever patient-specific boundary conditions are not available, that information can be obtained from the literature (e.g., hip joint force during normal walking can be obtained as in Bergmann et al. [39]).

22.4.4 Material Properties

The assignment of tissue material properties is a fundamental step in the generation of finite element models. The mechanical properties of skeletal tissues have been measured and registered in many experimental studies [40−47].

Subject-specific material properties are not always available and generally cannot be determined for each tissue (bones, ligaments, soft tissues, etc.) as a standard preliminary procedure when running finite element simulation. When those data are not available, material properties should be derived from experimental data from the literature. This practice may be appropriate in instances in which the integrity of these data can be ensured (e.g., cases in which details regarding the loading and boundary conditions are given during experimental session). Biological soft tissues have been generally represented with isotropic hyperelastic constitutive models [40]. A complete description of the constitutive equation and its FE implementation can be found in Weiss et al. [40]. The other tissues are generally hypothesized to be homogeneous, isotropic, and linearly elastic. Young's modulus and Poisson's ratio for the bony structures can be estimated by averaging the elasticity values of cortical and trabecular bones in terms of their volumetric contribution. However, when deriving these parameters from the literature, it should be taken into account that bone mechanical properties have been shown to depend on the subject, anatomic location, orientation, biological processes, and time. Furthermore, it has been proven that material distribution strongly influences the predicted mechanical behavior of biological tissues. When developing subject-specific finite element models, mechanical properties can be derived from CT data.

In this case, the skeletal tissue is considered to be an inhomogeneous material. In Carter and Hayes, a method to derive the distribution of mechanical properties in the bone tissue from CT data has been reported, and a linear correlation between CT numbers and the apparent density of skeletal tissue has been demonstrated over a large range of density values [41].

The same authors reported empirical relationships established experimentally between bone density and mechanical properties. In this context, mechanical properties derived from CT data are averaged over the CT scalar field for each element volume. In Taddei et al. [42], an advanced method for assigning the material properties to finite element models starting from a CT dataset can be found. The method accounts for the different types of tissue from the CT data, in order to overcome the limitations in the majority of the studies that assign a uniform elastic module to each element of the mesh. The model then averages the radiological densities of the voxels that fall inside the element. In Taddei et al. [42], a fully automatic method was provided that does not require any *a priori* information on the mesh topology, therefore it can be applied to either unstructured or structured meshes.

Finally, because of biologic variability, definition of material properties almost always involves some idealization. In this regard, it is usually suggested that one conduct preliminary sensitivity trials to gather information about the sensitivity of the finite element simulation with respect to the material input, especially in the case of nonlinear problems. In this regard, once the finite element mesh has been generated, the material properties relative to each element should be defined (i.e. when a subject-specific model is considered mechanical properties should be derived from CT data [42].)

22.5   Postprocessing

Postprocessing is an important step in the FEA process. For optimal postprocessing, the FE mesh must be built taking into account the needed results and the aim of the FE simulation. If the needed results are forces or moments, as in the case of ligaments, one-dimensional (1D) elements, such as trusses or beams, are suitable. Conversely, if we are interested in the stress and strain fields, 2D or 3D elements are the most appropriate.

The primary variables of most biomechanical analyses are the displacements of the nodes of the FE mesh; also node rotations are primary variables if beam elements are used.

In the case of beams, the bending moments, M1 and M2, the torsional moment M3 (3 is the local longitudinal axis of the beam), and shear and normal forces, T1, T2, and N, are also given by the analysis (M1, M2, M3, T1, T2, N, and their evolutions along the element are available for each beam). If truss elements are used (e.g., for ligaments), only the axial force N in each truss element together with the displacement vector in the nodes are given as results. The stress and strain fields in beams and trusses are then obtained by the FE code during the postprocessing phase, usually by applying the linear theory of Navier. Note that for beams, this theory is not valid in the proximity of points where forces or moments are applied; therefore, in these areas, the stress and strain fields obtained from the postprocessing phase may not be consistent with the real ones.

When 2D or 3D elements are used, the stress and strain fields are computed from the gradients of the displacements via geometry and the material properties. Depending on the gradient of the displacement vector, the quality of the results in terms of stresses and strains is of a lower level than that of the displacement. These considerations can be extended to other physical domains; some examples are given:

- In a thermal FEA of a biological system, the primary variable of the problem is the temperature, while the heat fluxes are computed from the gradient of the temperature and are hence more affected by modelling errors.
- In an FEA of diffusion of a chemical species within a biological tissue, the primary variable is the concentration of the chemical species, while the diffusive velocities are computed from the gradient of concentration usually using a Fick-type equation.
- In an FEA of flow of a fluid within a porous medium (for instance in the analysis of the pulpous nucleus of an intervertebral disc or in the modelling of the heel pad), the primary

variable is the pressure of the fluid, while fluxes are computed from the pressure, usually using a Darcy-type law.

- For flow in an artery, the usual output of CFD is obtained (i.e., pressures and velocities).

Different techniques are available and often used during the FE computation to correct the obtained flows and to respect the "divergence-free" property (conservation of mass and energy). Alternatively, other numerical methods, such as finite differences (FDs), finite volumes (FVs), or mixed-hybrid finite elements (MHFEs), can be used if the major aim is the precision of the computed fluxes.

Error estimation gives a measure of the accuracy obtained with a chosen mesh or time discretization (e.g., see Ref. [22] for space or Ref. [48] for time). Refinement in particular points or in time may be needed to increase the accuracy. It is good practice to perform more than one FEA using different spatial and time discretizations to check the possible influence of these aspects on the computed solution.

Stresses and strains have to satisfy failure criteria. Common criteria for bones can be found in Refs. [4,43,44]. Criteria for ligaments and soft tissues in general are often strain-driven criteria based on viscohyperelastic damage models [45−47].

In the postprocessing phase, very often the obtained numerical solution must be compared with experimental data to validate the model. The experimental data may be available as spatial distribution of a quantity (temperature, pressure, stress, etc.) at a particular time: in this case, results at a given time versus position must be plotted. When a transient problem is also studied, the evolution through time of a quantity in a specific point or area of the FE mesh can be available experimentally: in the second case, results for a given point or area versus time must be plotted. For validation procedures see the following section.

22.6 Validation

By considering that finite element modelling is an approximation technique, it is important to demonstrate the reliability of the obtained solutions. When FEA is applied to biomechanics, the usual methodologies for assessing validity of the selected mesh (convergence test) are not applicable in a straightforward manner. The so-called h- and p-convergence tests are rarely performed in biomechanical models, either because high-order elements generally are not appropriate or because of the geometrically irregular structures that characterize biological models. In this case, a general convergence test is carried out that involves several alternative meshes with progressively increasing resolution, avoiding uniform systematic subdivision that is generally characteristic of simple geometries.

Alternatively, what is called "model validation" can be performed [25]. In this case, it is crucial to identify the sources of information that are appropriate for model validation. If we consider that strain gauges can give direct numerical data as output, surface strain measurements can be used as a standard for comparison. Alternatively, a kinematic validation can be chosen, which is based on comparison between point displacements or velocities in the model with corresponding ones experimentally measured by means of stereophotogrammetric systems or fluoroscopic methodologies. Another alternative frequently adopted in the case of finite element problems involving contact stress and fluid pressures is the comparison with pressure sensors (e.g., Emed (Novel GmbH), Tekscan, Loran pressure sensors) or other commercially available transducers.

No matter which methods have been chosen for physical validation, it is mandatory to be aware of the sources of errors that are involved in the finite element modelling analysis when applied to solve biomechanical problems. In this context, the work of Anderson et al. [24] offers an accurate description of the most common source of errors that may affect musculoskeletal finite element models. In their work, errors are classified as either numerical errors or modelling errors. Numerical errors are referred to as those occurring when mathematical equations are solved using computational techniques (e.g., discretization error, incomplete grid convergence, and computer rounding errors [24]). Modelling errors are related to the simplifications applied either to the physical problem or to the physiological system representation in performing the finite element model analysis (e.g., any sort of approximations about geometries, boundary and loading conditions, material properties, or constitutive equations [24]).

Although any source of error is generally associated with a lack of modelling accuracy, the sources should be addressed separately. Some errors can be prevented by adopting subject-specific approaches in defining material data (e.g., assigning the material properties to finite element models starting from CT dataset), boundary conditions (e.g., subject-specific kinematics, ground reaction forces, and contact pressure distributions, muscle and tendon forces). On the other hand, physical approximation errors (e.g., rigid body assumption in defining bones in joint models) or iterative convergence tolerances are rarely avoidable. Finally human error should also be kept in mind whenever considering any computer—human interaction procedure. However, a statistically meaningful comparison between finite element modelling results and experimental measurements within the range of usability of the model could be an acceptable use compromise [24].

22.7   Case Study: FEA Foot Biomechanics

A wide variety of finite element models of the foot are available in the literature [49—69] and were developed under certain simplifications (e.g., 2D models, isotropic material behavior). For instance, some models developed partial foot segment [61,63,64] finite element models or simplified 2D models; or assumed linear material properties, infinitesimal deformation, and linear boundary conditions.

Several recent models have used the FEA to predict the loading of the foot's components during standing and gait as they relate to foot disorders and therapeutic footwear [51,52,55—57,59—67]. The various foot finite element models have been

Figure 22.2 Example of 2D finite element model of the heel coupled with the force plate used for recording the ground reaction forces (Abaqus, Simulia, v.6.12). The line passing through the malleoli is also highlighted.

generally developed in order to provide estimates of quantities that cannot be directly measured or to perform simulations that would be onerous or dangerous for human subjects [65]. For instance, an interesting field of application is diabetic foot pathology. This condition is of interest biomechanically because custom foot insoles and foot surgical interventions are generally the most-used approach for coping with its main consequences (e.g., plantar ulcers and amputations). An example of FEA applied to a diabetic foot either in the 2D or 3D case can be found in Figures 22.2 and 22.3, respectively. The former represents an attempt to describe differences in diabetic foot biomechanics with respect to healthy subjects by means of FEA [62,63]. Two subject-specific 2D models of the heel [63] and two 3D foot finite element models of both a diabetic neuropathic and a healthy subject's feet [62] were developed for this purpose. The models were subject specific in terms of foot geometry, *in vivo* kinematics, and kinetics measured data. The former were developed by segmenting bones, cartilage, and skin from MRI and drawing a horizontal plate as ground support. Four different phases of the stance phase of gait (heel strike, loading response, midstance, and push off) were simulated. Validity of the models was assessed by comparing the experimental plantar pressures data with the simulated data.

In this context, among the several finite element foot models available in the literature, an important role is played by the model of Cheung and Zhang [52], which developed a 3D model of the human foot and ankle. The geometry was obtained using a 2-mm coronal MRI of the right foot of a normal male subject by using Mimics software for automatic segmentation. The solid models of 28 foot bones and encapsulated soft tissue structure models were first established in Solidworks software and then imported into ABAQUS. A tetrahedral finite element mesh was adopted. The plantar fascia and 72 ligaments were defined by connecting the corresponding attachment points on the bones by means of tension-only truss elements,

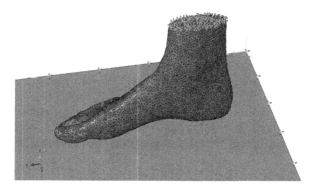

Figure 22.3 Example of 3D foot finite element model (Abaqus, Simulia, v.6.12). In this formulation, the model was developed by segmenting bones, cartilage, and skin from MRI. Bones, ligaments, and plantar fascia were assumed to have a linear isotropic behavior, while the surrounding soft tissue was defined as a formal 3D material nonlinear region using an isotropic, incompressible, hyperelastic second-order polynomial formulation.

and contact interactions among the major joints were thus defined, allowing relative bone movements. A solid model of the foot orthosis was also defined in Solidworks software and imported into ABAQUS. In this case, hexahedral finite element mesh was chosen. The encapsulated soft tissue and orthotic material were defined as hyperelastic, while other tissues were modeled under a simplified assumption of homogeneous, isotropic, and linearly elastic material. The ground reaction forces component as well as that generated by the muscles were applied at the inferior ground support and at their corresponding points of insertion by defining contraction forces via axial connector elements, respectively. The stance phase of gait was simulated. Validity of the FEA was assessed by comparison with experimental measurements conducted on cadavers and on the same subject who volunteered.

Additionally, the work of Yarnitzky et al. [68] should be mentioned, which developed a hierarchical modelling system integrating a 2D foot model (which represents the high-order model) together with local finite element models of the plantar tissue padding, the calcaneus, the medial metatarsal heads (that represents the low-order models). The high-order 2D foot model enables a real-time analytical evaluation of the time-dependent plantar fascia tensile forces during the stance phase of gait. These force evaluations are transferred together with foot—shoe local reaction forces (measured in real time on specific sites of the foot, namely: the calcaneus, medial metatarsals, and hallux), to the low-order finite element models. The former represent the boundary conditions for the analyses of local deformations and stresses in the plantar pad. Information provided by this original model allows for the measurement of internal deformations and stresses in the plantar pad in real time, thus allowing identification of elevated deformation and stress exposures of the foot during gait.

Another important contribution is represented by the work of Tadepalli et al. [69], in which the issue of choosing an appropriate mesh when dealing with particular shapes, as in the case of foot and foot orthosis, is extensively discussed. When dealing with an anatomical structure such as the foot, hexahedral mesh generation is rarely used due to geometric complexities. On the contrary, tetrahedral meshing, which can be more easily automated, is the most commonly used. In their contribution, Abaqus (Simulia, Providence, RI) is used to compare tetrahedral and hexahedral elements under compression and shear loading, material incompressibility, and frictional contact conditions, which are the conditions that characterize foot and footwear FEA.

The present description is not meant to be an exhaustive one of the application of FEA to foot biomechanical analysis, because this would be beyond the scope of this chapter. Indeed, it should be considered an example of the application of FEA methodology to solve a biomechanics problem.

22.8 Conclusion

It has now been almost 30 years since the first application of FEA in biomechanics appeared. There has been a great acceleration in the development of mathematical models, software, and tools that has resulted in more sophisticated and faster simulations. These improvements have been empowered by the parallel improvements of computing capabilities, thanks to the new generations of both computer technology and image processing techniques. FEA is a growing field of research in the area of biomechanics, whose possibilities of improvement are still open.

Acknowledgment

We acknowledge the contribution of Annamaria Guiotto in preparing the figures reported in this chapter.

References

[1] Zienkiewicz OC, Chan AHC, Pastor M, Schrefler BA, Shiomi T. Computational geomechanics. Chichester: Wiley; 1999.

[2] Holzapfel GH, Ogden RW, editors. Biomechanics of soft tissue in cardiovascular systems. CISM courses and lectures 2003, 441. New York, NY: Springer Wien; 2003.

[3] Holzapfel GH, Ogden RW, editors. Biomechanical modelling at the molecular, cellular and tissue levels. CISM courses and lectures 2009, 508. New York, NY: Springer Wien; 2009.

[4] Cowin SC. Bone mechanics handbook. Boca Raton, FL: CRC Press; 2001.

[5] Li X, von Holst H, Kleiven S. Influence of gravity for optimal head positions in the treatment of head injury patients. Acta Neurochir 2011;153:2057—64.

[6] Natali AN, Carniel EL, Pavan PG. Numerical analysis of biomechanical response of a dental prosthesis with regard to bone—implant adhesion phenomena. J Adhes Sci Technol 2009;23:1187—99.

[7] Schmid H, Wang W, Hunter PJ, Nash MP. A finite element study of invariant-based orthotropic constitutive equations in the context of myocardial material parameter estimation. Comp Methods Biomech Biomed Eng 2009;12(6):691—9.

[8] Argente dos Santos HAF, Auricchio F, Cont M. Fatigue life assessment of cardiovascular balloon-expandable stents: a two-scale plasticity-damage model approach. J Mech Behav Biomed Mat 2012;15:78—92.

[9] Calo VM, Brasher NF, Bazilevs Y, Hughes TJR. Multiphysics model for blood flow and drug transport with application to patient-specific coronary artery flow. Comp Mech 2008;43(1):161—77.

[10] Schroeder Y, Huyghe JM, van Donkelaar CC, Ito K. A biochemical/biophysical 3D FE intervertebral disc model. Biom Mod Mechanobiol 2010;9(5):641—50.

[11] Shim VB, Mithraratne K, Anderson IA, Hunter PJ. Simulating *in vivo* knee kinetics and kinematics of tibio-femoral articulation with a subject-specific finite element model. World Congress Med Phys Biomed Eng IFMBE Proc 2010;25(4):2315—8.

[12] Fernandez JW, Haque MZU, Hunter PJ, Mithraratne K. Mechanics of the foot. Part 1: A continuum framework for evaluating soft tissue stiffening in the pathologic foot. Int J Numer Methods Biomed Eng 2012;28(10):1056—70.

[13] Mithraratne K, Ho H, Hunter PJ, Fernandez JW. Mechanics of the foot. Part 2: A coupled solid—fluid model to investigate blood transport in the pathologic foot. Int J Numer Methods Biomed Eng 2012;28(10):1071—81.

[14] Hawkins-Daarud A, van der Zee KG, Oden JT. Numerical simulation of a thermodynamically consistent four-species tumor growth model. Int J Numer Methods Biomed Eng 2012;28:3—24.

[15] Sciumè G, Shelton SE, Gray WG, Miller CT, Hussain F, Ferrari M, et al. A multiphase model for three dimensional tumor growth. New J Phys 2013;15(35p):015005.

[16] Rybicki EF, Simonen FA, Weis EB. On the mathematical analysis of stress in the human femur. J Biomech 1972;5:203—15.

[17] Brekelmans WAM, Poort HW, Sloof TJJH. A new method to analyze the mechanical behavior of skeletal parts. Acta Orthop Scand 1972;43:301—7.

[18] Cottrell JA, Hughes TJR, Bazilevs Y. Isogeometric analysis: toward integration of CAD and FEA. Chichester: Wiley; 2009.

[19] Hughes TJR, Reali A, Sangalli G. Efficient quadrature for NURBS-based isogeometric analysis. Comp Methods Appl Mech Eng 2010;199(5—8):301—13.

[20] Lewis RW, Schrefler BA. The finite element method in the static and dynamic deformation and consolidation of porous media. Chichester: Wiley; 1998.

[21] Schenke M, Ehlers W. On the analysis of soils using an Abaqus—PANDAS interface. Proc Appl Math Mech 2011;11:431—2.

[22] Zienkiewicz OC, Zhu JZ. The superconvergent patch recovery (SPR) and adaptive finite element refinement. Comp Methods Appl Mech Eng 1992;101:207—24.

[23] FEAP-8.2.k. Finite element analysis program. University of California; 2008.

[24] Anderson AE, Ellis BJ, Jeffrey Weiss A. Verification, validation and sensitivity studies in computational biomechanics. Comput Methods Biomech Biomed Eng 2007;10 (3):171—84.

[25] Brown TD. Finite element modeling in musculoskeletal biomechanics. J Appl Biomech 2004;20:336−66.

[26] Kim J, Peruski B, Hunley C. Influence of surrounding tissues on biomechanics of aortic wall. Int J Exp Comput Biomech, in press.

[27] Murray WM, Buchanan TS, Delp SL. Scaling of peak moment arms of elbow muscles with upper extremity bone dimensions. J Biomech 2002;35:19−26.

[28] Arnold AS, Blemker SS, Delp SL. Evaluation of a deformable musculoskeletal model: application to planning muscle-tendon surgeries for crouch gait. Ann Biomed Eng 2001;29:263−74.

[29] Fernandez JW, Pandy MG. Integrating modelling and experiments to assess dynamic musculoskeletal function in humans. Exp Physiol 2006;91:371−82.

[30] Besier TF, Beaupre GS, Delp SL. Subject specific modeling to estimate patellofemoral joint contact stress. Med Sci Sports Exerc 2004;36(5 Suppl):S1−2.

[31] Sazonov I, Nithiarasu P. Semi-automatic surface and volume mesh generation for subject specific biomedical geometries. Int J Numer Methods Biomed Eng 2012;28:133−57.

[32] Tremel U, Deister F, Hassan O, Weatherill NP. Automatic unstructured surface mesh generation for complex configurations. Int J Numer Methods Fluids 2004;45 (4):341−64.

[33] Wang DS, Hassan O, Morgan K, Weatherill NP. EQSM: an efficient high quality surface grid generation method based on remeshing. Comp Methods Appl Mech Eng 2006;195:5621−33.

[34] Loehner R. Regridding surface triangulations. J Comp Phys 1996;126:1−10.

[35] Borouchhaki H, Laug P, George P-L. Parametric surface meshing using a combined advancing-front generalized Delaunay approach. Int J Numer Methods Eng 2000;49:233−59.

[36] Banks SA, Hodge WA. Accurate measurement of three-dimensional knee replacement kinematics using single-plane fluoroscopy. IEEE Trans Biomed Eng 1996;43:638−49.

[37] Zuffi S, Leardini A, Catani F, et al. A model-based method for the reconstruction of total knee replacement kinematics. IEEE Trans Med Imag 1999;18(10):981−91.

[38] Erdemir A, Guess TM, Halloran J, Tadepalli SC, Morrison TM. Considerations for reporting finite element analysis studies in biomechanics. J Biomech 2012;45:625−33.

[39] Bergmann G, Deuretzbacher G, Heller M, Graichen F, Rohlmann A, Strauss J, et al. Hip contact forces and gait patterns from routine activities. J Biomech 2001;34:859−71.

[40] Weiss JA, Maker BN, Govindjee S. Finite element implementation of incompressible, transversely isotropic hyperelasticity. Comput Methods Appl Mech Eng 1996;135:107−28.

[41] Carter DP, Hayes WC. The compressive behaviour of bone as a two-phase porous structure. J Bone Joint Surg Am 1977;59(7):954−62.

[42] Taddei F, Pancanti A, Viceconti M. An improved method for the automatic mapping of computed tomography numbers onto finite element models. Med Eng Phys 2004;26:61−9.

[43] Cowin SC, He Q-C. Tensile and compressive stress yield criteria for cancellous bone. J Biomech 2005;38:141−7.

[44] Fritsch A, Hellmich C. Universal microstructural patterns in cortical and trabecular extracellular and extravascular bone materials: micromechanics-based prediction of anisotropic elasticity. J Theor Biol 2007;244(4):597−620.

[45] Peña E. Damage functions of the internal variables for soft biological fibred tissues. Mech Res Commun 2011;38(8):610−5.

[46] Peña E, Alastrué V, Laborda A, Martínez MA, Doblaré M. A constitutive formulation of vascular tissue mechanics including viscoelasticity and softening behaviour. J Biomech 2010;43(5):984−9.

[47] Natali AN, Carniel EL, Pavan PG, Sander FG, Dorow C, Geiger M. A visco-hyperelastic-damage constitutive model for the analysis of the biomechanical response of the periodontal ligament. J Biomech Eng 2008;130:1−8.

[48] Secchi S, Schrefler BA. A method for 3-D hydraulic fracturing simulation. Int J Fract 2012;178:245−58.

[49] Bandak FA, Tannous RE, Toridis T. On the development of an osseo-ligamentous finite element model of the human ankle joint. Int J Solids Struct 2001;38:1681−97.

[50] Camacho DLA, Ledoux WR, Rohr ES, Sangeorzan BJ, Ching RP. A three-dimensional, anatomically detailed foot model: a foundation for a finite element simulation and means of quantifying foot-bone position. J Rehabil Res Dev 2002;39:401−10.

[51] Chen WP, Ju CW, Tang FT. Effects of total contact insoles on the plantar stress redistribution: a finite element analysis. Clin Biomech 2003;18:S17−24.

[52] Cheung JT, Zhang MA. 3D finite element model of the human foot and ankle for insole design. Arch Phys Med Rehabil 2005;86:353−8.

[53] Gefen A, Megido-Ravid M, Itzchak Y, Arcan M. Biomechanical analysis of the three dimensional foot structure during gait: a basic tool for clinical applications. J Biomech Eng 2000;122:630−9.

[54] Giddings VL, Beaupre GS, Whalen RT, Carter DR. Calcaneal loading during walking and running. Med Sci Sports Exerc 2000;32:627−34.

[55] Goske S, Erdemir A, Petre M, Budhabhatti S, Cavanagh PR. Reduction of plantar heel pressures: insole design using finite element analysis. J Biomech 2006;39:2363−70.

[56] Jacob S, Patil MK. Stress analysis in three-dimensional foot models of normal and diabetic neuropathy. Front Med Biol Eng 1999;9:211−27.

[57] Lemmon D, Shiang TY, Hashmi A, Ulbrecht JS, Cavanagh PR. The effect of insoles in therapeutic footwear: a finite-element approach. J Biomech 1997;30:615−20.

[58] Nakamura S, Crowninshield RD, Cooper RR. An analysis of soft tissue loading in the foot—a preliminary report. Bull Prosthet Res 1981;18:27−34.

[59] Actis RL, Ventura LB, Smith KE, Commean PK, Lott DJ, Pilgram TK, et al. Numerical simulation of the plantar pressure distribution in the diabetic foot during push-off stance. Med Biol Eng Comput 2006;44(8):653−63.

[60] Gefen A. Stress analysis of the standing foot following surgical plantar fascia release. J Biomech 2002;35(5):629−37.

[61] Gefen A. Plantar soft tissue loading under the medial metatarsals in the standing diabetic foot. Med Eng Phys 2003;25(6):491−9.

[62] Sawacha Z, Guiotto A, Boso D, Sciumè G, Guarneri G, Scarton A, et al. Development of a foot multiscale model for diabetic foot prevention. J Biomech 2013, in press.

[63] Scarton A, Guiotto A, Sawacha Z, Guarneri G, Avogaro A, Cobelli C. Gait analysis driven 2D finite element model of the neuropathic hindfoot. J Biomech 2013, in press.

[64] Budhabhatti SP, Erdemir A, Petre M, Sferra J, Donley B, Cavanagh PR. Finite element modeling of the first ray of the foot: a tool for the design of interventions. J Biomech Eng 2007;129(5):750−6.

[65] Cavanagh P, Erdemir A, Petre M, Owings T, Botek G, Chokhandre S, et al. Biomechanical factors in diabetic foot disease. J Foot Ankle Res 2008;1(Suppl. 1):K4.

[66] Oosterwaal M, Telfer S, Trholm S, Carbes S, Rhijn LWv, Macdu R, et al. Generation of subject-specific, dynamic, multisegment ankle and foot models to improve orthotic design: a feasibility study. BMC Musculoskelet Disord 2011;12(1):256.

[67] Cheung JT-M, Zhang M. Parametric design of pressure-relieving foot orthosis using statistics-based finite element method. Med Eng Phys 2008;30(3):269—77.

[68] Yarnitzky G, Yizhar Z, Gefen A. Real-time subject-specific monitoring of internal deformations and stresses in the soft tissues of the foot: a new approach in gait analysis. J Biomech 2006;39(14):2673—89.

[69] Tadepalli SC, Erdemir A, Cavanagh PR. Comparison of hexahedral and tetrahedral elements infinite element analysis of the foot and footwear. J Biomech 2011;44 (12):2337—43.

23 Modelling for Synthetic Biology

*Lorenzo Pasotti[a,b], Susanna Zucca[a,b]
and Paolo Magni[a,b]*

[a]Dipartimento di Ingegneria Industriale e dell'Informazione, Università degli Studi di Pavia, Pavia, Italy, [b]Centro di Ingegneria Tissutale, Università degli Studi di Pavia, Pavia, Italy

23.1 Background

23.1.1 Synthetic Biology

Synthetic biology is an emerging field of bioengineering that combines the expertise of biologists and engineers to design and implement novel biological functions *in vivo* [1]. Key concepts from the engineering world, such as abstraction, standardization, modularity, and datasheet specification, are used to design, build up, and characterize biological systems that can quantitatively work as intended [2]. DNA sequences, herein called *biological parts*, are the main building blocks that can be assembled to obtain a *genetic program* that can carry out the function of interest [3]. Biological parts include genes and regulatory sequences that can be tuned to optimize system behavior. A genetic program can work once introduced into a host organism (a *biological chassis*), which can use its transcriptional/translational machinery to decode the program and carry out the encoded function.

Synthetic biology extends the potential of genetic engineering and may benefit many application fields, such as bioenergy (e.g., waste-to-fuel conversion), environment (e.g., biosensors), medicine (e.g., production of drugs), or bioremediation (e.g., degradation of pollutants) through biological pathway design and optimization in living systems like microbes [4]. Recent successes include a microorganism able to produce the antimalarial drug precursor Artemisinin at low cost [5], a whole-cell biosensor to detect arsenic in drinking water [6], and a microorganism for the production of bioethanol from algal biomass [7]. While synthetic biology has demonstrated its potential to solve high-impact problems, a plethora of basic research studies are still ongoing to discover the engineering limits for biological systems. Genetic programs inspired by electronic engineering have led to the creation of living systems that behave like oscillators [8], bistable switches [9], waveform generators [10], logic functions [11], or high/low/band-pass filters [12,13]. Their study contributes to the understanding of biological mechanisms.

Modelling Methodology for Physiology and Medicine. DOI: http://dx.doi.org/10.1016/B978-0-12-411557-6.00023-9

In an ideal process for the creation of a new biological function, design specifications are met by choosing well-characterized parts from datasheets and parts are combined together in a genetic program that is finally incorporated into a chassis. This bottom-up strategy is commonly used in all of the fields of engineering, but is feasible only in a modular framework, where designers can predict the behavior of the composite system from the knowledge of basic parts functioning. In a biological context, modularity is not always valid for several reasons that are to be considered during a design process, such as energy demand of genetic programs in a chassis with finite resources, intrinsic biological noise [14], crosstalk, or context-dependent behavior of parts [15].

23.1.2 Biological Parts

According to the basic flow of information in biological systems, defined by the central dogma of molecular biology, protein-coding DNA sequences (herein called genes) are transcribed into RNA molecules, which are converted into proteins at the ribosome level, and, finally, DNA sequences can be replicated in living cells to propagate the encoded function to the progeny.

Proteins can perform a wide range of specific functions: from transcriptional control (transcription factors) to biochemical reactions (enzymes). At the regulatory level, transcription rate is controlled by *promoters*, which are DNA sequences upstream of a gene or a set of genes (operon). *Constitutive* promoters have a constant strength, while *regulated* promoters can be activated or inhibited by specific factors. For each produced transcript, translation rate is controlled by *ribosome binding sites* (RBSs), which are small sequences downstream of promoters and upstream of genes. RBSs are transcribed together with the downstream gene and regulate translation by affecting the binding of transcript molecules to ribosomes. Finally, *terminators* are sequences that determine the transcription stop. In order to incorporate synthetic DNA into a microbial chassis, *plasmids* containing the program are widely used as vectors. A replication origin is included in plasmids, which are replicated in parallel to genomic DNA at a specific copy number. Commonly used molecular chassis to host artificial systems are gram-negative bacteria, gram-positive bacteria, and yeasts. The most widespread organism used in synthetic biology is the gram-negative bacterium *Escherichia coli*.

Genetic programs can be conceptualized as genetic circuits or networks for a wide range of synthetic biology studies: in a pathway for a biofuel or drug production, a set of enzymes, synthesized by the host organism at a suitable level, can perform the biochemical transformation of substrates to a final compound. In contrast, in a synthetic oscillator circuit, different transcription factors are produced and degraded dynamically to actuate oscillations in gene expression or protein production.

23.1.3 Mathematical Models

Mathematical models are widely used in many areas of engineering to support the bottom-up building of systems from the early design steps to the final debugging.

Likewise, they can also drive and support synthetic biology studies. In recent research studies, different models have been exploited for disparate aims: a biophysical model has been developed to predict the translation initiation rate of RBSs given the surrounding DNA sequences [16]; ordinary differential equation (ODE) models have been used to capture the behavior of genetic circuits [9,13,17], estimate nonobservable parameters, and predict the effects of parameter changes (e.g., temperature, cell growth rate, DNA copy number, promoter/RBS tuning) [8,18]; a linear model of gene expression has been recently used to study context-dependent effects on protein levels [19].

This chapter focuses on models based on ODEs, which can support the study and bottom-up design of a wide range of genetic circuits. Section 23.2 illustrates the modelling concepts that can be exploited toward this goal.

ODE models describe system behavior in a deterministic framework. However, in antithesis to this paradigm, stochastic models can also be exploited to account for randomness affecting the process. Species are not described by deterministic values, but rather by probability distributions. Taking into account the fluctuations that exist in living systems, the investigation of the effects of biological noise on the system behavior is enabled.

23.2   Models of Genetic Circuits

Although the dynamics of simple gene networks can sometimes be inferred without the aid of mathematical models, the behavior of more complex reaction networks can be difficult to predict without the aid of modelling tools, used in combination with experimental data [20]. A key advantage in this process is that it partially avoids the use of time- and resource-consuming trial-and-error approaches for the fine-tuning of the network [21].

23.2.1   Mechanistic and Empirical Models

The most suitable model to describe the investigated phenomenon can be difficult to identify and the use of models of a different nature (i.e., empirical [18,22] or mechanistic [23,24] models) might be necessary to individuate the best candidate [20].

When dealing with simple gene networks, composed for example by a regulated promoter expressing a target protein, empirical laws based on Hill-like functions can be used to describe the activation characteristic of the promoter in terms of protein expression as a function of input signal intensity. The main feature of these models is their ease of use, which contrasts with the lack of biological significance of the parameters and with the impossibility of deepening the knowledge about every single reaction. When empirical models are inadequate to study the phenomena of interest, mechanistic models derived according to the law of mass action kinetics can be more properly formulated [21]. These models represent each elementary biomolecular interaction with biochemical reactions.

Mechanistic models are characterized by a higher number of parameters and intermediary steps. In contrast with what happens for the empirical models, here the parameters have a physiological meaning, representing, for example, association and dissociation rates, dimerization and dedimerization rates, or translation rates [25]. Such parameters might be expensive and difficult to measure or even not experimentally detectable.

23.2.2 Model of a Simple Gene Network

Simple regulatory networks composed by a promoter expressing a target protein can be described by an empirical mathematical model that accounts for transcriptional and translational processes and for protein maturation [26,27]. Reporter genes encoding fluorescent/luminescent/pigment-producing proteins are widely used by researchers to indirectly measure protein levels in real time. For example, a widely used reporter gene is *gfp*, encoding for the green fluorescent protein (GFP). The mathematical model describing GFP synthesis rate in a simple network is here reported.

$$\frac{d[M]}{dt} = n \cdot \text{PoPS} - \gamma_M \cdot [M] \tag{23.1}$$

$$\frac{d[I]}{dt} = \rho \cdot [M] - (\alpha + \gamma_I) \cdot [I] \tag{23.2}$$

$$\frac{d[G]}{dt} = \alpha \cdot [I] - \gamma_G \cdot [G] \tag{23.3}$$

Equation (23.1) describes the transcription of the *gfp* gene into mRNA, where $[M]$ is the mRNA concentration per cell. The transcription rate depends on the PoPS signal (polymerase per second: defined as the number of times that an RNA polymerase molecule passes a specific point on DNA per time unit), multiplied by the promoter copy number n. The γ_M parameter is the mRNA extinction rate.

Equations (23.2) and (23.3) represent the translation of mRNA into the immature GFP protein $[I]$ and the maturation of $[I]$ into the mature (fluorescent) protein $[G]$, respectively. ρ is the protein synthesis rate, α is the GFP maturation rate, and γ_I and γ_G are the immature and mature GFP extinction rates, respectively.

The mature GFP synthesis rate per cell (S_{cell}) can be expressed as

$$S_{cell} = \alpha \cdot [I] \tag{23.4}$$

By using this model, it is possible to cope with the difficult task of predicting intracellular species. As an example, the definition of the relationship between promoter transcriptional activity (represented by the hard to measure PoPS species) and the experimentally detectable GFP synthesis rate per cell (S_{cell}) is allowed.

Assuming the steady state of the system (i.e., all of the time derivatives are zero) the following relations can be determined:

$$S_{cell}^{SS} = \gamma_G \cdot [G]^{SS} \tag{23.5}$$

$$PoPS_{cell}^{SS} = \frac{\gamma_M \cdot (\alpha + \gamma_I) \cdot S_{cell}^{SS}}{\rho \cdot \alpha \cdot n} \tag{23.6}$$

Since GFP is a stable protein, the degradation rates of its mature and immature forms are negligible compared to its dilution rate due to cell growth (this is true only for GFP variants without fast-degradation tags). For this reason, the assumption $\gamma_I = \gamma_G = \mu$, where μ is the cellular growth rate, persists. Growth rate is highly affected by many factors, such as cell strain, growth medium composition (e.g., minimal or rich media), and metabolic burden for the host cell (e.g., plasmid copy number, toxicity of the DNA-encoded function). On the other hand, mRNA degradation typically occurs faster than cell doubling [27−29], thus γ_M can be assumed to be equal to the mRNA degradation rate, independent of cell growth.

γ_M, μ, ρ, n, and α can be estimated through *ad hoc* experiments (see Section 23.3 for details). Given a plasmid and an mRNA encoding a reporter device with RBS and terminator, these parameters should be independent of the promoter used and its activity can be estimated via PoPS computation. S_{cell}^{SS} can be estimated from experimental data of population fluorescence (F_{GFP}) and optical density (OD) measured over time in cell cultures in the exponential growth phase. F_{GFP} is proportional to the total amount of GFP molecules in the culture, while OD is proportional to the number of cells. Conversion factors to compute the number of GFP molecules and colony forming units (CFUs) from F_{GFP} and OD, respectively, can be experimentally determined [26,27]. GFP synthesis rate per cell can be estimated as

$$S_{cell}^{SS} = \frac{dF_{GFP}}{dt} \cdot \frac{1}{OD} \tag{23.7}$$

The described model is suitable for studying the expression of *gfp* driven by a constitutive promoter. In the more general case where a regulated promoter drives the expression of the *gfp* gene, an empirical Hill-like function can be adopted to describe the steady-state input−output characteristic X-$PoPS_{cell}^{SS}$, where X is the intensity of the input signal, triggering promoter activation or repression.

$$PoPS_{cell}^{SS} = \delta + \frac{V_{max} \cdot X^\eta}{K^\eta + X^\eta} \tag{23.8}$$

$$PoPS_{cell}^{SS} = \delta + V_{max} \cdot \left(1 - \frac{X^\eta}{K^\eta + X^\eta} \right) \tag{23.9}$$

In this function, δ represents the basic activity of the regulated promoter in its OFF state, V_{max} is the maximum activation rate that can be achieved, K is the input that achieves an output of $\delta + V_{max}/2$, and η is the Hill coefficient. Equation (23.8) describes the behavior of an inducible promoter, when the activity is an increasing function of X, while Eq. (23.9) represents the behavior of a repressible promoter, when the activity decreases for increasing values of X.

Although these empirical relations have been used to describe the steady-state behavior of the gene network, the model is also suitable for studying its dynamic characteristics. In a dynamic framework, PoPS(t) is a function that could be included as a state variable. Models like this can be defined for each expression cassette of an arbitrarily large network and static/dynamic functions can be estimated and used for prediction, debugging, and tuning.

Because sometimes it is difficult to estimate the parameters involved in transcription in a large gene network, transcription dynamics can be neglected when faster than other dynamics. As an example, in the previously studied case, Eqs. (23.1)−(23.3) become

$$\frac{d[I]}{dt} = P - (\alpha + \gamma_I) \cdot [I] \tag{23.10}$$

$$\frac{d[G]}{dt} = \alpha \cdot [I] - \gamma_G \cdot [G] \tag{23.11}$$

P is the protein production rate that can be described by a Hill function as illustrated above for PoPS. As before, an additional equation for P could be included that empirically summarizes the dynamics in protein production rate.

In summary, the dynamic behavior of arbitrarily complex networks can be captured by a set of ODEs and, under the steady-state hypothesis, the static behavior can be studied through algebraic equations. Transcription dynamics can be implicitly incorporated into translation-describing functions or described by an empirical dynamic equation.

23.2.3 Models of Interconnected Networks

In this section, two examples that use modelling tools to predict and study the behavior of complex networks are provided. In the first example, a logic inverter is quantitatively characterized and reused in another context to respond to a different stimulus. In the second example, an oscillator network is studied.

23.2.3.1 Logic Inverter

Figure 23.1A shows the structure and functioning of a genetic logic inverter (also called NOT gate) based on the TetR-Ptet system. The inverter is composed by a promoter-less repressor gene (tetR)-expression cassette and a promoter (Ptet) that can be repressed by the specific repressor protein (TetR) [30].

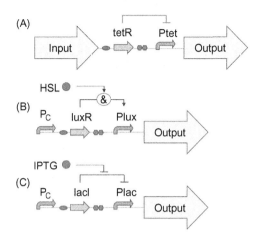

Figure 23.1 Structure of a genetic logic inverter. (A) An inverter can be implemented with a promoter-less repressor-encoding gene and a specific repressible promoter. Here, a tetR/Ptet-based inverter is shown: when the input signal is high, TetR represses Ptet and the output is low, while if the input signal is low, Ptet is not repressed anymore and its output is high. This implements the functioning of a logic inverter. The inverter input is a transcriptional signal that can be provided by a promoter, while the output is a protein of interest (e.g., GFP) encoded by the gene that is placed downstream of Ptet. (B and C) Examples of input devices that can be interconnected to the logic inverter to characterize its transfer function. (B) It consists of a luxR expression cassette driven by a constitutive promoter and the lux promoter (Plux). When LuxR transcription factor is expressed and the inducer N-3-oxohexanoyl-L-homoserine lactone (HSL) is added to the medium, Plux activity is triggered in an HSL concentration-dependent fashion. (C) It consists of a lacI repressor expression cassette driven by a constitutive promoter and the LacI-repressible promoter (Plac). When the inducer isopropyl β-D-1-thiogalactopyranoside (IPTG) is added to the medium, it inactivates LacI and, as a result, Plac activity is triggered in an IPTG concentration-dependent fashion. Symbols: Curved arrows represent promoters, straight arrows represent coding sequences, ovals represent RBSs, octagons represent transcriptional terminators, circles represent chemical inducers, P_C indicates a constitutive promoter, and & indicates that two elements are required to activate a promoter.

In order to predict the steady-state behavior of a composite system including a sensor (input) device upstream of the inverter, a static transfer function of the inverter needs to be measured. This can be carried out as described in Figure 23.2: (i) the transfer function of the input device has to be measured. This task can be achieved by assembling a reporter device (e.g., including GFP) downstream of the input device. The obtained measurement system is ready for the characterization to quantitatively measure the signal that it can generate (Figure 23.2A); (ii) the input device has to be assembled upstream of the inverter and a reporter device has to be assembled downstream of the inverter (Figure 23.2B); finally, (iii) the output of the inverter is measured as a function of the signal generated by the input device.

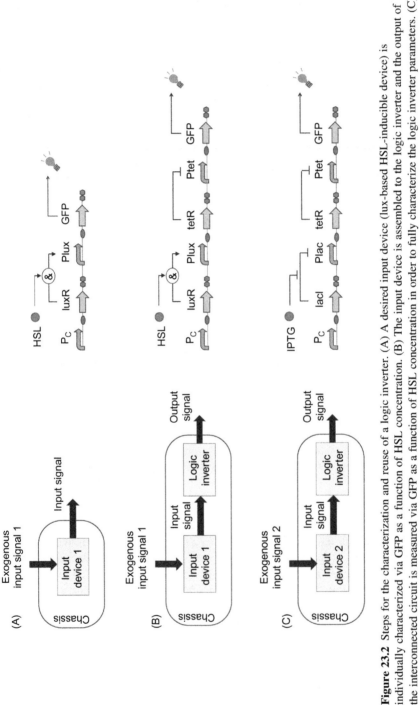

Figure 23.2 Steps for the characterization and reuse of a logic inverter. (A) A desired input device (lux-based HSL-inducible device) is individually characterized via GFP as a function of HSL concentration. (B) The input device is assembled to the logic inverter and the output of the interconnected circuit is measured via GFP as a function of HSL concentration in order to fully characterize the logic inverter parameters. (C) The characterized logic inverter can be reused by assembling a different pre-characterized input device upstream (in this example, a lac-based IPTG-inducible device) and the output of the interconnected circuit can be predicted via mathematical model. The input signal is the transcriptional signal, generated by an input device, that drives the NOT gate. All symbols are consistent with Figure 23.1.

First, the input module has to be characterized. Here, an HSL-inducible device is considered (Figure 23.1B). During its characterization stage, the GFP reporter is used as output of the device. Its static transfer function can be measured through the following model:

$$\frac{d[I]}{dt} = \delta_{\text{lux}} + V_{\text{max,lux}} \cdot \frac{1}{1 + (K_{\text{lux}}/H)^{\eta_{\text{lux}}}} - (\alpha + \mu) \cdot [I] \tag{23.12}$$

$$\frac{d[G]}{dt} = \alpha \cdot [I] - \mu \cdot [G] \tag{23.13}$$

where H is the HSL input concentration, the lux subscript indicates that the parameter is relative to the Plux activation function, and the other parameters are defined as before. I and G degradation rate is assumed to be much lower than the growth rate μ. Protein production rate (described by Hill function) is assumed to be a static function of its input H without considering the transcription dynamics. Given an S_{cell} time series at the steady state of all species, the parameters of the Hill function can be easily estimated (assuming that α and μ are known). In order to carry out experimental measurements, bacterial cultures bearing this circuit have to be induced with different amounts of HSL to yield the desired final concentrations and then cell growth and fluorescence have to be assayed over time.

According to Eqs. (23.4) and (23.12), the steady-state equations are

$$S_{\text{cell}} = \alpha \cdot [I] = \frac{\alpha}{\alpha + \mu} \cdot \left(\delta_{\text{lux}} + \frac{V_{\text{max,lux}}}{1 + (K_{\text{lux}}/H)^{\eta_{\text{lux}}}} \right) \tag{23.14}$$

$$\text{PS}_{\text{in}} = \frac{\alpha + \mu}{\alpha} \cdot S_{\text{cell}} = \delta_{\text{lux}} + \frac{V_{\text{max,lux}}}{1 + (K_{\text{lux}}/H)^{\eta_{\text{lux}}}} \tag{23.15}$$

In particular, PS_{in} is the synthesis rate per cell of the protein encoded by the gene downstream of Plux, thus representing the protein synthesis rate generated by the input device. Assuming that only the RBS affects the translation of an mRNA into a protein, the synthesis rate can be rescaled to describe the effect of different RBS sequences upstream of the same gene. For this reason, the PS_{in} can be rescaled to take into account possible different RBSs upstream of the reporter gene (during the input characterization stage) and of tetR (in the inverter characterization stage).

Second, the interconnected circuit is considered. The input device characterized above is assembled upstream and a GFP reporter device is assembled downstream (Figure 23.2B). The following ODE model describes genetic circuit including input device, logic inverter, and output reporter device:

$$\frac{d[T]}{dt} = \delta_{\text{lux}} + V_{\text{max,lux}} \cdot \frac{1}{1 + (K_{\text{lux}}/H)^{\eta_{\text{lux}}}} - (\gamma_T + \mu) \cdot [T] \tag{23.16}$$

$$\frac{d[I]}{dt} = \delta_{tet} + V_{max,tet} \cdot \left(1 - \frac{1}{1 + (T/K_{tet})^{\eta_{tet}}}\right) - (\alpha + \mu) \cdot [I] \tag{23.17}$$

$$\frac{d[G]}{dt} = \alpha \cdot [I] - \mu \cdot [G] \tag{23.18}$$

In this model, T is the TetR repressor, while the other species are those introduced before. γ_T represents the TetR degradation rate, while the Hill function parameters δ, V_{max}, η, and K are described above. An important assumption of this model is that cells bearing this genetic circuit and bearing the circuit containing only the input device have comparable growth rates. Otherwise, the steady-state level of the LuxR protein would change and the input device activity would not be black-box summarized by a simple Hill function (this situation is not considered here).

Assuming the steady state of all of the species, the circuit output (S_{cell}) can be computed as a function of HSL:

$$S_{cell} = \alpha \cdot [I] = \frac{\alpha}{\alpha + \mu} \cdot$$

$$\left(\delta_{tet} + V_{max,tet} \cdot \left(1 - \frac{1}{1 + ((1/(\gamma_T + \mu)) \cdot (\delta_{lux} + V_{max,lux}/(1 + (K_{lux}/H)^{\eta_{lux}})))/K_{tet})^{\eta_{tet}}}\right)\right)$$

$$\tag{23.19}$$

Equation (23.19) describes the HSL-S_{cell} static transfer function of the whole circuit. However, this relation does not directly enable the reuse of the NOT gate circuit with other input or output devices, as it explicitly depends on the specific input (HSL) and output (GFP) molecules. Conversely, the steady-state behavior of the logic inverter can be described as

$$PS_{out} = \delta_{tet} + V_{max,tet} \cdot \left(1 - \frac{1}{1 + ((PS_{in}/(\gamma_T + \mu))/K_{tet})^{\eta_{tet}}}\right) \tag{23.20}$$

where PS_{in} and PS_{out} are the synthesis rates per cell of TetR and of the protein of interest downstream of the Ptet promoter, respectively (in case of GFP, it is its immature form). Equation (23.20) well summarizes the static transfer function of the logic inverter, as it represents the synthesis rate per cell of any protein synthesized by the module that can be assembled as output and is a function of the synthesis rate of TetR that can be driven by any input device.

For prediction purposes, Eq. (23.20) can be further simplified under the two following assumptions: (i) the growth rate of cells bearing this circuit is similar to that of cells bearing the inverter with different input devices, whose function has to be predicted and (ii) all input devices are characterized in the same conditions (i.e., same reporter gene, RBS, and plasmid). The simplified relation is

$$PS_{out} = \delta_{tet} + V_{max,tet} \cdot \left(1 - \frac{1}{1 + (S_{cell,in}/K_{tet})^{\eta_{tet}}}\right) \tag{23.21}$$

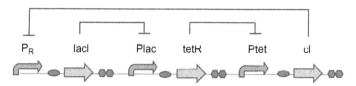

Figure 23.3 Biological oscillatory network. The circuit is composed by three interconnected NOT gates with feedback, thus obtaining a ring oscillator. All of the species involved in the system (i.e., mRNA and protein concentrations) can oscillate when the regulatory elements are properly tuned.

where $S_{cell,in}$ is the S_{cell} of the input device. The main advantages of using this formula are that the variations due to different RBSs between the input device and logic inverter do not need to be modeled since they are hidden in the S_{cell} signal; input device functions can be measured in arbitrary units by their S_{cell}; finally, a hard to estimate parameter like γ_T does not need to be determined. Then, the simple Hill function model of Eq. (23.21) can be effectively used to quantitatively predict the output of the logic inverter when driven by other quantitatively characterized input genetic devices (Figures 23.1C and 23.2C). A major difference between Eqs. (23.20) and (23.21) is that in the latter case the K_{tet} parameter is expressed in S_{cell} units.

Although not discussed in this chapter, these types of ODE models can also be used to predict the dynamic behavior of genetic circuits [31,32].

23.2.3.2   Oscillator Network

As a second example, a genetic circuit implementing an oscillator is illustrated. This circuit is called *repressilator* and is one of the milestones of synthetic biology, appeared in a paper in 2000 [33]. While the purpose of the previous example was the characterization and reuse of a genetic logic function, here the goal is to describe modelling tools to design a complex circuit from the early steps. Figure 23.3 shows the circuit structure. Briefly, it is composed of three repressor genes and three repressible promoters: the first repressor protein (LacI) inhibits the transcription of the second repressor gene (tetR), whose protein product (TetR) in turn inhibits the expression of a third gene (cI). Finally, its protein product (CI) inhibits lacI expression, completing the cycle. This negative feedback loop, essentially caused by three logic inverters, leads to temporal oscillations in mRNA and protein levels.

A simple model of transcriptional regulation has been used by Elowitz and Leibler [33] to design the device: transcription and translation dynamics can be expressed as in Eqs. (23.1) and (23.2), and promoter repression is assumed to follow a Hill function as in Eqs. (23.8) and (23.9).

In the early design steps of the repressilator, the primary goal is to assess *in silico* whether such a genetic configuration can exhibit oscillations of its elements. For this reason, model analysis can be performed on dimensionless equations, by using reasonable parameter values to numerically simulate the system and to identify the parameters with major impact on oscillatory behavior. Dimensionless equations are

$$\frac{dm_i}{dt} = \alpha_{0_i} + \frac{\alpha_i}{1 + p_j^{n_i}} - m_i \tag{23.22}$$

$$\frac{dp_i}{dt} = \beta_i \cdot (m_i - p_i) \tag{23.23}$$

$$\begin{pmatrix} i = \text{lacI}, & \text{tetR}, & \text{cI} \\ j = \text{cI}, & \text{lacI}, & \text{tetR} \end{pmatrix}$$

where m and p are the mRNA and protein levels, rescaled by translation rate (ρ) and $1/K$, respectively; α is the maximum activity of promoter and $\alpha 0$ its basic activity; time is rescaled by mRNA degradation rate; n is the Hill coefficient; and, finally, β is the ratio of protein degradation rate and mRNA degradation rate. The described model has been studied in a symmetric framework, where amounts and parameters (α, β, $\alpha 0$, and n) of the three repressors are identical, except the specificity toward their corresponding promoter.

The inclusion of transcription equations is essential to get an oscillatory behavior for the circuit, as oscillations can only occur when protein production is delayed (e.g., by transcription dynamics). Otherwise, model analysis shows that there is no parameter value compatible with self-sustained oscillations.

The equilibrium points of the ODE model can be computed by setting the time derivatives at zero. Because p/α and $1/(1 + p^n)$ have only one intersection, the system has a single equilibrium point p^*, solution of $p^* = \alpha 0 + \alpha/(1 + p^{*n})$. The stability of the equilibrium point can be studied by linearizing the equation system:

$$\frac{d\delta m_i}{dt} = -\delta m_i - \frac{\alpha \cdot n \cdot p_j^{n-1}}{(1 + p_j^n)^2}\Big|_{p*} \cdot \delta p_j \tag{23.24}$$

$$\frac{d\delta p_i}{dt} = \beta \cdot \delta m_i - \beta \cdot \delta p_i \tag{23.25}$$

where the δ indicates the variation of the state variable from the equilibrium point p^*. The full linearized system has the form:

$$\frac{d\delta y}{dt} = A \cdot \delta y \tag{23.26}$$

where δy is the column vector of the state variables of linearized system $[\delta m_1\, \delta p_1\, \delta m_2\, \delta p_2\, \delta m_3\, \delta p_3]^T$. From the 6×6 matrix A, it is possible to study the stability boundaries of the system, compatible with oscillations, as a function of the model parameters. In particular, the equilibrium point must not be asymptotically stable to be compatible with the desired oscillatory behavior, so at least one

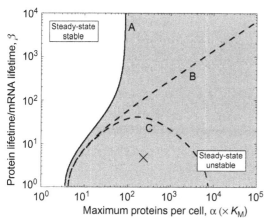

Figure 23.4 Stability diagram for the repressilator model. Boundary curves divide the regions where positive real part eigenvalues might exist for three different parameter configurations. A–C correspond to the boundary curves for different parameter values: A, $n = 2.1$, $\alpha_0 = 0$; B, $n = 2$, $\alpha_0 = 0$; C, $n = 2$, $\alpha_0/\alpha = 10^{-3}$. The shaded region corresponds to the unstable region of curve A.
Source: From Ref. [33] with permission of Nature Publishing Group.

eigenvalue of A must have a positive real part. Briefly, by obtaining the eigenvalues of A, it is possible to find an expression that describes the boundary curve dividing regions where positive real part eigenvalues might exist:

$$\frac{(\beta+1)^2}{\beta} = \frac{3 \cdot X^2}{4 + 2 \cdot X} \tag{23.27}$$

where

$$X = -\frac{\alpha \cdot n \cdot p^{n-1}}{(1+p^n)^2} \tag{23.28}$$

This expression is a function of all of the model parameters and it enables the identification of critical parameters and values that allow the system to possibly show an oscillatory behavior. All of the parameters have a crucial role for obtaining an unstable equilibrium point (Figure 23.4): high α, β, and n values, and a low α_0/α ratio are required. Such specifications correspond to strong promoters with low basic activity and high cooperativity, while proteins should have a high decay rate (for a fixed mRNA decay rate). In particular, given β, n, and α_0, the equilibrium point cannot be unstable for α values under a specific threshold. The unstable region becomes larger for increasing values of n: by comparing curves A and B, it is possible to see that a slight increase (from 2 to 2.1) of n value results in a much larger unstable region, so that the protein decay rate becomes less important in oscillatory behavior design.

This study highlights the importance of mathematical modelling in synthetic biology to support the design of an artificial function whose behavior was not trivial to predict as a function of the involved parameters. Such model-driven design allowed the creation of a biological network capable of an oscillatory behavior successfully implemented in *E. coli* in a low copy plasmid [33]. However, if this network had to be reused in a higher order interconnected circuit, it would require quantitative characterization steps similar to those followed in the first example.

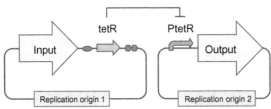

Figure 23.5 Copy number change of the Ptet promoter in the logic inverter. The input-tetR device and the Ptet-GFP device are placed in two plasmids with different replication origins, yielding different DNA copy numbers in the cell.

23.2.3.3   Mechanistic and Stochastic Models to Expand the Mathematical Modelling Tools for Synthetic Biology

Although the modelling approach just presented is quite general and allows for the description of a large number of synthetic biology design problems, some difficulties could arise when a designer wants to modify network elements (e.g., copy number of regulatory DNA or proteins) or to explain an observed behavior that is not predicted by a deterministic model. In this section, as an example, mechanistic models are introduced with the goal of studying DNA copy number changes and retroactivity in genetic circuits; finally, the advantages of stochastic models over deterministic ones are also briefly introduced. The three following questions can help explaining the previously mentioned problems.

1. How does one predict the output of the logic inverter of Figure 23.1A in a modified circuit where the Ptet promoter copy number is increased by fivefold (e.g., by placing it on a different plasmid, see Figure 23.5)?

    Because DNA copy number does not explicitly appear in Eqs. (23.16)–(23.18), such equations cannot be used for prediction purposes without the performance of a new steady-state experimental characterization of the logic inverter with the Ptet promoter in the new copy number context. Conversely, a mechanistic model can support this study *in silico* by explicitly describing copy number changes of promoter DNA. Mechanistic models are mainly based on the law of mass action and for this reason such models support an arbitrarily high level of detail. While mechanistic models are able to describe all molecular interactions of interest, their use is limited by the limited knowledge of biological systems and their parameter values. In fact, such kind of models often include a large number of parameters that require complex (and sometimes even unfeasible) experiments to be fully estimated.

2. How does the repressilator behavior change when a reporter gene is included downstream of a second Ptet promoter on a different plasmid (see Figure 23.6, which corresponds to the repressilator design used by Elowitz and Leibler [33] to detect oscillations via fluorescence)?

    Even if the core of the circuit is unvaried, an additional part of the circuit, not independent of the oscillatory network, is present. In fact, the TetR repressor can bind the Ptet promoter in the repressilator and also the Ptet promoter upstream of the reporter. Because there are additional DNA binding domains for TetR that compete with the Ptet DNA in the repressilator core circuit, the DNA needs to be explicitly considered in the model to predict quantitative behavior changes. This phenomenon, where circuit behavior can change with interconnections or the addition of supplementary modules, is called *retroactivity* and is under study by many research labs [30,34,35].

3. Why do sibling cells bearing the repressilator circuit exhibit non-synchronized oscillation patterns (Figure 23.7)?

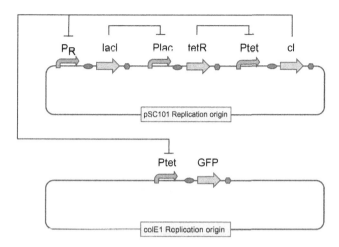

Figure 23.6 Addition of a reporter device to the repressilator system. The core of the repressilator (i.e., the three interconnected NOT gates with feedback) is in a low copy number plasmid with pSC101 origin, while a Ptet-GFP reporter is placed in a high copy number plasmid with ColE1 origin to enable oscillation measurements via fluorescence.

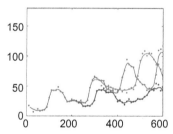

Figure 23.7 Experimental results of the repressilator. Fluorescence (y-axis, arbitrary units) is shown over time (x-axis, minutes) for three single sibling cells bearing the plasmids of Figure 23.6. The cells maintain a synchronized behavior for a limited time period, after which they go out of phase ($t = \sim 200$ min).
Source: From Ref. [33] with permission of Nature Publishing Group.

Deterministic models cannot explain the pattern shown in Figure 23.7, in which GFP is monitored in single cells. Only in a stochastic framework de-synchronization of oscillations emerges. Noise in biological systems is a key feature that in certain circuits becomes essential to explain the observed behavior [14]. Mathematical models can help handle noisy components at different levels, from cell doubling to gene expression [14,36].

23.3   Experimental Measurements for Parameter Identification

This section briefly describes how model parameter values can be estimated from *in vivo* experiments involving a given genetic circuit incorporated into a microbial chassis. Not all of the existing measurement techniques are listed, but the main currently used methods are illustrated.

23.3.1   Cell Growth

Given a cell population growing in a liquid medium, the number of cells per unit of volume can be measured by spreading a properly diluted aliquot of the culture on a solid medium (e.g., an agar plate). The number of colonies that appear on the plate corresponds to the number of cells in the spread aliquot called *CFUs* in the hypothesis that a single colony is generated by a single cell. Although it is reliable, the described procedure is hard to automate and provides offline results because colonies usually appear after a 12- to 24-h incubation. An easier procedure estimates cell density via turbidity measurements with a spectrophotometer: absorbance (OD at 600 nm, OD_{600}) is widely used and provides highly consistent results with the CFU method in a specific OD_{600} range [27,37]. This methodology provides real-time measurements, suitable for high-throughput assays in multiwell plates, and allows the precise estimation of cell growth rate.

23.3.2   DNA Copy Number

The number of genetic circuits in a cell depends on the number of copies of the plasmid vector that contains the circuit. The number of copies of the plasmid (i.e., the number of plasmids per genomic equivalent) can be measured by real-time qPCR: specific primers are used that anneal in plasmid and in genome. To carry out this procedure, samples are taken from the culture of interest, DNA is extracted, and the reaction is performed for each desired time point.

23.3.3   Transcription

After reverse transcription of mRNA into cDNA, real-time qPCR can be used to quantify the number of transcripts in a cell population. Specific primers anneal in the cDNA of interest and in control reactions used for calibration. If the number of cells in the population is known, the mRNA molecules per cell can be measured. Similar to the DNA copy number quantification, for transcript quantification, sampling, RNA extraction, reverse transcription, and qPCR, reactions are required for each desired time point.

23.3.4   Translation

In general, Western blot can be used to quantify protein level, provided that a specific antibody is available. Samples are taken from the culture of interest, which is disrupted, and the lysate is run on a protein gel. Western blot is then performed on the gel and image analysis is exploited to quantify protein level against control reactions used for calibration. While protein measurement is generally expensive, hard to automate, and limited to expert laboratory staff, fluorescent/luminescent/ color-producing reporter proteins are easily detectable in real time. Fluorimeter, luminometer, and spectrophotometer are used to measure reporter protein levels in a cell population. Such methodology can be automated for high-throughput assays in multiwell plates via a nonexpensive detection procedure. When working with fluorescent proteins, flow cytometry is a widely used technique for the measurement of single-cell fluorescence.

In summary, techniques are available to quantify the number of cells in a population and the per-cell copy number of genetic circuits, transcripts, and proteins. However, the quantification of DNA, transcript, and protein copies are expensive and hard to automate, as a sampling and reaction are required for each sample and time point to be measured. Moreover, a protein requires a specific antibody for quantification and its availability is not always granted. Among the cited methodologies, only absorbance for cell density quantification and reporter protein activity represent real-time techniques to monitor cell growth and protein levels, respectively.

23.4   Conclusion

This chapter has shown the basic mathematical tools to support the design, characterization, and reuse of genetic circuits in synthetic biology. Such tools are essential for biological engineers from the early design steps of a biological function to subpart and whole system characterization, debugging, and reuse. In particular, a network encoding the complex function of interest generally needs to be studied before its physical construction in order to evaluate if the desired behavior can occur. As shown in the repressilator case study, this early analysis allowed the identification of high-impact parameters for correct functioning. Because in the early design steps parameters may not be available, this process represents a qualitative support tool, while an in-depth analysis is required to fully characterize the circuit to improve the knowledge for simulation and reuse purposes. In fact, the potential of modelling in all the engineering fields is the capability of predicting behaviors never experimentally tested and of identifying hard-to-measure parameters, if a sufficient knowledge of the processes is available. Such advantages have been shown in the measurement of the transcription rate parameter (PoPS), which is normally expensive to estimate from experimental measurements, while a model-based approach can infer it from less-expensive fluorescence and absorbance measurements, given the other model parameters (e.g., culture doubling time, protein decay rates, etc.). Moreover, models of genetic circuits can be used to predict the behavior of parts upon interconnection, a key aspect of synthetic biology. Modularity is a major issue that has to be valid for this prediction step: if the behavior of a composite system cannot be predicted from the behavior of its basic parts, designers cannot exploit the power of the modelling process. Recent research studies have been focused on modularity, in order to experimentally evaluate the conditions in which biological systems can behave predictably [30,32]. The concept of retroactivity has been introduced to describe the quantitative behavior changes in a system occurring upon interconnection of other circuits. This has been recently faced by using mathematical modelling approaches to describe, for example, the competition in the binding of a transcription factor to a DNA binding site in a given circuit or in a newly interconnected circuit [38].

However, both modelling and experimental studies are needed to fully understand the modularity boundaries of biological systems.

The concept of datasheet has also been introduced to facilitate the characterization and reuse steps of biological parts. Even if technological advances have been made in the analysis of biological systems, variability is still a major issue affecting the characterization of parts. For this reason, standard units of measurement have been introduced for different biological parts to express their activity (in a reproducible unit). In particular, the relative promoter unit (RPU) approach has been proposed to measure the activity of promoters relative to an *in vivo* standard reference promoter [27]. This has been demonstrated to decrease the variability between promoter strength measurements performed by different instruments and different labs, so parameter estimation can be highly robust.

The considerations done so far are relative to empirical ODE models for genetic circuits, which are widely used because they are in general amenable to analytical studies and easy to identify. In contrast, mechanistic ODE models are harder to study and identify, but they are required when explicit molecular interactions need to be included. When properly identified, they can show superior predictive performance compared to empirical models. Biological noise, in some cases, dramatically affects the system behavior and for this reason the definition of stochastic models is rapidly spreading in the synthetic biology community.

References

[1] Serrano L. Synthetic biology: promises and challenges. Mol Syst Biol 2007;3:158.

[2] Endy D. Foundations for engineering biology. Nature 2005;438:449−53.

[3] Andrianantoandro E, Basu S, Karig D, Weiss R. Synthetic biology: new engineering rules for an emerging discipline. Mol Syst Biol 2006;2.

[4] Anderson JC, Dueber JE, Leguia M, Wu GC, Goler JA, Arkin AP, et al. BglBricks: a flexible standard for biological part assembly. J Biol Eng 2010;4(1):.

[5] Paddon CJ, Westfall PJ, Pitera DJ, Benjamin K, Fisher K, McPhee D, et al. High-level semi-synthetic production of the potent antimalarial artemisinin. Nature 2013;10:10.1038/nature12051.

[6] de Mora K, Joshi N, Balint BL, Ward FB, Elfick A, French CE. A pH-based biosensor for detection of arsenic in drinking water. Anal Bioanal Chem 2011;400(4):1031−9.

[7] Wargacki AJ, Leonard E, Win MN, Regitsky DD, Santos CN, Kim PB, et al. An engineered microbial platform for direct biofuel production from brown macroalgae. Science 2012;335:308−13.

[8] Danino T, Mondragon-Palomino O, Tsimring L, Hasty J. A synchronized quorum of genetic clocks. Nature 2010;463:326−30.

[9] Gardner TS, Cantor CR, Collins JJ. Construction of a genetic toggle switch in *Escherichia coli*. Nature 2000;403:339−42.

[10] Basu S, Mehreja R, Thiberge S, Chen MT, Weiss R. Spatiotemporal control of gene expression with pulse-generating networks. PNAS 2004;101(17):6355−60.

[11] Tamsir A, Tabor JJ, Voigt CA. Robust multicellular computing using genetically encoded NOR gates and chemical 'wires'. Nature 2010;469:212−5.

[12] Basu S, Gerchman Y, Collins CH, Arnold FH, Weiss R. A synthetic multicellular system for programmed pattern formation. Nature 2005;434:1130−4.

[13] Tabor JJ, Salis HM, Simpson ZB, Chevalier AA, Levskaya A, Marcotte EM, et al. A synthetic genetic edge detection program. Cell 2009;137(7):1272−81.

[14] Elowitz MB, Levine AJ, Siggia ED, Swain PS. Stochastic gene expression in a single cell. Science 2002;297(5584):1183−6.

[15] Kwok R. Five hard truths for synthetic biology. Nature 2010;463:288−90.

[16] Salis HM, Mirsky E, Voigt CA. Automated design of synthetic ribosome binding sites to control protein expression. Nat Biotechnol 2009;27(10):946−50.

[17] Pasotti L, Quattrocelli M, Galli D, Cusella De Angelis MG, Magni P. Multiplexing and demultiplexing logic functions for computing signal processing tasks in synthetic biology. Biotechnol J 2011;6(7):784−95.

[18] Zucca S, Pasotti L, Mazzini G, Cusella De Angelis MG, Magni P. Characterization of an inducible promoter in different DNA copy number conditions. BMC Bioinformatics 2012;13:S11.

[19] Mutalik VK, Guimaraes JC, Cambray G, Mai QA, Christoffersen MJ, Martin L, et al. Quantitative estimation of activity and quality for collections of functional genetic elements. Nat Methods 2013;10:347−53.

[20] Endler L, Rodriguez N, Juty N, Chelliah V, Laibe C, Li C, et al. Designing and encoding models for synthetic biology. J R Soc Interface 2009;6(Suppl. 4):S405−17.

[21] Chandran D, Copeland WB, Sleight SC, Sauro HM. Mathematical modeling and synthetic biology. Drug Discov Today Dis Models 2008;5(4):299−309.

[22] Rohwer J, Hanekom AJ, Hofmeyr JH. A universal rate equation for systems biology p. 175−87 Proceedings second international symposium on experimental standard conditions of enzyme characterizations (ESEC'06). Frankfurt am Main, Germany: Beilstein Institute; 2007

[23] Goryachev AB, Toh DJ, Lee T. Systems analysis of a quorum sensing network: design constraints imposed by the functional requirements, network topology and kinetic constants. BioSystems 2006;83(2):178−87.

[24] Stamatakis M, Mantzaris NV. Comparison of deterministic and stochastic models of the lac Operon Genetic Network. Biophys J 2009;96(3):887−906.

[25] Kaznessis YN. Models for synthetic biology. BMC Syst Biol 2007;1(1):.

[26] Canton B, Labno A, Endy D. Refinement and standardization of synthetic biological parts and devices. Nat Biotechnol 2008;26:787−93.

[27] Kelly JR, Rubin AJ, Davis JH, Ajo-Franklin CM, Cumbers J, Czar MJ, et al. Measuring the activity of BioBrick promoters using an *in vivo* reference standard. J Biol Eng 2009;3:4.

[28] Selinger DW, Saxena RM, Cheung KJ, Church GM, Rosenow C. Global RNA half-life analysis in *Escherichia coli* reveals positional patterns of transcript degradation. Genome Res 2003;13(2):216−23.

[29] Bernstein JA, Khodursky AB, Lin P-H, Lin-Chao S, Cohen SN. Global analysis of mRNA decay and abundance in *Escherichia coli* at single-gene resolution using two-color fluorescent DNA microarrays. PNAS 2002;99:9697−702.

[30] Pasotti L, Politi N, Zucca S, Cusella De Angelis MG, Magni P. Bottom-up engineering of biological systems through standard bricks: a modularity study on basic parts and devices. PLoS One 2012;7:e39407.

[31] Braun D, Basu S, Weiss R. Parameter estimation for two synthetic gene networks: a case study. IEEE ICASSP 2005;5:769−72.

[32] Moon TS, Lou C, Tamsir A, Stanton BC, Voigt CA. Genetic programs constructed from layered logic gates in single cells. Nature 2012;491(7423):249–53.

[33] Elowitz MB, Leibler SA. Synthetic oscillatory network of transcriptional regulators. Nature 2000;403(6767):335–8.

[34] Del Vecchio D, Ninfa AJ, Sontag ED. Modular cell biology: retroactivity and insulation. Mol Syst Biol 2008;4:161.

[35] Sauro HM. Modularity defined. Mol Syst Biol 2008;4:166.

[36] Tan C, Marguet P, You L. Emergent bistability by a growth-modulating positive feedback circuit. Nat Chem Biol 2009;5:842–8.

[37] Pasotti L, Zucca S, Lupotto M, Cusella De Angelis MG, Magni P. Characterization of a synthetic bacterial self-destruction device for programmed cell death and for recombinant proteins release. J Biol Eng 2011;5:8.

[38] Jayanthi S, Nilgiriwala KS, Del Vecchio D. Retroactivity controls the temporal dynamics of gene transcription. ACS Synth Biol 2013;10.1021/sb300098w.

Printed in the United States
By Bookmasters